Applications of Statistical and Field Theory Methods to Condensed Matter

NATO ASI Series

Advanced Science Institutes Series

A series presenting the results of activities sponsored by the NATO Science Committee, which aims at the dissemination of advanced scientific and technological knowledge, with a view to strengthening links between scientific communities.

The series is published by an international board of publishers in conjunction with the NATO Scientific Affairs Division

A	Life Sciences	Plenum Publishing Corporation
B	Physics	New York and London
C	Mathematical and Physical Sciences	Kluwer Academic Publishers Dordrecht, Boston, and London
D	Behavioral and Social Sciences	
E	Applied Sciences	
F	Computer and Systems Sciences	Springer-Verlag
G	Ecological Sciences	Berlin, Heidelberg, New York, London,
H	Cell Biology	Paris, and Tokyo

Recent Volumes in this Series

Volume 213—Interacting Electrons in Reduced Dimensions
edited by Dionys Baeriswyl and David K. Campbell

Volume 214—Science and Engineering of One- and Zero-Dimensional Semiconductors
edited by Steven P. Beaumont and Clivia M. Sotomayor Torres

Volume 215—Soil Colloids and their Associations in Aggregates
edited by Marcel F. De Boodt, Michael H. B. Hayes, and Adrien Herbillon

Volume 216A—The Nuclear Equation of State, Part A:
Discovery of Nuclear Shock Waves and the EOS
edited by Walter Greiner and Horst Stöcker

Volume 216B—The Nuclear Equation of State, Part B:
QCD and the Formation of the Quark-Gluon Plasma
edited by Walter Greiner and Horst Stöcker

Volume 217—Solid State Microbatteries
edited by James R. Akridge and Minko Balkanski

Volume 218—Applications of Statistical and Field Theory Methods to Condensed Matter
edited by Dionys Baeriswyl, Alan R. Bishop, and José Carmelo

Series B: Physics

Applications of Statistical and Field Theory Methods to Condensed Matter

Edited by
Dionys Baeriswyl
University of Fribourg
Fribourg, Switzerland

Alan R. Bishop
Los Alamos National Laboratory
Los Alamos, New Mexico

and
José Carmelo
University of Évora
Évora, Portugal

Plenum Press
New York and London
Published in cooperation with NATO Scientific Affairs Division

Proceedings of the NATO Advanced Study Institute on
Applications of Statistical and Field Theory Methods
to Condensed Matter,
held May 22–June 2, 1989,
in Évora, Portugal

Library of Congress Cataloging in Publication Data

NATO Advanced Study Institute on Applications of Statistical and Field
Theory Methods to Condensed Matter (1989:Évora, Portugal)
   Applications of statistical and field theory methods to condensed matter
/ edited by Dionys Baeriswyl, Alan R. Bishop, and José Carmelo.
      p.   cm.—(NATO ASI series. Series B, Physics; vol. 218)
   "Published in cooperation with NATO Scientific Affairs Division."
   "Proceedings of the NATO Advanced Study Institute on Applications of
Statistical and Field Theory Methods to Condensed Matter, held May
22–June 2, 1989 in Évora, Portugal"—T.p. verso.
   Includes bibliographical references.
   ISBN 0-306-43526-8
   1. Condensed matter—Statistical methods—Congresses. 2. Field theory
(Physics)—Congresses. I. Baeriswyl, D. (Dionys), 1944-   . II. Bishop, A. R.
(Alan R.), 1947-   . III. Carmelo José. IV. North Atlantic Treaty
Organization. Scientific Affairs Division. V. Title. VI. Series: NATO ASI
series. Series B, Physics; v. 218.
QC173.4.C65N372   90-7153
530.4'1—dc20   CIP

© 1990 Plenum Press, New York
A Division of Plenum Publishing Corporation
233 Spring Street, New York, N.Y. 10013

All rights reserved

No part of this book may be reproduced, stored in a retrieval system, or transmitted
in any form or by any means, electronic, mechanical, photocopying, microfilming,
recording, or otherwise, without written permission from the Publisher

Printed in the United States of America

PREFACE

There is no doubt that we have, during the last decade, moved into a "golden age" of condensed matter science. The sequence of discoveries of novel new states of matter and their rapid assimilation into experimental and theoretical research, as well as devices, has been remarkable. To name but a few: spin glasses; incommensurate, fractal, quasicrystal structures; synthetic metals; quantum well fabrication; fractional quantum Hall effect; solid state chaos; heavy fermions; and most spectacularly high-temperature superconductivity.

This rapid evolution has been marked by the need to address the reality of materials in "extreme" conditions -- disordered, nonlinear systems in reduced dimensions, restricted geometries and at mesoscopic scales, often with striking competitions between several length and frequency scales, and between strong electron-phonon and electron-electron interactions.

In such new territory it is not surprising that very interdisciplinary approaches are being explored and traditional boundaries between subjects and disciplines re-defined. In theory, this is evident, for instance, in attempts: (1) to advance the state of the art for electronic structure calculations so as to handle strongly interacting many-body systems and delicate competitions for collective ground states (spin models or many-electron Hamiltonians, field theory, band structure, quantum chemistry and numerical approaches); or (2) to understand pattern formation and complex (including chaotic) dynamics in extended systems. This demands close involvement with applied mathematics, numerical simulations and statistical mechanics techniques.

In condensed matter, then, the rapidly changing research environment is very different from the existing textbook world of ordered, linear, isotropic, 3-dimensional, weakly-interacting materials and models. It is essential that young researchers are introduced to the new points of view. However, this is not yet available in textbooks, certainly not in any systematic fashion. Therefore Advanced Study Institutes play a very important role in disseminating the new information and promoting an appreciation for the necessary interdisciplinary research.

The Evora Institute focused on current theoretical *techniques* in statistical mechanics and field theory, and on their *application* to specific condensed matter and solid state materials -- important contributions were made by experimentalists (Monceau and Mydosh) giving perspective to our knowledge of spin glasses, high-temperature superconductors and charge-density-wave materials. Some 15 experts from 9 countries gave tutorial and in-depth introductions and ideas on future directions to approximately 80 attendees from 18 countries. In addition there were round-table discussions and a series of poster presentations by junior attendees. The major topics covered at the Workshop were:

Quantum Coherence, Tunneling and Dissipation

Solitons, Patterns and Incommensurate Structures

Conformal Invariance and Phase Transitions

Chaos

Disorder and Localization

Spin Glasses and Heavy Fermions

Many-Body Techniques

Superconductivity

Charge- and Spin-Density Waves

The Institute received, besides a substantial NATO grant, generous support from various organizations: the Commission of European Communities, the Portuguese Convention for Scientific and Technological Investigations (JNICT), the Portuguese Institute for Scientific Investigations (INIC), the Gulbenkian Foundation (Portugal), the Spanish Ministry of Education and the American National Science Foundation.

Evora was a delightful town in which to hold this Institute. Facilities for accommodations, lectures and informal discussion were excellent. The attendees were welcomed with great generosity by the Chancellor and his University, by the Mayor, and by the Contessa (and her vineyard!). The Institute Secretary, Mrs. Fátima Loureiro handled every emergency with equanimity and made life easy for the organizers. Finally, the local organizing committee (José Camelo, Francisco Guinea, João dos Santos, and Vitor Viera) foresaw every difficulty and formulated excellent scientific and logistic plans.

Dionys Baeriswyl
Alan Bishop
José Camelo
November 1989

PARTICIPANTS

Alonso, Juan J., Dep. Fisica Aplicada, Univ. Granada, Spain
Alvarellos, J.E., Departamento de Fisica Fundamental, Madrid, Spain
Amaral, Vitor B. de Sequeira, Centro de Fisica da Univ. do Porto, Portugal
Assunçao, Manuel A.C. de, Dept. of Physics, Univ. de Aveiro, Portugal
Baeriswyl, Dionys, IBM Research Laboratory, Rüschlikon, Switzerland
Barone, Paulo, Phys. Instit., UNICAMP, Campinas SP, Brazil
Bartkowiak, Miroslav, Instit. of Phys., A. Mickiewicz Univ., Poznan, Poland
Bishop, Alan R., Los Alamos National Laboratory, USA
Bjelis, Alex, Institute of Physics, Univ. of Zagreb, Yugoslavia
Bobbert, Peter A., Technical University of Delft, The Netherlands
Bohr, Tomas, Niels-Bohr Institute, Copenhagen, Denmark
Brajczewska, Marta, Dept. de Fisica, Univ. of Coimbra, Portugal
Bruinsma, Robijn, Dept. of Phys., Univ. of California, Los Angeles, USA
Caldeira, Amir O., Phys. Inst., UNICAMP, Campinas SP, Brazil
Capellmann, Herbert, Institut Laue-Langevin, Grenoble, France
Carlos, Luis A.D., Dept. of Physics, Univ. of Evora, Portugal
Carmelo, José M.P., Dep. de Fisica, Univ. of Evora, Portugal
Castillo, Jr., Victor M., Dept. of Phys., San Jose State Univ., USA
Chatterjee, Ram, Physics Dept., University of Calgary, Canada
Chatterjee, Rupak, Physics Dept., University of Calgary, Canada
Conde, Jesús Pérez, Lab. de Physique des Solides, Orsay, France
Conrado, Claudine V., Niels-Bohr Institute, Copenhagen, Denmark
Costa, Antonio A. da, Dept. of Phys., Univ. of Evora, Portugal
Doty, Curtis Alan, Dept. of Physics, Princeton University, USA
Dzierzawa, Michael, Theory of Cond. Matter, Univ. Karlsruhe, FRG
Eckern, Uli, Theory of Cond. Matter, Univ. Karlsruhe, FRG
Efetov, Konstantin B., Max-Planck-Inst. f. Festkörperphys., Stuttgart, FRG
Etxebarria, Inigo E., Phys. Dept. of Cond. Matt., Univ. del Pais Vasco, Spain
Evangelista, Luiz R., UEM/CCE/DFI, Maringà PR, Brazil
Evans, Sarah M.M., Dept. of Theor. Phys., Oxford University, UK
Extremera, Alfonso, Depto. de Fisica Moderna, Universidad de Granada, Spain
Fernandez, V. F. Javier, Univ. Nacion. de Educacion a Dist., Madrid, Spain
Ferreira da Silva, Antonio, Univ. Federal da Bahia, Salvador BA, Brazil
Ferreira, Antonio L.C. de Sousa, Dept. of Phys., Univ. of Aveiro, Portugal
Ferrer, Jaime, Dept. of Cond. Matter Phys., Univ. Auton., Madrid, Spain
Floria, Luis M., Instit. of Mater. Science, Univ. of Zaragoza, Spain

Gabas Perez, Merche, Instit. of Mater. Science, Univ. of Zaragoza, Spain
Gawlinski, Edward T., Dept. of Phys., Temple University, Philadelphia USA
Gedik, Mehmet Zafer, Dept. of Phys., Bilkent University, Ankara, Turkey
Gooding, Robert James, Dept. of Phys., Cornell University, Ithaca, USA
Gottlieb, David, Facultad de Ciencias, Universid. de Chile, Santiago, Chile
Guinea, Francisco, Faculdad de Ciencias, Univ. Autonoma, Madrid, Spain
Hadley, Peter, Dept. of Phys., Stanford University, USA
Hebel, Wolfgang, CEC, DG XII, Brussels, Belgium
Hekking, Frank, Technical University of Delft, The Netherlands
Horbach, Marcus, Technical University of Delft, The Netherlands
Jain, Sudhir, Dept. of Chemistry, University of Durham, UK
Jelcic, Dajana, Dept. of Phys., Univ. of Zagreb, Yugoslavia
Kleban, Peter H., Dept. of Phys. and Astron., Univ. of Maine, Orono, USA
Lage, Eduardo J.S., Lab. of Phys., Univ. of Porto, Portugal
Lantwin, Christian J., Dept. of Theor. Phys., Oxford University, UK
Leggett, Anthony J., Dept. of Phys., Univ. of Illinois, Urbana, USA
Lopez, Maria J., Dept. of Theor. Phys., Univ. of Valladolid, Spain
Lucheroni, Carlo, Dip. di Fisica, Univ. di Perugia, Italy
Maki, Kazumi, Dept. of Phys., Univ. of Southern California, Los Angeles, USA
Marathe, Yatin, Dept. of Physics, Indian Inst. of Science, Bangalore, India
Marino, Eduardo C., Dept. de Fisica, PUC, Rio de Janeiro, RJ, Brazil
Martin Moreno, Luis, Depto. de Fisica de la Mat. Cond., Univ. Auton., Madrid, Spain
Martinez, Gerardo, Max-Planck-Institut f. Festkörperf., Stuttgart FRG
Martins, Paulo M. da Costa, Center of Condensed Matter Physics, Lisbon, Portugal
Membrado Ibanez Manuel C., Dpt. of Theor. Phys., Univ. of Zaragoza, Spain
Mendes, José F. Ferreira, Dept. of Physics, University of Porto, Spain
Mendiratta, Sushil K., Dept. of Phys., Univ. of Aveiro, Portugal
Monceau, Pierre, Centre de Rech. sur les Basses Témp., CNRS, Grenoble, France
Montorsi, Arianna, Dip. di Fisica, Politecnico di Torino, Italy
Morgado, Jorge Manuel F., Depto. de Quím., Inst. Sup. Técn., Lisboa, Portugal
Mydosh, John A., Kamerling Onnes Laboratorium, Leiden, The Netherlands
Neofotistos, George, Purdue University, West Lafayette, USA
Nicopoulos, Vassilios N., Los Alamos National Lab., Los Alamos, USA
Öğüt, Ali S., Middle East Technical University, Ankara, Turkey
Olmsted, Peter D., Dept. of Phys., Univ. of Illinois, Urbana, USA
Özer, Nilgün, Phys. Dept., Istanbul Technical University, Turkey
Pérez-Pérez, Rubén, Dep. Fisica Mat. Cond., Univ. Auton., Madrid, Spain
Pimentel, Iveta R., Center of Cond. Matter Phys., Lisboa, Portugal
Ramalho, Joao P.C.A., Dept. Chimica, Lisboa, Portugal,
Ramsak, Anton, Institute Jozef Stephan, Llubljana, Yugoslavia
Reis, Antonio Heitor S., Phys. Dept., Univ. of Evora, Portugal
Rubio, Angel, Dept. of Theor. Phys., Univ. of Valladolid, Spain
Sanchez, Angel, Depto. de Fisica Teorica I, Univ. Complutense, Madrid, Spain
Santos, Maria A., Faculty of Science, Univ. of Porto, Portugal
Santos, Maria Cristina dos, Univ. Fed. de Pernambuco, Recife PE, Brazil
Santos, J. Lopes dos, Laboratorio de Fisica, Universidade do Porto, Portugal
Silva, Edison Z. da, Physics Institute, UNICAMP, Campinas SP, Brazil
Smithline, Shepard J., James Franck Institute, Univ. of Chicago, USA

Tagliacozzo, Arturo, Dip. di Scienze Fisiche, Univ. di Napoli, Italy
Tahir-Kheli, Raza A., Dept. of Phys., Temple University, Philadelphia, USA
Tatlipinar, Hasan, Phys. Dept., Yildiz University, Sisli-Istanbul, Turkey
Tognetti, Valerio, Dept. of Physics, University of Florence, Italy
Toz, Iskender E., Science Faculty, Phys. Dept., University of Istanbul, Turkey
Trugman, Stuart, Los Alamos National Lab., Los Alamos, USA
Vaia, Ruggiero, Istituto di Elettronica Quantistica, CNR. Florence, Italy
Vallat, André, Phys. Institute, Univ. of Neuchâtel, Switzerland
Ventriglia, Franco, Dept. Phys., Univ. di Napoli, Italy
Vieira, Vitor J.R., INIC, Centro de Fisica da Materia Condens., Lisboa, Portugal
Voruganti, Purushotham, Dept. of Physics, Stanford University, USA
Würger, Alois, Institut Laue-Langevin, Grenoble, France
Wiegmann, Paul B., Landau Institute for Theor. Phys., Moscow USSR
Zhang, Fei, Dept. of Theor. Phys., Universidad Complutense, Madrid, Spain
Zimmer, Michael F., Dept. of Phys., University of Illinois, Urbana, USA
Zotos, Xenophon, Instit. for Theory of Condensed Matt., Univ. of Karlsruhe, FRG
Zürcher, Ulrich, Physics Institute, University of Basle, Switzerland

CONTENTS

QUANTUM COHERENCE, TUNNELING AND DISSIPATION

Quantum Mechanics of Complex Systems I 1
   A. J. Leggett

Quantum Mechanics of Complex Systems II 11
   A. J. Leggett

Quantum Tunneling in Macroscopic Systems 27
   A. O. Caldeira

Ground State of a Non-Dissipative Josephson Junction 49
   A. Tagliacozzo

The "Cold Fusion" Problem 63
   A. J. Leggett

Unstable Behavior of a Superconducting Ring Containing a Josephson
     Junction 77
   A. Extrema

Influence of the Radiation Field on the Process of Electronic Interference 79
   P. M. V. B. Barone and A. O. Caldeira

Quantum Mechanical Harmonic Chain Attached to Heat Baths 81
   U. Zürcher and P. Talkner

CONFORMAL INVARIANCE AND PHASE TRANSITIONS

Conformal Invariance – a Survey of Principles with Applications to Statistical
     Mechanics and Surface Physics 83
   Peter Kleban

A Possible Field-Theoretical Model for the Nematic-Isotropic Phase Transition
     in Liquid Crystals 117
   Luiz R. Evangelista and M. Simões

Lattice Unstabilities of Magnetic Origin 119
   D. Gottlieb and M. Lagos

SOLITONS, PATTERNS, INCOMMENSURATE STRUCTURES

Dual Quantization of Solitons 121
   E. C. Marino

Variational Approach to Quantum Statistical Mechanics 141
   Riccardo Giachetti, Valerio Tognetti and Ruggero Vaia

Pattern Changes in Electrodeposit of $CuSO_4$ 151
   V. M. Castillo, R. D. Pochy and L. Lam

Linear Dynamics of Modulated Spin Magnets.................................. 153
   Christian J. Lantwin

The Stochastic ϕ^4 Atomic Chain 155
   Angel Sánchez and Luis Vázquez

CHAOS

Chaos and Turbulence... 157
   Tomas Bohr

DISORDER AND LOCALIZATION

Electron Localization in Disordered Systems 187
   K. B. Efetov

SPIN GLASSES AND HEAVY FERMIONS

An Introduction to the Dynamics of Quench-Disordered Spin Systems 209
   R. Bruinsma

Experimental Studies of Spin Glasses and Heavy Fermions: their Magnetism
   and Superconductivity... 225
   J. A. Mydosh

Heavy Fermions: Theoretical Aspects.. 239
   Herbert Capellmann

MANY-BODY TECHNIQUES

General Many-Body Systems 253
   S. A. Trugman

Numerical Methods for Many-Body Problems
   A) Exact Diagonalization of Small Systems 265
   X. Zotos
   B) Quantum Monte Carlo Methods 273
   M. Dzierzawa and X. Zotos

High-Density Expansion for Electron Systems................................ 281
   M. Bartkowiak, P. Münger, K. A. Chao and R. Micnas

Perturbative Results using the Cumulant Expansion in the Anderson Lattice .. 283
   Gerardo Martinez and Mario E. Foglio

Application of Gutziller's Correlated Method to the Electronic Effective
   Mass of Degenerate N-type Silicon 285
   A. Ferreira da Silva and F. de Brito Mota

Optical Absorption in Disordered Semiconductor Systems: Application to
   Correlated Phosphorus-Doped Silicon 287
   A. Ferreira da Silva

SUPERCONDUCTIVITY

High T_c Superconductivity: Lessons to be Learned from Neutron Scattering ... 289
   H. Capellmann

Superconducting Networks in a Magnetic Field: Exact Solution of the
   J^2 Model 299
   L. M. Floria and R. B. Griffiths

Vortex Dynamics in Networks of Josephson Junctions 311
   Ulrich Eckern

Comparison of Effective Models for CuO_2 Layers in Oxide Superconductors ... 323
   R. Ramšak and P. Prelovšek

CHARGE- AND SPIN-DENSITY WAVES

Charge-Density Waves in Quasi-one-Dimensional Systems 325
   Aleksa Bjeliš

Recent Developments in Charge-Density Wave Systems 357
   Pierre Monceau

Spin-Density Waves in Organic Conductors 379
   Kazumi Maki

Poster Contributions 401

Contributors 403

Index 405

QUANTUM MECHANICS OF COMPLEX SYSTEMS, I

A. J. Leggett

Department of Physics
University of Illinois at Urbana-Champaign
1110 West Green Street
Urbana, IL 61801 USA

Very often in physics and chemistry we are interested in the motion of some variable which is described by quantum mechanics and is strongly coupled to an "environment," that is, one or (more likely) many other variables whose behavior is of no particular interest to us in its own right, but only because of the effect it may exert on the motion of the "system," that is, the variable of primary interest. Some familiar examples of such "systems," with the "environment" indicated in parentheses, are an electron participating in a chemical reaction (the vibronic modes), a paraelectric defect in a solid, such as OH^- in KCl, (phonons), a μ-meson in a metal (conduction electrons) and the K_0-$\bar{K}_0$ system (surrounding matter). All these are old problems. A class of problem which is of rather more recent interest is when the "system" variable is in some sense <u>macroscopic</u>: examples include charge density waves (normal electrons), the magnetization of small ferromagnetic particles (magnons, phonons), the early Universe (pre-existing mesons) and, par excellence, the phase of the Cooper pairs in devices incorporating one or more Josephson junctions (normal electrons, phonons, radiation field, nuclear spins ...). The last example has been studied intensively both experimentally and theoretically over the last few years and provides probably the best test-bed for ideas in this area. One further example which is of great current interest is the quantum tunnelling motion of a pair of deuterium nuclei in a matrix of a metal such as Pd or Ti; in this case the "environment" is the conduction electrons, the nuclei of the metal and any "third-party" deuterons which may be present.

It should be stressed that the business of applying quantum mechanics to this kind of situation is not at all trivial. Indeed, the context in which the theory was originally developed and tested experimentally in the late 20's was that of <u>isolated</u> microscopic systems, such as the atoms in a beam, whose interactions with their environment were so weak as to be negligible for most purposes. It is, actually, not even obvious a priori that the quantum formalism applies at all when the coupling to the environment is strong and the latter is complex (and, in fact, some of the writings of Nils Bohr are most naturally read as indicating that he thought it did not). Most physicists, being reductionist in temperament, assume that it does in principle still apply under these conditions, and for the purposes of these lectures I shall follow this view, for which there is at least plenty of circumstantial evidence (though see e.g. ref. (1) for some important reservations). Even if we believe this, the behavior of a quantum system interacting strongly with a complex environment may be qualitatively quite different from that of the same system in isolation. For example, if the system has available to it two degenerate potential walls, then in free space the ground state is split by tunnelling and we get the coherent oscillations well-known in the context of (e.g.) the NH_3 molecule; in a solid or liquid, on the other hand, we often find that in such a situation the motion is described by the classical "rate theory" familiar to chemists, which allows no such oscillations. A general rule of thumb is that interaction with a complex environment, particularly if the latter is dissipative, tends to make the motion of a quantum system appear more "classical"; we return below to this question. In view of this it is of particular interest to study the effects of the environment on these quantum phenomena which have no classical analogue or limit, such as tunnelling through a potential barrier and the coherent NH_3-type oscillations mentioned above.

In this lecture and the next I will discuss some general aspects of the way in which the quantum motion of a system is affected by a complex (and in general dissipative) environment, with illustrations from the exactly soluble problem of the damped harmonic oscillator and the "qualitatively soluble" spin-boson problem; in the third lecture I will study specifically one particular example of this type of problem which is currently in the news, namely the so-called "cold fusion" problem. Dr. Caldeira's lectures will discuss some related ideas in the context of the important problem of "macroscopic quantum tunnelling" and related problems.

The first question we need to ask is: Exactly how do we propose to specify the effect of the environment on the system? Let us denote the variable(s) of the system schematically by q, and those of the environment by x; q may, and x certainly will, represent more than one variable, in fact in most circumstances x will represent a macroscopic number of (microscopic) variables. Were we doing <u>classical</u> mechanics, life would be straightforward: in the simplest case we would write down for the classical variable q(t) a Langevin equation of the familiar form

$$M\ddot{q} + \eta\dot{q} + \partial V/\partial q = F(t) \tag{1}$$

where M is the mass associated with q, η a friction coefficient and F(t) a "random" force due to the environment, whose statistics are Gaussian and given by

$$\langle F(t) \rangle = 0, \qquad \langle F(t)F(t') \rangle = 2\eta kT\delta(t-t'). \tag{2}$$

Eqns. (1) and (2) correspond to the case of "white noise," where the correlation time of the forces exerted by the environment is negligibly short; in cases where the environment has a "memory" (finite correlation time) they are simply generalized, see e.g. ref. (2). The important point to note in the present context is that the environment coordinates x(t) never enter explicitly; their effect is totally encapsulated in the statistics of the "random" force F(t) as given in eqn. (2).

It is tempting to try to extend the Langevin formation to the quantum case, and in the literature one finds two different ways of doing this. The first, which is familiar in quantum optics under the name of the "quantum Langevin equation" consists in writing down a Schrödinger equation for the system including a full term due to the environment, with the force regarded as an <u>operator</u>, and then making certain approximations concerning the correlations of this operator. This technique appears to be equivalent to a special case (appropriate to the "weak-coupling" limit characteristic of quantum optics) of the more general technique I discuss below, so there is no need to discuss it explicitly here. The second method, which has been applied predominantly in condensed-matter contexts and is sometimes called the "quasiclassical (quantum) Langevin equation" consists in writing down eqn. (1) and the first of eqns. (2), but replacing the second of eqns. (2) by the "quantum correlation"

$$\langle F(t)F(t')\rangle = \int_{-\infty}^{\infty} d\omega \; \coth(\hbar\omega/2kT) \; \eta\hbar\omega \; e^{-i\omega(t-t')} \tag{3}$$

which reduces to (2) in the limit $\hbar \to 0$. As discussed in detail by Schmid[3], the quasiclassical Langevin equation may be a good description under certain circumstances, in particular for motion in the classically accessible regime with strong damping; however, there seems no a priori reason to believe it will give even qualitatively correct results for phenomena such as tunnelling or NH_3-type oscillations (though see ref. (4)). It should be stressed that the reason the QCLE is unreliable in these circumstances has very little to do with the effect of the environment as such, but is rather a consequence of the fact that we are trying to treat an intrinsically quantum system by a classical description. It is no accident that this works best precisely when the environment has effectively destroyed the essentially quantum features of the motion.

Given, then, that the effect of the environment cannot in general be specified entirely in terms of "random" forces exerted on the system, how shall we treat it? There is only one safe prescription, and that is to write down a complete quantum description of the "universe" (that is, the interacting complex of system (S) + environment (E)), solve for the wave function or density matrix of the universe, and then trace over the environmental degrees of freedom to obtain the dynamics of the system. Here it is crucial to note that it is a general feature of the quantum description of reality, in strong contrast to the classical description, that if we have two interacting quantum systems (here S + E), then in general there is <u>no</u> way of assigning each system definite properties in isolation; in particular, a general pure state of the "universe" is written $\Psi(S,E)$ and in general <u>cannot</u> be factorized into a product of pure states $\phi(S)$, $\chi(E)$ referring to "system" and "environment" separately. (This characteristically quantum-mechanical "nonseparability" has of course spectacular consequences in other contexts, as brought out by Bell's theorem.) In general, even if the quantum state of the "universe" is itself a pure state $\psi(q,x)$, the only description of the system is by a density matrix

$$\rho(q,q') \equiv \int \psi^*(qx)\psi(q'x)dx \tag{4}$$

where the notation $\int dx$ of course in reality represents integration over a large (often macroscopically large) collection of environmental coordinates. Moreover, inspection of the density matrix $\rho(q,q')$ by itself, without a knowledge of the environmental dynamics, may give substantially misleading results: see below. Thus the environment enters the problem in a more subtle and fundamental way than in classical physics.

At this stage it is convenient to make a small digression to discuss the connection of all this with the quantum theory of measurement. Suppose that q represents a reasonably macroscopic variable, such as the position of a cat (or pointer). Under certain (highly idealized) conditions it is possible to engineer a situation such that initially the density matrix of this variable corresponds to a pure state $\phi_0(q) \neq \delta(q-q_0)$, i.e. a superposition of different eigenstates of the position operator (for an example, see ref. (5)). On the other hand, classical experience tells us that if we look at a particular cat or pointer we always see a definite value of q (and can never exhibit any consequence of the superposition of different q-values). In order for this piece of everyday experience to be compatible with the predictions of the quantum formalism, it is usually argued that a necessary condition is that the density matrix rapidly evolves into an incoherent mixture of terms diagonal in q, i.e.

$$\rho(q,q') \to |c(q)|^2 \delta(q-q'). \tag{5}$$

It is clear that the interaction with the environment will cause just such a result, if it induces a "universe" wave function $\Psi(q,x)$ such that the right-hand side of eqn. (3) is zero for $q \neq q'$. Intuitively speaking, this means that different values of q are associated with mutually orthogonal states of the environment. To examine this point, let us suppose for simplicity that q is a discrete variable with eigenvalues q_n and associated wave functions (of S alone) $\phi_n(q) = \delta(q-q_n)$. Then initially we have

$$\phi(q) = \sum_n c_n \phi_n(q). \tag{6}$$

Let us assume for simplicity (it is certainly not the most general case) that when the system starts in a state $\phi_n(q)$ the interaction with the environment induces a product state of the universe of the form $\Psi_n(q,x) = \phi_n(q)\chi_n(x)$. Then, by linearity, if the initial state is (5) the final universe state will be

$$\Psi(q,x) = \sum_n \psi_n(q)\chi_n(x) \tag{7}$$

and the density matrix in the n-representation will be

$$\rho_{nn'} = \int \chi_n^*(n)\chi_{n'}(x)dx \quad c_n^* c_{n'}. \tag{8}$$

If the various χ_n are mutually orthogonal, then the density matrix becomes diagonal in the n-representation (and hence, automatically, the q-representation): $\rho_{nn'} = |c_n|^2 \delta_{nn'}$. This is essentially the same argument as used by von Neumann in his theory of measurement: in that context q represents the coordinate of a microscopic particle and x some coordinate(s) of a macroscopic measuring apparatus, and a particular form of interaction is assumed which induces the transition

$\phi_n(q) \to \phi_n(q)\chi_n(x)$ with the different χ_n orthogonal. The argument is then used to show that after a particle has interacted with a "measuring apparatus" it is impossible to recover interference between different values of the measured variable. In the case of interest to us, however, we may regard q as itself a macroscopic variable and x as the coordinates of a very complex environment; then the argument tends to show that interaction with such an environment will destroy any possibility of interference between appreciably different values of q.

This argument is partly right and partly wrong. As we will see, there are many cases in which we can convincingly show that the states of the environment, $\chi_n(x)$, associated with different values q_n of the system variable q are indeed, if not quite orthogonal, very nearly so. However, it should be strongly emphasized that, contrary to the above argument, this does <u>not</u> mean that the coupling to the environment necessarily destroys all possibility of exhibiting interference between states corresponding to different q_n. The reason lies in the crucial distinction between "adiabatic" and "irreversible" effects of interaction with the environment, and to explain this distinction in a qualitative way it is convenient to remind you at this stage about the <u>Born-Oppenheimer (adiabatic) approximation</u>, which will also play an important role in lecture 3.

Let us suppose, for the sake of the argument, that all the variables x of the environment are "fast" compared to the variables q of the system. Exactly what we mean by this needs to be defined self-consistently in the course of the argument, but in many cases our classical intuition gives us a good zero-order idea of whether this condition is fulfilled. For example, the classical and original application of the BO approximation is to the motion of the nuclei ("system") of a small molecule in the presence of the electrons ("environment"). Since the electrons are much lighter than the nuclei, one argues that they can follow the motion of the latter virtually instantaneously. The approximation is then formalized as follows: Consider the total Schrödinger equation of the "universe" (environment plus system), which we can write schematically in the form

$$\left\{ -\frac{\hbar^2}{2M}\frac{\partial^2}{\partial q^2} - \frac{\hbar^2}{2m}\frac{\partial^2}{\partial x^2} + V(q,x) \right\} \psi(q,x) = E\psi(q,x) \tag{9}$$

where M is the nuclear mass and m the electron mass, and the kinetic-energy terms are of course a shorthand for a much more complicated expression involving several different nuclei, etc. Write (9) in the form

$$\left\{ -\frac{\hbar^2}{2M}\frac{\partial^2}{\partial q^2} + \hat{H}_q(x) \right\} \psi(q,x) = E\psi(q,x) \tag{10}$$

where $\hat{H}_q(x)$ depends on q only as a parameter (i.e. it contains no differential operators with respect to q). Now solve the eigenvalue equation

$$\hat{H}_q(x)\chi_q(x) = E_q\chi_q(x) \tag{11}$$

for a normalized wave function $\chi_q(x)$ which depends parametrically on q. Suppose the lowest eigenvalue E_q is $U(q)$ and the corresponding eigenfunction $\chi_q^o(x)$. Then we take as an ansatz for the total ground state wave function of the "universe"

$$\Psi(q:x) = \chi_q^o(x)\phi_o(q) \quad (12)$$

where $\phi_o(q)$ is the ground state solution of the Schrödinger equation

$$\left[-\frac{\hbar^2}{2M}\frac{\partial^2}{\partial q^2} + U(q)\right]\phi_o(q) = E\phi(q). \quad (13)$$

In other words, we solve the energy eigenvalue problem in x for fixed q, and then use the result as an effective potential for the motion of q. The ansatz (12) constitutes the <u>lowest-order Born-Oppenheimer approximation</u> (LOBOA). It is convenient to generalize the ansatz to the case where neither the "fast" nor the "slow" subsystem is necessarily in its ground state: let $\chi_q^{(n)}$ be the n-th excited-state solution of (11), with eigenvalue $U_n(q)$; then the generalization is

$$\Psi_{jn}(q:x) = \chi_q^{(n)}(x)\phi_{jn}(q)$$

where $\phi_{jn}(q)$ is the j-th solution, with eigenvalue E_{jn}, of the Schrödinger equation

$$\left[-\frac{\hbar^2}{2M}\frac{\partial^2}{\partial q^2} + U_n(q)\right]\phi(q) = E\phi(q).$$

Note that the functions $\phi_{jn}(q)$ by themselves are not necessarily mutually orthogonal, since for $n \neq n'$ they are solutions of different equations (i.e. in general we do <u>not</u> have $(\phi_{jn}(q), \phi_{j'n'}(q)) = 0$ for $n \neq n'$); however, since for each value of the different $\chi_q^{(n)}(x)$ are mutually orthogonal, the $\Psi_{jn}(q,x)$ form a complete orthonormal set.

It is now intuitively obvious that <u>to the extent that the LOBOA is valid</u>, the environment can never really destroy the coherence that would have been present for the isolated system. As an example, consider the extreme case of a heavy atom diffracted by a Young's-slits apparatus. In this case the "system" is the nucleus, (so that q is simply the single nuclear coordinate) and the "environment" is the atomic electrons; the function $\chi_q(x)$ has the rather trivial form $\Psi_o(x-q)$, where $\Psi_o(x)$ is the atomic ground state wave function for a nucleus at $q = 0$. As the nucleus passes the screen with the two slits (positioned say at q_1 and q_2) the universe wave function is, schematically

$$\psi(q:x) = \frac{1}{\sqrt{2}}(\delta(q-q_1)\Psi_o(x-q_1) + \delta(q-q_2)\Psi_o(x-q_2)). \quad (14)$$

The off-diagonal element of the density matrix of the nucleus $\rho(q_1, q_2)$ is proportional to the quantity

$$\int \Psi_o^*(x-q_1)\Psi_o(x-q_2)\,dx \quad (15)$$

and provided the distance between the slits is much greater than the extent of the atomic wave functions $\Psi_0(x)$, this is exponentially small. Thus to all intents and purposes the density matrix $\rho(q,q')$ is an incoherent mixture at this stage. Does this mean, then, that we will see no Young's slits interference pattern? Of course not. What happens is that as the two beams from slits q_1 and q_2 converge again towards any particular point on the screen, the associated electronic wave functions also adiabatically move back into coincidence, until at the point where the interference is observed they are totally coincident and do nothing to destroy the coherence. Clearly this phenomenon of "adiabatic reconstruction of coherence" occurs much more generally in any problem for which the LOBOA is a good description. (For a more technical discussion of this point, see ref. (6), section 2.) It is amusing to note that were it not for this consideration, a standard neutron interferometer (see e.g. ref. (7)) could not work, since the states of the radiation field induced (because of the interaction with the neutron magnetic moment) by the neutron on its two possible paths are very nearly orthogonal! (You might want to work this out as a problem.)

The situation may be quite different if corrections to the LOBOA are important. By inserting the ansatz (12) into the complete Schrödinger equation (9) it is found that the terms in the latter which are left out in the LOBOA can be treated formally as a perturbation which in the basis of LOBOA functions $\phi_j^n(q)\chi_q^{(n)}(x) \equiv |n:j\rangle$ may be written

$$H'_{njn'j'} = \int \phi_j^n(q)^* \chi_q^{(n)*}(x) \left\{ -\frac{\hbar^2}{M} \frac{\partial}{\partial q} \phi_{j'}^{n'}(q) \frac{\partial}{\partial q}\chi_q^{(n')}(x) \right.$$

$$\left. - \frac{\hbar^2}{2M} \phi_{j'}^{n'}(q) \frac{\partial^2}{\partial q^2}\chi_q^{(n')}(x) \right\} dx\, dq. \quad (16)$$

Note that both terms can cause transitions in which both n and j change, corresponding to exchange of energy between the "level of excitation" of (e.g.) the electrons and the motion of the nuclei in the adiabatic potential. A crude estimate of the importance of the correction terms (16) in a typical molecule may be obtained by noting that the range of the dependence of $\chi_q(x)$ on q in this case is prima facie comparable to its range of dependence on x; since $-i\hbar\partial/\partial x$ is the electron momentum operator p, we can say that the expression in the last term in brackets in (16) is of the order of $p^2/2M$. On the other hand, in a small molecule the spacing of the electronic levels will be of the order of the mean kinetic energy $p^2/2m$. Thus we would conclude that the expansion parameter, $H_{njn'j'}/\Delta E$, is of order m/M, ($\lesssim 1/2000$). Actually this argument is a bit too simple, because the range of dependence on q of $\phi_0^n(q)$ in the first term of (16) turns out to be of order $(m/M)^{1/4}$ times that of $\chi_q(x)$, so the correct estimate of the corrections in a small molecule is $(m/M)^{3/4}$, which is still small. In a solid this argument does not go through, since the spacing of the electronic levels is very small (~1/volume), and a much more sophisticated consideration (essentially that embodied in Migdal's theorem[8]) is needed. It then turns out that the correct expansion parameter is $(m/M)^{1/4}$: this is still negligible for many purposes, but an important exception is for these phenomena, such as the absorption of ultrasound (nuclear vibrations) by the electrons in a metal, which vanish identically in the LOBOA. It turns out in fact that the ultrasound absorption is proportional to c_S/v_F (c_S = speed of sound, v_F = Fermi velocity), which for reasonably simple models of the metal is, by the Bohm-Staver relation, of order $(m/M)^{1/2}$. In a finite molecule, by contrast, there is strictly speaking never any real dissipation of the small-amplitude nuclear motion into the electronic degrees of freedom, since the electronic levels are discrete (For large enough amplitude, however, one can excite electrons into the continuum above ionization threshold, and then dissipation is possible.).

This distinction between (essentially) "adiabatic" and "dissipative" behavior is of crucial importance in the context of quantum measurement theory. So long as not only the LOBOA but the corrections to it generate only virtual transitions, the conclusions of the above discussion are not qualitatively altered. However, the moment real transitions are allowed the picture changes qualitatively. Suppose, for example, that in our Young's-slits example the atom is traveling fast enough that a collision with the edge of the slit can excite an electron into a continuum state. If this happens, then the electron is no longer tied to the nucleus and will not follow it adiabatically to the point of interference, but rather will propagate off into free space. Moreover, the electron wave packet corresponding to having been stripped off at slit 1 is likely to be nearly orthogonal to that resulting from stripping at 2. Thus, we should expect that on inspection of the state of the electron <u>at any subsequent time</u> should tell us which slit the nucleus went through; the electron has in effect "measured" the position of the nucleus, in a way which it certainly did not in the adiabatic case, and the interference at the final screen will certainly be destroyed.

More generally, it is important to emphasize that (contrary to the impression given in some papers on quantum measurement theory) the mere fact that the system has totally or approximately "orthogonalized" the environment at some intermediate stage of its motion (i.e. that different values of q have induced essentially orthogonal states of the environment) does <u>not</u> guarantee that no recovery of interference is possible; the crucial question is whether or not energy has been transferred <u>irreversibly</u> from the system to the environmental degrees of freedom, and this question cannot be decided without a detailed study of the dynamics of the whole "universe." We will see a specific example of this situation in the next lecture when we discuss the so-called spin-boson problem. I believe that neglect of these considerations may have led to a generally over-pessimistic estimate in the quantum measurement literature of the possibility of observing interference between macroscopically distinct states (another whole lecture's worth at least, see e.g. ref. (9)): it may also be important to take them into account explicitly when discussing the question of phase coherence in so-called mesoscopic systems (see e.g. ref. (10)).

<u>References</u>

1. A. J. Leggett, in *Quantum Implications*, ed. B. J. Hiley and F. D. Peat, Routledge and Kegan Paul, London (1987).

2. P. Hänggi, J. Stat. Phys. **42**, 105 (1986).

3. A. Schmid, J. Low Temp. Phys. **49**, 602 (1982).

4. U. Eckern, W. Lehr, A. Menzel-Dorwarth, F. Pelzer and A. Schmid, preprint.

5. A. J. Leggett, in *Directions in Condensed Matter Physics*, ed. G. Grinstein and G. Mazenko, World Scientific, Singapore, pp. 191-3 (1986).

6. A. J. Leggett, in *Chance and Matter*. (Proc. 1986 Les Houches Summer School), ed. J. Souletie, J. Vannimenus and R. Stora, North-Holland, Amsterdam (1987).

7. D. M. Greenberger, Revs. Mod. Phys. **55**, 875 (1983).

8. A. B. Migdal, Zh. Eksp. Teor. Fiz. **34**, 1438 (1958): translation, Soviet Physics JETP **34**, 996 (1958).

9. A. J. Leggett and Anupam Garg, Phys. Rev. Lett. **54**, 857 (1985).

10. Y. Imry, in *Directions in Condensed Matter Physics*, ed. G. Grinstein and G. Mazenko, World Scientific, Singapore, p. 101 (1986).

QUANTUM MECHANICS OF COMPLEX SYSTEMS, II

A. J. Leggett

Department of Physics
University of Illinois at Urbana-Champaign
1110 West Green Street
Urbana, IL 61801 USA

So far our treatment has been general and schematic. In order actually to calculate the effect of the environment quantitatively, we need an explicit model of the dynamics of the environment and its interaction with the system. At this point one of two situations can arise, corresponding roughly to whether our system is microscopic (and relatively simple) or macroscopic. In the first case, we often have a good a priori knowledge of the Hamiltonian governing the environment and its interaction with the system. This is the case, for example, in small molecules (when for most purposes the total Hamiltonian of the molecule is just the sum of the nuclear and electron kinetic energies and the various Coulomb interactions); it may also be approximately true in other situations of interest in chemical physics or solid-state physics, e.g. in muon diffusion. On the other hand, if we are dealing with the motion of a macroscopic variable, such as the angle made by a pendulum with the vertical, or the relative phase of the Cooper pairs in a Josephson junction, then it is unlikely that we know the details of the microscopic Hamiltonian with any confidence. What is much more likely, and indeed the norm, in this type of case is that we know the <u>effects of the environment on the classical motion of the system</u>, as instantiated for example in the classical friction coefficient. It is then an important question, to which we shall return below, whether a knowledge of such classical effects is sufficient to determine the effects of the environment also on the quantum dynamics of the system.

A model of the environment and its interaction with the system which has been widely used in the discussion of both microscopic and macroscopic systems is the so-called "oscillator bath" model. This introduces canonically conjugate variables x_j, p_j to describe the environment and writes the total Hamiltonian of the universe as

$$H = H_S(p,q) + H_{env} - \sum_j F_j(q) x_j + \frac{1}{2} \sum_j F_j^2(q)/m_j \omega_j^2 \qquad (1)$$

where H_S is the Hamiltonian of the isolated system

$$H_S(p,q) = \frac{p^2}{2M} + V(q) \qquad (2)$$

(p is the momentum canonically conjugate to q) and H_{env} describes a set of simple harmonic oscillators, that is

$$H_{env}\{p_j, x_j\} \equiv \frac{1}{2} \sum_j (p_j^2/m_j + m_j \omega_j^2 x_j^2). \tag{3}$$

(Strictly speaking, the parameter m_j is redundant, in the sense that we can always redefine the oscillator coordinates to get rid of it, but it is convenient to keep it to help our intuition). The last term in (1) is sometimes known as the "counterterm": its significance will be discussed below. Before we examine the justification for the Hamiltonian (1), a few general remarks about it are in order. (For further details, see appendix C of ref. (1)).

(1) The obvious advantage of the form of Hamiltonian (1) is that since it contains only terms linear and bilinear in the bath variables, the latter can be explicitly integrated out of the problem. This can be done by any of a variety of methods - equation of motion, Green's functions, functional integrals, etc.

(2) The simplest case is where the coupling function $F_j(q)$ has the form

$$F_j(q) = q C_j. \tag{4}$$

In that case the whole effect of the environment on the system is encapsulated in the single (q-independent) function

$$J(\omega) \equiv \frac{\pi}{2} \sum_j c_j^2 \, \delta(\omega - \omega_j) / m \omega_j. \tag{5}$$

For an environment with a finite number of degrees of freedom (such as the vibronic modes of a molecule) $J(\omega)$ is a discontinuous function. However, in the more common case of one with a very large number N of degrees of freedom (as is the case for any macroscopic variable, and also for example for a muon diffusing in a metal) we can let $N \to \infty$, and $C_j \to 0$ as $N^{-1/2}$, so that $J(\omega)$ tends to a continuous function which is independent of N in this limit. We will generally assume this case below.

(3) The simplest case of all is when $J(\omega)$ has the form

$$J(\omega) = \eta \omega, \qquad \omega < \omega_c \tag{6}$$

where ω_c is some large frequency (far above any characteristic frequencies of the system). In the literature this is sometimes called the "ohmic" case. In this case it is very easy to show that by writing down the *classical* equations of motion and eliminating the oscillator coordinates we obtain an equation of motion for the system which, neglecting terms of order ω_c^{-1}, reads

$$M\ddot{q} + \eta \dot{q} + \frac{\partial V}{\partial q} = 0 \tag{7}$$

so that the η appearing in eqn. (6) is nothing but the classical friction coefficient. It should be carefully noted (a) that the $V(q)$ appearing in the damped equation of motion (7) is just the original potential energy of the isolated system (see eqn. (2)), so that there is no renormalization of the potential by the environment, and (b) that this is only so because of the presence of the last term ("counterterm") in eqn. (6). Whether or not this counterterm should indeed be added must be decided by asking whether or not we expect the environment to provide not just dissipative effects as in eqn. (7) but also a renormalization of the reactive system parameters: in most cases of practical interest we do not, so the counterterm should be included.

(4) In the more general case described by (1), even with the counterterm included, it is impossible to prevent the environment inducing also some reactive effects on the classical motion of the system (the counterterm in effect ensures only that these will vanish in the static limit, cf. ref. (2)). However, there is one important general point we should note about eqn. (1): The presence of the counterterm automatically guarantees that the minimum value of the total potential energy attained on the hyperplane corresponding to given q is just the original $V(q)$. Thus, whenever q lies in a region which for the isolated system was classically forbidden it is intuitively obvious that coupling to the environment according to eqn. (1) cannot increase the probability of tunnelling into this region. Again, we emphasize that this is only true because of the inclusion of the "counterterm."

We now turn to the question of the justification of eqn. (1). Consider first the case where q represents a microscopic variable, so that we might hope to know the microscopic Hamiltonian of the "universe" a priori. In the first place, there are a number of instances where this Hamiltonian is indeed explicitly of the form (1): obvious examples are the case of an electron participating in a chemical reaction, where the "environment" is the vibronic modes (nuclear vibrations) of the molecule, and the case of a paraelectric defect tunnelling in an insulating solid (where the environment is the phonons). In each case, to the extent that we linearize the system-environment interaction in the environment coordinates, an expression of the form (1) immediately falls out: note however that whether or not we include the counterterm depends on our choice of initial potential $V(q)$.

A second class of "microscopic" cases is where the explicit microscopic Hamiltonian is not explicitly of the form (1), but can be recast into this form provided we are interested in the environment only for its effect on the system. An important case of this type is the interaction of (say) a diffusing muon with the electrons of a metal. It is clear that since the electrons are fermions, the microscopic Hamiltonian is certainly not of the form (1). However, in recent years it has been shown in a series of papers[3-5] of increasing generality that provided it is indeed only the effect of the environment on the system that is of interest, this effect may be mimicked by an effective Hamiltonian of the form (1) with suitably chosen values of the parameters. (It should be noted in particular that if the original interaction with the electrons was linear in q, then the interaction with the fictitious oscillators described by eqn. (1) will not be: see ref. (5)). A second example (this time involving a "macroscopic" variable) of a microscopic Hamiltonian which is not of the form (1) but can be converted into it for our purposes is the interaction of the phase variable in a Josephson junction with the electrons of the normal component: see refs. (6) and (7). Yet another example is a class of tunneling phenomena in solids where the "one-phonon" interaction which would correspond explicitly to eqn. (1) is forbidden for reasons of symmetry, and the first relevant interaction is bilinear in the phonon coordinates: once again this problem can be cast for our purposes into the form (1) by introducing a bath of oscillators whose parameters are explicitly temperature-dependent[8]. (Whether this somewhat Procrustean maneuver is helpful to our physical intuition may of course be legitimately regarded as a matter of opinion.)

It should be emphasized that although the effective Hamiltonian (1) is adequate for many cases involving a microscopic variable, it is not universally applicable. In particular, it cannot necessarily cope with the case of nuclear motion in molecules (or a fortiori in solids) once we need to go beyond the lowest order in the ratio m/M, as may be necessary for a discussion of "cold fusion," and quite different techniques may be needed for this problem (see lecture 3).

The question of the justification of eqn. (1) in the case of a macroscopic variable when the microscopic Hamiltonian is not known in detail is a good deal more delicate, and is discussed in some detail in appendix C of ref. (1) and in section (2) of ref. (2). In such a case our starting point has to be the classical equation of motion of the system, which incorporates the damping and reactive effects, if any, of the environment. It is then fairly easy to show (ref. (1), pp. 438-441) that <u>provided</u> any one degree of freedom of the environment is only weakly perturbed by the motion of the system, the latter can always be represented by a bath of oscillators and the coupling of the system to it by a set of terms linear in the oscillator coordinates and momenta. The question of whether, by a suitable choice of coordinates, it can always be put in the specific form (1), is much more difficult and a completely rigorous proof[2] is available only for the case that the coupling is known a priori to be linear in the <u>system</u> coordinates and moreover to be invariant under time reversal. In the more general case there may even be counter examples (see ref. (6), p.296) but fortunately they do not seem to be of much practical importance.

One very important general case in which we can justify more or less rigorously the use of the form of Hamiltonian (1), is the case of a macroscopic variable for which the "bare" Hamiltonian is that of the lowest-order Born-Oppenheimer approximation (LOBOA) reviewed in lecture 1. In view of the importance of this approximation in the context of these lectures, it is worth sketching the derivation (see ref. (1), appendix C for more details). As in lecture 1, we write the LOBOA wave functions in the form (changing the notation slightly)

$$\Psi_{jn}(q,x) = \phi_{jn}(q)\chi_n(x:q). \qquad (8)$$

The correction terms ΔH are of the form

$$\langle jn|\Delta H|j'n'\rangle = -\frac{\hbar^2}{2M}\int dq \int dx \left\{ \phi^*_{jn}(q)\chi_n(x:q)\left[2\frac{\partial}{\partial q}\phi_{j'n'}(q)\frac{\partial}{\partial q}\chi_{n'}(q:x)\right.\right.$$

$$\left.\left. + \phi_{j'n'}(q)\frac{\partial^2}{\partial q^2}\chi_{n'}(q:x)\right]\right\}. \qquad (9)$$

Note that this description is always formally valid, even when the correction terms ΔH are so large as to make the LOBOA a very poor approximation.

The following observation is now crucial: If the variable q is indeed macroscopic (and in some cases even where it is not) the states which are coupled together by ΔH differ only in the behavior of a small fraction ($\sim 1/N$) of the macroscopic number of environmental particles involved, and consequently the potential $U_n(q)$ which occurs in the Schrödinger equation for $\phi_{jn}(q)$ is essentially independent of the "microscopic" index, or more precisely can be separated into a part $U(q)$ which is independent of n and a term H_{env} which is a function only of the x's and p's and is independent of q (and hence cannot affect the form of $\phi_j(q)$). (This is one point at which the argument is not totally rigorous.) Thus, $\phi_{jn}(q)$ is itself essentially independent of n and can be written simply $\phi_j(q)$, and the terms in the expression for ΔH effectively factorize into a piece depending only on j,j' and a piece depending only on n,n'. In fact, if we define the Hermitian "momentum-like" operator $K(q)$ by its matrix elements

$$\langle n|\hat{K}(q)|n'\rangle \equiv \int \chi_n^*(x:q)\,(-i\hbar \frac{\partial}{\partial q})\,\chi_{n'}(x:q)\,dx \tag{10}$$

then after a bit of algebra we can write (since for given q the eigenfunctions $\chi_n(x:q)$ form a complete set)

$$\langle jn|\Delta H|j'n'\rangle = \frac{1}{2M}\langle j|pK_{nn'}(q) + K_{nn'}(q)p + K^2_{nn'}|j'\rangle \tag{11}$$

where p is just the usual momentum operator acting on the system wave functions $\phi_j(q)$. Since the other terms in the Hamiltonian were the "slow" kinetic energy $p^2/2M$, the n-independent potential $U(q)$ and the quantity H_{env} which depends only on the environmental variables, the total Hamiltonian can be written

$$\hat{H} = \frac{1}{2M}(p + \hat{K}(q))^2 + U(q) + \hat{H}_{env} \tag{12}$$

where the circumflex denotes operators with respect to the environmental variables. Note that this Hamiltonian is formally identical to that of a charged particle in a one-dimensional electromagnetic vector potential $-K(q)/e$.

We now argue as in the general case that provided any one environmental degree of freedom is only weakly perturbed (as we have already assumed) then H_{env} can be represented as the Hamiltonian of a set of harmonic oscillators, and the operator $K(q)$ can be expanded as a sum of terms linear in the oscillator coordinates x_j and the corresponding canonically conjugate momenta p_j. If we choose the oscillator coordinates in the "obvious" way, then the time-reversal invariance of ΔH implies that only terms in p_j can enter: thus,

$$\hat{H} = \frac{1}{2M}(p + \sum_j K_j(q)\,\hat{p}_j)^2 + U(q) + \hat{H}_{SHO}. \tag{13}$$

This is not in the "canonical" form (1), but it can be put into that form by a canonical transformation. The easiest way to see this is to note that since for an isolated oscillator the "coordinates" and "momentum" enter on an exactly footing, we may make the exchange $p_j \rightleftarrows m_j\omega_j x_j$, go to the Lagrangian description, add to the Lagrangian a total time derivative $d/dt \left[\sum_j m_j\omega_j x_j(t) \int_0^s K_j[q'(t)]\,dq'\right]$ which has no physical effect,* go back to the Hamiltonian description and finally reverse the exchange $p_j \rightleftarrows m_j\omega_j x_j$. The result of this rather inelegant series of operations is a Hamiltonian of precisely the form of eqn. (1), with the coupling constants $F_j(q)$ given by

$$F_j(q) = (m_j\omega_j)^2 \int_0^q K_j(q')\,dq'. \tag{14}$$

Note that the counter-term in eqn. (1) has arisen as an automatic result of this procedure. This is of some importance in connection with general arguments about the sign of effects of corrections to the LOBOA on expressions for tunneling, etc. (cf. lecture 3).

We conclude that for almost all practical purposes the description of the interaction of any macroscopic system, and many microscopic systems, with their environments is adequately modelled by the "canonical" Hamiltonian of eqn. (1), with the counter term included. From now on we shall assume this form without further comment.

An important further question is whether, given the general form (1), the form of the classical equation of motion uniquely fixes the parameters $F_j(q)$, etc. In the "strictly linear" case described by eqn. (4), the answer is yes, in the sense that while the classical motion does not uniquely fix the individual C_j, ω_j etc., it does fix the spectral function $J(\omega)$ (which, as we shall see, in turn uniquely determines the quantum dynamics). In the more general case a plausible but not rigorous answer, given in appendix C of ref. (1), is that a __complete__ knowledge of the classical dynamics will indeed fix the quantity corresponding to $J(\omega)$ uniquely, but that the kind of knowledge we are likely to have in practice will not do so in the absence of some a priori knowledge about the interaction.

We now turn to some simple applications of the "canonical" model. Let us first briefly discuss the exactly soluble problem of the damped harmonic oscillator (for more details, see appendix B of ref. (1)).** Suppose it is known that the _exact_ classical equation of motion is

$$M\ddot{q} + \eta\dot{q} + \omega_0^2 q = F(t) \tag{15}$$

where $F(t)$ is some prescribed external force. Then, according to the above considerations, there is an essentially unique choice of Hamiltonian for the "universe" given by eqn. (1) with the choice $F_j(q) = qC_j$, $J(\omega) \equiv \frac{\pi}{2}\sum_j C_j^2 \delta(\omega-\omega_j)/m_j\omega_j = \eta\omega$. As a result, the universe Hamiltonian is a bilinear form in the variables p, q, p_j, x_j and can be exactly diagonalized. Without going through this procedure explicitly, we can immediately deduce some general properties of the behavior of the system. Consider the reduced density matrix of the system at temperature $T \equiv 1/k_B\beta$:

---

* It should be carefully noted that this step is only valid so long as the region of q of interest is not multiply connected. In the case of a phase-like variable under conditions when the identity of 0 and 2π is relevant, extreme care is needed: cf. ref. (9).

** A detailed discussion of this system, and further references, may be found in ref. (10) where, however, the emphasis is somewhat different.

$$\rho(q,q':\beta) \equiv Z^{-1} \sum_n e^{-\beta E_n} \prod_j \int dx_j \, \psi_n^*(q,x_j) \psi(q',x_j)$$

$$(Z \equiv \sum_n e^{-\beta E_n}). \tag{16}$$

Now, it is very easy to show that for a single (undamped) harmonic oscillator with coordinate y_j the density matrix is of the general form const. $\exp-P(y_j, y_j')$, where P is bilinear and symmetric in the variables y_j, y_j'. The complete "universe" density matrix is of the form $\exp-\sum_j P_j(y_j, y_j')$: since the variables q and x_j are related to the y_j by a linear transformation and the x_j are integrated over in (16), it is bilinear and symmetric in q,q'. Thus, if we introduce the sum and difference variables X, ξ by

$$X \equiv \frac{1}{2}(q+q'), \quad \xi \equiv q-q', \tag{17}$$

the most general possible form of $\rho(X,\xi)$ is

$$\rho(X,\xi,\beta) = C_0 \exp - \frac{1}{2}(\lambda^{-1} X^2 + \mu\xi^2) \tag{18}$$

where in general λ, μ and C_0 will be functions of the parameters ω_0 and η and of the inverse temperature β.

We can immediately relate λ and μ to the mean-square fluctuations of the position and momentum of the damped oscillator:

$$\langle q^2 \rangle \equiv \int X^2 \rho(X,0:\beta) dX = \lambda \tag{19}$$

$$\langle p^2 \rangle \equiv \int dX \int p^2 dp \, \exp(ip\xi/\hbar)\rho(X,\xi:\beta) = \hbar^2 \mu, \tag{20}$$

and so ρ can be written

$$\rho(q,q') = C_0 \exp-\frac{1}{2}\left\{\frac{\frac{1}{2}(q+q')^2}{\langle q^2 \rangle} + \frac{\langle p^2 \rangle}{\hbar^2}(q-q')^2\right\} \tag{21}$$

To calculate $\langle q^2 \rangle$ and $\langle p^2 \rangle$ in terms of the parameters of the model, we use the fact that the position and momentum response functions for a single (undamped) quantum harmonic oscillator are identical to those for the corresponding classical oscillator, and that since the coordinate of the actual damped oscillator is a linear combination of those for the normal-mode oscillators, the same must hold for its response functions. Thus the (quantum) response functions of the system are given by the expressions which follow from the classical equations of motion (15), namely

$$\chi_q(\omega) = \frac{1}{M(\omega_0^2 - \omega^2 - 2i\gamma\omega)} \quad (\gamma \equiv \eta/2M), \quad \chi_p(\omega) = M^2\omega^2 \chi_q(\omega). \tag{22}$$

But the mean-square fluctuation $\langle q^2 \rangle$ is given in terms of the imaginary part of $\chi_q(\omega)$ by the fluctuation-dissipation theorem

$$\langle q^2 \rangle = \frac{\hbar}{\pi} \int_0^\infty \coth(\beta\hbar\omega/2) \; \text{Im} \; \chi_q(\omega) \, d\omega \qquad (23)$$

and similarly for $\langle q^2 \rangle$. Consequently we have

$$\langle q^2 \rangle = \frac{\hbar}{M\pi} \int_0^\infty \frac{\coth(\beta\hbar\omega/2) \cdot 2\gamma\omega}{(\omega_0^2 - \omega^2)^2 + 4\gamma^2\omega^2} d\omega \qquad (24)$$

$$\langle p^2 \rangle = \frac{M\hbar}{\pi} \int_0^\infty \frac{\coth(\beta\hbar\omega/2) \cdot 2\gamma\omega^3}{(\omega_0^2 - \omega^2)^2 + 4\gamma^2\omega^2} d\omega. \qquad (25)$$

To make the expression for $\langle p^2 \rangle$ finite, we need to impose some high-frequency cutoff ω_c, say, on the integral: apart from this, we see that $\langle q^2 \rangle$ and $\langle p^2 \rangle$ are functions only of the phenomenological parameters of the problem, ω_0 and γ (or η).

Without trying to evaluate $\langle q^2 \rangle$ and $\langle p^2 \rangle$ explicitly for the general case, we note some important qualitative features of the results:

(1) In the limit $T \to \infty$ ($\coth(\beta\hbar\omega/2) \to 2/\beta\hbar\omega$) we recover the classical "equipartition" results $\langle q^2 \rangle = kT/M\omega_0^2$, $\langle p^2 \rangle = MkT$, which are completely independent of the degree of damping γ. Thus in this limit the thermal equilibrium density matrix (16) knows nothing about the damping (although the dynamics of course remains sensitive to it). However, it should be noted that the "crossover temperature" T_0 at which $\langle q^2 \rangle$ approaches its classical form does depend on γ: for small γ, kT_0 is of order $\hbar\omega_0$, while for $\gamma \gg \omega_0$ we have $kT_0 \sim (\hbar\omega_0^2/\gamma) \ln(\omega_c/\gamma)$. The corresponding "crossover temperature" for $\langle p^2 \rangle$ is much higher, of order $\hbar\omega_c$.

(2) At zero temperature and in the limit of zero damping (which is of course a special case of the above formulae) we obtain $\langle q^2 \rangle = \hbar/2M\omega_0$, $\langle p^2 \rangle = M\hbar\omega_0/2$ as we would get from an elementary analysis. Since $\langle q \rangle = \langle p \rangle = 0$, the uncertainties Δq, Δp are equal to $\langle q^2 \rangle^{1/2}$, $\langle p^2 \rangle^{1/2}$ respectively, and so in this limit the uncertainty principle $\Delta p \Delta q \geq \hbar/2$ is fulfilled as an equality. If we now increase the damping (still at $T = 0$) analysis of eqns. (24-5) shows that Δq decreases while Δp increases; however, the increase of Δp is much more dramatic (e.g. for $\gamma \sim \omega_0$, we have Δq still $\sim \hbar/2M\omega_0$, but $\Delta p \sim (M\hbar\omega_0) \ln(\omega_c/\omega_0)$.) Thus, generally speaking, for a system with any appreciable degree of damping we have $\Delta p \Delta q \gg \hbar/2$.

(3) A qualitative interpretation of the result (2) is that any interaction of the type described by eqn. (1), where the coupling to the environment is through a function of position q, in effect acts as a "position measurement" and tends to decrease the superposition of states corresponding to different q and hence to contract the wave packet in q-space. However, it is essential to distinguish between the question of spread of the wave packet and that of coherence between different q-values. The first is essentially measured directly by $\langle q^2 \rangle^{1/2}$: the second is defined by asking how different q and q' have to be before the density matrix essentially decays to zero, i.e. it is effectively measured by the quantity $\sqrt{\hbar^2/\langle p^2 \rangle}$. For zero damping the two quantities are essentially identical, but we see that for any reasonable degree of

damping the "coherence length" is very much less then the "spread" (rms uncertainty in position).

(4) The above is related to a general point which I emphasized in lecture 1: a high degree of apparent "destruction of coherence" need not necessarily imply the total disappearance of quantum-mechanical effects provided that this destruction is adiabatic in nature. To be sure, in the present case it is not entirely trivial to say exactly what we mean by "quantum-mechanical effects," since as we already observed the quantum and classical dynamics of the damped harmonic oscillator is identical. What is clear, however, is that even a very large degree of destruction of the off-diagonal elements of the density matrix is compatible with an equation of motion which is only weakly damped and hence qualitatively similar to that of the isolated system. The reason is that most of the destruction comes from the environmental oscillators with $\omega \sim \omega_c \gg \omega_0$, i.e. is largely adiabatic in nature, whereas the dissipative effects on the dynamics come entirely from oscillators with $\omega \lesssim \omega_0$.

We go on to a second application, namely the well-known "spin-boson" problem. Suppose we have a system which has available to it two potential wells separated by a potential barrier: for simplicity we will assume that the potential is symmetric around the barrier top, which is located at q = 0, and we take the bottoms of the wells to lie at q = ±q$_0$/2. A well-known microscopic example of this situation is the NH3 molecule, but it can occur in other systems as well, including some where the variable q is macroscopic (e.g. the flux in an rf SQUID, see ref. (11)). As is well known, for an isolated system of this type, the ground state is split by tunnelling through the barrier and, if the system is started in one well, the probability of finding it there oscillates in time.

Suppose now we wish to introduce dissipation by coupling the system to its environment. For this purpose we can use the "canonical" Hamiltonian (1), and in any specific case the parameters entering it can be determined from the classical motion. Now, if we are interested in the behavior at sufficiently low temperatures (say, small compared to $\hbar$ times the frequency of small oscillations in the individual wells), then it is intuitively plausible that the behavior <u>within</u> each well is fixed and the only question of interest concerns the motion between the two wells. This idea is formally substantiated by truncating the possible states of the system to two "states" corresponding to positions ±q$_0$/2 respectively, and projecting the full Hamiltonian onto this subspace. The details of this procedure are straightforward but somewhat tedious, and I will not give them here: see appendix A of ref. (12). The upshot is that the low-temperature motion of the system is well described by the so-called "spin-boson" (two-state) Hamiltonian:

$$\hat{H} = -\frac{1}{2}\Delta_0\sigma_x + \frac{1}{2}\sum_j (p_j^2/2m_j + m_j\omega_j^2 x_j^2) - \frac{1}{2}q_0\sigma_z \sum_j C_j x_j \qquad (26)$$

where σ_x and σ_z are Pauli matrices (the "two wells" correspond to $\sigma_z = \pm 1$ respectively).* Note that this is a special and rather simple case of the general "canonical" Hamiltonian (1), corresponding to the "strict linearity" condition (4) (it may be shown that any more complicated behavior of $F_j(q)$ in the original problem disappears during the truncation procedure, see ref. (12)). Consequently, the effect of the environment is completely encapsulated in the spectral function $J(\omega)$ defined, as in the more general case, by

$$J(\omega) = \frac{\pi}{2} \sum_j c_j^2 \, \delta(\omega - \omega_j) / m_j \omega_j. \qquad (27)$$

In general we will assume that $J(\omega)$ is proportional to $\omega^n e^{-\omega/\omega_c}$, where ω_c is some arbitrary "microscopic" cutoff frequency introduced by the truncation procedure (in practice usually of the order of the small-oscillation frequency in the individual wells) which is $\gg \Delta$. It turns out that a consistent application of the truncation process means that the "bare tunnelling frequency" Δ which appears in eqn. (26) is itself a function of ω_c, so that the apparent dependence on ω_c which arises from its explicit occurrence in $J(\omega)$ is cancelled in the final answers. In the special case of so-called "ohmic" friction, where $J(\omega) = \eta\omega$ at small ω, the quantity

$$\frac{q_0^2}{2\pi\hbar} \left(\frac{J(\omega)}{\omega} \right) \equiv \frac{\eta q_0^2}{2\pi\hbar} \equiv \alpha \qquad (28)$$

is dimensionless and it is useful to have a special notation for it.

The particular interest of the deceptively simple-looking Hamiltonian (16) in the context of the general theory of quantum systems interacting with a complex environment is that it brings out with particular clarity the competition of two opposing tendencies: The term $\frac{1}{2}\Delta_0 \sigma_x$ is nondiagonal in the "natural" (σ_z-, two-well) representation and hence tends to produce linear combinations (even- and odd-parity ones) of the single-well eigenfunctions, i.e. to "delocalize" the system. On the other hand the third term is an interaction with the environment which is diagonal in the two-well representation, and therefore, according to the standard rule of thumb of quantum measurement theory, tends to force σ_z to take a definite value, i.e. to localize the system in one well or the other. The essential conclusions of the considerations which follow are (1) that the outcome of the competition between these two tendencies depends critically on the behavior of the environment spectral density at low frequencies ($\lesssim \Delta_0$), and (2) that a substantial (indeed near-complete) destruction of coherence at an intermediate stage of the motion need not imply disappearance of the characteristically quantum-mechanical oscillation phenomena.

Let us first consider the nature of the ground state of the system. Suppose we were to set $\Delta_0 = 0$ in eqn. (26). Then it is clear that for given σ_z the ground state is just obtained by shifting the oscillator wave functions to a new origin displaced by the amount

---

* An identical form of Hamiltonian occurs in quantum optics, when one considers transitions between a particular pair of atomic states in the presence of the radiation field. Note that in that case the two atomic levels correspond to the even-and-odd parity combinations of the single-well eigenfunctions (i.e. to eigenstates of σ_x, not σ_z in the above notation).

$$a_j = \frac{1}{2} q_0 \sigma_z \, c_j/m_j\omega_j^2 \tag{29}$$

with a resulting gain in energy (the "solvation energy" in the language of chemistry)

$$E_{solv} = -\frac{1}{2} m\omega_j^2 a_j^2 = -\frac{1}{8} q_0^2 \sum_j c_j^2/m_j\omega_j^2 = \frac{q_0^2}{4\pi} \int_0^\infty \frac{J(\omega)\,d\omega}{\omega} \tag{30}$$

which is finite for all reasonable forms of $J(\omega)$. In this limit, therefore, the ground state is doubly degenerate, corresponding to the possible choices $\sigma_z = \pm 1$, which are not connected by the Hamiltonian. Let the correspondingly eigenstates be written $\psi_\uparrow^{(0)}\{x_j\}|\uparrow\rangle$, $\psi_\downarrow^{(0)}\{x_j\}|\downarrow\rangle$. However, if we now let Δ_0 be finite, the states $\sigma_z = \pm 1$ are connected, and we might try to take this into account to a first approximation by simply forming the linear combination

$$\psi(\sigma_z, x_j) = \frac{1}{\sqrt{2}} (\psi_\uparrow^{(0)}\{x_j\}|\uparrow\rangle + \psi_\downarrow^{(0)}\{x_j\}|\downarrow\rangle). \tag{31}$$

Now it is obvious that by displacing the oscillators differently for the $\uparrow$ and $\downarrow$ states we have lost much of the "coherence" energy corresponding to the term $\Delta_0 \sigma_x$: in fact, we have

$$E_{coh} = -\langle \tfrac{1}{2}\Delta_0 \sigma_x \rangle = -\frac{1}{2}\Delta_0 \prod_j (\psi_\uparrow^{(0)}(x_j), \psi_\downarrow^{(0)}(x_j)). \tag{32}$$

We can work out the inner product $(\psi_\uparrow^{(0)}(x_j), \psi_\downarrow^{(0)}(x_j))$ by using the fact that the shifted ground state $\psi_\uparrow^{(0)}(x_j)$ may be written in the form $(\exp i\hat{p}_j a_j/\hbar)\psi_0(x_j)$ where ψ_0 is the unshifted ground state and $\hat{p}_j$ the momentum operator. Using standard results for Gaussian integrals we obtain

$$(\psi_\uparrow^{(0)}(x_j), \psi_\downarrow^{(0)}(x_j)) = \exp - 2a_j^2 \frac{\langle p_j^2\rangle_0}{\hbar^2} = \exp - m_j \frac{\omega_j a_j^2}{\hbar} \tag{33}$$

and so

$$E_{coh} = -\frac{1}{2}\Delta_0 \exp - \sum_j \hbar^{-1} m_j \omega_j a_j^2$$

$$= -\frac{1}{2}\Delta_0 \exp -\{q_0^2/2\pi\hbar \int_0^\infty \frac{J(\omega)}{\omega^2} d\omega\}. \tag{34}$$

Note that for $n \le 1$ the integral is divergent at the lower limit and E_{coh} hence zero. It is clear that losing all the coherence energy in this way is not necessarily the best way to achieve a good ground state, so the following procedure suggests itself (cf. ref. (13)): Let us continue to take our trial wave function, as above, to be of the general form

$$\psi(\sigma_z, x_j) = \frac{1}{\sqrt{2}} (\psi_\uparrow\{x_j\}|\uparrow\rangle + \psi_\downarrow\{x_j\}|\downarrow\rangle) \tag{35}$$

where $\psi_{\uparrow,\downarrow}(x_j)$ are products of shifted oscillator wave functions $\psi(x_j \pm a_j)$ for each oscillator, but now treat the shift a_j as a variational parameter. It is clear that with such a trial wave function we obtain $\langle E \rangle = E_{solv} + E_{coh}$, where

$$E_{solv} = \sum_j \frac{1}{2} m_j \omega_j^2 a_j^2 - q_o \sum_j C_j a_j \tag{36}$$

$$E_{coh} = -\frac{1}{2} \Delta_o \exp - \sum_j \hbar^{-1} m_j \omega_j a_j^2 \equiv -\frac{1}{2} \Delta\{a_j\}. \tag{37}$$

Minimizing $\langle E \rangle$ with respect to a_j gives

$$a_j = \frac{q_o C_j}{m_j \omega_j^2 + m_j \omega_j \Delta/\hbar} \tag{38}$$

and substituting this back into the expression for Δ gives the self-consistent equation

$$\Delta = \Delta_o \exp - \sum_j \frac{1}{m_j \omega_j} \frac{1}{(\omega_j + \Delta/\hbar)^2}$$

$$\equiv \Delta_o \exp - \left\{ \frac{q_o^2}{2\pi\hbar} \int_0^\infty \frac{J(\omega) d\omega}{(\omega + \Delta/\hbar)^2} \right\}. \tag{39}$$

We see that Δ has the physical significance of a <u>renormalized</u> tunnelling matrix element. From (38) we see that, if Δ is finite, the optimum displacement of oscillators with frequency ω is determined primarily by the coherence energy for $\omega < \Delta$ and by the solvation energy for $\omega > \Delta$.

The solution we have obtained is based on a variational wave function and cannot therefore be expected to be exact; however, many of the qualitative features, at least, appear to be reflected in more rigorous treatments of the <u>dynamics</u> of the problem (on which see below). One point deserves particular note: Let us write (39) in the form

$$\ln(\Delta_o/\Delta) = \frac{q_o^2}{2\pi\hbar} \int_0^\infty \frac{J(\omega) d\omega}{(\omega + \Delta/\hbar)^2}. \tag{40}$$

As a function of Δ, the left-hand side starts at zero when $\Delta = \Delta_o$ and tends to ∞ as $\Delta \to 0$, while the right-hand side is positive and finite for $\Delta = \Delta_o$. Thus, if the right-hand side is finite in the limit $\Delta \to 0$, there must exist a solution for finite Δ. This is the case whenever the power n in $J(\omega) \sim \omega^n$ is greater than 1, and then for small Δ_o/ω_c the value of Δ is approximately given by

$$\Delta \cong \Delta_o \exp - \int_0^\infty \frac{J(\omega)}{\omega^2} d\omega \equiv \Delta_o F \tag{41}$$

which is just the familiar Franck-Condon expression. Note that $\Delta = 0$ is <u>not</u> a possible solution in this case. The ground state is a nontrivial superposition of $|\uparrow\rangle$ and $|\downarrow\rangle$ correlated with appropriate shifted oscillator functions. For n < 1 the right-hand side of (40) diverges for $\Delta \to 0$ and it is clear that $\Delta = 0$ is a possible solution (and in fact generally speaking the only one). But the situation $\Delta = 0$ corresponds to total loss of coherence: the oscillators essentially ignore the coherence energy entirely and displace themselves so as to minimize the solvation energy. The density matrix of the 2-state system is then simply a mixture of the states $\psi_\uparrow^{(o)}\{x_j\}|\uparrow\rangle$ and $\psi_\downarrow^{(o)}\{x_j\}|\downarrow\rangle$, which tells us that there are in fact two degenerate ground states, corresponding to localization in one or the other well.

Thus, at zero temperature we have two qualitatively different outcomes: For n > 1 the system is delocalized in the ground state, while for n < 1 it is localized. What happens for the "marginal" case n = 1? In this case explicit evaluation of the right-hand side of (40) gives $\alpha[\ln(\omega_c/\Delta)+c]$ where the constant c is of order 1. Since we have assumed $\omega_c \gg \Delta_o$, for $\alpha > 1$ eqn. (40) has only the solution $\Delta = 0$, while for $\alpha < 1$ it has the nontrivial solution

$$\Delta = \Delta_o \left(\frac{\Delta_o}{\omega_c}\right)^{\frac{\alpha}{1-\alpha}}. \tag{42}$$

(It actually turns out that Δ_o is itself proportional to $\omega_c^{-\alpha}$ (see ref. (12), appendix A), so that the apparent dependence on the arbitrary cutoff ω_c drops out of the final result.) A result analogous to (42) is familiar in the Kondo problem. Thus, as regards the statics the "ohmic" case with $\alpha = 1$ is the point of cross over between two qualitatively different types of behavior. More sophisticated treatments (e.g. refs. (14,15)) confirm this result.

We should expect that this pattern of behavior would be reflected in the dynamics of the spin-boson problem, and indeed it is (see ref. (12)). Actually, the dynamics is in some sense an easier problem than the statics, at least if one starts from "natural" initial conditions (e.g. σ_z known to be +1, oscillators relaxed under that condition) and is not interested in the behavior on exceedingly long time scales. The reason for this somewhat surprising circumstance is rather trivial: to obtain the dynamics under these conditions one can, in principle at least, do finite order perturbation theory in Δ_o, whereas to obtain the static behavior one needs to go to infinite order. The relevant calculations for the dynamics are presented in detail in ref. (12) and I shall just quote the results. One finds that for n > 1 the system behaves (at zero temperature) qualitatively like the undamped system, that is, it oscillates coherently between the two wells, with however a frequency $\Delta/\hbar$ which is reduced from the "undamped" frequency $\Delta_o/\hbar$ by the Franck-Condon factor (41); the oscillation also undergoes a weak decay in time, with a decay rate of order $J(\Delta)$ which is in general much smaller than Δ itself. For n < 1 the system is completely localized at T = 0, and for T > 0 undergoes an <u>incoherent</u> relaxation between the wells as in the classical "rate theory" used in chemical physics. The "ohmic" case n = 1 shows a particularly rich pattern of behavior as a function of the dimensionless dissipation coefficient α and temperature T: for α and T both small we get underdamped oscillation, and for either large incoherent relaxation, while in a small region of low T and $\frac{1}{2} < \alpha < 1$ the behavior appears to be describable by no simple picture. Thus, the spin-boson problem shows with particular clarity that even the qualitative nature of the motion of a quantum system coupled to an environment is sensitive to the form of the environmental spectral density at and below the characteristic system frequencies.

One more point needs to be emphasized: In the case $n > 1$, the lowest relevant states are approximately of the form

$$\psi_\pm = \frac{1}{\sqrt{2}} (\psi_\uparrow^{(o)}(x_i)|\uparrow\rangle \pm \psi_\downarrow^{(o)}(x_i)|\downarrow\rangle) \tag{43}$$

and a general wave function in this subspace can therefore be written

$$\psi(t) = \alpha(t)\, \psi_\uparrow^{(o)}(x_i)|\uparrow\rangle + \beta(t)\, \psi_\downarrow^{(o)}(x_i)|\downarrow\rangle. \tag{44}$$

Consider now the reduced density matrix of the system ("spin") at a time when $\alpha(t)$ and $\beta(t)$ are both appreciable (say at $\frac{1}{4}$ of the cycle period). In the completely uncoupled case $c_i = 0$ (when, trivially, $\psi_\uparrow^{(o)}(x_i) = \psi_\downarrow^{(o)}(x_i) = \psi_o(x_i)$) we obviously would have

$$\hat{\rho} = \begin{pmatrix} |\alpha(t)|^2 & \alpha^*\beta \\ \alpha\beta^* & |\beta|^2 \end{pmatrix}. \tag{45}$$

In the actual case of interest we have

$$\hat{\rho} = \begin{pmatrix} |\alpha(t)|^2 & \alpha^*\beta F \\ \alpha\beta^* F & |\beta(t)|^2 \end{pmatrix} \tag{46}$$

where F is the Franck-Condon factor defined by eqn. (41). In a realistic case, F may well be very much smaller than unity, and it is therefore tempting to argue that since the off-diagonal elements of the density matrix are so small, the amplitude of any future coherence effects in the system must be reduced by a similar factor. Not at all! If we wait for a complete half-cycle or cycle (say) we will find the probability amplitude has oscillated in essentially the same way as in the isolated system with only a small amount of damping ($\propto J(\Delta)$, which in such cases is proportional to Δ^n and hence in general $\ll \Delta$). The <u>time scale</u> of the oscillations has been much lengthened by the interaction with the environment (mostly, with the high-frequency part of the latter) but the <u>amplitude</u> is barely affected. This brings out spectacularly the general point I have stressed throughout: The mere fact that at some intermediate stage the density matrix of the system is nearly diagonal does <u>not</u> necessarily prevent the recovery of coherence. In this case as in many similar ones, the reduction of the density matrix is mainly due to the <u>high-frequency</u> oscillators and is therefore mostly adiabatic in nature, while any true destruction of coherence has to come from the much smaller group of oscillators which have $\omega \sim \Delta$, i.e. which can absorb energy at the characteristic frequency of the system. It is clear that this is a special case of the general point made in lecture 1 in the context of the LOBOA: indeed, it is clear that eqn. (43) is precisely of the LOBOA form, and it is the small corrections to it which will give the damping. I make no apology for reiterating this point, which I believe is not nearly widely enough appreciated.

<u>References</u>

1. A. O. Caldeira and A. J. Leggett, Ann. Phys. **149**, 374 (1983).
2. A. J. Leggett, Phys. Rev. **B30**, 1208 (1984).
3. C. C. Yu and P. W. Anderson, Phys. Rev. **B29**, 6165 (1984).

4. L-D. Chang and S. Chakravarty, Phys. Rev. **B31**, 154 (1985).
5. Y. C. Chen, J. Stat. Phys. **47**, 17 (1987).
6. A. J. Leggett, in *Frontiers and Borderlines in Many-Particle Physics*, ed. R. A. Broglia and J. R. Schrieffer, Società Italiana di Fisica, Bologna (1988).
7. V. Ambegaokar and U. Eckern, Z. Phys. **B69**, 399 (1987).
8. H. Grabert, talk at 1986 Budapest Workshop on Quantum Tunnelling in Many-Dimensional Systems, unpublished.
9. W. Zwerger, A. T. Dorsey and M. P. A. Fisher, Phys. Rev. **B34**, 6518 (1986).
10. W. H. Zurek and W. G. Unruh, Phys. Rev. D, in press.
11. A. J. Leggett and Anupam Garg, Phys. Rev. Lett. **54**, 857 (1985).
12. A. J. Leggett, S. Chakravarty, A. T. Dorsey, Matthew P. A. Fisher, Anupam Garg and W. Zwerger, Revs. Mod. Phys. **59**, 1 (1987).
13. R. A. Silbey and R. A. Harris, J. Chem. Phys. **80**, 2615 (1984).
14. S. Chakravarty, Phys. Rev. Lett. **50**, 1811 (1982).
15. A. J. Bray and M. A. Moore, Phys. Rev. Lett. **49**, 1546 (1982).

QUANTUM TUNNELLING IN MACROSCOPIC SYSTEMS

A.O. Caldeira

Instituto de Física "Gleb Wataghin", Universidade Estadual
de Campinas, C.P. 6165, 13081 Campinas, SP, Brasil

ABSTRACT

In this lecture I shall introduce the concept of metastable states (or phases) in macroscopic systems at very low temperatures. I shall particularly be interested in two specific examples: superconducting devices and anisotropic magnetic films. By applying some techniques previously used in quantum field theory I will be able to describe the decay of these metastable states by quantum mechanical tunnelling within the WKB approximation.

1. INTRODUCTION

The concept of metastability in macroscopic systems has long been known from thermodynamics. Suppose we want to study the liquid-gas transition (at constant temperature) that takes place when one varies the pressure of a system described by the van der Waals equation

$$(p + \frac{a^2}{v})(v-b) = RT \quad . \tag{1.1}$$

Here, p is the pressure, v is the molar volume and a, b and R are constants. Furthermore, let us assume that the temperature $T < T_c$ (the critical temperature). The pv diagram for such a system can be found in any standard thermodynamics textbook (see, for example, Callen[1]) and looks like the one showed in fig. 1 below.

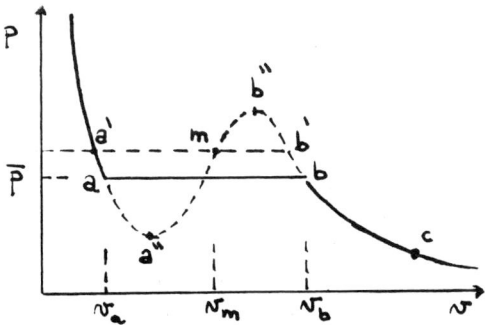

Fig. 1. pv diagram for the van der Waals equation when $T < T_c$.

Let us investigate the transition starting from the gaseous phase represented by point c in figure 1. By slowly increasing the pressure we can reach point b with the corresponding molar volume v_b for the gaseous phase. From there onwards, the theory of equilibrium phase transitions tells us that along the flat part of the curve there is coexistence of gas and liquid phases with molar volumes v_b and v_a, respectively. Once the complete transition is accomplished the whole system will be in the liquid state with molar volume v_a and the pressure can again be raised beyond the value $\bar{p} = p_a = p_b$ with the corresponding variation of the molar volume of the liquid phase.

If on the other hand, we are careful enough and protect our system from any sort of external perturbation, we can slowly bring it in its gaseous phase to a point b' along the dotted branch bb" of the pv diagram. This branch actually corresponds to a set of local minima of the Gibbs free energy. To each such a point there is another one (a'), with the same pressure and $v < v_a$ that has lower free energy. Consequently, the branch bb" consists of metastable states of the system. An analogous argument can be used to the branch aa" from the liquid to gas transition.

Once the system has been brought to one of those metastable states, say b', it stays there for a long time but eventually decays into the more stable phase (liquid state). The reason why such a transition takes place is the fluctuation of the state variable v. As we know v must fluctuate due to thermal effects about its equilibrium value and if we wait long enough a small part of the system may have $v \lesssim v_m$. When this happens, this small region, usually called "the droplet", shrinks (expands) whenever its surface energy is greater (smaller) than its bulk energy. When the droplet expands it converts the whole system into the liquid phase. This is the well-known phenomenon of homogeneous nucleation.[2]

Energetically we could imagine that the Gibbs free energy of the system has a potential barrier separating the two metastable phases, say a' and b' (as shown in fig. 2 below), and the droplet corresponding to the more stable phase must find its optimum form so as to overcome this free energy barrier.

The problems I wish to address in this lecture are basically equivalent to the one described above. The difference is that we shall be dealing with macroscopic systems at extremely low temperatures and therefore thermal fluctuations cannot be responsible for the decay of the metastable state.

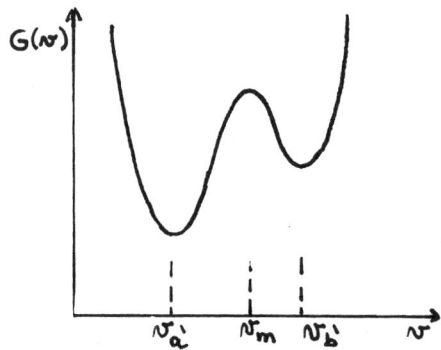

Fig. 2. The Gibbs free energy of the system.

On the other hand these macroscopic systems might have their physical behaviour described by a set of (or a single) collective dynamical variables and this very fact naturally leads us to ask about the possibility of a more sophisticated mechanism for driving the above mentioned transition, namely; quantum tunnelling.

In simple terms, let us suppose we have a particle subject to a potential $V(q) = Aq^2 - Bq^3$ where $A,B > 0$. We can easily convince ourselves that $q = 0$ is a metastable position for the particle that sees a potential barrier $V_o = 4A^3/27B^2$. The probability/time that the particle will leave this metastable state by thermal fluctuation is proportional to the factor[3] $\exp -(V_o/kT)$ which clearly vanishes when $T \to 0$. However, we know that this is not the only mechanism that takes the particle beyond the potential barrier. When we have very low temperatures, quantum mechanical effects are important and the particle can tunnel through the potential barrier.

I shall organize the remaining of these lecture notes as follows. In the next section I shall briefly discuss the physical systems of interest (superconducting devices and magnetic films). In section 3 I shall develop the path integral formulation for the quantum mechanical decay rate out of a metastable potential for a point particle and generalize it to other two cases: field theory and dissipative systems.

In section 4 I shall apply the previously obtained results to each case of interest while in section 5 I shall present the conclusions of this work.

2. SYSTEMS OF INTEREST

As mentioned above, I shall particularly be interested in two kinds of systems in these lectures; superconducting devices and magnetic films.

2a) Superconducting Devices

2ai) <u>Superconducting Quantum Interference Devices (SQUID's)</u>. Here I will deduce the equation of motion of the total electromagnetic flux inside the SQUID ring. This device consists of a superconducting ring closed by a weak link. The weak link can be either a point contact or even a junction made of a material in its normal state. The SQUID is shown in fig. 3 below.

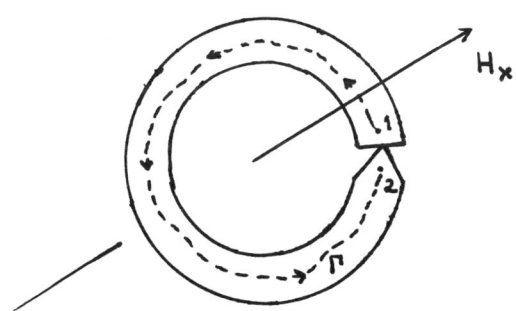

Fig. 3. The SQUID.

A well known fact from the theory of superconductivity is that there is a condensate formed by Cooper pairs which can be properly described by the so-called "macroscopic wave function" which is given by[4]

$$\psi(\vec{r}) = \sqrt{\rho(\vec{r})} \exp i\theta(\vec{r}) \quad , \tag{2a.1}$$

where $\rho(\vec{r})$ is the density of Cooper pairs and $\theta(\vec{r})$ is the phase of the wave function. The density of matter current, $\vec{J}_p$, is given by the expression

$$\vec{J}_p = \frac{\hbar}{2m_p i}(\psi^*\vec{\nabla}\psi - \psi\vec{\nabla}\psi^*) = \frac{1}{m_p} \mathrm{Re}[\psi^*(\vec{p}\psi)] \quad , \tag{2a.2}$$

which allows us to compute the supercurrent density, $\vec{J}_s$, if we remember that $m_p = 2m$ (m = electron mass) and $\vec{J}_s = 2e\vec{J}_p$ (e = electronic charge). Therefore, applying (2a.2) to (2a.1) we have

$$\vec{J}_s = \frac{\hbar e}{2mi}(\psi^*\vec{\nabla}\psi - \psi\vec{\nabla}\psi^*) = \frac{e\hbar}{m}\rho(\vec{r})\vec{\nabla}\theta(\vec{r}) \quad . \tag{2a.3}$$

In order to accomodate the external field H_x in our formalism, we must modify (2a.3). The standard procedure is to make the replacement

$$\vec{p} \rightarrow \vec{p} - 2e\vec{A} \tag{2a.4}$$

in (2a.2) (remember that our particles are now Cooper pairs of charge 2e) which gives us the new supercurrent density

$$\vec{J}_s = \frac{e\hbar}{m}(\vec{\nabla}\theta - \frac{2\pi}{\phi_o}\vec{A}) \quad , \tag{2a.5}$$

where $\phi_o = \frac{h}{2e}$ is the quantum of flux and $\vec{A}$ is the total vector potential.

The first step towards the obtainment of the equation of motion for the flux ϕ is to write down a new quantization condition when the superconducting ring has a constriction. This can be easily achieved if we evaluate the line integral of $\vec{J}_s$ along the contour Γ of fig. 3. Using (2a.5) this integral can be written as

$$\int_\Gamma \vec{J}_s \cdot d\vec{\ell} = \frac{e\hbar}{m}\int_\Gamma \vec{\nabla}\theta \cdot d\vec{\ell} - \frac{e\hbar}{m}\frac{2\pi}{\phi_o}\int_\Gamma \vec{A} \cdot d\vec{\ell} \quad . \tag{2a.6}$$

If we choose Γ to be a contour deep inside the ring, the Meissner effect tells us that $J_s = 0$ along Γ and consequently

$$\int_\Gamma \vec{\nabla}\theta \cdot d\vec{\ell} = \frac{2\pi}{\phi_o}\int_\Gamma \vec{A} \cdot d\vec{\ell} \quad . \tag{2a.7}$$

Now, making the approximation

$$\int_\Gamma \vec{A} \cdot d\vec{\ell} \sim \oint \vec{A} \cdot d\vec{\ell} = \phi \quad , \tag{2a.8}$$

(2a.6) can be cast into the form

$$\frac{2\pi\phi}{\phi_o} = \int_\Gamma \vec{\nabla}\theta \cdot d\vec{\ell} \quad . \tag{2a.9}$$

On the other hand we can rewrite the integral above as

$$\int_\Gamma \vec{\nabla}\theta \cdot d\vec{\ell} = \oint \vec{\nabla}\theta \cdot d\vec{\ell} - \int_2^1 \vec{\nabla}\theta \cdot d\vec{\ell} \qquad (2a.10)$$

and then use the fact that θ is defined modulo 2π (since it is the phase of ψ) to finally write (2a.9) as

$$\phi = n\phi_o + \frac{\phi_o}{2\pi}\Delta\theta \qquad (2a.11)$$

where $\Delta\theta \equiv \theta_2 - \theta_1$. Notice that for a uniform ring $\Delta\theta = 0$ and we recover the well-known flux quantization condition $\phi = n\phi_o$.

Now let us analyze the behaviour of the total electromagnetic flux inside the ring. As we already know from elementary classical electrodynamics the total flux in a circuit is given by

$$\phi = \phi_x + Li \qquad (2a.12)$$

where ϕ_x is the flux due only to H_x, L is the self inductance of the circuit and i is its total current. On the other hand, the latter is composite of three parts

a) Josephson Current; this component is due to the tunnelling of Cooper pairs through the junction and is written as

$$i_s = i_o \sin \Delta\theta \qquad (2a.13)$$

b) Normal Current; this is a result of the two-fluid model for superconductivity; Cooper pairs and normal electrons. The latter obey Ohm's law

$$i_n = \frac{V}{R} \qquad (2a.14)$$

where V is the induced e.m.f. and R is the normal resistance of the contact.

c) Polarization Current; this is due to the finite value of the capacitance of the contact. Its expression reads

$$i_c = C\dot{V} \qquad (2a.15)$$

where C is the above mentioned capacitance.

Assuming that the total current is given by the sum of these three components (resistively shunted junction model) we have

$$i = i_o \sin \Delta\theta + \frac{V}{R} + C\dot{V} \qquad (2a.16)$$

Substituting (2a.16) into (2a.12) and using (2a.11) together with the fact that $V = -\dot{\phi}$, we can write

$$\frac{\phi_x - \phi}{L} = i_o \sin \frac{2\pi\phi}{\phi_o} + \frac{\dot{\phi}}{R} + C\ddot{\phi} \qquad (2a.17)$$

which is the equation of motion of a "particle" of "coordinate" ϕ subject to a "potential" $U(\phi)$ given by

$$U(\phi) = \frac{(\phi-\phi_x)^2}{2L} - \frac{\phi_o i_o}{2\pi} \cos \frac{2\pi\phi}{\phi_o} \qquad (2a.18)$$

Using (2a.18) we can rewrite (2a.17) as

$$C\ddot{\phi} + \frac{\dot{\phi}}{R} + U'(\phi) = 0 \qquad . \qquad (2a.19)$$

Actually we should yet add a fluctuating current, $i_f(t)$, to the RHS of (2a.19) in order to correctly describe the thermodynamical properties of the system. With this extra term, (2a.19) is nothing but the well known Langevin equation.

In order to be able to appreciate the variety of physical phenomena in (2a.19) we had better analyze the potential (2a.18) in some detail.

The minima of $U(\phi)$ (stable solutions) are obtained as solutions ϕ_m of $U'(\phi) = 0$, subject to the condition $U''(\phi_m) > 0$. With the help of (2a.18) these two equations take us to two distinct cases:

i) $2\pi L i_o/\phi_o > 1$ several minima
ii) $2\pi L i_o/\phi_o \leq 1$ only one minimum.

In fig. 4 below we sketch the form of $U(\phi)$ in case (i) for three different values of the external flux, ϕ_x.

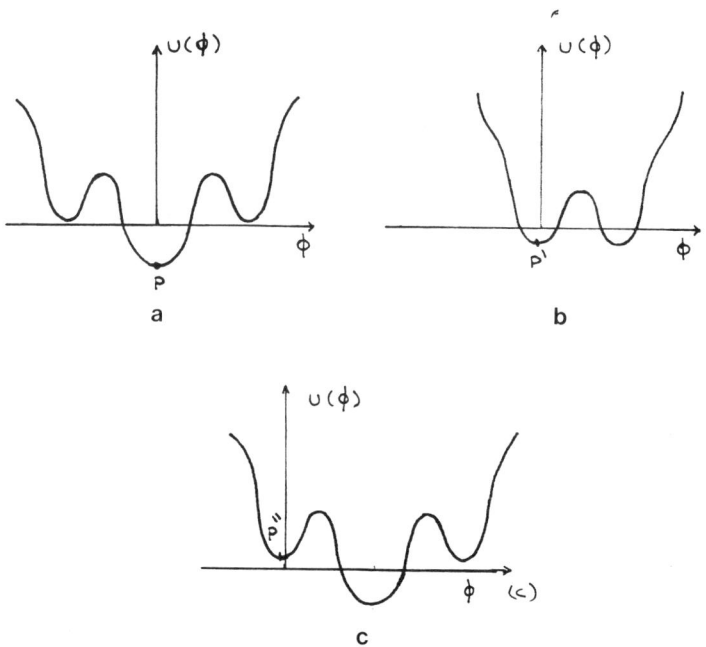

Fig. 4. The potential $U(\phi)$. a) $\phi_x = 0$; b) $\phi_x = \phi_o/2$; c) $\phi_x = \phi_o$.

Now, let us suppose that at t = 0 there is no external field ($H_x = 0$) and no superconducting current. In this case, the equilibrium value for the total flux inside the ring is $\phi = 0$ which is represented by the point P in fig. 4a. If we now change H_x very slowly the form of the potential $U(\phi)$ also changes and the equilibrium value of ϕ adiabatically follows its initial potential well until this minimum becomes a inflection point of $U(\phi)$. In fig. 4 we clearly see this change as the equilibrium point moves as P → P' → P". We should notice that this adiabatic approximation is only valid when dH_x/dt is much smaller than any other frequency of the system;

$$\frac{1}{C}\left.\frac{d^2U}{d\phi^2}\right|_{\phi_m} \quad \text{or} \quad \frac{1}{2RC}.$$

The verification that the dynamics of the total flux inside the ring is really the same as of a Brownian particle in a potential $U(\phi)$ was done in[5] where the decay of the metastable state by thermal fluctuation was studied.

2aii) <u>Current Biased Josephson Junction (CBJJ)</u>. In this case the variable of interest is the difference in the phase of the macroscopic wave function across the Josephson junction. The equation of motion for $\Delta\theta$ can easily be obtained from our previous analysis if we regard the CBJJ as being a SQUID with $L \to \infty$ and $\phi_x \to \infty$ in such a way that $\phi_x/L = I_x$ (the external current). Therefore, making the replacement

$$\phi \to -\frac{\phi_o}{2\pi}\Delta\theta \equiv \frac{\phi_o \varphi}{2\pi} \tag{2a.20}$$

in (2a.17) we have

$$\frac{\phi_o}{2\pi}C\ddot{\varphi} + \frac{\phi_o}{2\pi R}\dot{\varphi} + U'(\varphi) = 0 \tag{2a.21}$$

where $U(\varphi)$, the so-called "washboard potential", is given by

$$U(\varphi) = -I_x\varphi - i_o\cos\varphi \tag{2a.22}$$

which is sketched in fig. 5 below.

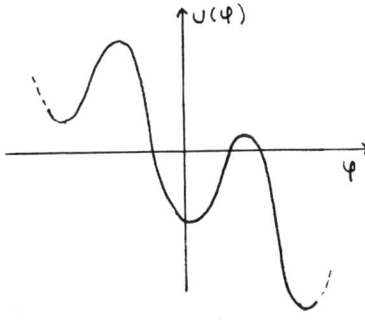

Fig. 5. The washboard potential.

Although this equation for φ is suitable for our purposes, one should be aware of the fact that its physical meaning is not so simple as in the case of the SQUID. The reason for this is that φ is defined modulo 2π and consequently one has to be careful with the boundary conditions for the wave function in the quantum mechanical version of the problem. However, since we shall be dealing with values of I_x close to i_o (where the minima of $U(\varphi)$ become inflection points) only "distances" much less than 2π will be involved in the problem and we can safety rely on the usual interpretation of the dynamics of a particle in the washboard potential.

In both examples, we have seen that the variable of interest obeys an equation of motion that is analogous to the one of a particle subject to a given potential and to a dissipative force. Actually, we should have added a term corresponding to a fluctuation current in equations (2a.19) and (2a.21) in order to properly describe the fluctuations about the equilibrium value of the dynamical variable. Since these are the equations of motion of truly dynamical quantities we can immediately try to study the quantum mechanics of these systems and ask very interesting questions such as the possibility of quantum mechanical tunnelling of this electromagnetic analogue of a point particle when $T = 0$.

2b) Magnetic Systems

We consider the following two-dimensional spin Hamiltonian

$$\mathcal{H} = -J \sum_{<ij>} \vec{S}_i \cdot \vec{S}_j - \Delta \sum_{<ij>} S_i^{(z)} S_j^{(z)} - \mu H \sum_i S_i^{(z)} \tag{2b.1}$$

which represents a ferromagnetic 2D lattice placed in the xy plane. Here, $J\hbar^2$ (> 0) is the exchange energy, $\Delta\hbar^2$ (> 0) is the uniaxial anisotropy energy, μ is the magnetic moment at each site of the lattice and H is the magnitude of the applied field which is pointing in the $\hat{z}$ direction. The brackets $<\;>$ mean we are summing over nearest neighbours only.

Writing down the equations of motion for the spin variable

$$\dot{\vec{S}}_k = \frac{1}{i\hbar} [\vec{S}_k, \mathcal{H}]$$

and expressing them in terms of its cylindrical components leads us to the following equations

$$\dot{S}_k^{(\rho)} = -4Js' \sum_i (S_i^{(\theta)} S_k^{(z)} - S_i^{(z)} S_k^{(\theta)}) + 4\Delta s' \sum_i S_k^{(\theta)} S_i^{(z)} + 2s'\mu H S_k^{(\theta)} \tag{2b.2a}$$

$$\dot{S}_k^{(\theta)} = -4Js' \sum_i (S_i^{(z)} S_k^{(\rho)} - S_i^{(\rho)} S_k^{(z)}) - 4\Delta s' \sum_i S_k^{(\rho)} S_i^{(z)} - 2s'\mu H S_k^{(\rho)} \tag{2b.2b}$$

$$\dot{S}_k^{(z)} = -4Js' \sum_i (S_i^{(\rho)} S_k^{(\theta)} - S_i^{(\theta)} S_k^{(\rho)}) \;. \tag{2b.2c}$$

Now, taking the continuum approximation in equations (2b.2) yields

$$\dot{S}^{(\rho)} = -a(S^{(z)} \nabla^2 S^{(\theta)} - S^{(\theta)} \nabla^2 S^{(z)}) + bS^{(z)} S^{(\theta)} + cS^{(\theta)} \tag{2b.3a}$$

$$\dot{S}^{(\theta)} = -a(S^{(\rho)}\nabla^2 S^{(z)} - S^{(z)}\nabla^2 S^{(\rho)}) - bS^{(z)}S^{(\rho)} - cS^{(\rho)} \qquad (2b.3b)$$

$$\dot{S}^{(z)} = -a(S^{(\theta)}\nabla^2 S^{(\rho)} - S^{(\rho)}\nabla^2 S^{(\theta)}) \qquad (2b.3c)$$

where the constants a, b and c are given by

$$a = 4Js'zd^2 \qquad b = 4\Delta s'z \qquad c = 2s'\mu H \qquad (2b.4)$$

z is the number of nearest neighbours at each site, s' is related to the spin quantum number, $s' \equiv [s(s+1)]^{1/2}$ and ∇^2 is the Laplacian operator in the xy plane.

Since we are going to treat the problem semiclassically, we introduce at this level the classical approximation in the sense that we treat the spin as an ordinary vector (of length S) specified by two angles φ and ψ[6,7].* In cylindrical coordinates S is writen as

$$\vec{S}(\rho,\theta,t) = (S^{(\rho)}, S^{(\theta)}, S^{(z)})$$

$$= S(\cos\varphi(\rho,\theta,t), \sin\varphi(\rho,\theta,t)\sin\psi(\rho,\theta,t)\sin\varphi(\rho,\theta,t)\cos\psi(\rho,\theta,t))$$

$$(2b.5)$$

where S = s'ℏ and (φ,ψ) are local spherical angles (relative to the ρ axis) of the spin vector at the position (ρ,θ) at time t. In figure 6 we indicate the local angles (φ,ψ) relative to the (ρ,θ,z) coordinate system.

We expressed the equations (2b.3) in terms of the angular variables and ψ. Equation (2b.3a) multiplied by $(\sin\varphi)^{-1}$ gives us the equation for $\dot{\varphi}$

$$-\dot{\varphi} = 2aS\cos\varphi \frac{\partial\varphi}{\partial\rho}\frac{\partial\psi}{\partial\rho} + aS\sin\varphi \frac{\partial^2\psi}{\partial\rho^2} + aS\sin\varphi \frac{1}{\rho}\frac{\partial\psi}{\partial\rho}$$

$$+ \frac{2aS}{\rho^2}\cos\varphi \frac{\partial\varphi}{\partial\theta}\frac{\partial\psi}{\partial\theta} + \frac{aS}{\rho^2}\sin\varphi \frac{\partial^2\psi}{\partial\theta^2} + \frac{bS}{2}$$

$$\times \sin\varphi \sin 2\psi + c \sin\psi \quad . \qquad (2b.6a)$$

The combination $(\dot{S}^{(\theta)}\cos\psi - \dot{S}^{(z)}\sin\psi)$ of equations (2b.3b) and (2b.3c) leads us to the equation for $\dot{\psi}$

$$\dot{\psi}\sin\varphi = -\frac{aS}{2}\sin 2\varphi(\frac{\partial\psi}{\partial\rho})^2 + aS\frac{\partial^2\varphi}{\partial\rho^2} + aS\frac{1}{\rho}\frac{\partial\varphi}{\partial\rho} + \frac{aS}{\rho^2}\frac{\partial^2\varphi}{\partial\theta^2}$$

$$- \frac{aS}{2\rho^2}\sin 2\varphi(\frac{\partial\psi}{\partial\theta})^2 - \frac{bS}{2}\sin 2\varphi\cos^2\psi - c\cos\varphi\cos\psi. \qquad (2b.6b)$$

---

*Such a treatment is expected to make sense for large values of spin (S>>1), and the equations of motion that follow are the same as those one could have obtained classically using Poisson brackets.

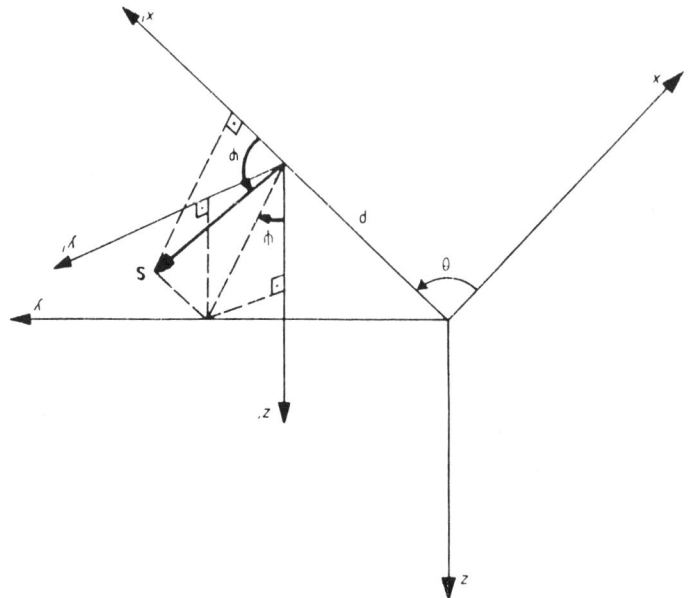

Figure 6. Local angular variables $\varphi(\rho,\theta)$ and $\psi(\rho,\theta)$ of the spin vector $S(\rho,\theta)$ at the position (ρ,θ) in the xy plane at a given time t: φ is the angle the vector S makes with the x' axis; ψ is the angle the projection S $\sin\varphi$ of vector S onto the y'z' plane makes with the z' axis.

Now we make an approximation based on experimentally known facts[8], i.e. that the walls of the magnetic bubbles observed in thin films are of two types: Bloch walls (ferromagnetic case) or Néel walls (antiferromagnetic case). The former has $\varphi \simeq \pi/2$ and $0 \leq \psi \leq \pi$ while the latter has $\psi \simeq 0$ and $0 \leq \varphi \leq \pi$. Since we are dealing with a ferromagnetic model it is convenient to start with the following hypothesis:

$$\varphi = \pi/2 - \varphi' \quad \text{and} \quad \varphi' \ll \pi/2 \quad . \tag{2b.7}$$

Under such an approximation, we can drastically simplify equations (2b.6). Then, it is possible to eliminate the φ dependence and write an equation for the ψ variable as

$$\frac{1}{F(\psi)} \ddot\psi - aS\nabla^2\psi + \frac{bs}{2} \sin 2\psi + c \sin \psi = 0 \tag{2b.8}$$

where

$$F(\psi) = -aS(\vec{\nabla}\psi)^2 + bS \cos^2\psi + c \cos \psi \quad . \tag{2b.9}$$

At this point we are going to argue in favour of the approximation

$$F(\psi) \simeq K^2 \tag{2b.10}$$

(K^2 independent of position and time). First, considering the situation where the field H is negative and small ($\mu|H|\hbar \ll \Delta\hbar^2$), the only possibility for a bubble to survive is to be formed with large radius and thin wall[9].

36

In this situation $\psi(\rho,\theta,t)$ is either zero (metastable vacuum) or π (inner region of the bubble). Hence, except for a small region of space we can consider a constant 'velocity'

$$K_o \simeq (bS)^{1/2} \qquad \text{(thin-wall regime)} \quad . \qquad (2b.11)$$

With this approximation, (2b.8) can be seen as an equation of motion for a fictitious particle with 'position' ψ in a potential

$$V(\psi) = -\tilde{U}(\psi) = -\frac{bS}{2} \sin^2\psi - c(1-\cos\psi) \quad . \qquad (2b.12)$$

Now, if one starts to increase the value of H this argument is no longer valid. The profile of the quantity K becomes very strongly configuration dependent. Physically this is an expected result since the stiffnesses of spin waves about the metastable and stable configurations are very different. Therefore, one would need to take the complete expression (2b.9) into account, which is obviously a very hard task. In order to overcome this problem we shall insist on a constant -K approximation but give up the goal of obtaining even an approximate value for the decay rate. The quantity we can compute within this model with the specific choice

$$F(\psi) = K^2 \simeq bS + c = K_o^2(1-|c|/bS) \qquad \text{(if } c < 0\text{)} \qquad (2b.13)$$

(which is the minimum of K) is actually a lower bound for the decay rate as we shall see later. The equation of motion (2b.8) finally reduces to

$$(1/K^2)\ddot{\psi} - |a|S(\nabla^2\psi) + d\tilde{U}/d\psi = 0 \quad . \qquad (2b.14)$$

A 2D Lagrangian density that yields the equation of motion (2b.14) can be written as

$$= \frac{\hbar aS}{d^2}[\frac{1}{2v^2}(\frac{\partial\psi}{\partial t})^2 - \frac{1}{2}(\nabla\psi)^2 - g\sin^2\psi - \alpha(1-\cos\psi)] \qquad (2b.15)$$

d being the lattice spacing, and the constants v^2, g and α defined as

$$v^2 \equiv aSK^2 \qquad g \equiv b/(2a) \qquad \alpha \equiv c/(aS) \quad . \qquad (2b.16)$$

In terms of these parameters we define the function

$$U(\psi) = g\sin^2\psi + \alpha(1-\cos\psi) \quad . \qquad (2b.17)$$

Let us analyze the behaviour of $U(\psi)$ as we change the external field H (which is contained in α). In figure 7 we sketch the form of this function for values of α belonging to five different regions, namely $\alpha > \alpha_c$, $\alpha_c > \alpha > 0$, $\alpha = 0$, $0 > \alpha > -\alpha_c$ and $-\alpha_c > \alpha$, where

$$\alpha_c = 2g > 0 \quad . \qquad (2b.18)$$

This potential has extrema at $\psi_n = n\pi$ (n integer) and at

$$\psi_m = \cos^{-1}(-\alpha/2g) \quad . \qquad (2b.19)$$

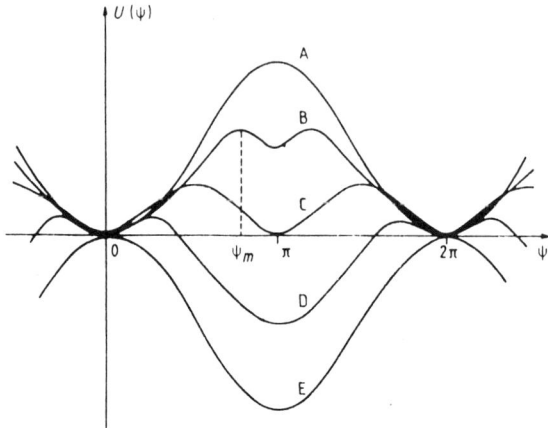

Fig. 7. The potential energy density for different values of α: A, α > α_c; B, α > α_c > 0; C, α = 0; D, 0 > α > -α_c; E, -α_c > α.

The latter are always maxima of $U(\psi)$ (see figure 7) while the former can be either maxima or minima depending on values of α. If n is even, ψ_n are (classically) stable configurations for α > 0, metastable ones for $0 \geq \alpha > -\alpha_c$ and unstable ones for $-\alpha_c \geq \alpha$. On the other hand if n is odd, ψ_n are unstable for $\alpha \geq \alpha_c$, metastable for $\alpha_c > \alpha \geq 0$ and stable for α < 0.

What makes this problem interesting is the fact that unlike high-energy problems[10], the energy density difference between the non-degenerate minima of $U(\psi)$ can be controlled by the experimenter via the external field (α is proportional to H).

3. THE PATH INTEGRAL FORMULATION

In this section I shall introduce the path integral formulation for tunnelling problems in three cases; particle mechanics, dissipative systems and field theory. Although these lectures are intended to be introductory to this subject I shall not develop the whole theory of tunnelling via path integrals in detail. Instead, I think it will be much more interesting to refer the reader to texts where this has been carefully done[10,11,12] and only describe the general strategy herein.

3a) Particle Mechanics[10,11]

Our main goal in this section is to introduce the method of path integrals in order to solve the problem of tunnelling of a particle out of a metastable minimum of a given potential.

Suppose that a particle of mass m is at rest at the local minimum (q=0) of the potential sketched below. Its Hamiltonian is given by the usual expression

$$H = \frac{p^2}{2m} + V(q) \qquad (3a.1)$$

where $V(q)$ can be approximated by the form $V(q) = \frac{1}{2} m\omega_0^2 q^2 - \lambda q^3$.

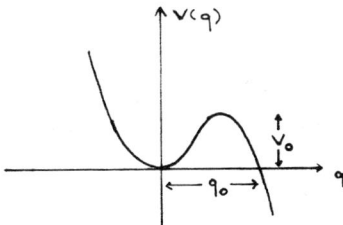

Fig. 8. The potential V(q).

In order to compute the tunnelling rate of this particle we shall develop a method that allows us to calculate the imaginary part of the energy of the particle in a metastable state within the potential well. In this way we could write

$$\psi(q,t) \propto e^{-i(E_R + iE_I)t/\hbar} \qquad (3a.2)$$

which implies that (for $E_I < 0$)

$$\psi^*(q,t)\psi(q,t) \propto e^{-2|E_I|t/\hbar} \qquad (3a.3)$$

and then

$$\Gamma = \frac{2|\text{Im}E|}{\hbar} \qquad (3a.4)$$

would be the tunnelling rate for which we are looking.

The starting point to obtain ImE is the expression for the density operator of the particle at temperature $(k\beta)^{-1}$

$$\rho(x,y,\beta) = \langle x|e^{-\beta H}|y\rangle = \int_y^x \mathcal{D}q(\tau) \exp - \frac{S_E[q(\tau)]}{\hbar} \qquad (3a.5)$$

where $S_E[q(\tau)]$ is the Euclidean action of the particle[10]

$$S_E[q(\tau)] = \int_0^{\hbar\beta} \{\frac{1}{2} m \left(\frac{dq}{d\tau}\right)^2 + V(q)\} d\tau \qquad (3a.6)$$

Our second step is to find another representation for $\rho(x,y,\beta)$ relating the solution of the path integral (3a.5) with Γ in (3a.4). To do so, let us imagine that the potential in fig. 8 is a function of a parameter ε in such a way that for $\varepsilon > 0$ $V_\varepsilon(q)$ be a potential with a single minimum. As we vary ε the potential is continuously deformed. Furthermore, let us assume that for $\varepsilon < 0$ the absolute minimum of $V_\varepsilon(q)$ becomes the local minimum in fig. 8. Therefore, when $\varepsilon > 0$ we can write

39

$$\rho_\varepsilon(x,y,\beta) = \langle x | e^{-\beta H_\varepsilon} | y \rangle = \sum_{n=0}^{\infty} \psi_n^{(\varepsilon)}(x) \psi_n^{(\varepsilon)*}(y) e^{-\beta E_n^{(\varepsilon)}} \tag{3a.7}$$

where we have employed the completeness relation $\sum_n |\psi_n^{(\varepsilon)}\rangle\langle\psi_n^{(\varepsilon)}| \equiv I$ in passing from the second to the third term in (3a.7). Now, with the help of (3a.5) one has ($x=y=0$)

$$\int_0^0 \mathcal{D}q(\tau) \exp -\frac{1}{\hbar} S_E^{(\varepsilon)}[q(\tau)] = \sum_{n=0}^{\infty} |\psi_n^{(\varepsilon)}(0)|^2 e^{-\beta E_n^{(\varepsilon)}} \tag{3a.8}$$

This expression is extremely useful if we wish to compute the energy of the ground state of the system (still for $\varepsilon > 0$). It is enough to take the limit $\beta \to \infty$ ($T \to 0$) in (3a.8) to get

$$\lim_{\beta \to \infty} \int_0^0 \mathcal{D}q(\tau) \exp -\frac{1}{\hbar} S_E^{(\varepsilon)}[q(\tau)] \underset{\sim}{\sim} |\psi_0^{(\varepsilon)}(0)|^2 e^{-\beta E_0^{(\varepsilon)}} \tag{3a.9}$$

where we have replaced the series in (3a.8) by its leading term when $\beta \to \infty$. Consequently all one has to do to determine $E_0^{(\varepsilon)}$ is to evaluate the path integral in (3a.9).

Expression (3a.9) is only plausible when $\varepsilon > 0$. However, our problem is to use it when $\varepsilon < 0$ although this procedure seems, at first sight, to be wrong. The way out is to make an appropriate analytic extension of (3a.9) to the region of the complex plane where $\varepsilon < 0$. In this way, besides proceeding in a mathematically rigorous way, we can get an imaginary part for the energy of the metastable state. This procedure was carefully studied in references[9,10] and I shall only quote the final results here, reminding the reader to consult the above mentioned references.

The decay rate can be computed as

$$\Gamma_0 = \frac{2 \mathrm{Im} E_0}{\hbar} = \left(\frac{B_0}{2\pi\hbar m}\right)^{1/2} \left|\frac{\det(-m\partial_t^2 + m\omega_0^2)}{\det'(-m\partial_t^2 + V''(q_c))}\right|^{1/2} \exp -\frac{B_0}{\hbar} \tag{3a.10}$$

where $\omega_0^2 = \frac{1}{m} V''(0)$, q_c is a non trivial solution ($q_c \neq 0$) of

$$\left.\frac{\delta S_E}{\delta q}\right|_{q_c} = -m\ddot{q}_c + V'(q_c) = 0 \tag{3a.11}$$

with $q_c(-\infty) = q_c(\infty) = 0$, and

$$B_0 = S_E[q_c] . \tag{3a.12}$$

In (3a.10), det' means the determinant has been computed omitting the zero eigenvalue.

3b) Dissipative Systems[13]

As we have seen in the lectures by A.J. Leggett[14] the quantum dynamics of a dissipative system of the form of those introduced in section 3a can be studied by coupling the system to a bath of non-interacting oscillators with a given spectral function. The total Hamiltonian reads

$$H = \frac{p^2}{2m} + V(q) + \sum_k C_k q_k q + \sum_k \left\{ \frac{p_k^2}{2m_k} + \frac{1}{2} m_k \omega_k^2 q_k^2 \right\} + \sum_k \frac{C_k^2}{2m_k \omega_k^2} q^2 \tag{3b.1}$$

and the spectral function, defined as

$$J(\omega) = \frac{\pi}{2} \sum_k \frac{C_k^2}{m_k \omega_k} \delta(\omega - \omega_k) \quad, \tag{3b.2}$$

must assume the form

$$J(\omega) = \begin{cases} \eta\omega & \text{if } \omega < \Omega \\ 0 & \text{if } \omega > \Omega \end{cases} \tag{3b.3}$$

In this way, the composite system particle-reservoir is subject to a N+1 dimensional potential which has a metastable minimum at $(q,\vec{R}) = (0,\vec{0})$ ($\vec{R} = (q_1, q_2, \ldots, q_N)$) and a saddle point at $(q,\vec{R}) = (\frac{m\omega_o^2}{3\lambda}, \ldots, -\frac{C_k}{3} \frac{m}{m_k} \frac{\omega_o^2}{\omega_k^2}, \ldots)$. Therefore, in semiclassical terms, there is an optimal path going through the saddle point along which the N+1 dimensional variable could tunnel. The decay rate can, once again, be found by computing the imaginary part of the energy of the metastable state about $(q,\vec{R}) = (0,\vec{0})$. Since the only dangerous direction in this N+1 dimensional space is associated with the potential V(q) we can interpret the imaginary part of the energy as being related to the decay rate of the particle influenced by the environment.

The strategy to compute the ground state energy of the composite system is very similar to that we have developed in the previous section. Let us consider the density operator for the complete system in equilibrium at temperature $(k\beta)^{-1}$. This reads

$$\rho_{eq}(x,\vec{R};y,\vec{Q},\beta) = \sum_{\{n\}} \psi_{\{n\}}(x,\vec{R}) \psi_{\{n\}}^*(y,\vec{Q}) \exp{-\beta E_{\{n\}}} \tag{3b.4}$$

where {n} is a set of quantum numbers $n_o, n_1, \ldots, n_N$.

The reduced density operator of the particle is then

$$\tilde{\rho}_{eq}(x,y,\beta) = \int d\vec{R} \, \rho_{eq}(x,\vec{R};y,\vec{R},\beta) = \int d\vec{R} \sum_{\{n\}} \psi_{\{n\}}(x,\vec{R}) \psi_{\{n\}}^*(y,\vec{R}) *$$

$$* \exp{-\beta E_{\{n\}}} \quad . \tag{3b.5}$$

Now, using the path integral representation, we can rewrite (3b.5) as

$$\tilde{\rho}_{eq}(x,y,\beta) = \int d\vec{R} \langle x\vec{R}|e^{-\beta H}|y\vec{R}\rangle = \int d\vec{R} \int_{y,\vec{R}}^{x,\vec{R}} \mathcal{D}q(\tau')\mathcal{D}\vec{R}(\tau') \exp -\frac{S_E}{\hbar}[q(\tau'),\vec{R}(\tau')] \quad (3b.6)$$

where $S_E[q(\tau'),R(\tau')]$ is the Euclidean action corresponding to the Hamiltonian (3b.1). Then, proceeding as in the previous section we can calculate the energy of the metastable configuration of the system. Defining $\tau \equiv \hbar\beta$, using the completeness relation $\sum_{\{n\}} |\psi_{\{n\}}\rangle\langle\psi_{\{n\}}| = I$ and taking the limit $\tau \to \infty$ we have

$$\int d\vec{R}|\psi_{\{0\}}(0,\vec{R})|^2 e^{-\tau E_o/\hbar} \underset{\sim}{\times} \int d\vec{R} \int_{0,\vec{R}}^{0,\vec{R}} \mathcal{D}q(\tau')\mathcal{D}\vec{R}(\tau')\exp -\frac{S_E}{\hbar}[q(\tau'),\vec{R}(\tau')] \quad (3b.7)$$

The trace over the oscillator coordinates can be easily evaluated[13] and in so doing we get

$$\tilde{\rho}(0,0,\beta) = \tilde{\rho}_o(\beta) \int_0^0 \mathcal{D}q(\tau') \exp -\frac{S_{eff}}{\hbar}[q(\tau')] \quad (3b.8)$$

where

$$S_{eff}[q(\tau')] = \int_{-\infty}^{\infty} \{\frac{1}{2}m\dot{q}^2 + V(q)\}d\tau' + \frac{\eta}{4\pi}\int_{-\infty}^{\infty}d\tau'\int_{-\infty}^{\infty}d\tau'' \frac{\{q(\tau')-q(\tau'')\}^2}{(\tau'-\tau'')^2} \quad (3b.9)$$

In order to reach (3b.9) we have replaced $\sum_k \to \int d\omega$ with the help of (3b.2) and (3b.3).

The explicit expression for E_o comes from the evaluation of the path integral in (3b.9) that can be done by the saddle point method as in section 3a. The imaginary part of E_o allows us to write the expression for the decay rate as

$$\Gamma = \left(\frac{\|q_c\|}{2\pi\hbar}\right)^{1/2} \left|\frac{\det \hat{D}_o}{\det' \hat{D}}\right|^{1/2} \exp -\frac{B}{\hbar} \quad (3b.10)$$

where q_c is a non-trivial solution ($q_c \neq 0$) of

$$\left.\frac{\delta S_{eff}}{\delta q}\right|_{q_c} = m\ddot{q}_c - \frac{\partial V}{\partial q_c} - \frac{\eta}{\pi}\int_{-\infty}^{\infty}d\tau'' \frac{[q_c(\tau')-q_c(\tau'')]}{(\tau'-\tau'')^2} = 0 \quad , \quad (3b.11)$$

with boundary conditions $q_c(-\infty) = q_c(\infty) = 0$ and the operator D is defined through its application on $q(\tau')$ as

$$\hat{\mathbb{D}}q(\tau') = -m\frac{d^2}{d\tau'^2}q(\tau') + V''(q_c)q(\tau') + \frac{\eta}{\pi}\int_{-\infty}^{\infty}\frac{q(\tau')-q(\tau'')}{(\tau'-\tau'')^2} \qquad (3b.12)$$

The norm of $\dot{q}_c$ is given by

$$\|\dot{q}_c\|^2 = \int_{-\infty}^{\infty}\dot{q}_c^2(\tau')d\tau' \qquad (3b.13)$$

and $\hat{\mathbb{D}}_o$ is the same as $\hat{\mathbb{D}}$ replacing $V''(q_c)$ by $m\omega_o^2$. Finally, the factor B is the effective action computed at q_c

$$B = S_{eff}[q_c] \quad . \qquad (3b.14)$$

3c) Field Theory[11,12]

The decay rate in field theory is a straightforward generalization of the one computed for a point particle. Let us imagine that we have a field theory in D dimensions whose dynamics is described by the equation of motion

$$\frac{1}{c^2}\frac{\partial^2\varphi}{\partial t^2} - \nabla_D^2\varphi + U'(\varphi) = 0 \qquad (3c.1)$$

where ∇_D^2 is the D-dimensional Laplacian operator.

Now, the important quantity to be computed is the imaginary part of the energy of the metastable state per unity volume in this D-dimensional space. Generalizing what we have done for a point particle[11] we reach

$$\frac{\Gamma}{V_D} = A\,e^{-B/\hbar} \qquad (3c.2)$$

where

$$B = S_E[\varphi_c] = \int_{V_D}\int_{-\infty}^{\infty} d\tau\, d^Dx\,\{\frac{1}{2c^2}(\frac{\partial\varphi_c}{\partial\tau})^2 + \frac{1}{2}|\vec{\nabla}_D\varphi_c|^2 + U(\varphi_c)\} \quad , \qquad (3c.3)$$

$$A = (\frac{B}{2\pi\hbar})^{(D+1)/2}\,c\,\left|\frac{\det(-\nabla_{(D+1)}^2 + U''(\varphi_m))}{\det'(-\nabla_{(D+1)}^2 + U''(\varphi_c))}\right|^{1/2} \quad , \qquad (3c.4)$$

φ_m is the metastable classical configuration and φ_c is a nontrivial solution ($\varphi_c \neq \varphi_m$) of

$$\frac{d^2\varphi}{d\rho^2} + \frac{D}{\rho}\frac{d\varphi}{d\rho} - U'(\varphi) = 0 \quad . \qquad (3c.5)$$

with boundary conditions $\varphi_c(\rho = \infty) = 0$ and $\left.\frac{d\varphi_c}{d\rho}\right|_{\rho=0} = 0$. The new variable ρ is defined as

$$\rho = \sqrt{r^2 + c^2\tau^2} \qquad (3c.6)$$

with $\vec{r} = (x_1 \ldots x_D)$.

4. RESULTS

Now that we know the expressions of the quantum mechanical decay rate for those cases of interest presented in section 2 we will apply them for each specific example.

4a) Particle Mechanics[11]

In this case, the decay rate (3a.10) becomes

$$\Gamma_o = A_o \exp - \frac{B_o}{\hbar} \qquad (4a.1)$$

where

$$B_o = \frac{36}{5} \frac{V_o}{\omega_o}, \qquad (4a.2)$$

$$V_o = \frac{m^3 \omega_o^6}{54\lambda^2} \qquad (4a.3)$$

is the height of the potential barrier and

$$A_o = 6\omega_o \sqrt{\frac{6}{\pi} \frac{V_o}{\hbar\omega_o}} \qquad (4a.4)$$

which is the usual WKB result for the tunnelling rate out of a cubic potential.

Let us apply this result to the case of the SQUID without damping. In this example the capacitance C plays the role of the mass m and $U(\phi)$ (2a.18) replaces the potential $V(q)$. Expanding $U(\phi)$ up to third order in the deviation from the metastable value of the flux we have the desired form for the potential $U(\phi)$. With the numerical values $C \sim 10^{-12}$F, $L \sim 10^{-10}$H and $i_o \sim 10^{-5}$A we can find a approximate value for $B_o/\hbar$ as

$$\frac{B_o}{\hbar} \simeq 7.2 \, \phi_o i_o \frac{\sqrt{LC}}{2\pi\hbar} \, \epsilon(\frac{\phi_x}{\phi_c}) \qquad (4a.5)$$

where the function $\epsilon(\frac{\phi_x}{\phi_c})$ is a smoothly varying function of the ratio ϕ_x/ϕ_c which tends to zero when the critical value of the flux is reached ($\phi_x = \phi_c$). Actually, ϵ is a measure of the barrier height. The decay rate (4a.1) then becomes

$$\Gamma_o \sim 10^{13} \sqrt{\epsilon} \exp - 10^3 \epsilon \qquad (4a.6)$$

From this expression one easily concludes that there are appropriate values of ϕ_x (or ε) for which the decay rate is appreciable. However one should be very careful in choosing these values within a given range where the approximations we have made make any sense at all. For example, $10^{-2} < \varepsilon < 10^{-1}$ because for $\varepsilon < 10^{-2}$ the potential barrier is too small and this would make the WKB approach not reliable. On the other hand, for $\varepsilon > 10^{-1}$ the barrier is too high and the decay rate would be negligible.

4b) Dissipative Systems[13]

This case will be separated in two distinct regimes; weakly damped and strongly damped. Defining $\gamma \equiv \eta/2m$, the former is such that $\omega_o \gg \gamma$ while for the latter $\gamma \gg \omega_o$.

For both the weakly damped as for the strongly damped regime one can still show that

$$\Gamma = A \exp - B/\hbar \qquad (4b.1)$$

where A and B are presented below for each case separately.

In the weakly damped regime:

$$A = A_o + \mathcal{O}(\gamma) \qquad (4b.2)$$

and

$$B = B_o + \frac{12}{\pi^3} \zeta(3)\eta q_o^2 \qquad (4b.3)$$

where $\zeta(3) = \sum_{n=1}^{\infty} \frac{1}{n^3}$ and $q_o = \frac{m\omega_o^2}{2\lambda}$ is the exit point of the particle (see fig. 8) in a cubic potential.

In the strongly damped regime:

$$\frac{A}{A_o} \propto \gamma^{7/2} \omega_o^{-5/2} \qquad (4b.4)$$

and

$$B = \frac{2\pi}{9} \eta q_o^2 + \mathcal{O}(\frac{\omega_o}{\gamma}) \qquad (4b.5)$$

In both cases as the exponential factor is the most important contribution to Γ we can say that the effect of ohmic dissipation is always to supress quantum mechanical tunnelling ($\Gamma < \Gamma_o$).

Now if we want to apply these results to the SQUID or to the CBJJ we only need to replace γ by $1/2$ RC and use the appropriate expression for Γ. For this particular application there is experimental support by Clarke et.al.[16] who reported measurements on the onset of voltage for damped and undamped CBJJ's at extremely low temperatures even before the critical current is reached.

4c) <u>Field Theory</u>[15]

Now, I wish to apply the results of section 3c to the problem of quantum nucleation of magnetic bubbles in the magnetic system introduced in section 2b.

Defining a parameter ε as

$$\varepsilon \equiv 1 - \frac{|\alpha|}{2g} = 1 - \frac{H}{H_c} \tag{4c.1}$$

(ε controls the height of the potential barrier) we can write (2b.17) as

$$U(\psi) = -4g \sin^4(\frac{\psi}{2}) + 4g\varepsilon \sin^2(\frac{\psi}{2}) \tag{4c.2}$$

Using (4c.2) directly in (3c.2-3-4) we can compute the probability of nucleation of the stable phase per unity area per second.

Unfortunately, an analytic solution to this problem is only available when $\varepsilon \sim 1$ or $H \sim 0$ which means that the potential barrier is very high and consequently, $\Gamma \to 0$. The decay rate when $\varepsilon \to 0$ can only be computed numerically. Its final form is

$$\frac{\Gamma}{\text{area}} = 8s(s+1)z\, R(\varepsilon)\, (\frac{J\hbar^2}{\hbar d^2})(\frac{\Delta}{J})^{1/2} (\frac{2I(\varepsilon)}{\varepsilon})^{3/2} \exp(-\frac{4\pi J}{\Delta}\frac{I(\varepsilon)}{\varepsilon}) \tag{4c.3}$$

where $R(\varepsilon)$ is a numerical factor that results from the detailed evaluation of the ratio of determinants in (3c.4) ($1 < R < 10$ in general) and $I(\varepsilon)$ is given by[15]

$$I(\varepsilon) = \int_0^\infty dx\, x^2\, [-\psi_c \sin(\frac{\psi_c}{2})\cos(\frac{\psi_c}{2}) + \frac{2\psi_c}{\varepsilon}\sin^3(\frac{\psi_c}{2})\cos(\frac{\psi_c}{2})$$
$$- \frac{2}{\varepsilon}\sin^4(\frac{\psi_c}{2}) + 2\sin^2(\frac{\psi_c}{2}) \tag{4c.4}$$

where ψ_c is the solution of

$$\frac{d^2\psi}{dx^2} + \frac{2}{x}\frac{d\psi}{dx} = \frac{1}{\varepsilon}\sin\psi(\cos\psi - 1) + \sin\psi \tag{4c.5}$$

with boundary conditions $\psi(\infty) = 0$ and $\psi'(0) = 0$.

This solution, ψ_c, is only available numerically. Once it is computed we can evaluate (4c.4) for different values of ε (see Table 1 below).

Table 1. Dimensionless action as a function of ε.

ε	0.1	0.2	0.3	0.4	0.5	0.6	0.7	0.8	0.9
$\dfrac{4\pi I(\varepsilon)}{\varepsilon}$	45.8	59.3	78.5	106.8	152.7	234.6	405.5	883.6	3280.0

This table allows us to determine a range of values of the external field for which the decay rate by quantum tunnelling is appreciable. For example, if we consider a moderately anisotropic system ($\Delta \approx J$) with the transition temperature ($T_c \approx J\hbar^2 k^{-1}$) ranging between 1-10 °K the prefactor of (4c.3) will be close to $10^{31} m^{-2} s^{-1}$. Therefore, we can easily conclude that for $\varepsilon \lesssim 0.3$ the prefactor and the exponential will combine to give us $\Gamma \sim 1\ m^{-2} s^{-1}$. The smaller the factor ε, the greater the probability of quantum decay of the metastable configuration. However, one should keep in mind that when $|H| \lesssim H_c$ ($\varepsilon \approx 0$), zero point fluctuations of the spins about their requilibrium positions will tend to destroy the metastable configuration before any tunnelling event can take place. Therefore, we should introduce a lower bound for ε, below which the WKB approach we have been employing is no longer valid.

5) CONCLUSIONS

Throughout these lecture notes I have been investigating some very interesting systems that might exhibit unusual behaviour at very low temperatures. If this behaviour can be confirmed the only natural explanation would be that these systems undergo quantum mechanical tunnelling out of a metastable state.

This expectation has actually been fulfilled with the experimental results[16] to which I refer above in the text. Nowadays, it is the general belief that the appearence of the finite voltage state in CBJJ's before the critical current is reached is due to quantum tunnelling of the phase variable ψ. Of course I am referring to the very low temperature region only.

The evidence for this tunnelling event is a signature that tunnelling between fluxoid states in a SQUID will also take place. This phenomenon clearly increases the expectations that quantum coherence[14] is quite likely to be observed (or at least tested) in this same system and this experiment will be of fundamental relevance to the foundations of quantum mechanics.

On the other hand the other kind of problem I have treated in these lectures (quantum nucleation) is of a much harder experimental access. The reason for this is that I have only addressed the problem of homogeneous nucleation which is a very ideal situation. In a realistic situation, impurities and boundaries will play a very important role in the nucleation process because they are good nucleation centers. My intention was only to show that we can find magnetic systems such that the numerical values of their parameters would allow them to be good candidates to exhibit quantum nucleation.

Finally, I would like to stress the difference between macroscopic quantum tunnelling and quantum nucleation. In the former, there must be a coherent decay of a macroscopic number of particles (electrons in the case of SQUID's or CBJJ's) from a metastable state into a stable state. On the other hand, in the latter, this is a matter of collective decay of a huge number of correlated microscopic variables, however, not necessarily in a coherent way.

REFERENCES

1. H.B. Callen, "Thermodynamics", John Wiley & Sons Inc. (1960).
2. L.D. Landau and E.M. Lifshitz, "Statistical Physics", Pergamon (1977).
3. H.A. Kramers, Physica 7:284 (1940).
4. M. Tinkham, "Introduction to Superconductivity", McGraw-Hill, Inc. (1975).
5. J. Kurkijarvi, Phys. Rev. B6:832 (1972).
6. H.J. Mikeska, J. Phys. C11:L29 (1978).
7. A.R. Bishop, J. Phys. C13:L67 (1980).
8. A.H. Eschenfelder, "Magnetic Bubble Technology", Springer Series in Solid State Sciences, Vol. 14, ed. H.J. Queisser (Springer; Berlin).
9. J.S. Langer, Ann. Phys. 41:108 (1967).
10. S. Coleman, Phys. Rev. D15:2929 (1977).
11. C.G. Callan and S. Coleman, Phys. Rev. D16: 1762 (1977).
12. M. Stone, Phys. Lett. 67B:186 (1977).
13. A.O. Caldeira and A.J. Leggett, Ann. Phys. 149:374 (1983).
14. A.J. Leggett, Lecture Notes delivered at the NATO-ASI in Evora (in this same volume).
15. A.O. Caldeira and K. Furuya, J. Phys. C21:1227 (1988).
16. J. Clarke, A.N. Cleland, M.H. Devoret, D. Esteve and J.M. Martins, Science 239:992 (1988).

GROUND STATE OF A-NON DISSIPATIVE JOSEPHSON JUNCTION

A. Tagliacozzo

Dipartimento di Scienze Fisiche, Universita' di Napoli e.
G.N.S.M. (CNR)
Mostra d'Oltremare Pad.19
I-80125 Napoli, Italy

Josephson junctions display quantum properties at a macroscopic level, due to many-body quantum mechanics and superconductive coherence. In fact they allow for a classical description at a macroscopic level in terms of equations of motion which include $\hbar$ in a constitutive way.

Recently, it has been suggested that they are good candidate systems in search of evidence for direct quantum behavior of some macroscopic degree of freedom (Macroscopic Quantum Phenomena)[1]. One of the difficulties that immediately arises, when dealing with quantum behavior of macroscopic bodies, is how to describe their interaction with the environment. Basically the interaction with the environment is twofold:

a) interaction with the "outside" thermal bath, i.e. dissipation;

b) interaction with the external sources.

In my opinion, while the first topic is by now widely studied, starting from the pioneering work of Caldeira and Leggett[2], the second is often misconsidered.

Inclusion of external sources in a coherent quantum picture is the main subject of this manuscript.

Take a non dissipative Current Biased Josephson Junction (CBJ) as an example. If we can disregard magnetic fields, both external and due to the supercurrent itself, the junction can be described by a phase difference of the order parameter Φ, which is constant on the area of the insulating layer. The system is in this respect one-dimensional.

The by far most common view in studying a current bias is to take into account the presence of the current generator in the circuit, by coupling the phase to the external current. This is done adding a term in the potential energy where the external current appears as a "dead" parameter controlled from the outside. Thus the potential energy usually employed in describing CBJ's is:

$$U_J(\Phi) = - E_J(\cos \Phi - 1) - \hbar I_{ext} \Phi/2e . \qquad (1)$$

Here E_J is the Josephson coupling energy.

The main justification of the second term on the r.h.s. is the fact that the minimum of the potential energy, describing classical equilibrium for the phase Φ_{cl} is found for $I_{ext} = 2eE_J \sin \Phi_{cl}/\hbar = I_{cl}^J$. In fact superconductors allow for current carrying equilibrium states, what is a special feature of these systems.

The drawback that I see in this procedure is that the bias is included in the system as a totally classical variable, in such a way that deviations from its value (in our case fluctuations of the phase), are immediately deviations from equilibrium. On the other hand, direct quantization of the phase implies that, at $T = 0$, its zero point motion is part of the ground state properties in a fundamental way and consequently of the quantum equilibrium itself.

A contrasting perspective when quantizing the system is the following.

One allows for fluctuations of the dynamical variables and imposes at the end given expected values for the relevant observables which classically characterize its state. These are the values which are supposed to describe the operating conditions. In practice, in the case of a CBJ, one imposes as a constraint that, e.g., the state is a zero voltage state and that the expectation value for the current is the one which is supposed to be given by the classical external generator.

Let us first include harmonic quantum fluctuations of a zero voltage state for a CBJ with a given current, quantizing as usual by means of the correspondence principle. To quantize the model of eq. (1) we have to introduce the kinetic energy term:

$$H = \frac{Q_J^2}{2C} + U_J(\Phi). \qquad (2)$$

Here C is the capacitance of the junction and we have introduced the charge difference at the junction Q_J, which is the canonical conjugate variable of the phase difference at the junction Φ. If the Josephson coupling energy E_J is much larger than the capacitive energy $E_C = (2e)^2/2C$ (limit of large capacitance), the fluctuations of the phase induced by the charging of the capacitance are very small and we are close to the classical description.

It is useful to define the deviations φ of the phase from its 'classical' value: $\Phi = \Phi_{cl} + \varphi$. The potential energy appearing in eq. (1) can be expanded in powers of φ and the smallness of the oscillations of the phase around the minimum allows us to retain just the quadratic term:

$$U_J(\Phi) \sim U_J(\Phi_{cl.}) + \frac{1}{2}E_J \cos\Phi_{cl.}\varphi^2 + \frac{\hbar}{2e}(I_{cl.}^J - I_{ex})\varphi . \qquad (3)$$

We define: $I^* = I_c \cos \Phi_{cl.} = (I_c^2 - I_{cl.}^{J\,2})^{\frac{1}{2}}$ with $I_c = 2eE_J/\hbar$ and we leave aside the linear term in φ for the moment. Thus, if is $I_{cl}^J = I_{ext}$, we are lead to the following hamiltonian:

$$H_{osc} = \frac{e^2}{2C}\left(\frac{Q_J}{e}\right)^2 + \frac{1}{2}\frac{\hbar}{2e}I^*\varphi^2 + U_J(\Phi_{cl.}) . \qquad (4)$$

The classical contribution $U_J(\Phi_{cl.})$ will be omitted in the following discussion. The charge accumulated at the junction capacitor plates, Q_J, is the conjugate variable to φ due to the fact that $\dot{\Phi}_{cl.} = 0$. (That is $\Phi_{cl.}$ refers to the zero voltage state). It follows that in a phase representation, Q_J is expressed, as $Q_J = -i2e\partial/\partial\varphi$. The approximation involved in eq. (3) is consistent if $<\varphi^2>_{osc}\ll 1$, which in fact occurs if $E_J \gg E_C$ and $I^J_{cl.} \ll I_c$.

We second quantize H_{osc} according to:

$$\frac{Q_J}{e} = \left(\frac{I^*}{e\omega_J}\right)^{\frac{1}{2}}(a + a^\dagger)$$

$$\varphi = i\left(\frac{e\omega_J}{I^*}\right)^{\frac{1}{2}}(a - a^\dagger) \tag{5}$$

with $[a, a^\dagger] = 1$.

Here ω_J, is the plasma frequency

$$\omega_J = \left(\frac{2eI^*}{\hbar C}\right)^{\frac{1}{2}}. \tag{6}$$

It is straightforward to verify that classical electrostatics is consistent with the second Josephson equation, due to the operator identity:

$$V = \frac{\hbar}{2e}\dot{\Phi} = \frac{Q_J}{C}, \tag{7}$$

with $i\hbar\dot{\Phi} = [\Phi, H_{osc}] = i\hbar\omega_J^2 Q_J/I^*$.

According to eq. (3), the operator for the extra current flowing in the junction should have the expectation value:

$$<\delta I> = <\dot{Q}_J> = -I^*<\varphi> = I_c\sin\Phi_{cl} - I_{ext}. \tag{8}$$

The state of the system is characterized by an average value of the phase difference $<\varphi>$ and of the charge difference $<Q_J>$. Due to eqs. (7,8), the obvious choice for a CBJ in the zero voltage state is that they both vanish.

The generating functional for the hamiltonian is:

$$Z[\eta(\tau), \eta^*(\tau)] = \mathrm{Tr}\left\{e^{-\beta H_{osc}}\mathrm{T}_\tau\exp\left\{1/\hbar\int_0^{\hbar\beta}dr[\eta(\tau)a(\tau) + \eta^*(\tau)a^\dagger(\tau)]\right\}\right\} \tag{9}$$

from which all the relevant average quantities can be calculated. An auxiliary time dependent source $\eta(\tau)$ has been introduced. The result is [3]:

$$Z = [2\sinh u]^{-1}\exp 1/\hbar^2 \int_0^{\hbar\beta}\int_0^{\hbar\beta} d\tau\, d\tau'\, \eta(\tau)g(\tau - \tau')\eta^*(\tau^*), \tag{10}$$

where $u = \beta\hbar\omega_J/2$ and $g(\tau) = [N + \vartheta(\tau)]e^{-\omega\tau}$ is the equilibrium Green's function for the oscillator. Here $N = [e^{2u} - 1]^{-1}$ and $\vartheta(\tau)$ is a ϑ-function. We can displace the source by a time independent quantity ξ to take care of the constraints

51

$$<\delta I> = -\frac{e}{\hbar\beta}\left(\frac{I^*}{e\omega_J}\right)^{\frac{1}{2}}\frac{\delta \ln Z[\xi]}{\delta \,\Im m\, \eta(\tau)}\bigg|_{\eta=0}, \quad (11)$$

and

$$<\frac{Q_J}{e}> = \frac{1}{\beta\hbar\omega}\left(\frac{I^*}{e\omega_J}\right)^{\frac{1}{2}}\frac{\delta \ln Z[\xi]}{\delta \,\Re e\, \eta(\tau)}\bigg|_{\eta=0}. \quad (12)$$

If we want to describe a zero voltage state with an extra current $<\delta I>$, we fix $\xi = \xi' + i\xi''$ (ξ', ξ'' real) as follows:

$$<Q_J> = 0,$$
$$<\delta I> = <\dot{Q}_J> = -2\xi''\left(\frac{I^*}{e\omega_J}\right)^{\frac{1}{2}} e\omega_J, \quad (13)$$

giving: $\xi' = 0$ and $\xi'' = -\left(\frac{e\omega_J}{I^*}\right)^{\frac{1}{2}} <\delta I>/(2e\omega_J)$.

If we substitute back this result into eq. (9), the hamiltonian appearing in the partition function incudes an extra term which reads:

$$H_{bias} = -\hbar\omega_J \xi'' i(a - a^\dagger) \sim \frac{\hbar}{2e} <\delta I> \varphi, \quad (14)$$

with $<\delta I>$ given by eq. (8). This is in fact the term that appears in eq. (3), in the expansion around Φ_{cl}.

Within this picture of quantum harmonic oscillations of the phase, described by the hamiltonian of eq. (4) it is clear that a reduction of the capacitance C keeping the other parameters fixed (e.g. I^*) amounts to an increase in the quantum fluctuations of the phase and to a decrease of those of the charge, respectively. In fact eqs. (4,5) imply that $<\varphi^2>_{osc} = e\omega_J/I^* \coth u$, while $<Q_J^2>_{osc} = eI^*/\omega_J \coth u$.

However, the classical minimum is not a good starting point to describe the quantum ground state of a non dissipative junction at least in the unbiased case, due to the indistinguishability of φ and $\varphi + 2\pi$. At zero temperature quantum effects will be dominant except in a narrow range of values for the current close to the critical value. For a non dissipative system quantum fluctuations can induce tunneling of the phase between adjacent minima of the washboard potential, which are phase slips of 2π. As a result, a finite voltage would develop across the junction. It follows that, if tunneling occurs, quantum fluctuations in the electromotive force are an intrinsic characteristics of the ground state. Its expectation value will of course vanish in the zero voltage state, which has to be imposed as a constraint.

1. ONE-DIMENSIONAL WIRE

To show how all of this should be handled, we study a simpler reference system, consisting of a very long wire between two superconductors, which can be described within the Ginzburg Landau (G-L) approach. Prior to discussing the quantum picture, we will shortly review the properties of the G-L free energy, stressing its gauge properties.

The Ginzburg-Landau free energy density, which includes the magnetic field is[4]

$$\mathcal{F} = \frac{1}{8\pi}(B^2 - H_c^2) + \frac{\hbar^2}{2m}\left|(\vec{\nabla} + \frac{ie}{\hbar c}\vec{A})\psi\right|^2 + \frac{H_c^2}{8\pi}\left(\frac{1}{n_s}|\psi|^2 - 1\right)^2. \quad (15)$$

Here ψ is a complex scalar field, the order parameter describing the superconductive state. $|\psi|^2 = n_s$ is the equilibrium superfluid density for homogeneous systems and the energy gain in the superconductive state is $H_c^2/8\pi$, where H_c is the critical magnetic field. In view of the gauge invariance, it is useful to define the gauge covariant derivatives $D_\mu = \partial_\mu - i\bar{\phi}_o^{-1}A_\mu$ with (μ = 1,2,3). We have here defined $\bar{\phi}_o$ as the flux quantum unit $\phi_o = \hbar c/2e$ divided by 2π ($-e$ is the electron charge). The magnetic part can also be rewritten in terms of the electromagnetic field strength tensor ($F_{\mu\nu} = \partial_\mu A_\nu - \partial_\nu A_\mu$). A static Lagrangian density arises

$$\mathcal{L} = -\frac{1}{16\pi}F^{\mu\nu}F_{\mu\nu} + \frac{H_c^2}{4\pi n_s}\left\{\xi^2(D^\mu\psi)^*(D_\mu\psi) + |\psi|^2 - \frac{1}{2n_s}|\psi|^4\right\}, \quad (16)$$

where ξ is the coherence length. There is by now no difference between upper and lower indices (μ = 1,2,3), but they are introduced to include subsequently the time.

The classical equations of motion are readily obtained by functional minimization:

$$\xi^2 D^\mu D_\mu \psi + \frac{1}{n_s}|\psi|^2\psi - \psi = 0 \quad (17)$$

and

$$\frac{c}{4\pi}\partial_\nu F^{\nu\mu} = -\frac{\bar{\phi}_o c}{4\pi n_s \lambda^2}\left\{\frac{i}{2}(\psi\partial^\mu\psi^* - \psi^*\partial^\mu\psi) + \bar{\phi}_o^{-1} A^\mu|\psi|^2\right\} = j^\mu. \quad (18)$$

Eq. (18) defines the current via Maxwell's equations. Here two typical lengths appear which characterize the space variation of the scalar and the vector field, that is the coherence length ξ and the penetration length $\lambda^{-2} = 2H_c^2\xi^2/\bar{\phi}_o^2$.

In the absence of fields and currents the solutions of these equations for an homogenous system are:

$$D^\mu\psi(x) = 0 \text{ and } F^{\mu\nu}(x) = 0$$

that is $\psi(x) = n_s^{\frac{1}{2}}e^{i\eta(x)}$ with

$$\eta(x) = \eta_o + \bar{\phi}_o^{-1}\chi(x) \text{ and } A^\mu(x) = -i\, e^{-i\eta(x)}\partial^\mu e^{i\eta(x)}. \quad (19)$$

They can all be obtained via gauge transformation from the solution

$$\psi(x) = \psi_o = n_s^{\frac{1}{2}}e^{i\eta_o} \text{ and } A^\mu = 0,$$

and they are called pure gauges. These extrema of the functional $\mathcal{F}$ correspond to minima of the free energy: they are stable with respect to small perturbations. For this reason they are called "classical vacua." The gauge symmetry group $U(1)$ is the broken symmetry, i.e., order parameters differing by a constant phase describe different classical vacua.

When a static magnetic field is present, let's say along the z axis, there is an approximate solution of the motion equations of the vortex type[5]:

$$\psi = n_s^{\frac{1}{2}} e^{im\vartheta} f(r) \ , \quad \vec{A}(r) = A_\vartheta(r)\hat{\vartheta} \ . \tag{20}$$

Here cylindrical coordinates (r, ϑ, z) have been chosen and m is an integer. $f(r)$ is the solution of an equation derived from eq. (17) with limiting behaviors:

$$f(r) \sim r \quad \text{for} \quad r \to 0$$
$$ 1 \phantom{\quad \text{for} \quad} r \to \infty$$

and the vector potential is approximately of the form:

$$rA_\vartheta = \bar{\phi}_0 m \left(1 - \frac{r}{\lambda} K_1\left(\frac{r}{\lambda}\right)\right) \ , \tag{21}$$

where K_1 is the first order Bessel function. The vanishing of $F_{\mu\nu}$ far from the vortex core $(r \to \infty)$ implies that A_μ asymptotically becomes a gauge transform of zero:

$$A_\mu = g\partial_\mu g^{-1} + \mathcal{O}(e^{-\frac{r}{\lambda}}),$$

with $g(\vec{r}) = e^{im\vartheta}$. Also ψ can be viewed as gauge transformed:

$$\psi \to g\,\psi_0 + \mathcal{O}(e^{-\frac{r}{\lambda}}).$$

Far from the vortex core the presence of a flux line trapped at the origin is just a gauge transformation on the fields. This will correspond to the instanton solution of our quantum model in $1 + 1$ dimensions.

Let the system be a long wire of length l, $(l \to \infty)$ biased with a current I. By one dimensional, we mean that its diameter is much less than the coherence length ξ and the penetration depth λ, so that the order parameter and all the other quantities can be assumed as non-varying within its cross section σ. The physical state is chosen once the electromagnetic sources and the phase difference $\Delta\eta$ have been fixed between $x \to \pm\infty$. Of course, the free energy is a function of these given conditions. From the equation of motion we have [6]:

$$\psi(x) = n_s^{\frac{1}{2}} f(x') e^{i\eta(x')}$$
$$j = \frac{c}{\phi_o} H_c^2 \, \xi^2 \, f^2(x') D_1\eta = c\frac{H_c^2 \xi}{\phi_o} J \ , \tag{22}$$

which defines the dimensionless constant J. Here $x' = x/\xi$, but we will drop the prime in the following. If we consider the electromagnetic field generated by the current to be negligible and we ignore the current generator, we can choose the gauge in which the vector potential A_1 and the scalar potential A_0 both vanish. The solutions are:

$$\psi(x) = n_s^{\frac{1}{2}} f_\kappa e^{i(\eta_o + \kappa x)}$$
$$J = \kappa(1-\kappa^2) \text{ and } f_\kappa = (1-\kappa^2)^{\frac{1}{2}} . \qquad (23)$$

The maximum allowed value for κ^2 is 1/3 for a true minimum, which defines the critical current.

If we do not restrict the phase within $(0, 2\pi)$, there are various classical minima for given boundary conditions, labeled by integers m:

$$\Delta\eta = \kappa_m \frac{l}{\xi} + 2\pi m . \qquad (24)$$

The difference in energy between neighboring minima is:

$$\delta F = \frac{dF}{d\kappa} \delta\kappa = H_c^2 J \sigma \xi = \frac{\hbar}{2e} I . \qquad (25)$$

Again, they are stable with respect to small perturbations. The rate for thermally activated transitions over the energy barrier, $\kappa \to \kappa - 2\pi\xi/l$, has been studied and the resistive transition to the normal state for T close to T_c interpreted in this way.

The usual procedure when dealing with the classical description at constant current bias is to perform a Legendre transformation on the free energy:

$$F(\Delta\eta) \to G(I) = F - \frac{\hbar}{2e} I \Delta\eta . \qquad (26)$$

There is a crucial difference between a superconducting wire and a Josephson Junction, even in the case in which the latter can be treated as a quasi one-dimensional system. In the presence of the junction it is impossible to count the windings of the phase because this is not defined within the insulating layer. This implies that in a JJ $\Delta\eta$ is defined modulus 2π and the supercurrent will be periodic in 2π. This differs from eq. (24): This difference has no practical consequences, however, in the limit of infinite length of the wire.

Our aim is to discuss the ground state energy of the system, taking into account properly the constraint on the phase difference $\Delta\eta$.

The static problem cannot be disentangled from the dynamical one due to quantum fluctuations[7]. In place of the static free energy $\mathcal{F}$ of eq. (15) a Lagrangian density is needed:

$$\mathcal{L}[\psi, \dot\psi; \vec{A}, \dot{\vec{A}}] .$$

The time dependent terms for the electromagnetic fields giving rise to the appearance of the electric field and to Maxwell's equations are straightforward. A proper derivation of the dependence on time of the order parameter, on the contrary, requires the inclusion of the particle-particle interaction within a microscopic picture and will not be performed here[8]. The choice of an effective Lagrangian has been recently reconsidered, to describe the dynamics of a simple Nambu-Goldstone field, allowing also for variations in the magnitude of the order parameter[9]. We use the fact that at very low temperature the excitation modes are propagative, implying that a second derivative with respect to time will appear in the equation of motion for ψ.

As we analyze the zero temperature case, charge fluctuations involving plasma oscillations of frequency ω_p are highly improbable. We also neglect gradients in the chemical potential μ_c, which give rise to resistance-producing fluctuations. They decay on a very short time scale. In fact, let R_n be the resistance associated with them. Classically, neglecting induced fields, the current reads ($A_0 = A_1 = 0$):

$$I(x,t) = \phi_o^{-1}\sigma H_c^2 \xi c(1-\kappa^2)\xi \nabla \eta(x,t) + lR_n^{-1}\nabla \mu_c.$$

Formally extending the range of the greek labels also to the time (i.e. $\mu = (0,1\ldots d)$, where d is the space dimensionality), and including the chemical potential, we have:

$$-\frac{1}{4\pi}\vec{\nabla}\vec{E} = \frac{\phi_o}{4\pi c\lambda^2}\frac{|\psi|^2}{n_s}\left(\dot{\eta} + \frac{2e}{\hbar}\mu_c\right). \tag{27}$$

Time variations of the phase of the order parameter locally follow the electrochemical potential if the system is charge neutral:

$$\frac{\partial \vec{\nabla}\eta}{\partial t} = \frac{2e}{\hbar}\vec{\nabla}\mu_c, \tag{28}$$

which can be substituted in the second term of the current expression. Again, by charge neutrality $\nabla I(x,t) = 0$, so that

$$\nabla^2 \left(\frac{1}{\tau_r}\eta + \dot{\eta}\right) = 0,$$

where $\tau_r = 4\pi(e/I_o)\cdot(R_Q/R_n)$ with $I_o = J_o\sigma$ and J_o given by eqs. (23,24) for $m = 0$, and R_Q the quantum resistence $h/4e^2$.

If the phase fluctuation has to be localized in space, the content of the bracket has to vanish, which says that the time scale of the variations of the phase and of the electric field is τ_r. We assume $R_Q/R_n \ll 1$ and τ_r very small. Within this approximation the chemical potential can be taken as constant and the current density is still given by eq. (18). Eq. (27) is the Gauss law for this model, which suggests the definition of a charge density ρ by equating the r.h.s. to $-v^2\rho/c$. In fact the continuity equation $\partial^\mu j_\mu = 0$ implies that

$$\nabla^2 \eta - \frac{1}{v^2}\frac{\partial^2}{\partial t^2}\eta = \frac{1}{\phi_o v}\frac{\partial \mu_c}{\partial t} = 0.$$

Thus the velocity v is the propagation velocity of the collective modes: these would be sound waves of velocity $v_F/\sqrt{3}$ for an uncharged system. To keep notation simple I put $v = c$ which I think doesn't affect the core of the argument. Then it is enough to extend μ also to $\mu = 0$ and $j^\mu = (c\rho, \vec{j})$. Otherwise, anisotropy should be introduced in the definition of ξ appearing in eq. (16).

We want to add a piece to the Lagrangian density which implements a constraint on the electromotive force in the wire. The term to be added is:

$$\Delta \mathcal{L} = \theta \frac{e}{\pi c}\partial^\mu \epsilon_{\mu\nu} A^\nu. \tag{29}$$

Here θ is a scalar and $\epsilon_{01} = -\epsilon_{10} = -1, \epsilon_{\mu\mu} = 0$. This term is a total derivative and thus doesn't affect the equations of motion. It appears however in the action as the quantity

$$W(A)\theta = -\frac{1}{2\phi_o}\theta \int \epsilon_{\mu\nu}F^{\mu\nu}d^2x = -\frac{2e}{\hbar}\theta \int dt\, f_{em}\,. \tag{30}$$

Of course $W(A)$ is a gauge invariant quantity that can be rewritten as a closed contour integral at infinity in the space-time plane of the form[10]:

$$W(A) = \frac{1}{2\pi\phi_o} \oint \vec{A}\cdot\vec{dl}\,. \tag{31}$$

Written in this form its gauge invariance is not apparent but allows us to recognize that this quantity only depends on the asymptotic behavior of the vector potential and not on the details of the behavior at points inside the space-time plane. In a system which is always neutral the usual choice is $A_o = 0$ and $\vec{A} = A_t$, i.e., only the transverse component of the vector potential survives.

If one keeps $A_0 = 0$ all static gauge transformation are still possible

$$A_\mu \to A_\mu + \partial_\mu \chi(\vec{x}).$$

In a 1-d system, within fixed boundary conditions, the functions χ are classified by their behavior at infinity ($x \to \pm\infty$) into two types:

(i) gauge transformations or "small" transf.: $x \to \pm\infty$: $\chi(x) = \tilde{\chi}(x) \to 0$, and

(ii) "large" transf.: $\chi(x) \to 2\pi m_\pm \, \phi_o(m_\pm$ any integer$)$.

While by construction $W(A)$ does not change under a small transformation, under a large one, we have:

$$\delta\{W(A)\} = (\chi(+\infty) - \chi(-\infty))|_0^T = \nu \quad \nu \text{ integer}\,, \tag{32}$$

where $\nu = m_+ - m_-$ is the winding number of the large transformation. We have also chosen $\nu(t = 0) = 0$. This is equivalent to fixing the minimum of the free energy one is assuming as the initial condition at $t = 0$. Results are expected to be independent of this choice if all the minima are equivalent. To satisfy the Gauss law of eq. (27) at each point, the quantum physical state has to be a superposition of all small gauge transformed states.

We conclude that physical states that are classical vacua are homotopy classes for the field configurations $\{\psi, A\}$ characterized by the winding number N given by eq. (30) or eq. (31), the members of each class being related to one another by a small transformation. The point is that the classical "vacuum" is unstable with respect to "large" transformations which imply a change of $\Delta \eta$ by multiples of 2π. A "large" transformation $\mathcal{G}_\nu$ is an unitary operator, which, acting on the physical state $|N>_{phys}$ changes it into $|N + \nu>_{phys}$. On the other hand, given a scalar θ, states of the form:

$$|\theta> = \sum_{N=-\infty}^{+\infty} e^{iN\theta} |N>_{phys} \tag{33}$$

are eigenstates of $\mathcal{G}_\nu$ with eigenvalue $e^{-i\nu\theta}$.

The total derivative in eq. (31) can be eliminated from the Lagrangian just by adding a phase factor to the states $|\theta>$:

$$|\theta>_{phys} = e^{iW(A)\theta}|\theta> .$$

Both $|\theta>$ and $W\theta$ are left invariant by a small gauge transformation, while both change, under "large" ones by opposite phase factors $\pm i\nu\theta$. Therefore, this is the same as requiring that the physical state $|\theta>_{phys}$ is invariant under $\mathcal{G}_\nu$. This requirement is of course only conceivable in the absence of dissipation. Quantum coherence will be lost as soon as a dissipative coupling is included.

2. JOSEPHSON JUNCTION AT $T=0$ IN THE ZERO VOLTAGE STATE

Let us now discuss the quantum ground state of our model Josephson junction with zero external current. In this case, according to eq. (25), the minima of the static free energy are all equivalent.

Since $\mathcal{G}_\nu$ is a symmetry operation for the system, it commutes with the hamiltonian $\mathcal{H}$. Hence, we look for simultaneous eigenstates of the form (33). We evaluate the quantum correction to the ground state energy due to tunneling of the phase between adjacent minima in the semiclassical approximation. We perform a Wick rotation to imaginary times: $cr = ict = x_0, A_\mu = (iA_0, \vec{A}), j_\mu = (ic\rho, \vec{j})$. The propagator can be written, according to Feynman's prescriptions, in terms of the euclidean action. For large times, it will behave as

$$<\theta|e^{-HT/\hbar}|\theta'> = 2\pi\delta(\theta-\theta')e^{-E(\theta)T/\hbar} . \qquad (34)$$

In particular:

$$<\theta|e^{-HT/\hbar}|\theta> = 2\pi\delta(0)e^{-E(\theta)T/\hbar} = \sum_N \sum_\nu e^{-i\nu\theta} <N+\nu|e^{-HT}|N> .$$

Because the sum over ν is likely to be independent of N we have:

$$\lim_{T\to\infty} e^{-E(\theta)T/\hbar} = \lim_{T\to\infty} \mathcal{N} \sum_\nu \int \{\mathcal{D}A\mathcal{D}\psi\mathcal{D}\psi^*\}_\nu e^{-(S_E+S_{gf})/\hbar} e^{-i\nu\theta} . \qquad (35)$$

Here S_{gf} is a gauge fixing term to factor out the redundancy due to small gauge transformations and $\mathcal{N}$ is a suitable normalization. The r.h.s. of this equation can be rewritten without separating the configurations belonging to different homotopy classes if one deals with the action:

$$S_E/\hbar + S_{gf}/\hbar + i\frac{1}{4\pi\phi_o}\theta \int d^2x\, \epsilon_{\mu\nu}F_{\mu\nu} . \qquad (36)$$

The factor $-i$ is due to the continuation to imaginary times. This is another way to see how the added term arises.

To evaluate the ground state energy semiclassically, we note that there are saddle points of the action, whose field configuration, at infinity, corresponds to a large gauge transformation. These are vortex configurations in the euclidean space-time plane (τ, x). In fact, in our model, the equations of motion that describe a vortex in (τ, x) are the same as the ones for a static vortex in (2+1)-d. There is a sudden change of phase at $\tau = 0$ at the origin $x = 0$ where the phase difference jumps by 2π. These phase slip processes can occur because at that time the order parameter vanishes at the origin.

A single instanton of winding number ν can be viewed as the tunneling of the system in time T from one classical vacuum labeled by N to that labeled by $N + \nu$, spending most of the time in the two extremal configurations. Since the action of these field configurations is finite, they constitute the dominant contributions to the propagator. The energy of the ground state can thus be calculated semiclassically. If we denote the action of a single instanton of winding number n (or anti-instanton of number $\bar{n}$) S_o and we assume that they form a 'dilute gas' so that they do not interact with each other, we get:

$$< \theta | e^{-HT/\hbar} | \theta > \sim \sum_{n,\bar{n}} (K\, e^{-S_o/\hbar})^{n+\bar{n}} \frac{T^{n+\bar{n}}}{n!\bar{n}!} e^{i(n-\bar{n})\theta} = \exp\left(2KT e^{-S_o/\hbar} \cos\theta\right). \tag{37}$$

Here is $\nu = n - \bar{n}$ and K is the usual determinant factor arising from the evaluation of the path integral. This can be expressed in terms of the eigenvalues of differential operators (where the second functional derivative of the action around the minimum appears, evaluated at the instanton configuration, with the zero eigenvalue excluded in the denominator):

$$K = (S_o/2\pi\hbar)^{\frac{1}{2}} \left| \frac{\det\{-\partial_r^2 + \omega^2\}}{\det'\{-\partial_r^2 + V''(\{\psi, A\})\}} \right|^{\frac{1}{2}}. \tag{38}$$

Here, ω is the frequency of oscillations around the minima of the free energy. The energy of the vacuum defined in eq. (34), is given by:

$$E(\theta) = E_{class} + \frac{1}{2}\hbar\omega - 2\hbar K e^{-S_o/\hbar} \cos\theta. \tag{39}$$

The imposition that there is no electromotive force implies that the physical choice is $\theta = 0$.

In fact, for a time translationally invariant system we have:

$$<f_{em}> = \frac{1}{2cT} \int d^2x\, \epsilon_{\mu\nu} F^{\mu\nu} = -\frac{<W>}{cT} \phi_o.$$

Thus the electromotive force is equal to the expectation value of the number of flux quanta which cross the junction in unit time. Continuation to imaginary times yields:

$$\frac{1}{cT} \frac{<\theta|W|\theta>}{<\theta|\theta>} \rightarrow \frac{1}{icT} \frac{\int \mathcal{D}A\mathcal{D}\psi\, e^{-S/\hbar}\nu\, e^{i\nu\theta}}{\int \mathcal{D}A\mathcal{D}\psi\, e^{-S/\hbar} e^{i\nu\theta}} =$$

$$= -\frac{1}{cT} \frac{\partial}{\partial\theta} \ln \int \mathcal{D}A\mathcal{D}\psi\, e^{-S/\hbar} e^{i\nu\theta} = 2\frac{K}{c} e^{-S_o/\hbar} \sin\theta.$$

Hence:
$$<f_{em}> = -\frac{\hbar}{e}\hbar K e^{-S_o/\hbar}\sin\theta . \qquad (40)$$

Therefore, θ has to vanish.

A current bias can be taken into account if we allow θ to be time dependent. In fact the contribution due to the term added to the action $S/\hbar$ in the Weyl gauge is:

$$-\frac{1}{2\phi_o}\int d^2x\theta\epsilon_{\mu\nu}F^{\mu\nu} = -\frac{1}{\phi_o}\int dt(\partial_t\theta)\int A_1\,dx_1 + \\ +\frac{1}{\phi_o}\int dt\partial_t\left(\theta\int A_1\,dx_1\right) . \qquad (41)$$

The second term is a total derivative $\partial_t(W(A)\theta)$ added to the Lagrangian, as a generalization of that of eq. (29). It doesn't change the equations of motion. According to eq. (32) it gives:

$$\theta(t)\int dx_1\,A_1\Big|_0^T = \nu\,\theta(T) .$$

The first term is the change in the action due to the addition of the current

$$I = 2e\,\frac{\dot\theta}{2\pi} , \qquad (42)$$

corresponding to the energy term:

$$-\delta\mathcal{E} = \frac{1}{c}I\int \vec{A}\cdot d\vec{l} .$$

Therefore, this term allows us to rescale the free energies of all inequivalent minima, as we did in eq. (26).

In the case of a bulk one-dimensional wire, in which A is approximately a pure gauge one would have:

$$\int A_1\,dx_1 = \phi_o\Delta\eta ,$$

which gives back eq. (26). In the case of a current biased Josephson junction, when the e.m. field is eliminated by means of the classical equations of motion, the added term gives rise to the usual semiclassical Lagrangian for the phase difference $\Phi = \Delta\eta$:

$$\mathcal{L}_{JJ}(\Phi,\dot\Phi) = \frac{1}{2}C\left(\frac{\hbar}{2e}\dot\Phi\right)^2 - U_J(\Phi) - \frac{\hbar}{2e}\dot\Phi\,It . \qquad (43)$$

In fact, in the gauge $A_0 = 0$, the classical field satisfies:

$$\partial_1 E_1 + \frac{\phi_o}{\lambda^2 c}\dot\eta = 0 . \qquad (44)$$

For an electric field decaying exponentially from the junction at the origin we have:

$$\partial_1 E_1|_{0+} - \partial_1 E_1|_{0-} = -\frac{1}{\lambda^2} f_{em} . \quad (45)$$

It follows that:

$$-\frac{1}{\pi} \int dt \theta \, f_{em}(t) = -\frac{\hbar}{2e} \int dt It \frac{\partial \Delta \eta}{\partial t},$$

which defines θ in terms of the total charge q flowing in time T across the junction:

$$e\theta(t) = \int_0^t I \, dt' = q . \quad (46)$$

In eq. (43) the external bias has a form that preserves the invariance $\varphi \to \varphi + 2\pi$. The inclusion of damping is usually modelled by a normal conductor shunting the junction. It has been shown that, at zero temperature, if the shunting conductance has a non zero limit at $\omega \to 0$ (which is also the case of ohmic dissipation), the effects of the non trivial topology of the phase difference φ we have been discussing, are washed out[11].

The inclusion of magnetic fields, e.g. in the case of extended junctions, or in Josephson rings (SQUIDS), requires higher spatial dimensions.

Enlightening discussions with A. Leggett, P. Marchetti and F. Ventrigilia are gratefully acknowledged.

REFERENCES

[1] A. J. Leggett, Les Houches, XLVI, (1986), Elsevier Science Publ.; eds. J. Souletie, J. Vannimenus and R. Stora (1987), E. Ben Jacob, Y. Gefen Phys. Lett. **108 A**,289 (1985).

[2] A. O. Caldeira and A. J. Leggett, Ann. Phys. (NY) **153**, 445 (1984), H. Grabert, P. Olschowski and U. Weiss, Phys. Rev. **B 36**, 1931 (1987).

[3] C. Itzykson and J. B. Zuber, "Quantum field Theory," p. 612, McCraw-Hill Int. Book Company (1980).

[4] "Superconductivity," ed. by R. D. Parks, Dekker, New York (1969) (e.g. Chapter 6, page 321, by N. R. Werthamer).

[5] E. B. Bogomolnyi, Sov. J. Nucl. Phys. **24**, 449 (1979)
L. Jacobs and C. Rebbi, Phys. Rev. **B19**, 4486 (1979).

[6] J. S. Langer and V. Ambegaokar, Phys. Rev. **164**, 498 (1967)
D. E. McCumber, Phys. Rev. **172**, 427 (1968)
D. E. McCumber, Phys. Rev. **181**, 716 (1969)
D. E. McCumber and B. I. Halperin, Phys. Rev. **B 1**, 1054 (1970).

[7] J. A. Hertz, Phys. Rev. **B 14**, 1165 (1976).

[8] V. Ambegaokar and L. P. Kadanoff, Il Nuovo Cimento **XXII**, 914 (1961), V. Ambegaokar, in "Percolation, localization and Superconductivity," ed. by A. M. Goldman and S. A. Wolf (Plenum, New York, 1984), see also S. Doniach in the same volume
V. Ambegaokar, U. Eckern, G. Schön, Phys. Rev. Lett. **48**, 1745 (1982)
U. Eckern, G. Schön, V. Ambegaokar, Phys. Rev. **30**, 6419 (1984)

V. N. Popov, "Functional Integral Methods in Quantum Field Theory and Statistical Mechanics," (Reidel Publishing Co., 1983).

[9] M. Greiter, F. Wilczek, E. Witten, "Hydrodynamic Relations in Superconductivity," preprint.

[10] R. Rajaraman, "Solitons and Instantons," North Holland Publ. Co. Amsterdam (1982)
R. Jackiw, "Topological investigations of quantized gauge theories," in Les Houches, Session XL (1983), B. S. De Witt and R. Stora, eds., Elsevier Science Publ. B.V., 1984
S. Coleman, "The uses of instantons," in "'New Phenomena in Subnuclear Physics," ed. by A. Zichichi, Plenum Press, New York (1977).

[11] W. Zwerger, A. T. Dorsey, M. P. A. Fisher, Phys. Rev. **B34**, 6518 (1986).

THE "COLD FUSION" PROBLEM*

A. J. Leggett

Department of Physics
University of Illinois at Urbana-Champaign
1110 West Green Street
Urbana, IL 61801 USA

As I write this (early May, 1989) it is just over six weeks since the first claims[3,4] of observation of "cold fusion" burst upon the world, and it is still not entirely clear whether we are dealing with a potentially revolutionary new source of energy, a minor but intriguing new physical phenomenon or simply a catalog of experimental and statistical errors. No doubt the picture will have changed by the time this lecture is given, let alone by the time it is published; anyway, for present purposes I will take the view that there is sufficient circumstantial evidence that an unexpectedly high rate of nuclear fusion is taking place in deuterium trapped in metals such as palladium and titanium that it makes sense to ask what kinds of constraint theory can put on possible mechanisms for this phenomenon, if indeed it is genuine.

Consider two deuterium nuclei in free space which have succeeded in reaching small relative separation $\vec{r}$, so that the probability density at $\vec{r} = 0$ is some quantity $P(0)$ (units cm^{-3}). Having reached this point, they can undergo various fusion reactions. The rate of any particular reaction in free space, R, can be written in the form

$$R = AP(0) \tag{1}$$

where the constant A, which has dimensions $cm^3 \, sec^{-1}$, is usually approximately independent of energy at the energies of interest. For reactions which involve nuclear forces only we can make a crude order-of-magnitude estimate of A, in the absence of any special resonance effects, by multiplying a typical nuclear volume ($\sim 1 \, fm^3$) by a typical nuclear frequency ($\sim 10^{23} \, sec^{-1}$); this gives the estimate $A \sim 10^{-16} \, cm^3 \, sec^{-1}$. The two such reactions relevant in the "cold fusion" context are[5]

---

* The work reported here was done in collaboration with Gordon Baym, and parts of this lecture are reproduced more or less verbatim from our two papers[1,2] on the subject.

$$d+d \to {}^3\text{He}(0.82 \text{ MeV}) + n(2.45 \text{ MeV}), \quad A \cong 0.6 \times 10^{-16} \text{ cm}^3 \text{ sec}^{-1} \quad (2a)$$

$$d+d \to {}^3\text{H}(1.01 \text{ MeV}) + p(3.02 \text{ MeV}), \quad A \cong 0.9 \times 10^{-16} \text{ cm}^3 \text{ sec}^{-1} \quad (2b)$$

where the figures in brackets are the energies carried off by the decay products in the center-of-mass frame. A third reaction, $d + d \to {}^4\text{He}$, is forbidden in free space by the conservation of energy and momentum; the reaction

$$d + d \to {}^4\text{He} + \gamma \text{ (23.8 MeV)} \quad (3)$$

is possible, but has a much lower value of A ($\sim 10^{-23}$ cm^3 sec^{-1}) since it involves the electromagnetic interaction. Finally, we note the reaction

$$p + d \to {}^3\text{He} + \gamma \text{ (5.4 MeV)} \quad (4)$$

with a small rate constant ($A \sim 10^{-21}$ cm^{-3} sec^{-1}).

The reason why nuclear fusion is not an everyday phenomenon (on earth) lies in the factor $P(0)$, which is the probability density to find the two deuterons at relative separation zero. For any finite energy E, to reach this point requires tunnelling through a finite region where the Coulomb repulsion e^2/r is greater than E, and hence for low E the probability is exponentially small. In fact, if the wave function is normalized in a volume of order a^3, then an order-of-magnitude expression for $P(0)$ is (for two deuterons)

$$P(0) \sim a^{-3} \exp -B, \qquad B = \pi(E_d/E)^{1/2} \quad (5)$$

where E_d is the "deuteron rydberg", $M_d e^4 / 2\hbar^2 \sim 50$ keV. Thus, for example, if we take $a = 1\text{Å}$, $E \sim 1$ eV (an optimistic estimate) we find $P(0) \sim 10^{-300}$ cm^{-3} and hence a rate of fusion too low to be of any conceivable interest.

In the recent experiments of Jones et al.[4], rates of the reaction (2a) of the order of 10^{-23} sec^{-1} per deuteron pair have been claimed for deuterium trapped in the metals Pd and Ti. In the experiment of Fleischmann and Pons[3] rates about four orders of magnitude greater are said to have been directly detected, and if the anomalous heat production observed in these experiments is due to fusion then it would imply a rate of about 10^{-10}–10^{-11} sec^{-1} per deuteron pair. For the purpose of the present discussion we shall consider specifically the reaction (2a) and take the figure of 10^{-23} sec^{-1} per deuteron pair as a target to shoot at.

One possibility which one must of course consider is that the reported fusion rates are real and are due to some "extrinsic" perturbation such as cosmic-ray muons. A second is that the effect is intrinsic but nonequilibrium, being due perhaps to the high local electric fields which may arise in the process of electrolysis. I shall not consider here either of these possibilities explicitly, though the results to be derived will be seen to have some implications for the latter suggestion. We are then left with the hypothesis that the reported phenomenon is one which occurs in thermal equilibrium at room temperature when deuterium is trapped in metals such as Pd or Ti. What can theory usefully say about this hypothesis?

As already pointed out, if one simply takes the free-space expression (1) for the reaction rate of a single pair, treats this (at least to an order of magnitude) as the rate for a (neighboring) pair in the solid and estimates P(0) by putting a ~ 1Å, E ~ 1 eV, one gets a rate which is literally hundreds of orders of magnitude too small. How could the rate be enhanced to something of the order of 10^{-23} sec^{-1}? There are three obvious possibilities. First, it is conceivable that there is some exotic mechanism operating by which the rate is actually proportional to the _amplitude_ of the wave function at the origin rather than the probability density, and/or a large fraction of the pairs behave coherently (giving a factor ~ 10^{23}). I have seen no concrete proposal along these lines which seems at all plausible, so will not discuss it further here. Secondly, it is possible that the "intrinsic" nuclear rate constant A is somehow enhanced enormously by the solid-state environment. While it is just conceivable that this might happen for the free-space-forbidden reaction d+ d → ^{4}He (because the momentum and energy conservation laws are relaxed in the solid matrix) it seems extremely difficult to see how such an enhancement could take place for the free-space-allowed reaction (2a).

If one rejects these possibilities, then it follows that the rate R "per pair" (as defined in ref. (4)), i.e. total no. of fusions/2 × number of deuterons in system) is given as in free space by eqn. (1), where however P(0) is now the probability density of finding two deuterons at relative superation r = 0, averaged over the motion of their center of mass, the other N−2 deuterons and the remainder of the solid-state environment (electrons, metallic nuclei, etc.). Formally,

$$P(0) \equiv \frac{1}{N} \sum_{ij} P(\vec{r}_{ij})_{\vec{r}_{ij}=0}. \qquad (6)$$

It then follows from the above considerations that to explain a ratio of 10^{-23} sec^{-1} the quantity P(0) must be at least 10^{-7} cm^{-3}, i.e. some hundreds of orders of magnitude times the preliminary estimate made above.

At first sight this enormous enhancement is not all that implausible. To see why, let us start by considering the D_2 molecule. The calculated rates of fusion in this molecule, though low, (~ 10^{-53} sec^{-1})$^{(6)}$ are much higher then the crude estimate made above, and it is not difficult to see why: The "bare" Coulomb repulsion of the two deuterium nuclei is screened out and compensated, at distances of the order of atomic, by the two electrons. Indeed, if not only the nuclei but the electrons behaved classically then it is easy to see that we could take a given configuration in which the total Coulomb energy was negative (e.g. ⊕ ⊖ / ⊖ ⊕ with the nuclei and electrons arranged at the corners of a square as) and shrink it down to an arbitrarily small scale, always keeping the net interaction attractive. In this way the nuclei could approach arbitrarily closely without ever having to go into the classically forbidden region. In reality, needless to say, this does not work, because to confine the electrons in this way costs us kinetic energy T: if the characteristic scale is r, then we have to an order of magnitude

$$V_{coul} \sim \frac{e^2}{r}, \qquad T \sim \frac{(\Delta p)^2}{m} \sim \frac{\hbar^2}{mr^2} \qquad (7)$$

and so the total energy becomes positive when $r \gtrsim \hbar^2/me^2 \equiv a_0$, the Bohr radius, which therefore determines the order of magnitude of the molecular size. (In actual fact, the equilibrium bond length of the neutral D_2 molecule is 0.74 Å $\cong$ 1.5 a_0, and that of the D_2^+ ion 1·1 Å $\cong$ 2a_0). However, the screening effect of the electrons not only extends the classically accessible region of motion of the nuclei to smaller distances, it also to some extent compensates the direct Coulomb repulsion in the barrier region, and thereby enhances the value of P(0) and hence R by many orders of magnitude.

It is clear that if we could substitute for one or both of the electrons in the D_2 molecule a heavier negatively charged particle, we should be able to enhance the rate further (since the "effective Bohr radius" and hence the equilibrium size of the molecule is proportional to 1/mass). Just such an enhancement is achieved in muon-catalyzed fusion[7], where the electron is replaced by a muon; since the mass of the muon is about 200 times that of the electron, the equilibrium separation of the nuclei is reduced by a corresponding factor, with the result that the fusion rate is enhanced by around 80(!) orders of magnitude (a μ-mesic D_2^+ molecular ion actually undergoes fusion in a time $\sim 10^{-10}$ secs, which is $\sim 10^{-4}$ of the muon lifetime).

It is therefore natural to ask whether in the complicated conditions which may well prevail in the interior of a solid such as Pd or Ti, the conduction electrons could screen the direct D-D repulsion so effectively that a rate of order 10^{-23} sec^{-1} per pair (which is, after all, 33 orders of magnitude less then that achieved in muon-catalyzed fusion!) could be attained. In this context one red herring should be dismissed at once, namely the suggestion[4] that "quasi-electrons form in the deuterated metal lattice, with an effective mass a few times that of a free electron" (and presumably then play the role of the muon): The "effective masses" discussed in standard solid-state theory characterize the motion on the scale of (at least) the lattice cell, and have little or no relation to the ability to screen at the distances of the order of a Bohr radius or less which are relevant to cold fusion. Nevertheless, the complicated and highly correlated nature of the electron wave functions in solids – not to speak of possible effects of the motion of metal ions, other deuterons etc. – make the hypothesis of anomalously efficient screening quite attractive, and already by the time of writing numerous ingenious schemes have been concocted by theorists with a view to boosting P(0) to the required value of 10^{-7} cm^{-3}.

What I shall now demonstrate is that, given essentially known facts about the thermodynamics of D and ^{4}He in the metals in question and the applicability of standard nonrelativistic quantum mechanics, _all_ such schemes are quite simply unviable at zero temperature and, at the least, highly implausible at room temperature. This is a rigorous theorem, and no amount of clever maneuvering with electron correlations, violations of the LOBOA or anything else will get around it.

For pedagogic clarity I will start with a simple version[1] of the theorem which applies only to the extent that the LOBOA is valid; I will not emphasize rigor here since the result is essentially subsumed under the much stronger theorem I will prove subsequently. The essential point of the argument is that if the effective repulsion of two deuterons at short distance is substantially weakened by solid-state effects, then these effects should lead as well to greatly increased binding of an α particle to the metal. To make the argument precise, we define, within the lowest-order Born-Oppenheimer approximation, the effective "potential energy" $V(\vec{R},\vec{r})$ of two deuterons placed in the metal at $\vec{R} + \vec{r}/2$ and $\vec{R} - \vec{r}/2$ respectively. (We assume that two additional electrons are added to the metal as well, to preserve overall charge neutrality; since these electron energies cancel out in the argument we do not consider them explicitly.) We split V into three contributions: (a) the direct Coulomb interaction, e^2/r, of the two deuterons with each other, (b) the Coulomb interaction $V_c(\vec{R} + \vec{r}/2) + V_c(\vec{R} - \vec{r}/2)$, of the two deuterons with the "environment," i.e., with all other charges, electronic and ionic, and (c) everything else, i.e., the kinetic energies and mutual interactions of the environment, which we call K. Here

$$V_c(\vec{R}) = e \int d^3\vec{r} \; \frac{\rho(\vec{r})}{|\vec{r}-\vec{R}|^3}, \tag{8}$$

where $\rho(r)$ is the charge density (expectation value) of the environment.

Consider now the "chemical energy" $U(\vec{R} + \vec{r}/2)$ of an α particle placed at $\vec{R} + \vec{r}/2$, i.e., its potential energy relative to that of an α particle placed at infinite distance from the metal. If we use the exact environment wave function for the ground state of the two deuteron problem as a variational wave function for the α-particle problem, noting that the α has charge 2e, we see that $U(\vec{R} + \vec{r}/2) \leq K + 2V_c(\vec{R} + \vec{r}/2)$, with a similar inequality at $\vec{R} - \vec{r}/2$. Adding the two inequalities we find, for all $\vec{R}$ and $\vec{r}$, that

$$V(R,r) \geq \frac{1}{2} [U(\vec{R} + \vec{r}/2) + U(\vec{R} - \vec{r}/2)] + \frac{e^2}{r}. \tag{9}$$

In the limit of zero temperature, which we consider for the moment, the actual energy eigenvalue E_{2d} of the deuteron pair is bounded above by the energy of two deuterons at infinite separation in the metal, which is $-2(E_d + K_d)$, where $E_d = 1$Ry is the binding energy of the deuteron in free space (Ry is the Rydberg), and K_d is its affinity to the metal. Similarly, $U(\vec{R} \pm \vec{r}/2)$ is bounded below, for all R, r, by $-(E_4 + K_4 + E_{zp}^o)$, where $E_4 = 5.802$ Ry and K_4 are, respectively, the free-space binding energy and affinity of the ^{4}He atom, and E_{zp}^o is the zero-point energy of the α-particle in the metal, i.e., its energy relative to the minimum value of the potential $U(\vec{R})$. Inserting these inequalities into (9) we find

$$V(\vec{R},\vec{r}) - E_{2d} \geq \frac{e^2}{r} - \lambda \frac{e^2}{a_o}, \tag{10}$$

where $\lambda = \frac{1}{2} [E_4 + K_4 + E_{zp}^o - 2(E_d + K_d)]/$Ry, and a_o is the usual Bohr radius. The inequality (3) is rigorous and applies to arbitrary position of the two deuterons, whether in the bulk or on the surface.

Now clearly the exponent B for penetration to $\vec{r} = 0$ is bounded below by its value for the potential on the right side of (3). Further, it is adequate for the present purpose to evaluate the latter by the simple WKB approximation.[8] Thus, taking into account the reduced mass of the deuteron pair, we can write

67

$$B \geq B_O(\lambda) \equiv 2 \int_0^{a_0/\lambda} dr \, \left(\frac{M_d e^2}{\hbar^2}\right)^{1/2}$$

$$= \pi \, \left(\frac{M_d}{m_e}\right)^{1/2} \lambda^{-1/2} \simeq 189 \, \lambda^{-1/2}, \tag{11}$$

where m_e is the mass of the electron and M_d that of the deuteron. We will show below that a generous estimate for λ is ~ 1.78, so the WKB exponent is certainly not smaller than 141, and unless the prefactor in the expression for $P(0)$ (which is essentially the probability density near the boundary of the classically allowed region of motion, see below) has the quite incredible value $> 10^{30}$ Å^{-3} it is totally impossible to reproduce the required value of $P(0)$. It is clear that the above argument could be made rigorous along the lines set out below.

However, this argument, which assumes the validity of the LOBOA, is not totally satisfying. In the first place, even if the screening were entirely due to the electrons of the metal, the fact that the expansion parameter m/M_d (where m is the electron mass, and M_d the deuteron mass) of the B-O expansion is small, $\sim 3 \times 10^{-4}$, does not guarantee that corrections of high order in m/M_d need be unimportant in calculating exponentially small probabilities; the more so because in a metal, as distinct from say the D_2 molecule, the electronic excitation energies can be arbitrarily small. Secondly, the effects on the tunnelling of a particular deuteron pair of motion of "third-party" deuterons, let alone that of the nuclei of the metal itself, are not at all well described by the LOBOA, and some attempts to enhance the fusion rate (e.g. ref. (9)) have exploited this fact. Thus it is desirable to give a more general proof which does not assume the validity of the LOBOA, and I now do this.

At first sight it is tempting to argue that the results described in the previous lecture already prove that the actual tunnelling probability cannot be greater than that calculated in the LOBOA; if this is indeed so, then our desired result is proved. Now it is indeed true that once one has described the corrections to the LOBOA by the "canonical" Hamiltonian (1), then as pointed out in lecture 2 it immediately follows that these corrections can only lower the rate. However, the argument that the Hamiltonian (1) indeed adequately describes the corrections is not rigorous and in particular relies on the assumption that any one "environment" mode is only weakly excited. Thus, the hypothesis that the LOBOA indeed gives an upper limit on the tunnelling probability, while perhaps rendered plausible by the arguments of lecture 2, cannot be regarded as proved by them.

The argument I shall actually use proceeds in three stages. We first prove the quite general result that the exact rate of tunneling of a pair of nuclei into the classically forbidden region, in a condensed matter system, is bounded above by the rate calculated for any spherically symmetric lower bound on the lowest order BO potential, $U(\vec{r})$ (defined precisely below), for the motion of the relative coordinate $\vec{r}$ of the pair. Specifically, suppose that $U(\vec{r})$ is bounded below by some spherically symmetric potential $V_2(r)$ (where $r = |\vec{r}|$). For a given energy E of the many-body state, we define $r_0(E)$ as the largest distance for which $V_2(r) - E \geq 0$ for all $r \leq r_0(E)$. We define $\bar{\rho}(r)$ as the probability density angular-averaged in $\vec{r}$; $B(\mu, r, E)$ as $\ln[\bar{\rho}(r)/\rho(0)]$, where μ is the reduced mass of the pair; and $B_0(\mu, r, E)$ the value of B calculated using the lowest-order BO approximation with potential $V_2(r)$. Then for any $r \leq r_0(E)$ we have the general inequality

$$B(\mu,r,E) \geq B_0(\mu,r,E). \tag{12}$$

In the second stage we use an improved version of our earlier argument concerning ^{4}He to make a judicious choice of $V_2(r)$, and in the third stage exploit these results to obtain upper limits on $P(0)$.

Stage I. In the proof of (12) we collectively denote by ξ all the coordinates of the many-body system except the relative coordinate, $\vec{r}$, of the deuteron pair in question. We reserve the notation $\vec{\nabla}$ for gradients with respect to $\vec{r}$, and write $\int d\xi$ as a shorthand for the many-dimensional integration over ξ. Note that ξ includes not only all electron coordinates, but also the center-of-mass coordinate of the deuteron pair, and all coordinates of "third-party" deuterons and of the nuclei of the metal.

The Schrödinger equation for a stationary-state wave function $\psi(\vec{r},\xi)$ of the many-body system may be written as

$$\left[-\frac{\hbar^2}{2\mu}\nabla^2 + \hat{H}(\vec{r})\right]\psi(\vec{r},\xi) = E\psi(\vec{r},\xi). \tag{13}$$

The operator $\hat{H}$ is a function of the coordinates ξ and their conjugate momenta, and depends on $\vec{r}$ through the interaction of the two deuterons with each other and with the rest of the system. The lowest eigenvalue, $U(\vec{r})$, of $H(\vec{r})$ at given $\vec{r}$, is the potential for relative motion in the lowest BO approximation. (It should be carefully noted that since the latter is defined by letting all other coordinates, including the "slow" ones, adjust to the instantaneous value of $\vec{r}$, this potential may differ somewhat from the one derived in the approximation in which all the nuclei are held fixed.) Since $U(\vec{r})$ minimizes the expectation value of $H(\vec{r})$ for a given normalization of the wave function over ξ, we can write:

$$\int d\xi\, \psi^*(\vec{r},\xi)\hat{H}(\vec{r})\psi(\vec{r},\xi) \geq U(\vec{r}) \int d\xi |\psi(\vec{r},\xi)|^2 \equiv U(\vec{r})\rho(\vec{r}), \tag{14}$$

where $\rho(\vec{r})$ is the total probability density to find the deuterons at $\vec{r}$; hence from (13) we have

$$\frac{\hbar^2}{2\mu} \frac{\int d\xi \psi^*(\vec{r},\xi)\nabla^2\psi(\vec{r},\xi)}{\rho(\vec{r})} \geq U(\vec{r}) - E. \tag{15}$$

By using the Schwarz inequality, and the fact that differentiation with respect to $\vec{r}$ commutes with integration over ξ, we may show that for any real wave function f (here $|\psi(\vec{r},\xi)|$) one has $\int d\xi\, f\nabla^2 f \leq g\nabla^2 g$, where $g^2 \equiv \int d\xi\, f^2$ (a formal proof is given in Appendix A). Thus from (15) we obtain

$$\frac{\hbar^2}{2\mu} \frac{\nabla^2\chi(\vec{r})}{\chi(\vec{r})} \geq U(\vec{r}) - E, \tag{16}$$

where now $\chi(\vec{r}) \equiv \rho(\vec{r})^{1/2}$. We may formally regard the quantity $\chi(\vec{r})$ as the solution of the Schrödinger equation for a particle of mass μ moving in a potential $V_1(\vec{r}) \equiv (\hbar^2/2\mu)(\nabla^2\chi(\vec{r})/\chi(\vec{r})) + E$. Then if $V_2(r)$

is any direction-independent lower bound on $U(\vec{r})$, we have $V_1(\vec{r}) \geq V_2(r)$.

Let us define $\phi(r)$ to be the solution of Schrödinger's equation that is regular at the origin, for a particle of mass μ in the potential $V_2(r)$, at energy E. Then since both $\chi(\vec{r})$ and $\phi(r)$ can be taken to be positive, finite and continuous for all $r \leq r_0(E)$, where $r_0(E)$, as above, is the classical turning point in the potential $V_2(r)$, a simple application of Green's theorem to the quantity $\chi\nabla\phi - \phi\nabla\chi$ gives the result that as r decreases, $\ln\chi(r)$ decreases faster than $\ln\phi(r)$ for all $r \leq r_0(E)$, where, as above, $\chi(r)$ denotes the angular average of $\chi(\vec{r})$. Noting that B_0 is defined in terms of ϕ^2 rather than ϕ, that $\chi(\vec{r})_{\vec{r}=0}$ is isotropic and that in general $[\chi(r)]^2 \leq \bar{\rho}(r)$, we immediately derive the inequality (12).

Stage II. We now obtain a suitable potential $V_2(r)$ for constructing a useful bound on the tunneling rate. We write the quantity $H'(\vec{r}) \equiv H(\vec{r}) - \frac{e^2}{r}$ in the form

$$\hat{H}'(\vec{r}) = -\frac{\hbar^2}{2M_4}\nabla_R^2 + [V_C(\vec{R} + \vec{r}/2; \bar{\xi}) + V_C(\vec{R} - \vec{r}/2; \bar{\xi})] + \hat{K}(\bar{\xi}), \quad (17)$$

where $\vec{R}$ denotes the center-of-mass coordinate of the deuteron pair; $\bar{\xi}$ denotes all coordinates of the environment, other than $\vec{R}$ and $\vec{r}$; $V_C(\vec{a})$ is the Coulomb interaction between a deuteron at point a and the environment, and $M_4 = 2M_d$ is the mass of the α particle (up to the nuclear binding energy, which it is consistent with our nonrelativistic formulation to neglect). We let $\psi_0(\vec{R}\xi;r)$ be the lowest eigenfunction of $H'(r)$, for fixed r, with corresponding eigenvalue $E'(r)$. Then choosing, for the state $\psi_0(\vec{R},\xi;0)$, a trial density matrix that is a mixture of $\psi_0(\vec{R}+\vec{r}/2,\xi;r)$ and $\psi_0(\vec{R}-\vec{r}/2,\xi;r)$ with equal weights, we easily demonstrate that $E'(\vec{r}) \geq E'(0)$. But $E'(0)$ is by definition minus the binding energy of an α particle in the metal. (The complications regarding the two electrons are as in the earlier version of the argument.) Thus using the fact that the ground-state energy relative to that without the two deuterons cannot be greater than (minus) twice the binding energy of a single deuteron to the metal, we may write a lower limit $V_2(r)-E$ on $U(\vec{r})-E$ as

$$V_2(r) - E = \frac{e^2}{r} - \lambda(E)\frac{e^2}{a_0}, \quad (18)$$

where $\lambda(E) \equiv \lambda_0 + E$; here E is measured from the many-body ground state energy in units of the Hartree (twice the Rydberg, i.e., 27.2 eV), and

$$\lambda_0 = E_4 + K_4 - 2(E_d + K_d). \quad (19)$$

i.e., λ_0 is the λ defined above without the α-particle zero-point energy. The energies in (19), all measured in Hartrees, are as defined earlier, i.e. E_4, K_4, E_d and K_d are respectively the free-space binding energy of the ^{4}He atom (2.901), its affinity to the metal in question, the free-space binding energy of the deuterium atom (0.5) and its affinity to the metal.

The deuterium affinity, K_d, is[10] 0.087 Hartree in Pd and presumably of similar size in Ti; to the best of our knowledge there is no published direct measurement of the helium atom affinity, K_4, for Pd or Ti, but the fact that at room temperature ^{3}He desorbs[11] from Ti and forms bubbles[12] in Pd suggests that, as for other metals, K_4 is either negative or very small (<< 1 eV). A conservative estimate is therefore $\lambda_o \simeq 1.78$. (Strictly speaking, K_4 and K_d should be evaluated in the actual (possibly deuterium-soaked) ground state, but since it would take a quite extraordinary and unprecedented dependence on deuterium concentration, c, to affect the results appreciably, we use c = 0 values. Any corrections are easily incorporated.)

Stage III. We write the inequality (1) in the form $\rho(r)/\rho(0) \geq \phi^2(r)/\phi^2(0)$, where $\phi(r)$ is defined as in stage I (i.e. it is the solution of Schrödinger's equation in the potential $V_2(r)$ of eqn. (7)), and integrate both sides over a sphere of radius $r_0(E)$. Introducing the standard notation for the Coulomb problem,[13]

$$z \equiv \frac{2\eta(E)}{r_0(E)} r, \qquad \eta(E) \equiv \left(\frac{M_d}{4m\lambda(E)}\right)^{1/2} \qquad (20)$$

(z is the variable ρ in ref. (13), a notation we avoid for obvious reasons), we obtain the inequality

$$\rho(0;E) \leq \frac{1}{4\pi}\left(\frac{2\eta}{r_0}\right)^3 \int_{r \leq r_0} \rho(\vec{r})d^3r \times \left[\frac{(F_0'(0))^2}{\int_0^{2\eta} F_0^2(z)dz}\right], \qquad (21)$$

where $F_0(\eta,z)$ is the L = 0 radial Coulomb wave function (ref. (13), sec. 14.1). Noting that $F''_0(z) \geq 0$ for z less than the turning point 2η, we find that a lower bound on the integral in the denominator is $F_0(2\eta)/3F'_0(2\eta)$. Using the explicit formulae (14.6.2), (14.1.8), (14.5.10) and (14.5.11) of ref. 13, we can write the result in the form

$$\rho(0;E) \leq \left(\frac{3}{4\pi r_0^3(E)} \int_{r \leq r_0} \rho(\vec{r})d^3r\right) Q(E)\eta^{10/3}(E)e^{-2\pi\eta(E)}, \qquad (22)$$

where Q(E) is a number which depends weakly on E, and for E = 0, $\lambda_o \simeq 1.78$, is approximately 60.

For a single pair of deuterons, for which $P(0) \equiv \rho(0;E)$, the integral in (11) is clearly bounded above by unity. Then, putting in the numbers, we find that at zero temperature the equilibrium value of P(0) for dd fusion cannot exceed $2 \times 10^{-31}/cm^3$ and hence the rate of fusion R cannot exceed 2×10^{-47} secs^{-1}, a limit more than 23 orders of magnitude smaller than the value quoted in ref. (4). Similar evaluations lead, for the reaction p + d → ^{3}He + γ, to an upper bound on P(0) of $3 \times 10^{-20}/cm^3$, and a reaction rate $R_{pd} \leq 10^{-41}$/sec; and for the reaction d + t → ^{4}He + n, to P(0) $\leq 4 \times 10^{-37}/cm^3$, and $R_{dt} \leq 10^{-50}$/sec. We reemphasize that these results are rigorous, to within small numerical uncertainties arising from the poorly known value of K_4 and its variation (and that of K_d, etc.) with hydrogenic concentration.

In the physically more relevant case of N deuterons, we define the dimensionless correlation function $G(\vec{r};E) \equiv (N\rho_o)^{-1} \sum_{ij} \rho(\vec{r}_{ij};E)_{\vec{r}_{ij}=\vec{r}}$, here ρ_o is the average deuteron density. Then if $\bar{G}(E)$ denotes the average of G over the sphere $r \leq r_0(E)$, an upper limit on P(o) $\equiv \frac{1}{N}\sum_{ij} \rho(\vec{r}_{ij})_{\vec{r}_{ij}=0}$, is clearly given by

$$P(0) \leq \rho_o \bar{G}(E) Q(E) \eta^{10/3}(E) e^{-2\pi\eta(E)}. \qquad (23)$$

Estimating the maximum deuterium concentration as two per metal ion, we find that at zero temperature $P(0) \leq 4 \times 10^{-33}$ $\bar{G}/cm^3$, where $\bar{G}$ is the dimensionless correlation function averaged over the classically forbidden region of relative motion. To obtain a value of $P(0) \geq 10^{-7}/cm^3$ requires the totally incredible value $\bar{G} > 10^{25}$. (Such a value would *inter alia* imply partial Coulomb interaction energies well beyond TeV per deuteron, and could very likely be rigorously refuted by an argument similar to that of stage II above, but we shall not do this here.)

The case of finite temperature is slightly trickier, since in order to get rigorous results we have to take E to be the total energy of the many-body state measured from the ground state, a quantity proportional to the total number of particles in the system. It is clear from (22), (20) and the definition of $\lambda(E)$, that for two isolated deuterons, to obtain a value of $P(0;E)$ of the order of $10^{-7}/cm^3$ requires at least an energy of the order of 80 eV (and a similar number for the many-deuteron case as long as $\bar{G}(E)$ is not highly anomalous). If we use a microcanonical ensemble and take into account that the thermal energy per atom at room temperature of Pd (and D) is of order the Dulong-Petit value (0.075 eV), we see that such an energy value cannot be achieved at room temperature in an assembly of fewer than ~1000 atoms. Correspondingly, in a macroscopic system at least this number of atoms would have to collaborate to achieve the required enhancement. While we have not been able rigorously to rule out such an exotic long-range effort operative only at finite temperature, it seems extraordinarily implausible.

One technical loophole, which I would however regard as very minor, in the above argument should be pointed out: It is just conceivable that while the conclusion that any enhancement mechanism which produces the required fusion rate would also imply a helium affinity of the order of 80 eV is technically correct, this is not in conflict with the observed behavior of helium in the relevant metals because the sites which give this enormous affinity are few and far between (perhaps associated with a particular type of impurity or defect), so that in practice most helium atoms never find them. (It is clear that this is the situation in the case of muon-catalyzed fusion in the metal, which would otherwise in effect be excluded by our argument.) The only comments I have on this hypothesis are (a) that in this case the actual rate of fusion per "active" site would, of course, have to be much greater than the rate ~ 10^{-23} $secs^{-1}$ inferred in ref. (4), (b) that few if any of the theoretical attempts I have seen so far to account for the data of refs. (3) and (4) exploit this possibility, and (c) that since at least a few He atoms should get trapped at the "active" sites, it should be possible to test the hypothesis by tritium implantation experiments.

In view of the above results, it seems one is left with the following choice of conclusions:

(1) Relativistic effects, neglected above, save the day by increasing the upper limit on the allowed tunneling probability by (at least) the 23 orders of magnitude required.

(2) An "exotic" effect of the type described above is operative (so that the fusion state is proportional (a) to amplitude rather than probability, and/or (b) to the total number of deuterons present).

(3) Quantum mechanics fails for very tiny probabilities, of the order of magnitude of those relevant in the cold fusion problem.

(4) The effect is intrinsically a non-equilibrium one, so that the above calculation is of limited relevance.

(5) The experiments are wrong.

Regarding (1), I would regard this as highly improbable, since relativistic effects on the kinetic energy of relative motion of the deuterons are tiny on the scale (~ 80 eV) which is apparently relevant; however, since the relativistic calculation has not yet been done, it must be regarded as a possible loophole.

As to (2), I should say, first, that I have seen no concrete proposal along these lines which seems at all plausible. However, quite irrespective of this, it should be pointed out that effect (a) by itself is actually insufficient to give the required rate, and even the combination of (a) and (b) only succeeds by a few orders of magnitude. To see this, we note that for a coherent transition of a single pair into a single state the transition rate R_{coh} is at most of order $M/\hbar$, where M is the relevant matrix element, while for the "incoherent" mechanism considered above the rate R is $(2\pi/\hbar)M^2\rho(E)$, where $\rho(E)$ is the density of states in the volume of the sample. Thus the effect of mechanism (a) is to replace the rate R calculated above by (at most) the expression

$$R_{coh} \equiv (R/2\pi\hbar\rho(E))^{1/2}.$$

For the reaction (2a) in a 1 cm^3 sample, the quantity $\rho(E)$ is of order 10^{50} in SI units, so using the upper limit on R calculated above we find $R_{coh} \lesssim 10^{-32}$ secs^{-1}, still nine orders of magnitude below the rate quoted in ref. (4). If we assume also mechanism (b) and therefore multiply by a factor ~ 10^{23}, the rate is comfortably above that of ref. (4) but only barely above that required to account for the anomalous heat production claimed in ref. (3).

Opinions may legitimately differ concerning the plausibility or otherwise of hypothesis (3); all I would point out, here, is that the evidence that quantum mechanics works for probabilities as small as those relevant here is distinctly circumstantial. If one seriously considers the possibility that the standard quantum formalism may fail for very improbable events, then the most natural hypothesis would seem to be that the failure occurs as a function of the WKB exponent B. As we have seen, crudely speaking the theoretical (i.e. quantum-mechanical) lower limit on B is about 140, whereas the "required" value is about 80. Probably the most improbable phenomenon for which quantum mechanical calculations have given reasonable agreement with experiment is the radioactive decay of U^{238}, for which the value of B is about 90; however, in view of the complexity of the theoretical problem it can hardly be claimed that the agreement obtained, such as it is, is a rigorous test of the validity of the quantum formalism at this level.

In connection with (4), it should be stressed that while the rigorous results obtained above cannot rigorously exclude any particular type of nonequilibrium mechanism, they can make some of them distinctly implausible. For example, it has been speculated that a crucial role may be played by the intense electric fields developed in the electrolytic process. However, from the above results we see that it would be necessary to "channel" an energy of the order of 80 eV into the relative motion of two deuterons, and it is difficult to see how this could have been achieved in the original experiments[3,4] where the maximum voltage applied across the whole sample seems to have been[4] 25 V. However, there are other types of possible nonequilibrium mechanism about which the above arguments have little to say, in particular those involving cosmic-ray muons (either in the form of conventional muon-catalyzed fusion or by some kind of "knock-on" chain reaction).

Finally, as regards hypothesis (5), no doubt by the time this lecture is published the situation will be clearer than it is at the time of writing!

Appendix

We consider a real function $f(\underline{r},\xi)$, where the notation is as in the main text, and define, as there, the quantity $g(\underline{r})$ by

$$g^2(\underline{r}) \equiv \int f^2(\underline{r},\xi) \, d\xi. \tag{1}$$

Then we have the theorem

$$\int d\xi \, f(\underline{r},\xi) \, \nabla^2 f(\underline{r},\xi) \leq g(\underline{r}) \, \nabla^2 g(\underline{r}). \tag{2}$$

Proof of theorem (2).

$$\nabla^2 g^2(\underline{r}) \equiv \underline{\nabla} \cdot \underline{\nabla} \int f^2(\underline{r},\xi) \, d\xi = 2\underline{\nabla} \, \{\int f(\underline{r},\xi) \underline{\nabla} f(\underline{r},\xi) \, d\xi\} \tag{3}$$

(where we used the fact that integration over ξ commutes with the gradient operator $\underline{\nabla}$ with respect to $\underline{r}$). Hence,

$$(\underline{\nabla} g)^2 + g\nabla^2 g = \int (\underline{\nabla} f)^2 \, d\xi + \int f \nabla^2 f \, d\xi. \tag{4}$$

But we also have

$$(\underline{\nabla} g)^2 \equiv (\underline{\nabla} (\int f^2(\underline{r}\xi) d\xi)^{1/2})^2 = \frac{1}{\int f^2 d\xi} \, [\int f(\underline{\nabla} f \, d\xi]^2. \tag{5}$$

We put $f \equiv \phi$, $(\underline{\nabla} f)_\alpha \equiv \chi_\alpha$ and use the Cauchy-Schwarz inequality (for real ϕ, χ_a)

$$\int \phi^2 \, d\xi \int \chi_\alpha^2 \, d\xi \geq (\int \phi \chi_a \, d\xi)^2. \tag{6}$$

Substituting this in the right-hand side of (5), we find

$$(\underline{\nabla} g)^2 \leq \int \underline{\nabla} f)^2 \, d\xi. \tag{7}$$

Combining the inequality (7) with the equality (4), we immediately obtain the inequality (2), Q.E.D.

References

1. A. J. Leggett and G. Baym, Nature **340**, 45 (1989).
2. A. J. Leggett and G. Baym, Phys. Rev. Lett., **63**, 191 (1989).

3. M. Fleischmann and S. Pons, J. Electroanal. Chem. **261**, 301 (1989).
4. S. E. Jones, E. P. Palmer, J. B. Czirr, D. L. Decker, G. L. Jensen, J. M. Thorne, S. F. Taylor and J. Rafelski, Nature **338**, 737 (1989).
5. W. A. Fowler, G. R. Caughlan and B. A. Zimmermann, Ann. Rev. Astron.Astrophys. **5**, 525 (1967).
6. S. E. Koonin and M. Nauenberg, Nature **339**, 690 (1989).
7. S. E. Jones, Nature **321**, 127 (1986).
8. The WKB approximation can be shown to give an upper limit on the tunnelling probability in one dimension, but unfortunately not in three, the case of present interest. Fortunately we can bypass this difficulty by using the exact results for the Coulomb potential, see below. The WKB approximation gets the order of magnitude right.
9. S. E. Koonin, submitted to Phys. Rev. Lett.
10. R. Lässer and G. L. Powell, Phys. Rev. **B34**, 578 (1986).
11. P. Bach, Radiation Effects (G8) **78**, 77 (1983).
12. C. J. Thomas and J. M. Mintz, J. Nuc. Mat. **116**, 336 (1983).
13. M. Abramowitz and I. A. Stegun, *Handbook of Mathematical Functions* (National Bureau of Standards, Wash. D.C., Ch. 14 (1964)).

UNSTABLE BEHAVIOUR OF A SUPERCONDUCTING RING CONTAINING A JOSEPHSON JUNCTION

A. Extremera

Departamento de Física Moderna
Universidad de Granada
E-18071 Granada, Spain

The dynamic decay process for a SQUID ring is analyzed within the framework of the flux mode regime. Under the equilibrium flux condition, the quantum rates for the low-lying states are computed by using a simpler form for the Anderson-Josephson potential. This condensed matter model has acquired some fame since it describes collective quantum effects in thin magnetic films[1] and sine-Gordon systems[2]. Functional integral techniques specifically adapted to the general features of superconductivity[3] are used to derive WKB-type expansions. Hence, the probability in the dual charge ("momentum") mode is easily obtained.

REFERENCES

1. A. O. Caldeira and K. Furuya, Quantum nucleation of magnetic bubbles in a two-dimensional anisotropic Heisenberg model, J. Phys. C21:1227 (1988).
2. K. Furuya and A. O. Caldeira, Quantum tunnelling in a double sine-Gordon system near the instability, Physica A154:289 (1989).
3. A. Tagliacozzo and F. Ventriglia, Ordinary superconductivity and path integrals, Nuovo Cimento D11:141 (1989).

INFLUENCE OF THE RADIATION FIELD ON THE PROCESS OF ELECTRONIC INTERFERENCE

P.M.V.B. Barone* and A.O. Caldeira

Instituto de Física "Gleb Wataghin", UNICAMP

Caixa Postal 6165, CEP 13081 Campinas, SP, Brasil

ABSTRACT

We develop a non-relativistic formulation for the quantum dynamics of an electron coupled to its own radiation field. For this purpose, we have applied the Feynman-Vernon approach to the composite system in order to obtain the reduced density operator of the electron. In the classical limit, some well known results are reproduced such as the Abraham-Lorentz equation of motion. We have applied the resulting formalism to the problem of electronic interference in order to investigate the possible effects of the incoherent modes of the electromagnetic radiation on the interference fringes. The results allow us to conclude that the coupling to the radiation field destroys the interference fringes in a time scale that can be controlled by the parameters involved in the preparation of the initial state of the system.

---

*Permanent address: Departamento de Física, ICE, UFJF – Campus UFJF CEP 36100 Juiz de Fora, MG, Brasil

QUANTUM MECHANICAL HARMONIC CHAIN ATTACHED TO HEAT BATHS

U. Zücher and P. Talkner

Institut f. Physik, Universität Basel
Klingelbergstrasse 82, CH-4056 Basel, Switzerland

We investigate a finite linear chain of N equal particles connected by equal harmonic springs whose left and right ends are in contact with independent stochastic heat baths at temperatures T_1 and T_N, respectively. The heat baths itself can be modelled as a system of coupled oscillators that introduce both fluctuations and dissipation in the chain [1]. In its classical version this model has been studied in [2]. Our starting point are the corresponding quantum Langevin equations [1] for the operators $x_n(t)$, $p_n(t)$ of the displacement of the n-th particle out of its equilibrium position and its conjugate momentum, respectively. Exploiting the linearity of the system we derive the equations of motion for the equal time correlation functions of the operators x_n, p_n, i.e. $<x_n(t)x_m(t)>$, $<x_n(t)p_m(t)>$, etc. For the stationary state we have explicitely determined the covariance matrix in the limit $N \to \infty$, from which all statistical properties of the chain can be inferred. If both temperatures are equal, $T_1 = T_N$, these correlation functions are determined for vanishing damping constant by the standard weak coupling expression $<o_n o_m>$ $= Z^{-1} tr o_n o_m exp(-H/k_B T)$, where o_n is an operator of the particle at site n, H is the Hamiltonian of the chain and Z denotes the partition function. Classically these expressions are exact for all damping constants and yield e.g. equipartitioning of the kinetic energy. In the quantum case finite coupling corrections to the weak coupling expressions are still small well inside the chain but grow towards the end of the chain, where $<p_n^2>$ for $n = 1, N$ diverges logarthmically with an upper cutoff on the frequencies of the heat baths. If the heat baths are at different temperatures, $T_1 \neq T_N$, the chain is in a stationary nonequilibrium state, in which the quantum mechanical heat flux is decreased compared to the classical case. We shall discuss the equilibrium and nonequilibrium properties of the harmonic chain in greater detail elsewhere [3].

References

[1] G. W. Ford, M. Kac, and P. Mazur, J.Math.Phys.6,504 (1965)
[2] Z. Rieder, J L. Lebowitz, and E.H. Lieb, J.Math.Phys.8,1073 (1967)
[3] U. Zürcher, P. Talkner, to be published

CONFORMAL INVARIANCE - A SURVEY OF PRINCIPLES WITH APPLICATIONS
TO STATISTICAL MECHANICS AND SURFACE PHYSICS

Peter Kleban

Laboratory for Surface Science and Tech. and Dept. of
Physics and Astronomy, Univ. of Maine, Orono ME 04469
USA* and Materials Science Division, Argonne National
Laboratory

1. INTRODUCTION

These lectures are a brief survey of the basic principles of conformal invariance with examples and applications to statistical mechanical models, mainly in two-dimensions, and real surface systems. Given the amount of time available, we cannot even hope to be complete--note that the introductory lectures by Ginsparg [1] took up more than 13 hours! We will attempt only to outline the basic theory and present some consequences and examples of interest to the author. Hopefully this will suffice to convey a feeling for the subject, some appreciation of its implications, and more importantly stimulate the reader to further study. To this end, more exhaustive treatments may be found in various review articles [1-6]. A few comments on the approach and level of these papers are given below.

We assume, in what follows, familiarity with the theory of phase transitions and certain aspects of statistical mechanics, as well as some knowledge of surface systems and scattering theory. Sections 2 through 6 introduce the basics of conformal invariance and work out a few simple consequences. The following two sections (7 and 8) briefly review the more formal theory in two dimensions. This part of the lectures was necessarily rather hurried, and the treatment here is essentially an extended outline. The final section, on applications to surface phase transitions, in principle depends on the foregoing in some essential ways, however it may be read independently if one or two facts are accepted without proof.

---

*Permanent Address

2. COVARIANCE OF CORRELATION FUNCTIONS

2.1 Scaling Transformations

Consider an infinite system at a second-order phase transition. This will be characterized, in general, by scaling densities or "operators" $\phi_i(\vec{r})$ with anomalous dimensions x_i. The corresponding renormalization group eigenvalue $y_i = d - x_i$, where d is the spatial dimensionality. If a scale transformation $\vec{r}' = b^{-1}\vec{r}$ is performed, the multi-point correlations will satisfy

$$<\phi_1(\vec{r}_1)\phi_2(\vec{r}_2)\cdots> = b^{-x_1} b^{-x_2} \cdots <\phi_1(\vec{r}'_1)\phi_2(\vec{r}'_2)\cdots>. \tag{2.1}$$

In a physical system, (2.1) is valid in the scaling limit, i.e. all distances large compared to any microscopic length. Most of the theory discussed below assumes, equivalently, that we are already in the field-theoretic limit, so there are no microscopic lengths. In particular the lattice spacing a is taken to be zero. The existence of well-defined dimensions x and (2.1) follow since we evaluate the thermal average at the fixed point, and the ϕ are renormalized operators.

The relation to operators defined on a lattice is

$$\phi(\vec{r}) \sim ca^x \phi_{lattice}(\vec{r}) \tag{2.2}$$

where a is a lattice spacing and c a non-universal constant. In correlation functions either side of (2.2) may be used when all distances are $\gg$ a--this connects the long-distance properties of the physical system with field theoretic quantities.

For an isotropic, translation invariant system (2.1) implies

$$<\phi(r_1)\phi(r_2)> = |r_1-r_2|^{-2x} \tag{2.3}$$

when the operators ϕ are normalized in the usual way. Note that (2.3) implies that $<\phi> = 0$, i.e. any expectation value of the operator is subtracted off in its definition. Thus all correlation functions are automatically connected.

2.2 Conformal Transformations

Scale invariance, as expressed in (2.1), is believed to hold quite generally at critical points (see [2] for exceptions). In most systems, translation and rotational invariance are equally valid, as assumed in (2.3). Then one expects that the scale factor b may be replaced by a slowly varying function of position $b(\vec{r})$ so that

$$\langle\phi_1(\vec{r}_1)\phi_2(\vec{r}_2)\cdots\rangle = b(\vec{r}_1)^{-x_1}b(\vec{r}_2)^{-x_2}\cdots\langle\phi_1(\vec{r}'_1)\phi_2(\vec{r}'_2)\cdots\rangle \ . \quad (2.4)$$

For (2.4) to hold, b(r) must look *locally* like a combination of a translation, rotation and (uniform) scale transformation, as illustrated in Fig. 1. These are just the conditions for a *conformal* transformation, i.e. a coordinate transformation that does not change angles, shear not being allowed. (2.4) is the basic expression of covariance of correlations under such an operation, and the starting place for further development of the theory.

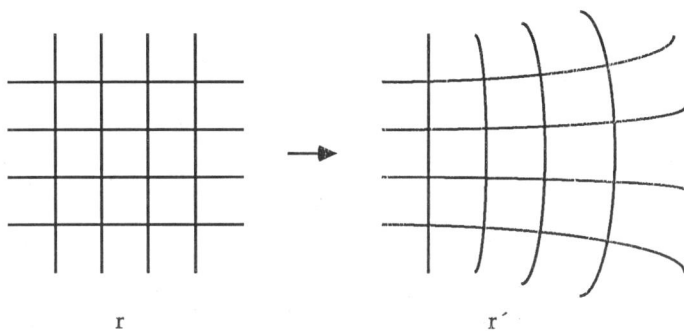

r   r´

Fig. 1.   *Action Of A Conformal Transformation*

2.3 *Complex Coordinates*

In what follows it will often be useful to employ complex coordinates z and $\bar{z}$ in two-dimensional systems. For most of these lectures $\bar{z}$ will be the complex conjugate of $z = x + iy$, however development of the full theory is greatly facilitated by taking them as independent complex variables, and only restoring the consequences of $\bar{z} = z^* = x - iy$ after most of the analysis has been completed. In this language conformal transformations may be expressed via $z = w(z)$, where w is an analytic function, and (2.4) becomes

$$\langle\phi_1(z_1,\bar{z}_1)\cdots\rangle = |w'(z_1)|^{-x}\cdots\langle\phi_1(w_1,\bar{w}_1)\cdots\rangle, \quad (2.5)$$

where the w′ is the derivative of w. If we let $z_{12} = z_1 - z_2$, etc., note that (2.3) may be written

$$\langle\phi_1(z_1,\bar{z}_1)\cdots\rangle = |w'(z_1)|^{-x}\cdots\langle\phi_1(w_1,\bar{w}_1)\cdots\rangle, \quad (2.6)$$

already suggesting the utility of using two independent complex coordinates. More generally, (2.6) may be extended by introducing the real numbers h and $\bar{h}$, which are called the "complex" scaling dimensions of ϕ. Then the separation is complete, giving

$$\langle\phi(1)\phi(2)\rangle = z_{12}^{2h}\bar{z}_{12}^{-2\bar{h}} . \tag{2.7}$$

Here $x = h + \bar{h}$ is the scaling dimension of ϕ and $s = h - \bar{h}$ its spin. The numbers h and $\bar{h}$ are not necessarily complex conjugates (in all cases herein they are in fact both real numbers).

3. CONFORMAL GROUP

For general dimension, the conformal group is composed of dilatations, rotations, translations, and "special" conformal transformations

$$\vec{r}' = \lambda\vec{r}$$

$$\vec{r}' = R\vec{r} + \vec{c}$$

$$\frac{\vec{r}'}{r'^2} = \frac{\vec{r}}{r^2} + \vec{a} , \tag{3.1}$$

so that one has a finite parameter group. The special transformation may be regarded as composed of an inversion and a translation. In two dimensions, as mentioned, (3.1) is extended to include all analytic functions w(z) (at points where w'(z) ≠ 0). Since an analytic function may be expressed as a power series with arbitrarily many terms, it follows immediately that the conformal group is infinite dimensional. This is responsible for the profound consequences of conformal invariance in two dimensions.

At this point, the reader is encouraged to show, as an exercise, that the form (2.3) of the correlation function is preserved under an arbitrary "bi-linear" transformation w(z) = (az+b)/(cz+d), ad-bc ≠ 0.

4. SIMPLE CONSEQUENCES OF CONFORMAL INVARIANCE-INFINITE SYSTEMS

4.1 Two-point Correlation Functions (G)

As a first application, consider [7] an infinitesimal special transformation

$$r'^2 = r^2(1 - 2\vec{a}\vec{r} + \cdots) . \tag{4.1}$$

Note that the scalar produce $\vec{a}\vec{r}$ ensures the transformation stretches space isotropically at each point $\vec{r}$, so that angles are preserved. In addition,

$$\delta\,\phi(\vec{r}_i) = (2\vec{a}\vec{r}_i)x\phi(\vec{r}_i)$$
$$\delta\,\ln(r_{12}) = -\vec{a}(\vec{r}_1 + \vec{r}_2) + \cdots. \qquad (4.2)$$

Since our transformation is conformal everywhere, there are only two changes induced--in the scaling operators and in the coordinates. Applying (4.2) to a two-point correlation function

$$G(r_{12}) = <\phi_i(\vec{r}_1)\phi_j(\vec{r}_2)> \qquad (4.3)$$

we find that

$$-\frac{\partial G}{\partial \ln r_{12}}\vec{a}(\vec{r}_1+\vec{r}_2) = 2(x_i\vec{a}\vec{r}_1 + x_j\vec{a}\vec{r}_2)G \qquad (4.4)$$

where we have assumed that G is evaluated in an infinite, uniform, and isotropic system. It follows from (4.4) that G must have the standard scaling form of (2.3) if the scaling dimensions $x_i = x = x_j$, while G must vanish if $x_i \neq x_j$ (unless the difference is an integer, see [2, 8]). The former statement already follows from scale invariance, but the orthogonality of critical operators does not.

4.2 Three-point Correlation Function (G_3)

A similar analysis of the three-point function [7]

$$G_3(\vec{r}_1,\vec{r}_2,\vec{r}_3) = <\phi_a(\vec{r}_1)\phi_b(\vec{r}_2)\phi_c(\vec{r}_3)> \qquad (4.5)$$

shows that

$$G_3 \propto r_{12}^{(x_c-x_a-x_b)} r_{13}^{(x_b-x_c-x_a)} r_{23}^{(x_a-x_b-x_c)} \qquad (4.6)$$

while scale, translation and rotational invariance alone result in [1]

$$G_3 = \sum_{a,b,c} \frac{C_{abc}}{r_{12}{}^a r_{23}{}^b r_{31}{}^c} \qquad (4.7)$$

with $a+b+c = x_a + x_b + x_c$. The extra requirement of invariance under the (finite) conformal group allows only a single term in (4.7). This result holds in any spatial dimension d.

4.3 Four and Higher Point Function (G_n)

It has also been shown [7] that the n-point function G_n, for $n \geq 4$, is determined up to a function of $n(n-3)/2$ variables if d is sufficiently large (or fewer if d is less). The four-point function G_4, which plays an important role in the theory, is given up to an unknown function of one variable for d = 2, as we will see. Further analysis determines this function in some important cases.

4.4 *G with a Surface*

Now consider correlations in a geometry with a (flat) surface, e. g. the upper half plane in two dimensions. In this situation, the two-point function satisfies

$$G = G(y_1, y_2, |\vec{x}_1 - \vec{x}_2|) \tag{4.8}$$

where the variables in (4.8) are illustrated in Fig. 2. Analysis similar to the above shows [2]

$$G = (y_1 y_2)^{-x} \Psi\left(\frac{|\vec{r}_1 - \vec{r}'_2|}{|\vec{r}_1 - \vec{r}_2|}\right) \tag{4.9}$$

(4.9) is valid for any d and any operator, and holds at the special or ordinary transitions. The unknown function Ψ can be determined in some important cases, as we will see. The appearance of an image point $\vec{r}'_2$ is characteristic of correlations in this geometry. It may (loosely) be thought of as necessary to preserve the surface. It is no accident that, for d = 2, the two-point function is determined only up to a function of one variable, the same as the four-point function in the infinite plane. This will be explained in Section 8.1.

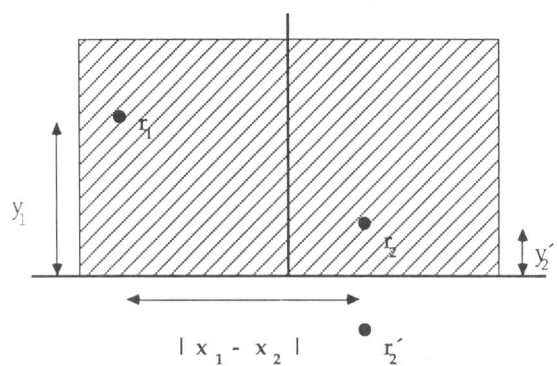

Fig. 2.   *Variables In The Half Plane*

5. THE OPERATOR PRODUCT EXPANSION

5.1 *Statement and Physical Basis*

The idea of the operator product expansion [9] is that as two operators approach each other, they may be expressed as a linear combination of operators

$$\phi_i(\vec{r}_1)\phi_j(\vec{r}_2) = \sum_k C_{ijk} \frac{1}{r^{x_i+x_j-x_k}} \phi_k(\vec{R}) \qquad (5.1)$$

where $r = |\vec{r}_1-\vec{r}_2|$ and $\vec{R} = (\vec{r}_1+\vec{r}_2)/2$. (5.1) is to be used in correlation functions when r is much less than the distance to any other operator, for example in (4.5) when r_{13}, $r_{23} \gg r_{12}$. Correction terms to the r.h.s. of (5.1) appear as higher inverse powers of r. These involve correction to scaling operators. In general, there will be infinitely many. Note also that (2.3) implies, when $i = j$, the presence of a term involving the unit operator for which $C_{ii1} = 1$.

One can understand the origin of relations like (5.1) by considering correlations at a critical point. In general, there will be some finite set $\{\phi_i(\vec{r}_i)\}$ of most relevant scaling operators. Therefore, any product of two these, in an expectation value with some other $\phi_j(\vec{r}_j)$, will have one or more terms that decay most slowly as $r_j \to \infty$. But these must just be the expectations of the r.h.s of (5.1) times $\phi_j(\vec{r}_j)$.

The coordinate dependence of the operator product expansion coefficients follows from scale invariance. The C_{ijk} are universal. In general, the operator product expansion is expected to be asymptotic; however at a critical point it is summable in at least some cases, as we will see in Section 9. Thus the restrictions on distances may be removed. At a critical point (or in a conformal field theory), the absence of a length scale provides a heuristic argument for convergence of the operator product expansion [1].

The Ising universality class provides a simple example of the operator product expansion. At each point there are three operators, the local order parameter (or "spin") $\sigma(\vec{r})$, the local energy $\epsilon(\vec{r})$ and the unit operator, with scaling dimensions $x = 1/8$, 1, and 0, respectively. For the simple Ising model on a lattice, the first two are proportional to the scaling limit of the spin s_i and bond energy $s_i s_{i+1} - \langle s_i s_{i+1}\rangle$. Proportionality constants and subtraction are necessary to fulfill (2.3). Taking the scaling limit preserves only the most relevant part of each lattice operator, so that the operator with the (single) most relevant scaling dimension remains. The operator product expansion is (omitting all coefficients)

$$\sigma\sigma = 1 + \epsilon$$
$$\sigma\epsilon = \sigma$$
$$\epsilon\epsilon = 1 \qquad (5.2)$$

Note that all terms allowed by up-down symmetry ($\sigma \to -\sigma$, $\epsilon \to +\epsilon$) and duality ($\epsilon \to -\epsilon$) appear in (5.2).

5.2 Application to G_3

Applying (5.1) and the orthogonality of critical operators mentioned above to the three-point function gives

$$<\phi_i(\vec{r}_1)\phi_j(\vec{r}_2)\phi_k(\vec{r}_3)> = C_{ijk} r_{12}^{(x_k-x_i-x_j)} r_{13}^{(x_j-x_k-x_i)} r_{23}^{(x_i-x_j-x_k)} \quad (5.3)$$

Thus the coefficient appearing in the three-point function (of properly normalized operators) is exactly the operator product expansion coefficient.

6. SIMPLE CONSEQUENCES OF CONFORMAL INVARIANCE-OTHER GEOMETRIES

6.1 Infinite Strips

6.1.a Correlation length ξ. A system in an infinite strip of width L is at criticality when its temperature and interactions are at a critical point of the corresponding infinite system. In this situation, the correlation length along the strip satisfies

$$\xi = AL \quad (6.1)$$

A variety of exact and numerical results on specific two-dimensional systems showed that the proportionality constant in (6.1) is given by

$$A = 1/\pi\eta \quad (6.2)$$

for periodic boundary conditions across the strip, where $\eta = 2x$. Why η should appear in this relation was rather mysterious. Now the conformal transformation

$$w = \ln z \quad (6.3)$$

maps the z-plane into a strip of width 2π, as shown in Fig. 3. It is important that the strip boundary conditions are periodic. Applying (6.3) to (2.5) shows, for points far apart along the strip (see also (6.5) below) that

$$<\phi(w_1)\phi(w_2)> \to e^{-x|u_1-u_2|} \quad (6.4)$$

which, by scaling to a strip of width L (or by suitably modifying (6.3)) establishes (6.2) [10]. A similar result also holds for fixed or free boundary conditions across the strip.

6.1.b Structure of the transfer matrix. We now examine the results of (6.3) more closely. Using (2.5), the full form of the two-point function in a strip of width L is seen to be

$$<\phi(w_1)\phi(w_2)> = \frac{(2\pi/L)^{2x}}{\left[2\cosh\frac{2\pi}{L}(u_1-u_2) - 2\cos\frac{2\pi}{L}(v_1-v_2)\right]^x} \quad (6.5)$$

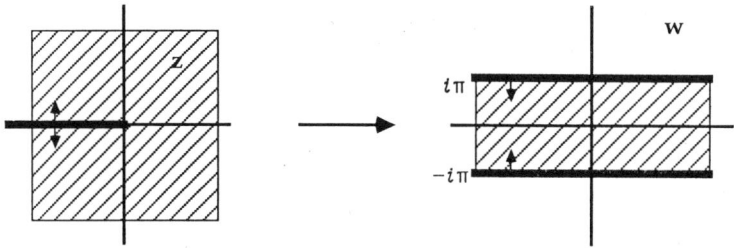

Fig. 3.   *The Transformation $w = \ln z$.*

Note that (6.4) follows for $|u_1-u_2| \to \infty$. On the other hand, as $w_1 \to w_2$, the "bulk limit" of (2.3) is recovered, since the two points are too close to each other for the finite width to have an effect. For $u_1 > u_2$ we can expand (6.5)

$$<\phi\phi> = \left(\frac{2\pi}{L}\right)^{2x} \sum_{m,\bar{m}=0}^{\infty} a_m a_{\bar{m}}\, e^{-\frac{2\pi}{L}(x+m+\bar{m})(u_1-u_2)}\, e^{\frac{2\pi i}{L}(m-\bar{m})(v_1-v_2)} \qquad (6.6)$$

where $a_m = \frac{\Gamma(x+m)}{\Gamma(x)m!}$ [2]. Considerable information can be obtained by computing the two-point function in a transfer matrix formulation and comparing the result with (6.5). The transfer matrix $\hat{T}$ may be expressed as $e^{-a\hat{H}}$, where a is a lattice spacing and $\hat{H}$ the Hamiltonian. With periodic boundary conditions, the eigenstates of $\hat{H}$ may be labelled by momentum (across the strip) eigenvalues k, quantized in units of $2\pi/L$, so that

$$\hat{H}\,|n,k\rangle = E_n\,|n,k\rangle \qquad (6.7)$$

If $\hat{\phi}$ is the lattice operator corresponding to ϕ, we find

$$<\phi\phi>_{strip} = \sum_n <0|\hat{\phi}(v_1)|n,k\rangle e^{-(E_n-E_0)(u_1-u_2)} <n,k|\hat{\phi}(v_2)|0> \qquad (6.8)$$

Comparison with (6.6) then shows that for each operator ϕ in the plane there is a doubly infinite tower of states $|n,k\rangle$ with energy and momentum eigenvalues

$$E_n = E_0 + \frac{2\pi}{L}(x+m+\bar{m})$$

$$P_k = \frac{2\pi}{L}(m-\bar{m}) \qquad (6.9)$$

where m and $\bar{m}$ are integers. Note, however, that the degeneracies of these states are not specified.

Let $|\phi_i\rangle$ denote the state of lowest energy ($m = \bar{m} = 0$) created by an operator $\hat{\phi}$ acting on the ground state $|0\rangle$. Since $|\phi_i\rangle$ is unique, it follows that

$$\langle 0|\hat{\phi}_i|\phi_i\rangle = \left(\frac{2\pi}{L}\right)^{x_i} \tag{6.10}$$

as long as the operators are properly normalized. If one calculates G_3 using the transfer matrix and compares with the general form (4.6) transformed to the strip, one finds

$$\langle \phi_i|\hat{\phi}_j|\phi_k\rangle = \left(\frac{2\pi}{L}\right)^{x_j} C_{ijk} ,$$

$$C_{ijk} = \frac{\langle \phi_i|\hat{\phi}_j|\phi_k\rangle}{\langle 0|\hat{\phi}_j|\phi_j\rangle} . \tag{6.11}$$

6.1.c Susceptibilities χ. The integral of G in (6.5) over the strip is the generalized susceptibility

$$\chi = \iint \langle \phi \phi \rangle \, du \, dv$$

$$= B^{(2)} L^{2-2x} \tag{6.12}$$

where the second line follows by scaling. The coefficient $B^{(2)}$ may be obtained exactly [11] as

$$B^{(2)} = -(2\pi)^x \left(\frac{4\pi}{x}\right) \frac{B(x/2, x/2)}{B(-x/2, -x/2)} \tag{6.13}$$

where B is the beta function [12]. Note that (6.13) (and (6.5)) is valid for any scaling operator whatever. In the small x limit

$$B^{(2)} \xrightarrow{x \to 0} (2\pi)^x \left(\frac{4\pi}{x}\right) \tag{6.14}$$

which is exact for $\langle \phi \phi \rangle \sim e^{-u}$, i.e. when the correlation function behaves one-dimensionally. This form for $B^{(2)}$ already agrees with the exact value for the Ising spin-spin function ($x = 1/8$) to within 0.12%.

6.1.d Correlation lengths and structure function. One can also obtain closed form expressions for the correlation lengths ξ_u and ξ_v defined as moments of $\langle \phi\phi \rangle_{strip}$ [11]. In addition, the structure factor

$$S(\vec{k}) = \frac{1}{L} \int <\phi\ \phi> e^{i(k_u u + k_v v)} du dv_1 dv_2 \ , \quad v = v_1 - v_2 \tag{6.15}$$

has been computed. It is noteworthy that S approaches the bulk limit $1/k^{(2-\eta)}$ only very slowly with increasing k,

$$S(\vec{k}) \sim \frac{1}{k^{2-\eta}} \left(1 \pm \frac{x(x-1)(x-2)}{3} \frac{1}{k^2} + \mathcal{O}(\frac{1}{k^4}) \right) \tag{6.16}$$

In (6.16) $\eta = 2x$, where the second term has a plus (minus) sign for $\hat{k}$ in the u(v) direction. This same slow convergence occurs for S in fully finite regions, a fact of interest for diffraction experiments at surface phase transitions.

6.2 Surface Critical Behavior-Systems with an Edge

A system with an edge may induce a non-zero value for $<\phi>$, even when this quantity is defined to vanish in the bulk. In this case, scaling suggests [13]

$$<\phi> = \frac{A}{y^x} \tag{6.17}$$

where y is the distance to the edge (cf. Fig. 2). Using (6.3) and (2.5) we find, for the infinite strip [12]

$$<\phi> = \frac{A}{\left[\frac{L}{\pi} \sin\left(\frac{\pi v}{L}\right) \right]^x} \tag{6.18}$$

Similarly, one can transform to the interior of a hypersphere, where [14]

$$<\phi> = \frac{A}{\left[\frac{1}{2}R(1-\frac{r^2}{R^2}) \right]^x} \tag{6.19}$$

In (6.19), R is the hypersphere radius and r the distance to the center. This formula is valid in any number of dimensions.

It is interesting to rewrite (6.17) in complex coordinates

$$<\phi> \sim \frac{1}{(z-\bar{z})^x} \tag{6.20}$$

Comparing (6.20) with (2.6), we see that the z, $\bar{z}$ dependence of the former is exactly the same as the z_1, z_2 dependence of the later. Thus the one-point function in the half plane and the two-point function in the infinite plane have the same functional dependence. This resembles the two-point and four-point function similarity mentioned above. These connections will be explained in Section 8.

7. USING THE STRESS TENSOR

7.1 *Conformal Ward Identity*

Considerable further progress follows on consideration of the transformation law (2.4) or (2.5) for an *infinitesimal* coordinate transformation. In doing this, it is very convenient to treat the z and $\bar{z}$ dependence as independent. Thus, for instance, we will take any operator to be a function of two complex variables

$$\phi = \phi(z,\bar{z}) \qquad (7.1)$$

Only when most of the analysis is complete are the consequences of setting $\bar{z}=z^*$ considered (see [5] for details). Next we consider an infinitesimal coordinate transformation

$$\begin{aligned} w(z) &= z + \alpha(z) \\ \bar{w}(\bar{z}) &= \bar{z} + \bar{\alpha}(\bar{z}) \end{aligned} \qquad (7.2)$$

If we require α (and $\bar{\alpha}$) to be non-trivial and infinitesimal everywhere, they cannot be analytic everywhere (This might seem to contradict the argument in Section 4.1, but note that the "infinitesimal" transformation (4.1) actually becomes arbitrarily large with r). Thus, to apply (7.2) to (2.5) we take α (and $\bar{\alpha}$) analytic in some region including all the arguments z_i ($\bar{z}_i$) of the operators in the expectation value, but non-analytic outside that region, as shown in Fig. 4. These functions are further chosen to be sufficiently smooth across the boundary and to vanish at infinity; otherwise they are arbitrary. The non-conformal parts of these functions will change the fixed point Hamiltonian H^*, giving rise to a term

$$\delta H^* = -\frac{1}{2\pi} \int \partial^i \alpha^j \, T_{ij} \, d^2 r \qquad (7.3)$$

where T_{ij} defines the stress tensor. The integration in (7.3) extends over the non-analytic region. The fact that H^* is conformally invariant means that the r.h.s. of (7.3) vanishes for shearless transformations, hence

$$\begin{aligned} T_{ij} &= T_{ji} \\ T_{ii} &= 0 \end{aligned} \qquad (7.4)$$

In complex coordinates, (7.4) becomes

$$T_{z\bar{z}} = T_{\bar{z}z} = 0 \qquad (7.5)$$

We further define $T = T_{zz}$, $\bar{T} = T_{\bar{z}\bar{z}}$. In what follows, we will see that

$$\begin{aligned} T &= T(z), \\ \bar{T} &= \bar{T}(\bar{z}) \end{aligned} \qquad (7.6)$$

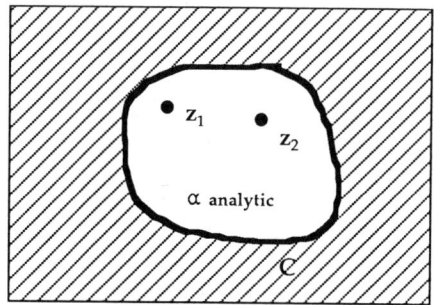

Fig. 4. *The Contour C*

i.e., T ($\bar{T}$) is an analytic (antianalytic) function of z. However, $\bar{T}$ is in general not the complex conjugate of T.

In applying (7.2) to (2.5) we must take account of (7.3). Hence the covariance equation will have an extra term, beyond the two encountered in (4.2), for example. This new term is due to the stress tensor. Expanding all three contributions to leading order gives

$$\int \partial^u \alpha^v(\vec{r}) <T_{uv}(\vec{r})\phi_1 \cdots> d^2r = \sum_j [\alpha(z_j)\Delta_j + \alpha(z_j)\frac{\partial}{\partial z_j} + \overline{\alpha(z_j)\Delta_j + \alpha(z_j)\frac{\partial}{\partial \bar{z}_j}}] \quad (7.7)$$

where Δ and $\bar{\Delta}$ (a.k.a. h,$\bar{h}$) are the complex dimensions of the operator ϕ and the integral extends over the non-analytic region. Next we consider the l.h.s. of (7.7). This may be transformed into an integral over the boundary C of the analytic region (see Fig. 4) plus an integral of $\alpha^v \partial^u T_{uv}$ over the non-analytic region. Since α ($\bar{\alpha}$) is arbitrary in this region, the divergence of T must vanish

$$\partial^u T_{uv} = 0 \qquad (7.8)$$

or, in complex coordinates

$$\frac{\partial}{\partial \bar{z}} T_{zz} = 0 = \frac{\partial}{\partial z} T_{\bar{z}\bar{z}} \qquad (7.9)$$

which establishes the analyticity properties (7.6). The integral on C may be further transformed to

$$\frac{1}{2\pi i} \int_C dz \alpha <T\phi \cdots> - \frac{1}{2\pi i} \int_C d\bar{z} \bar{\alpha} <\bar{T}\phi \cdots>. \qquad (7.10)$$

Equating (7.10) and the r.h.s. of (7.7), and using Cauchy's theorem gives

$$<T(z)\phi_1(z_1,\bar{z}_1)\cdots> = \sum_j \left[\frac{\Delta_j}{(z-z_j)^2} + \frac{1}{(z-z_j)}\frac{\partial}{\partial z_j} \right] <\phi_1(z_1,\bar{z}_1)\cdots> \quad (7.11)$$

A similar equation involving $\bar{T}$ and the $\bar{\Delta}_j$ follows since we take $\bar{\alpha}$ independent of α. (7.11), the conformal Ward identity, is the basis for further development. Notice that in contrast to (2.5) it is a local relation.

7.2 $T\phi$ Operator Product Expansion

Next we Laurent expand T acting on an operator

$$T(z)\phi(z_1,\bar{z}_1) = \sum_{-\infty}^{+\infty} \frac{1}{(z-z_1)^{n+2}} L_n \phi(z_1,\bar{z}_1) \quad (7.12)$$

where the L_n are defined by

$$L_n\phi(z_1,\bar{z}_1) = \frac{1}{2\pi i}\oint dz(z-z_1)^{n+1} T(z)\phi(z_1,\bar{z}_1) \quad (7.13)$$

$\bar{T}$ acts on the $\bar{z}$ dependence of ϕ; hence the analogous expansion of $\bar{T}\phi$ gives rise to the $\bar{L}_n$. The L_n and $\bar{L}_n$ are "pieces" of T corresponding to different conformal transformations. If we write

$$\alpha(z) = \sum_{-\infty}^{+\infty} \epsilon_n z^{n+1} \quad (7.14)$$

then for

$$\begin{aligned} n &= -1: & \alpha &= \epsilon_{-1} = \text{Const.} \\ n &= 0: & \alpha &= \epsilon_0 z \\ n &= +1: & \alpha &= \epsilon_1 z^2 \end{aligned} \quad (7.15)$$

so that L_{-1} arises in infinitesimal translations, L_0 rotations or dilatations, and L_{+1} the "special" transformation.

Further examination of the conformal Ward identity (7.11) and (7.13) shows that the action of the L_{-n} in correlation functions may be expressed via differential operators

$$<(L_{-n}\phi_1)\phi_2\cdots> = \mathcal{L}_{-n} <\phi_1\phi_2\cdots>$$

$$\mathcal{L}_{-n} = \sum_{j\neq 1} \left\{ \frac{(1-n)\Delta_j}{(z-z_j)^n} - \frac{1}{(z-z_j)^{n-1}} \frac{\partial}{\partial z_j} \right\}, \quad n \geq 2 \quad (7.16)$$

The representations of the conformal group in two dimensions include primary operators. These are operators with definite anomalous dimension, which are annihilated by the L_n

$$L_n \phi = 0, \quad n \geq 1$$
$$L_0 \phi = \Delta \phi \tag{7.17}$$

Acting on a primary operator with the L_{-n}, $n \geq 1$ gives rise to a conformal tower of new operators, called descendants of ϕ

$$L_{-1} \phi = \frac{\partial}{\partial z} \phi$$
$$\left. \begin{array}{l} L_{-2} \phi = \ldots \\ L_{-3} \phi = \ldots \end{array} \right\} \text{ new operators} \tag{7.18}$$

The $L_{-n}\phi$ have dimensions $\Delta + n$. The complete conformal tower of ϕ consists of all operators

$$L_{-k_1} \ldots L_{-k_m} \bar{L}_{-\bar{k}_1} \ldots \bar{L}_{-\bar{k}_{\bar{m}}} \phi \tag{7.19}$$

which have dimension

$$x_t = \Delta + \bar{\Delta} + \sum_{i=1}^{m} k_i + \sum_{i=1}^{\bar{m}} \bar{k}_i$$
$$= x + n + \bar{n}; \quad n,\bar{n} = \text{positive integers} \tag{7.20}$$

as anticipated in (6.9).

7.3 TT Operator Product Expansion and the Virasoro Algebra

The operator product expansion of T with itself is

$$T(z)T(z') = \frac{c/2}{(z-z')^4} + \frac{2}{(z-z')^2}T(z') + \frac{1}{(z-z')}\frac{\partial}{\partial z'}T(z') + \ldots \tag{7.21}$$

The first term on the r.h.s of (7.21) depends on the central charge c (also called the conformal or trace anomaly number). This term therefore depends on the representation (for phase transitions, the universality class). The central charge also may be seen to give rise to a finite size correction to the free energy [15]; additionally it governs the response of the system to curvature of the underlying space. The factor two in the second term of (7.21) arises because the complex dimension of T is (2,0). If T were a primary operator the first term would be absent (cf. 7.17, 7.18). In fact T is a descendant of the unit operator ($T = L_{-2}1$). The form of the first term is also inconsistent with (2.3) since the normalization of T is fixed by (7.3) and cannot be adjusted.

At this point it is convenient to realize the conformal symmetry in a true operator formalism, e.g. by using the transfer matrix as in Section 6.1.b. We again denote corresponding quantities with a carat, e.g. $T \to \hat{T}$. Since the theory is conformally invariant, we can further make use of the transformation to introduce "radial quantization", as illustrated in Fig. 5. Translating (7.21) into this language and making use of (7.13) results (after some computation) in the Virasoro algebra

$$[\hat{L}_m, \hat{L}_n] = (m-n)\hat{L}_{m+n} + \frac{c}{12} m(m^2-1)\delta_{m,-n} \quad (7.22)$$

Note that the commutation relations of the $\hat{L}_n$ depend on the representation via the central charge. The anomalous term vanishes for $m = 0, \pm 1$, since these terms correspond to finite (Möbius or bilinear) coordinate transformations that map the infinite plane onto itself, and do not "rip up" space. The study of conformal invariance in two dimensions is the problem of classifying the representations of (7.22).

7.4 Unitarity and the Kac Formula

Now consider all descendants with dimension $\Delta + N$ of a primary state $|\phi\rangle$ of dimension Δ. These are

$$\hat{L}_{-k_1} \ldots \hat{L}_{-k_m} |\phi\rangle \quad (7.23)$$

where $N = \sum_{i=1}^{m} k_i$. N is called the "level" of this set of states. There are $P(N)$ such states, where $P(N)$ is the partition of N. Next consider the $P(N) \times P(N)$ matrix M_N of all possible inner products at level N

$$\langle\phi| \hat{L}^+_{-k_m} \ldots \hat{L}^+_{-k_1} \hat{L}_{-k'_1} \ldots \hat{L}_{-k'_{m'}} |\phi\rangle \quad (7.24)$$

The determinant of M_N is known as the Kac determinant, and may be shown to satisfy

$$\det M_N = \alpha_N \prod_{p,q \leq N} \left(\Delta - \Delta_{p,q}(c)\right)^{P(N-pq)} \quad (7.25)$$

where p and q are positive integers, α_N is independent of p, q, and c, and

$$\Delta_{p,q} = \frac{[(m+1)p - mq]^2 - 1}{4m(m+1)} \quad (7.26)$$

with the central charge c parameterized by

$$c = 1 - \frac{6}{m(m+1)} \quad (7.27)$$

Requiring that the representation be unitary [16] imposes further constraints, since then all eigenvalues of M_N must be non-negative. The result of this analysis is that for $c \geq 1$ there is no constraint, but the only allowed values for $c < 1$ are given by (7.27) with $m = 3, 4, 5, \ldots$ and the Kac formula (7.26) with $1 \leq p \leq m-1$ and $1 \leq q \leq p$. Thus the latter provides a rational expression for critical exponents! The universality class is determined by the value of the central charge; e.g. m = 3 corresponds to the Ising class, m = 4 the tricritical Ising, m = 5 the three state Potts model or a tetracritical point, etc. c = 1 is the Gaussian class.

7.5 Differential Equations for Correlation Functions

Other important conclusions concerning the unitary representations follow from this analysis as well. By (7.25), det(M_N) vanishes for $\Delta = \Delta_{p,q}$. Thus there is a linear combination of the $\hat{L}_{-n}$ at level N = pq that annihilate $|\phi_{p,q}\rangle$, i.e. this state is degenerate. Further, (7.16) implies that any correlation function involving $\phi_{p,q}$ satisfies a partial dif-

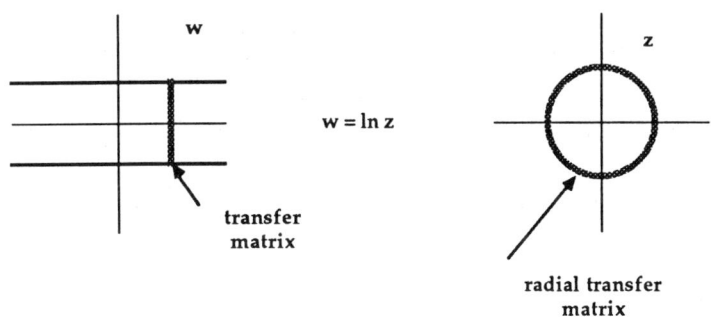

Fig. 5.

Radial Quantization. The Horizontal (Vertical)"Time" ("Space") Direction In The Strip Becomes The Radial (Azimuthal) "Time" ("Space") Direction In The Plane.

ferential equation of order pq. These differential equations also imply a closure relation on the operator product expansion (5.1); for each representation there is a finite number of primary operators that form a closed algebra under the operator product expansion. Additionally, one can solve (at least in principle) the differential equation for any four-point correlation function G_4. By taking appropriate limits, the operator product expansion coefficients C_{ijk} may be extracted. So the theory is completely specified--the anomalous dimensions, correlation functions, and operator product expansion are all determined.

8. SURFACE CRITICAL BEHAVIOR

8.1 Conformal Ward Identity Revisited

We next consider infinitesimal coordinate transformations for systems with an edge. The analysis is carried out in the upper half plane [17], which, using (2.4) or (2.5) is equivalent to any fully finite compact region by a conformal mapping. In this case the functions α and $\bar{\alpha}$ in (7.2) are no longer independent -- since $\alpha = 0$ on the real axis we require

$$\alpha(\bar{z}) = \overline{\alpha(z)} \tag{8.1}$$

As a result, we no longer have separate conformal Ward identities involving T and $\bar{T}$. However, T and $\bar{T}$ must coincide on the real axis, and $\bar{T}$ can thus be regarded as the analytic continuation of T in the lower half plane. It follows that the Ward identity has the form

$$<T(z)\phi(z_1,z_2)\cdots> = \sum_j \left[\frac{\Delta_j}{(z-z_j)} + \frac{1}{(z-z_j)} \frac{\partial}{\partial z_j} \right] <\phi(z_1,z_2)\cdots> \tag{8.2}$$

Note that summation in (8.2) now runs over all arguments of the operators, in contrast to (7.11) (and the corresponding Ward identity for $\bar{T}$). Thus, when the L_n and corresponding differential operators are introduced, as in (7.13) and (7.16), each operator is acted on twice. As a result, as already anticipated in Sections 4.4 and 6.2, the n-point correlation function in the half plane, taken as a function of all its variables, satisfies the same differential equation as the 2n-point function in the infinite plane, taken as a function of half its variables. Only the boundary conditions distinguish the two cases.

8.2 Ising Model

For the Ising class the order parameter-order parameter correlation function in the half plane is given by (see Fig. 2)

$$<\sigma(1)\sigma(2)>_{\frac{1}{2}} = \text{const.} \ (y_1 y_2)^{-\frac{1}{8}} \Psi(\tau)$$

$$\Psi(\tau) = \sqrt{\tau^{1/4} + \tau^{-1/4}}$$

$$\tau \equiv \frac{\rho^2 + (y_1+y_2)^2}{\rho^2 + (y_1-y_2)^2}$$

$$\rho \equiv |\vec{x}_1 - \vec{x}_2| \tag{8.3}$$

In (8.3) the $-$ sign corresponds to free boundary conditions ($<\sigma> = 0$ on the real axis) while the $+$ sign applies to fixed boundary conditions ($<\sigma> = \pm \infty$ on the real axis). This result is very useful in comparing with diffraction experiments on real surfaces, as we will see in Section 9.

9. APPLICATIONS TO SURFACE PHASE TRANSITIONS

9.1 Scattering Lineshapes at Surface Critical Points

Consider a surface system that undergoes a second-order phase transition. For instance, the Au(110) surface, at sufficiently low temperatures, exhibits a (1x2) "missing row" [18] reconstruction, as shown schematically in Fig. 6.a. Every other row in the topmost layer is missing (and there is some small rearrangement of the atoms in the next few layers as well). Thus the symmetry of the surface layer is different from the bulk, and extra diffraction features result, as shown in Fig. 7. At low temperatures, the reconstructed layer is well-ordered. Thus starting on an occupied row and moving perpendicular to the row by an even number of lattice spacings,

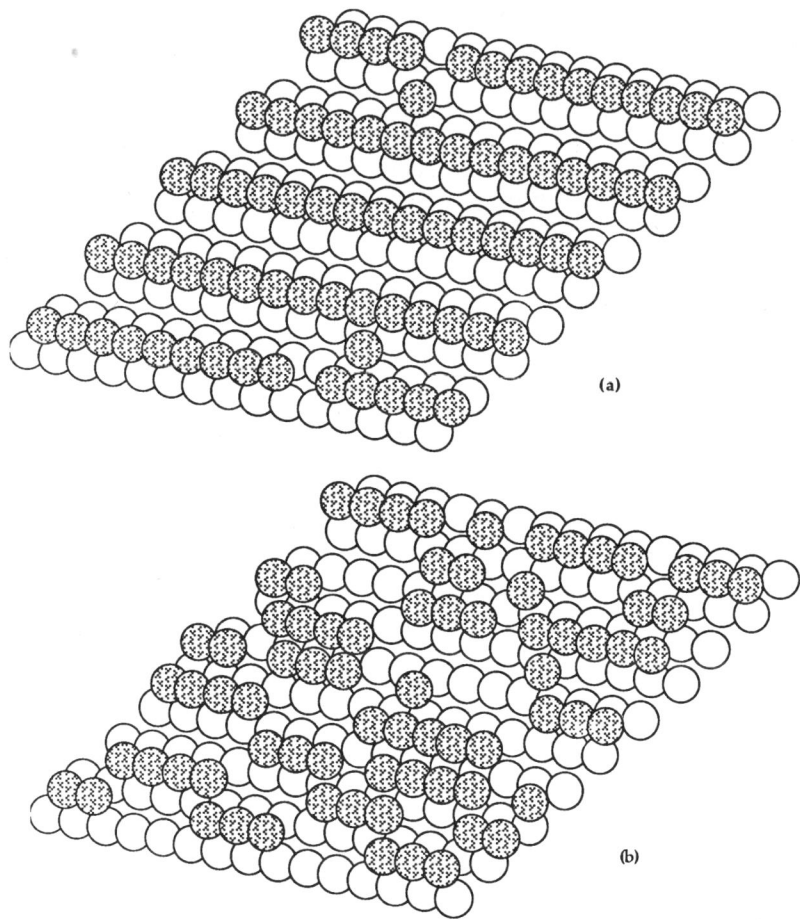

Fig. 6. *Schematic of the Au(110) surface below (a) and above (b) the phase transition. The shaded circles represent top-layer atoms.*

one will encounter another occupied row with probability near one. There is also long-range order in the perpendicular direction. However, as the temperature increases, disorder sets in. This is believed to occur by Au atoms in occupied rows moving to nearby unoccupied sites. For temperatures T less than the phase transition temperature T_c, the long-range (1x2) order is weakened but not destroyed. Thus the extra (superlattice) Bragg spots (Fig. 7) will decrease in intensity, but still be observable. For $T > T_c$, the long-range order will be gone (Fig. 6.b). If one moves perpendicular to an occupied row, there is an equal probability of encountering either an occupied or unoccupied row after moving by a sufficiently large even number of lattice spacings. Also, the rows themselves shorten. As a result, diffraction at the superlattice spots will be weak or non-existent, depending on how much short-range order remains, and there will be no true superlattice Bragg spots.

These effects can be seen in Low-Energy Electron Diffraction (LEED) data for the Au(110) surface system [19]. The superlattice feature has a lineshape that broadens and weakens with increasing temperature. This is due to the thermal disorder described above in combination with finite size effects. Even at low temperatures, where the reconstructed regions are well-ordered, the beam has a finite linewidth. This comes about since the reconstructed domains are of finite size--they are limited by surface defects such as steps. On the Au(110) surface, the domain size depends on sample preparation [20]. For the results discussed below, the largest domains were about 150 A across. This was determined by measuring the linewidth at low T.

In order to fix ideas, it is useful to picture the experimental situation as in Fig. 8. This illustrates elastic diffraction from a domain of size L. Variables of interest are the parallel momentum transfer $k = (k_f - k_0)_{||} - Q$, where Q is the momentum transfer of the superlattice beam, and the incident energy E_i, which we will allow to vary (at fixed k) in order to characterize multiple scattering effects. Fig. 8, and the theoretical treatment of the lineshape given below, actually apply to any type of elastic diffraction, e.g. electron (LEED), neutron, x-ray, ion or He atom scattering. However, for definiteness, we will give explicit results for the LEED case only.

The scattering lineshape of the superlattice beam is our main concern in this subsection. As mentioned, it is influenced by both thermal and finite size effects. The calculation of the thermal effects in finite regions is at best difficult, and certainly depends on the details of the system under consideration. At a second-order phase transition, however, by use of conformal field theory, one can determine these effects in a universal (model-independent) way. Thus it is not necessary to know the forces responsible for the phase transition. This is a considerable advantage, since the energies involved are difficult or impossible to determine, being generally very small compared to adsorption, cohesion, diffusion barrier or other energies in the problem.

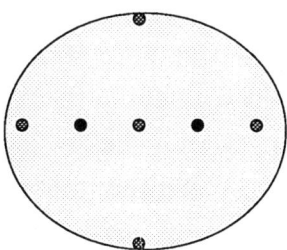

Fig. 7. Schematic Diffraction Pattern for the (1x2) Structure of Fig. 6. Dotted Circles-Bulk (Substrate Features; Solid Circles-Surface Layer (Superlattice) Beams.

When the domain is undergoing a second-order phase transition, it has no true long-range range order because of thermal fluctuations. Hence the superlattice beam, strictly speaking, is diffuse scattering rather than a true Bragg peak. However, in two dimensions, the fluctuations are generally large, and the scattering therefore substantial. The lineshape of this scattering is given by the long-range correlations at the phase transition. Thus it depends on the universality class of the transition, which in general may be determined by observation of the symmetry of the ordered state relative to the disordered state, and also the size, shape and boundary conditions of the domain. We emphasize that this diffraction occurs in a partially ordered, fluctuating medium where finite size effects are important--not the usual situation at all!

Now the lineshape of the scattering is the variation of the observed intensity I(k) at small k. This, as mentioned, depends on the long-range range thermal correlations. At a phase transition, the range in question is L, the domain size. For definiteness, consider a transition in the Ising class. Then, as explained in Section 5.1, there are three such quantities: the local order parameter $\sigma(\vec{r})$, local energy $\epsilon(\vec{r})$ and unit operator. In the language of the theory of phase transitions, these are the most relevant operators--the operators whose correlations decay most slowly with distance, and whose Fourier transform therefore varies most rapidly with k at small k. Thus, at the transition, the part of the scattering amplitude that determines the lineshape must be

$$T = \sum_i \left[A\sigma(\vec{r}_i) + B\epsilon(\vec{r}_i) + C \right] \quad (9.1)$$

where the sum is over the possible scattering sites in the domain, and A, B and C depend on the type of scattering, scattering conditions (especially E_i), the substrate, etc. A more detailed derivation of (9.1) will be given in Section 9.4. The scattered intensity is proportional to the Fourier transform of the thermal average of the square of the matrix element of T. For an Ising transition, for symmetry reasons ex-

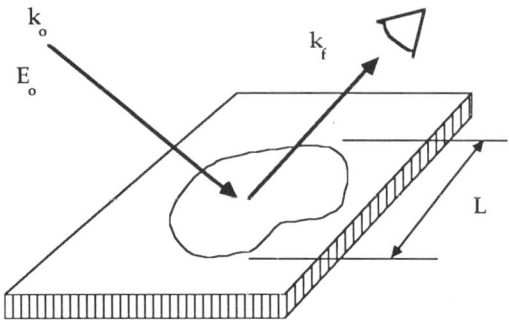

Fig. 8. Diffraction from a Surface Domain of Size L.

plained in Section 9.4, at a superlattice beam only the first term on the r.h.s. of (9.1) contributes and therefore

$$I(\vec{k}) = <|<\vec{k}_f|T|\vec{k}_o>|^2>$$
$$= A \sum_{i,j} e^{i(\vec{k}_f-\vec{k}_o)\vec{r}_{ij}} <\sigma(\vec{r}_i)\sigma(\vec{r}_j)> \qquad (9.2)$$

where $A = |A|^2$.

Two important consequences follow from (9.2). First, the lineshape is determined by the order parameter-order parameter correlation function, which can be calculated using conformal field theory, as explained in the preceding Sections. The results of this calculation [21] will be discussed shortly. Secondly, as shown in Fig. 9, the lineshape is independent of multiple scattering effects, even when strong multiple scattering occurs. This is a special feature of multiple scattering at a second-order phase transition in the Ising class. For other phase tranitions, the intensity is the sum of a few terms with this feature. However, there is no such simplification in general for scattering from disordered systems. These predictions have not been tested experimentally, to our knowledge. Additionally, the variation of A with multiple scattering conditions, e.g. E_i (the I-V curves) can differ substantially from the variation in the same system at low T, since the different orders of multiple scattering are weighted differently. We will examine these points more closely in Section 9.4, after we consider the scattering lineshape.

One may re-express (9.2) to emphasize the dependence of the intensity on the domain size. This gives

$$I(\vec{k}) = A L^{4-\eta} S(\vec{k}L) \qquad (9.3)$$

where L is the size of the domain scattered from (Fig. 8), S the scattering function (the Fourier transform of $<\sigma\sigma>$ for a region of unit size), and η a critical exponent ($\eta = 2x$). (9.3) may be derived using (2.1). For an ordered region the total intensity I at the superlattice peak ($\vec{k} = 0$) would grow with the size of the domain as L^4, since it is proportional to the square of the number of scatterers in the domain. At a second-order phase transition, critical fluctuations reduce the power of L by η ($\eta = 1/4$ for the Ising class). This is a reflection of the fact that the scattering at the transition is diffuse rather than true Bragg scattering.

At a second-order phase transition the correlation length ξ is proportional to the size of the system L, as illustrated in Fig. 10. This is a fundamental property of the transition. Since ξ measures the distance over which a disturbance is felt, a finite fraction of the domain always "feels" the edge, no matter how large it is (technically, the thermodynamic limit is not unique at a second-order critical point). Thus one expects the scattering function S to be strongly influenced by finite size effects. This is in fact the case, as will be borne out below.

To actually calculate S, we begin with the result (8.3) for $\langle\sigma\sigma\rangle$ in the half plane with free boundary conditions. Then, using standard conformal mapping techniques and (2.4),

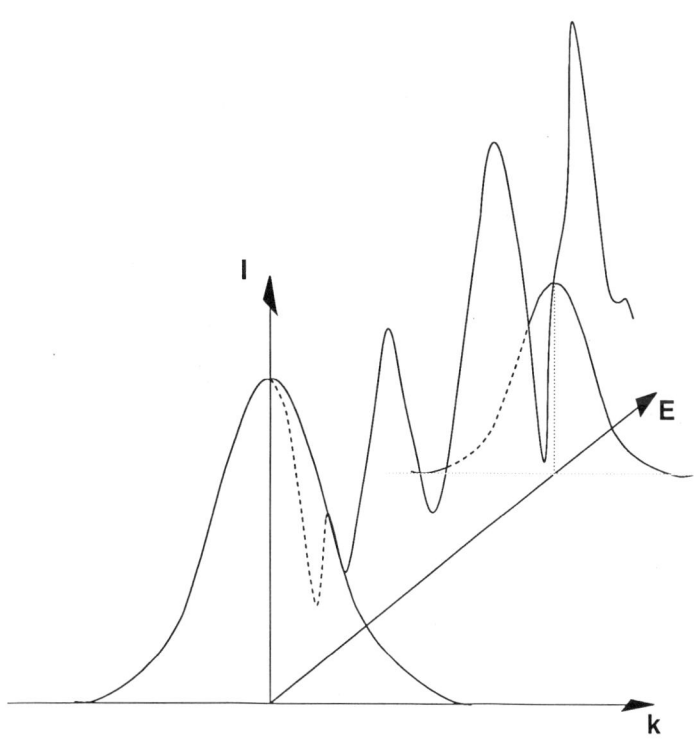

Fig. 9. *The lineshape is independent of multiple scattering at an Ising critical point.*

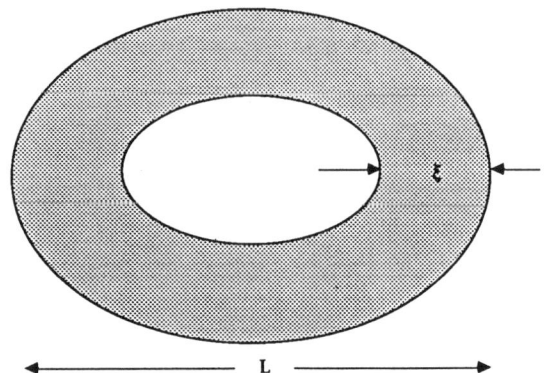

Fig. 10. *For A Finite System At Criticality, The Correlation Length Is Proportional To The System Size L.*

we can calculate $\langle \sigma\sigma \rangle$ in any (compact) domain. The Fourier transform of the result gives $S(\vec{k})$. The scattering function for the simplest plane shape, a circle, is shown in Fig. 11 [21]. Other domain shapes and boundary conditions have been considered in the literature [22], but for reasons to be explained we do not need to include them here. The lineshape for a circle, to good approximation, may be represented as a Gaussian plus a Lorentzian

$$S(k) = A_G \, e^{-(k/k_G)^2} + A_L \, \frac{1}{1+(k/k_L)^2} \qquad (9.4)$$

with specified height and width ratios $A_G/A_L = 0.669$, $k_G/k_L = 1.28$. (It should be noted that the numerical values quoted are sensitive to the fitting routine used, so the same routine should be employed for any experimental data). The Lorentzian part of (9.4) gives rise to a long "tail" as seen in Fig. 11. This has the property that it grows with domain size, i.e. $S \sim L^{2-\eta}/k^2$ at large k. Thus when one scatters from a distribution of domains, as occurs on real surfaces, even the scattering at large k is dominated by the largest domain. This is important for comparison with experiment, as we will see. By contrast, the lineshape for a well-ordered domain is approximately Gaussian at large k. This means that the scattering at large k depends on the smaller domains. Therefore, it is necessary to know the distribution of domain sizes to understand the lineshape, which greatly complicates the analysis.

Strictly speaking, the scattering at large k is dominated by the correlation function for small distances, which goes as (2.3), the "bulk" limit. Thus

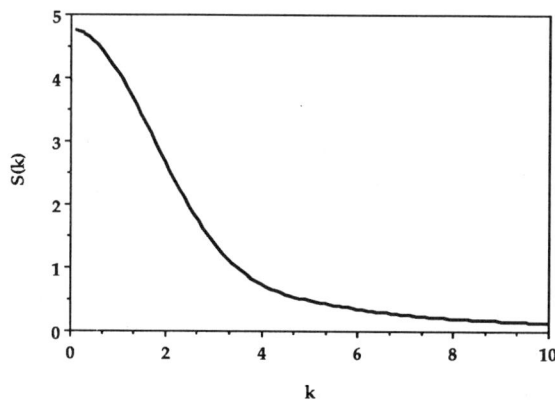

Fig. 11. Scattering Function at an Ising Transition in a Circular Domain

$$S(k) \to 1/k^{2-\eta} \tag{9.5}$$

However, this behavior only sets in at very large k values [22], as anticipated in (6.16), where (unless L is very large) S and hence the intensity are small and difficult to separate from the noise in any comparison with experiment.

We have seen that the diffraction lineshape is determined by the two-point correlation function in the half plane. As described in Section 8.1, this quantity satisfies the same differential equation as the four-point function in the infinite system. Thus it is fundamental in the field theory. A lineshape measurement is therefore a significant test of the theory. It is conceivable, for instance, that at a transition in a given surface system the critical exponents, which follow from scale invariance alone, are correct, but the lineshape, which is determined by conformal invariance, is not.

9.2 Comparison with Experiment--the Au(110) (1x2)-to-(1x1) Transition

As mentioned, the LEED scattering lineshape has been measured for the Au (110) surface system at various temperatures. Previous analysis of this data [19] and other measurements on this system [23] establish that it undergoes a second-order phase transition with critical exponents of the Ising class, as suggested by symmetry considerations [24]. Since the transition temperature T_c is not exactly known, we exhibit in Fig. 12 [25] a comparison of measured and predicted lineshapes at several temperatures near the expected value of T_c. Since neither the absolute magnitude nor linewidth of the scattering was exactly known, these were fit to the theory at each temperature. Thus deviations of predicted and observed intensities are due only to lineshape differences. A good fit is obtained for T = 661 ± 26 K. This is within the range of expected values for the transition temperature T_c. For temperatures below this range, the experimental lineshape is more Gaussian than the theory, consistent with increased order. For higher temperatures, the data is more Lorentzian, consistent with an increased importance of thermal fluctuations. These trends are already evident in Fig. 12.

The good agreement of theory and experiment exhibited in Fig. 12 is very gratifying. Considering the situation further brings up several questions that must be dealt with, however. These include the effects of any coherent scattering from more than one domain, other boundary conditions, the domain shapes, and the domain size distribution. In addition, there are other determinations of T_c, and the instrumental resolution. Consider the last effect first. It will alter the lineshape unless the instrumental transfer width t exceeds the domain size (t >> L). This condition implies, somewhat ironically, that for a given diffraction apparatus, a "bad" sample is best--one with L not too large! It was satisfied for the data analyzed in Fig. 13. Fixed boundary conditions give rise to a lineshape with very small "tails" that does not fit the data. The linewidth at low T is approximately isotropic, indicating that the domains are not greatly elongated. The difference in lineshape between, e.g., a circle and a square is within the experimental error. These and the other questions mentioned are discussed in more detail elsewhere [25].

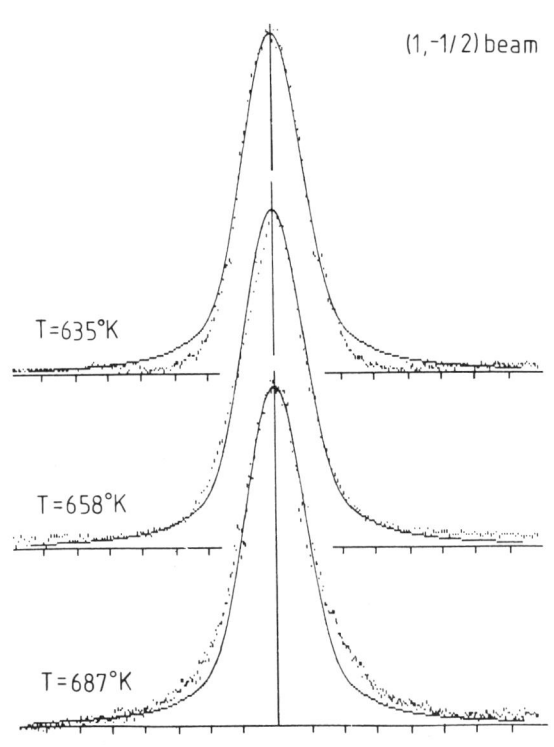

Fig. 12. *Lineshapes Near the Au(110) Phase Transition*

9.3 Lineshapes for Other Phase Transitions

Similar predictions of lineshapes can be made for phase transitions in universality classes other than Ising. The Gaussian class is characterized by continuously varying exponents. Realizations of this transition include the O/Ru(0001) system, which exhibits a 4-state Potts transition [26], and various surfaces that apparently show roughening transitions [27]. The predicted lineshape is a non-trivial function of the exponent. Fig. 13 illustrates this by showing the HWHM as a function of exponent. Analysis of data from systems with domains of the proper size, similar to that described in Section 9.2, would be very interesting. Fitting observed lineshapes to theoretical predictions may be a good way to determine the exponent and could help more convincingly establish whether or not a given system exhibits roughening, a controversial question [28]. The model independence of the conformal predictions should be especially helpful here.

Other cases of interest include the tricritical Ising class, which exhibits supersymmetry [2]. The consequences of this for the lineshape have yet to be fully determined, however the vanishing of certain operator product expansion coefficients may lead to observable consequences. The 3-state Potts model is realized in several surface systems [29]. Here the differential equation for the two-point function in the half plane is sixth order; however solutions for the four-point function are available [30].

9.4 Multiple Scattering at Surface Critical Points

Conformal invariance may also be used to calculate multiple scattering at surface critical points. The main tool here is the operator product expansion, reviewed in Section 5. Because of this, a good part of the argument is valid at any critical point. Fig. 14 illustrates the experimental setup-- the scattered probe interacts several times with the surface or substrate before being detected. The distance over which this occurs is the mean free path λ, which measures the average distance the probe travels before losing too much energy to be considered elastically scattered. For LEED and typical surfaces, $\lambda \ll L$. The resulting small parameter λ/L makes possible the simplification of the expression for multiple scattering.

The scattered probe will generally couple to the density $n(\vec{r})$ of atoms in the surface system. Multiple scattering terms will involve coupling to products of densities, at points separated by λ, or less. Now $n(\vec{r})$ generally includes a term proportional to the (local) order parameter $\sigma(\vec{r})$. Then the multiple scattering series for the transition amplitude can be rearranged into a series of products of σ. The σ in each product are at points within λ of each other. Using the operator product expansion, one can extract the most relevant part of each of these. The result is (9.1). Correction terms will affect the intensity to $\mathcal{O}(\lambda/L)^2$. Additionally, the part of $n(\vec{r})$ proportional to σ has a coefficient with the spatial symmetry of the ordered state.

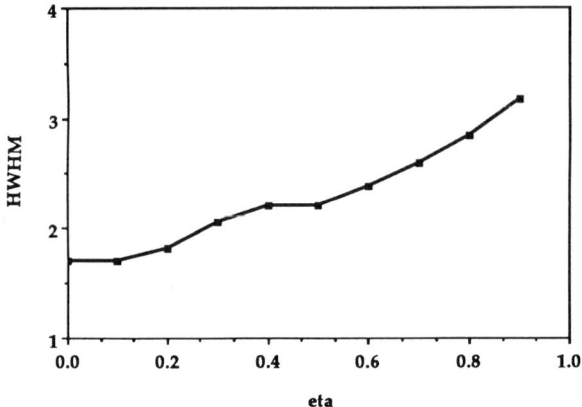

Fig. 13.   HWHM for the Gaussian Class

Therefore the contribution to the intensity at a superlattice beam from a given product of σ has the symmetry of the corresponding product of coefficients--combined with the operator product expansion, this can eliminate certain terms in the intensity at a superlattice beam, giving rise to (9.2), for instance. These points will become clearer as we consider a specific example.

It should be recognized that (9.2) describes the intensity variation $I(\vec{k})$ at small k. Thus the change in I with multiple scattering parameters (the I-V curve shown in Fig. 9) is the change in the intensity above the background. It is possible that the background intensity itself has a very different dependence on multiple scattering. This cannot be described in a universal way, since it depends on short-range effects.

For definiteness, in the following we consider multiple scattering of low-energy electrons. The results are comparable with experiment, and the methods employed to treat this specific problem can be adapted more generally if required. For simplicity, we will explicitly treat scattering from a

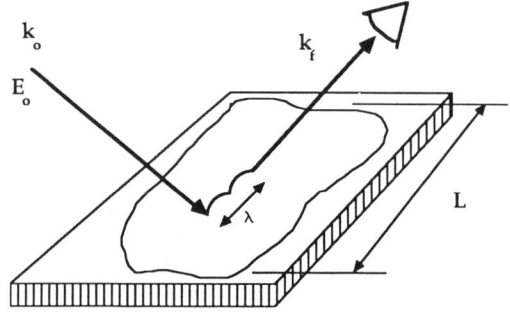

Fig. 14.   *Multiple Scattering.*

single layer only. The additional terms that appear due to the presence of an ordered substrate can be included by using renormalized propagators (and removing summation restrictions) [31].

The first step is to write a Born series for the scattering amplitude from an ordered layer

$$T = \sum_i t_i + \sum_{i \neq j} t_i G_{0,ij} t_j + \sum_{\substack{i \neq j \\ j \neq k}} t_i G_{0,ij} t_j G_{0,jk} t_k + \cdots \quad (9.6)$$

In (9.6) the t_i are single-ion scattering vertex operators that give the (total) scattering amplitude from a single atom at position i. For a uniform layer, they will in fact be independent of i. The G_{0ij} are electron propagators, which generally are taken to include the mean free path λ via a complex wavevector. The successive terms correspond to increasingly high orders of multiple scattering, as shown in Fig. 15. The summations extend over all atoms in the domain (see Fig. 14), with restrictions so that an electron does not interact with the same site twice in succession. In principle, the expression should be altered to account for edge effects. These are generally ignored in dynamical LEED studies, since they are small by factors of λ/L. At a phase transition, they will be further suppressed if the correlation function vanishes at the edge of the domain due to free boundary conditions. For an ordered layer, (9.6) may be summed to give useful expressions for comparison with measured I-V curves [32]. The structure of these curves depends sensitively on atomic arrangements, and extensive numerical work, compared with measurements, has provided much useful information on surface structure.

For a layer with substitutional disorder, such as the Au(110) surface (Fig. 6), the scattering will depend on the arrangement of atoms in the layer. Thus in (9.6) we replace

$$t_i \to t \, n_i \quad (9.7)$$

where the occupation numbers $n_i = 1$ (0) if site i is occupied (unoccupied) by an atom and we assume the vertices are site independent. To calculate the intensity we must then take the matrix element of T, square it, and then average the $\{n_i\}$ over an equilibrium ensemble. The result is

$$I = \sum_{i,j} e^{i(\vec{k}+\vec{Q}) \cdot \vec{r}_{ij}} t^2 \langle n_i n_j \rangle + \sum_{\substack{i,j,k \\ j \neq k}} e^{i(\vec{k}+\vec{Q}) \vec{r}_{ij}} t^3 G_{0,ij} \langle n_i n_j n_k \rangle + \cdots \quad (9.8)$$

where the brackets $\langle \ldots \rangle$ denote a thermal average. Note that the total momentum transfer $\vec{k} + \vec{Q} = \vec{k}_f - \vec{k}_0$ appears in (9.8). For a disordered system, in general, each term in (9.8) will have a different lineshape, and its contribution will change with multiple scattering conditions. Thus the overall line-

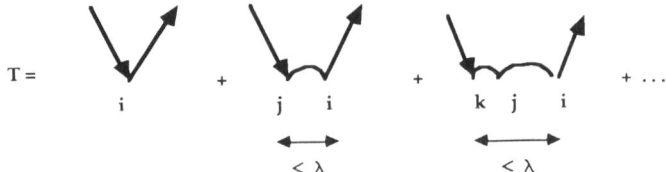

Fig. 15. *Born series for multiple scattering.*

shape will depend on multiple scattering, and the separation illustrated in Fig. 9 will not hold. There are two important exceptions to this rule. For a well-ordered system, all $n_i=1$, so the Fourier transforms in (9.8) will be identical term-by-term. The other exception is a critical point. Here there will be a small number of important terms on the r.h.s. of (9.8), and in some important cases only one term will appear.

For a transition in the Ising class, the ordered state has two sublattices, distinguished by the sublattice phase factor

$$\epsilon_i = e^{i\vec{Q}\cdot\vec{r}_i} = \pm 1 \tag{9.9}$$

We are now in a position to make the connection to field theory, since

$$n_i = \frac{1}{2}\left(1+\epsilon_i c_s a^x \sigma(\vec{r}_i)\right) \tag{9.10}$$

In (9.10) a is a lattice constant, and c_s a non-universal constant that depends on the specific system. There may be corrections to (9.10) due to field mixing effects and higher order terms in the conformal tower of σ. These will generally have only a small effect, as discussed in [32]. Substituting (9.10) into (9.8) and rearranging gives rise to a sum of terms involving multi-point correlations of the σ. Each of these consists of the expectation value of two products of operators σ multiplied together. Each σ in a product is within λ of each other σ in the product. For small k, the important part of each correlation is from distances of $\mathcal{O}(L)$, so any σ in one product is far from any σ in the other product. In this circumstance, it is appropriate to use the operator product expansion to extract the leading (most relevant) term in each correlation function.

Now note that at $\vec{k} = 0$ (at the superlattice spot center), there is a factor $\epsilon_i \epsilon_j$ appearing in each Fourier transform in (9.8). If any additional dependence on i or j in the function being transformed is smooth, the sum on these alternating variables will be very small. However, additional sublattice phase factors will enter via (9.10). Thus, since $\epsilon^2 = 1$, unless each product of operators in the correlation function has an odd number of σ the sum will be negligible. This argument also applies to small k. Now it follows from (5.2) that the operator product expansion of an odd number of order parameter operators (for the Ising class) involves the order parameter operator only. Thus only the <σσ> correlation function contributes to the intensity at a superlattice beam. This justifies (9.2). Note that a similar result is expected for Ising transitions in other dimensions as well.

In order to find the contribution of multi-σ correlation functions to the intensity, it is not sufficient to use (5.2) Higher order terms in the conformal tower must be included. Consider, for instance, the projection of three σ on one

$$\sigma(1)\sigma(2)\sigma(3) = F(123)\sigma(1). \qquad (9.11)$$

We can calculate F from the four-point function since

$$<\sigma(1)\sigma(2)\sigma(3)\sigma(4)> \xrightarrow[r_4 \to \infty]{} \frac{1}{r_4^{1/4}} F(123) \qquad (9.12)$$

and expressions for any correlation of σ are available for the Ising class [33]. For multiple scattering with LEED, it is in general not necessary to include terms past fifth order [34]. Combining the expression for F (and similar functions for higher order terms) with (9.8), re-expressed using (9.10) as described, then gives rise to an explicit formula for (9.2).

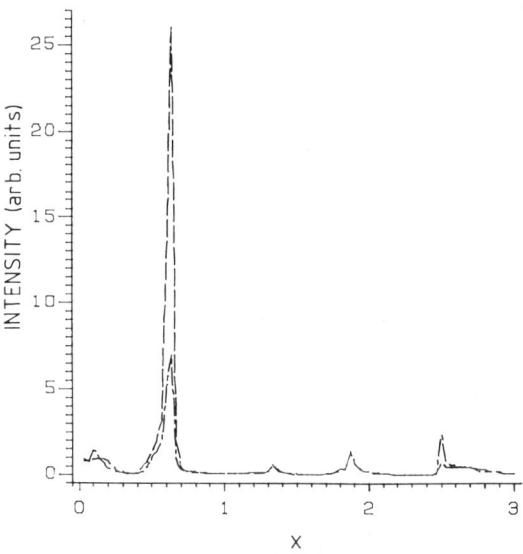

Fig. 16. *Intensity at a Bragg Peak vs. Electron Energy.*

It is now possible to examine some corrections to (9.2). The r.h.s. of (9.11) omits higher order (less relevant) contributions from the conformal tower of σ. Because they are less relevant, these contribute terms to the intensity that grow less rapidly with L than (9.2), as may be seen from (9.3). On the other hand, they must increase with the mean free path λ. A simple power counting argument indicates that the leading correction is $\mathcal{O}((\lambda/L)^2)$. This will be very small for typical LEED conditions and surfaces.

A full calculation of the I-V characteristics at a phase transition has yet to be carried out. However, an approximate treatment [32] shows (Fig. 16) that substantial differences from the well-ordered (low temperature) case can occur. In the figure, the uniformly dashed curve corresponds to the phase transition and the short-long dashes to low temperatures. In this computation, the single-ion scattering vertices t were approximated with s-wave phase shifts only, and the F coefficients of (9.12) were also simplified. It should be noted that the differences between low T and critical I-V curves begin in third order in multiple scattering from the top layer. Therefore proper scattering conditions, e.g., grazing incidence, will maximize the effect.

Finally we reiterate that the I-V curves are very sensitive to surface structure. Thus the treatment described here provides a means of understanding the effects of temperature on surface structure. In addition, it is one of the few cases in which an exact treatment of multiple scattering from disordered systems is possible.

10. REMARKS ON REFERENCES

There is a profusion of articles and reviews on conformal invariance. A recent and useful reprint volume [6] runs to over 1000 pages. We offer here a few comments that may help the uninitiated to get started in the subject, having in mind those schooled in statistical mechanics or many-body theory rather than field theory. The seminal article of Belavin, Polyakov and Zamolodchikov [5] is quite extensive but assumes considerable knowledge of this somewhat arcane branch of quantum field theory. Perhaps the most accessible comprehensive treatment for the neophyte is Cardy's review [2]. This is updated in [3], which includes more recent developments. Affleck's article in the same volume [4] considers the application of conformal methods to spin chains. The review by Ginsparg [1] offers to our knowledge the most complete treatment of the basics, filling in many of the steps not included in [2] and [5]. For this reason it makes for slower reading; additionally the treatment is somewhat more field theoretically oriented than in [2].

ACKNOWLEDGEMENTS

We thank J. Cardy for a helpful comment. This work was supported in part by the U.S. DOE, BES-Materials Science Contract No. W-31-109-ENG-38.

REFERENCES

1. P. Ginsparg, in *Champs, Cordes, et Phénomènes Critiques/Fields, Strings, and Critical Phenomena*, edited by E. Brézin and J. Zinn-Justin, Elsevier Publishers B.V. (to appear).
2. J. L. Cardy, in *Phase Transitions and Critical Phenomena*, edited by C. Domb and J. L. Lebowitz (Academic, London, 1986) Vol. 11.
3. J. L. Cardy, in *Champs, Cordes, et Phénomènes Critiques/Fields, Strings, and Critical Phenomena*, edited by E. Brézin and J. Zinn-Justin, Elsevier Publishers B.V. (to appear).
4. I. Affleck, in *Champs, Cordes, et Phénomènes Critiques/Fields, Strings, and Critical Phenomena*, edited by E. Brézin and J. Zinn-Justin, Elsevier Publishers B.V. (to appear).
5. A. A. Belavin, A. M. Polyakov and A. B. Zamolodchikov, Nucl. Phys. B **241**, 333 (1984).
6. Conformal Invariance and Applications to Statistical Mechanics, editors C. Itzykson et al., World Scientific Publishing (1988).
7. A. M. Polyakov, Sov. Phys. JETP Letters **12**, 381 (1970).
8. L. Schaefer, J. Phys. A **9**, 377 (1976).
9. L. P. Kadanoff, Phys. Rev. Letters **23**, 1430 (1969); A. M. Polyakov, Sov. Phys. JETP **30**, 151 (1969); K. G. Wilson, Phys. Rev. **179**, 1499 (1969).
10. J. L Cardy, J. Phys. A **17**, L385 (1984).
11. R. Hentschke, P. Kleban and G. Akinci, J. Phys. A **19**, 3353 (1986).
12. M. Abramowitz and I. A. Stegun, editors, *Handbook of Mathematical Functions*, (Dover, 1972).
13. M. E. Fisher and P. G. deGennes, C. R. Acad. Sci. Paris **287**, B-207 (1978).
14. T. W. Burkhardt and E. Eisenriegler, J. Phys. A **18**, L83 (1985).
15. H. W. J. Blöte, J. L. Cardy, and M. P. Nightingale, Phys. Rev. Letters **56**, 742 (1986); I. Affleck, Phys. Rev. Letters **56**, 746 (1986).
16. D. Friedan, Z. Qiu and S. Shenker, Phys. Rev. Letters **52**, 1575 (1984).
17. J. L. Cardy, Nucl. Phys. B **240** [FS12], 514 (1984).
18. I. Robinson, Phys. Rev. Letters **50**, 1145 (1983) and references therein; I. Robinson, Y. Kuk and L.C. Feldman, Phys. Rev. B **29**, 4762 (1984) and references therein.

19. J. C. Campuzano, M. S. Foster, G. Jennings, R. F. Willis and W. N. Unertl, Phys. Rev. Letters **54**, 2684 (1985); J. C. Campuzano, G. Jennings and R. F. Willis, Surface Sci. **162**, 484 (1985).
20. W. N. Unertl, private communication.
21. P. Kleban, G. Akinci, R. Hentschke and K. R. Brownstein, J. Phys. A **19**, 437 (1986); Surface Sci. **166**, 159 (1986).
22. P. Kleban and R. Hentschke, Phys. Rev. B **34**, 1980 (1986).
23. D. E. Clark, W. N. Unertl and P. Kleban, Phys. Rev. B **34**, 4379 (1986).
24. P. Bak, Solid State Commun. **32**, 581 (1979).
25. P. Kleban, R. Hentschke and J. C. Campuzano, Phys. Rev. B **37**, 5738 (1988).
26. P. Piercy and H. Pfnür, Phys. Rev. Lett. **59**, 1124 (1987).
27. See, for example, B. Salanon et al, J. Vac. Sci. Technol. A **6**, 655 (1988) or G. A. Held et al, Phys. Rev. Lett. **59**, 2075 (1987) and references therein.
28. S.G.J. Mochrie, Phys. Rev. Letters **59**, 3047 (1987); P. Zeppenfeld, K. Kern, R. David and G. Comsa, Phys. Rev. Letters **62**, 63 (1989); S. Thevusathan, thesis, Univ. of Maine (1989).
29. See W. N. Unertl, Comments Cond. Mat. Phys. **12**, 289 (1986); M. Schick, Prog. Surface Sci. **11**, 245 (1981).
30. A. B. Zamolodchikov and V. A. Fateev, Sov. Phys. JETP **62**, 215 (1985).
31. C. B. Duke and A. Liebsch, Phys. Rev. B **9**, 1126 (1974).
32. R. Hentschke and P. Kleban, Surface Sci. **202**, 533 (1988).
33. P. Di Francesco, H. Saleur and J. B. Zuber, Nuclear Phys. B **290** [FS20], 527 (1987).
34. S. Y. Tong, private communication.

A POSSIBLE FIELD-THEORETICAL MODEL FOR THE NEMATIC-ISOTROPIC PHASE TRANSITION IN LIQUID CRYSTALS

Luiz R. Evangelista[1]   and M. Simões[2]

[1]Departamento de Física, Universidade Estadual de Maringá
87020, Maringá, Paraná, Brazil
[2]Departamento de Física, Universidade Estadual de Londrina
86051, Londrina, Paraná, Brazil

ABSTRACT

We intend to study some aspects of the well-known nematic-isotropic phase transition in liquid crystals, by exploiting the nonlinear features of the system. Treating the molecules as hard rods and regarding the fact that the heads and the tails of the molecules cannot be distinguished in the nematic phase and using the translational invariance along the z-axis (which is parallel to the director $\hat{n}$ in the nematic phase), we arrive at a model that displays the essential features of the system. If we assume that the potential is periodic and write it as, for example, $-\alpha\cos\beta\phi$, the system undergoes a phase transition like that we found in the sine-Gordon model at $\beta^2 = 8\pi$[1]. Thus, the critical temperature (NI) is given by $T_c = 2\pi L K_\theta / K_B$ where L is the length of the system in the z-direction, K_θ is the elastic torsion constant and K_B is the Boltzmann constant.

It is necessary to ellaborate this general conjecture by continuing the development of the Statistical Mechanics of the system. But this is a hard problem (the model is in [2+1] dimensions). Despite of this, some interesting questions may arise. We mention, for example, the possibility of using the soliton solutions of the model and to investigate their role in the transition. On the other hand we may ask if it is possible to relate this picture of a liquid crystal to some manifestations of the Coulomb gas. The problem is how to do it if we attain a positive answer.

REFERENCES

1. J. Frohlich, Comm. Math. Phys., 47, 233 (1976); J. Frohlich and T. Spencer, J. Stat. Phys., 24, 4 (1981); Comm. Math. Phys., 81, 527 (1981). See also, M. Simões, The sine-Gordon Divergences, Preprint IFUSP, (1986).

LATTICE UNSTABILITIES OF MAGNETIC ORIGIN

D. Gottlieb and M. Lagos

Depto. de Física, Fac. de Ciencias, Universidad de Chile
Casilla 653, Santiago, Chile

Recent experiments [1-3] in the layered perovskites suggest that the spin lattice interaction may play an important role in the fenomenology of these compounds. Using a method which proved to be quite accurate for anisotropies towards the Ising limit, we solve the anisotropic Heisenberg antiferromagnet in which the spin–lattice interaction was modeled by adding to the antiferromagnetic coupling constant a linear dependence on the lattice configuration.

It was found that:
a) The spin–lattice interaction softens the elastic constants of the differents normal modes of the system. The softening is larger for the mode whose length is twice the lattice parameter.
b) At finite temperatures the lattice symmetry breaks and the system becomes dimerized, and
c) the dimerization vanishes exponentially as the temperatures approaches $T = 0$.

The calculation assumes a one dimensional model. However, the extension to higher dimensionalities is straitghtforward.

The calculation is a generalization for lattice deformation of arbitrary wave length of previous works done by the authors.[4-8]

References

1. D. Vaknin et al., Phys. Rev. Lett. **58**, 2802 (1987); T. Freltoft it et al., Phys. Rev. **B36**, 826 (1987).
2. G. Shirane et al., Phys. Rev. Lett. **59**, 1613 (1987).
3. G. Aeppli, D. J. Buttrey, Phys. Rev. Lett. **61**, 203 (1987).
4. D. Gottlieb, M. Lagos, Phys. Rev. **B39**, 2960 (1989).
5. D. Gottlieb, O. Maldonado, Solid State Comm. **70**, 495 (1989).
6. M. Lagos, G. G. Cabrera, Solid State Comm. **67**, 221 (1988.)
7. M. Lagos, G. G. Cabrera, Phys. Rev. **B38**, 659 (1988).
8. M. Lagos, M. Kiwi, E. R. Gagliano, G. G. Cabrera, Solid State Comm. **67**, 225 (1988.)

DUAL QUANTIZATION OF SOLITONS

E.C. Marino

Departamento de Física, Pontifícia Universidade Católica
C.P. 38071, 22452 Rio de Janeiro RJ, Brasil

INTRODUCTION

It is a remarkable fact that some systems contain states in their spectrum of excitations that cannot be created by the hamiltonian variables or polynomials of them. These states usually bear an identically conserved (topological) charge which characterizes some superselecting sectors of the Hilbert space. These excitations are usually called topological solitons.

In these lectures, we expose a method of quantization of topological excitations which relies on the observation that the soliton creation operator must be a disorder parameter of the system. Disorder variables were introduced long ago in Statistical Mechanics (SM) [1,2]. Later on it was realized [3] that the soliton operator sould be a disorder variable.

Consider, for instance a one dimensional system with hamiltonian $H = H[\sigma]$, $\sigma = \sigma(x,t)$. Suppose H has a classical symmetry $\sigma \to -\sigma$. Quantum mechanically there must exist an unitary operator $U = e^{iQ}$, such that $U\sigma U^{-1} = -\sigma$, $[H, U] = 0$. The expectation value of σ is an order parameter ($|0>$ ground state)

$$<\sigma> \equiv <0|\sigma|0> = <0|U^{-1}U\sigma U^{-1}U|0> = - <0|U^{-1}\sigma U|0> \qquad (1.1)$$

If the ground-state is unique, $U|0> = |0>$, it follows that $<\sigma> = 0$. If, on the other hand, $<\sigma> \neq 0$, then $U|0> = |0'>$, with $<0|\sigma|0> = - <0'|\sigma|0'>$. The ground-state is degenerate, indocating the occurrence of spontaneous symmetry breaking.

When $<\sigma> \neq 0$, $<0|\sigma|0> = \sigma_0$, $<0'|\sigma|0'> = -\sigma_0$, there may be static classical solutions such that $\sigma(x) \xrightarrow[x \to \pm\infty]{} \pm\sigma_0$. These solutions are called classical solitons. In order to describe the quantum states associated with these classical solutions we introduce a quantum oerator $\mu(x,t)$, such that $\mu|0> = |\text{soliton}>$. If $<\sigma> \neq 0$, then, $|\text{soliton}>$ is a genuine excitation and $<0|\text{soliton}> = 0$, implying that $<\mu> = 0$. If, on the other hand, $<\sigma> = 0$, there may be no solitons in the spectrum and $<0|\text{soliton}> = 1$. As a consequence, we must have $<\mu> \neq 0$ in this case. We see that, indeeed $<\mu>$ is a disorder parameter.

Kadanoff and Ceva determined the basic algebra that the order and disorder variables of a system must obey [2]. In the following chapters, we show how these methods were generalized for Quantum Field Theory (QFT) in $1+1$, $2+1$ and $3+1$ space-time dimensions to establish a general framework for the description of quantum solitons in a wide variety of systems, without resorting to any semiclassical approximation.

DUAL QUANTIZATION OF SOLITONS IN STATISTICAL MECHANICS

Consider the quantum $1d$ Ising Model in transverse field, described by

$$H = -J \sum_m \sigma_3(m)\, \sigma_3(m+1) - \sum_m \sigma_1(m), \qquad (2.1)$$

where $\sigma_i(m)$ are Pauli matrices acting on a Hilbert space $\mathcal{H}_n$. ($\mathcal{H} = \ldots \otimes \mathcal{H}_n \otimes \mathcal{H}_{n+1} \otimes \ldots$). The model possesses the symmetry $\sigma_3(n) \to -\sigma_3(n), \sigma_1(n) \to \sigma_1(n)$. This system has two phases [3], characterized by $<\sigma_3>$. For $J > 1 <\sigma_3> \neq 0$ and for $J < 1 <\sigma_3> \,= 0$. $J = 1$ is the critical point.

Let us define a dual lattice, for which the site n^* is placed between the sites n and $n+1$ of the original lattice.

Introduce now an operator $\mu_3(n^*)$, defined through the "Dual Algebra" [2,3]

$$\mu_3(n^*)\, \sigma_3(m) = \begin{cases} -\sigma_3(m)\, \mu_3(n^*) & n > m \\ \sigma_3(m)\, \mu_3(n^*) & n \leq m \end{cases} \qquad (2.2)$$

An operator realization for $\mu_3(n^*)$ is [3]

$$\mu_3(n^*) = \prod_{m > n} \sigma_1(m). \qquad (2.3)$$

It may be shown [3] that for $J < 1$, $<\mu_3> \neq 0$ while for $J > 1$, $<\mu_3> \,= 0$. We may see, further, that μ_3 is a soliton creation operator in the phase $J > 1$. Take one of the ground states for $J > 1$ (ordered phase), say all spins down: $|0_-\!>_{J \gg 1} = |\cdots \downarrow\downarrow\downarrow\downarrow\downarrow \cdots>$. It is clear that $\mu_3(n^*)|0_-\!>_{J \gg 1} = |\cdots \downarrow\downarrow\downarrow\uparrow\uparrow\uparrow \cdots>$. This state is a Bloch wall, the soliton of the Ising model.

We see that indeed the disorder operator of the system does create its topological excitations, namely, Bloch walls. In the ordered phase, we have $<\mu_3> \neq 0$ and the phase transition may be understood as a soliton condensation [3].

It was realized [3] that the model (2.1) possesses a classical discrete euclidean time version given by ($2d$ lattice)

$$H = -J \sum_{n,\mu} \sigma(n)\, \sigma(n + \hat{r}_\mu) \qquad (2.4)$$

where $\sigma(n) = \pm 1$ are spin variables, n are the lattice sites and $\hat{r}_\mu$, $\mu = 1,2$ are unitary vectors associated with the two directions in the lattice. The system possesses the symmetry $\sigma(n) \to -\sigma(n)$. The dual lattice is composed by sites n^*, which are placed at the center of each plaquette of the original lattice. The disorder variables $\mu(n^*) = \pm 1$ are then introduced on each site n^* of the dual lattice. As is shown in

[2], the disorder correlation function is given by

$$< \mu(n_1^*) \, \mu(n_2^*) > = \sum_{\{\sigma\}} e^{-H'[\sigma]} , \qquad (2.5)$$

where $H'[\sigma]$ has the same form as H in (2.1), except for the fact htat the coupling constant J is changed to $-J$ along an arbitrary curve C along the dual lattice sites connecting n_1^* and n_2^*. Observe that the curve C cuts the links of the original lattice and each link is associated with a pair of spin variables.

The disorder correlation-function may also be written in the form

$$< \mu(n_1^*) \, \mu(n_2^*) > = \sum_{\{\sigma\}} e^{-H[\sigma] - 2H_c[\sigma]} , \qquad (2.6)$$

where

$$H_c[\sigma] = \begin{cases} H[\sigma] & \text{for } \hat{r}_\mu \in C \\ 0 & \text{Otherwise} \end{cases} \qquad (2.7)$$

Observe that $H_c[\sigma]$ is a term being nonzero only on the curve C. As it is shown in [2], the disorder correlation functions (2.6) is independent of C. This can be easily seen by making the change of variable $\sigma(n) \to -\sigma(n)$ on sites n belonging to an arbitrary region R bounded by $\Gamma = C - C'$. As the sum $\sum_{\{\sigma\}}$ is invariant under this change of variable, one sees that $< \mu\mu >_C = < \mu\mu >_{C'}$.

For mixed functions, the prescription is:

$$< \mu(n_1^*) \, \sigma(m_1) \, \mu(n_2^*) \, \sigma(m_2) > = \sum_{\{\sigma\}} e^{-\{H[\sigma] - 2H_c[\sigma]\}} \sigma(m_1) \, \sigma(m_2) \qquad (2.8)$$

Path independence is attained now, except for a sign ambiguity which appears depending on the points m_1, and m_2 to belong or not the region R mentioned above.

In what follows, we will describe the generalization of the above ideas to continnum QFT in two, three and four spacetime dimensions, obtaining thereby a full quantum theory of solitons in each case.

DUAL QUANTIZATION OF SOLITONS IN $1+1\, d$ QUANTUM FIELD THEORY

Let us consider a QFT in $1+1$ space-time dimensions described by a lagrangian density $\mathcal{L} = (\phi, \partial_\mu \phi)$ which contains a local, complex scalar field $\phi(x)$. Let us suppose the theory possesses a global discrete symmetry $\phi(x) \to g\phi(x)$, where g belongs to some discrete group G. A remarkable fact about these kind of theories is that the Lagrangian field ϕ alone can not in general describe completely the theory. There may be excitations (sectors) of the Hilbert space which might never be created by the operation of polynomials of ϕ on the vacuum. The most well known examples are perhaps the Sine-Gordon solitons and ϕ^4-kinks in $1+1$ dimensions, the vortices of the abelian Higgs model in $2+1$ dimensions [4] and the magnetic monopoles in the non-abelian Higgs model in $3+1$ dimensions [5]. This phenomenon usually happens in spontaneously broken theories, where the vacuum has a non trivial structure.

Let us introduce, in the framework of the theory described by $\mathcal{L}(\phi, \partial_\mu \phi)$ a dual or disorder local field $\mu(x)$, through the dual algebra.

$$\mu(x)\,\phi(y) = \begin{cases} g\,\phi(y)\,\mu(x) & y^1 > x^1 \\ \phi(y)\,\mu(x) & y^1 < x^1 \end{cases} \quad (x-y)^2 < 0 \quad (3.1)$$

which is the generalization of the one described in the last section, appearing in the Ising model. Using our experience with statistical mechanical systems, we will assume that $\mu(x)$ is the field which creates the topological excitations of the theory. We will see that in well known models as the sine-Gordon for example, our assumption reproduces the standard results.

Based only on the algebra (3.1), one can demonstrate in the axiomatic QFT framework, very general theorems, regardless the form of $\mathcal{L}(\phi, \partial_\mu \phi)$ [6].

We showed for instance, that:
i) $<\phi><\mu>=0$
ii) $<\phi\mu>=0$
iii) whenever $<\phi>=0$ and $<\mu>=0$, the mass gap vanishes.

The first result shows that if $<\phi>\neq 0$, which means spontaneous symmetry breaking, then $<0|\mu|0>=0$, implying that μ acting on $|0>$ creates a state orthogonal to the vacuum, that is, μ is a true interpolating field in this case. When $<\mu>\neq 0$, on the other hand, which means that $\mu|0> \sim |0>$, in other words, that μ acting on the vacuum does not create a true particle state, the $<\phi>=0$ and the symmetry may not be broken.

The above results exhibit a deep relationship between the spectrum and the vacuum expectation values of order and disorder fields.

Let us discuss now how the ODD idea may be applied to the description of the quantum kinks of theories of the type $\mathcal{L} = \partial_\mu \phi^* \partial_\mu \phi - V(\phi, \phi^*)$, where ϕ is a complex local scalar field. Let us assume the theory has a $Z(N)$ symmetry $\phi(x) \to e^{i\alpha}\phi(x)$, with $\alpha = 2\pi/N$. Introducing the kink field operator through the dual algebra (3.1), we write the continuum generalization of the Kadanoff-Ceva formula (2.3) in euclidean space-time [7]:

$$<\mu(x)\,\mu^*(y)> = N^{-1} \int D\phi\,D\phi^* \exp - \int d^2z \{\mathcal{L} + \int_{x,c}^{y} \Psi_\mu \delta^2(z-\xi)\,\varepsilon^{\mu\nu} d\xi_\nu + \mathcal{L}_c\}, \quad (3.2)$$

where Ψ_μ is a functional of the fields and $\mathcal{L}_c$ is a path renormalization counterterm, both to be determined. N is the vacuum functional.

To find Ψ_μ and $\mathcal{L}_c$ we impose path independence, write $\int_C = \int_{C'} + \oint_{\Gamma = C - C'}$ and make a change of variables corresponding to the application of symmetry of the theory inside the region S, bounded by $\Gamma = C - C'$ [7]. V and $\mathcal{L}_c$ are assumed to be invariant under this change of variable. The orther terms transform as:

$$\partial_\mu \phi^* \partial^\mu \phi \to \partial_\mu \phi^* \partial^\mu \phi - i\alpha \phi^* \overleftrightarrow{\partial}_\mu \phi \partial^\mu \theta(S) + \alpha^2 \phi^* \phi \partial_\mu \theta(S)\,\partial_\mu \theta(S) \quad (3.3a)$$

$$\int_C \Psi_\mu \delta^2(z-\xi)\,\varepsilon^{\mu\nu}\,d\xi_\nu \to \left[\int_{C'} + \oint_\Gamma\right] \Psi_\mu\,\delta^2(z-\xi)\,\varepsilon^{\mu\nu}\,d\xi_\nu$$
$$+ \int_C \delta\Psi_\mu\,\delta^2(z-\xi)\,\varepsilon^{\mu\nu}\,d\xi_\nu \qquad (3.3b)$$

where $\delta\Psi_\mu$ represents the change of Ψ_μ under the above change of variable and the gradient of the Heaviside function is

$$\partial_\mu\,\theta(S) = \oint_{\Gamma=C-C'} \delta^2(z-\xi)\,\varepsilon_{\mu\nu}\,d\xi^\nu \qquad (3.4)$$

Imposing $<\mu\mu^*>_c = <\mu\mu^*>_{C'}$ and the cancelation of terms containing one closed contour integral, we get [7]:

$$\Psi_\mu = -i\alpha\phi^*\,\overleftrightarrow{\partial}_\mu\,\phi$$
$$\delta\Psi_\mu = 2\alpha^2\,\phi^*\phi\,\partial_\mu\,\theta(S) \qquad (3.5)$$

Inserting these expressions in the original equation, we find that

$$\mathcal{L}_S = \alpha\phi^*\phi \int_{x,c}^y d\xi^\mu \int_{x,c}^y d\eta_\nu\,\delta^2(z-\xi)\,\delta^2(z-\eta) \qquad (3.6)$$

ensures path independence.

The akward expression for $<\mu\mu^*>$ can be put in a nice form by introducing the external field.

$$A_\mu(z,C) = \int_{x,c}^y \varepsilon_{\mu\nu}\,\delta^2(z-\xi)\,d\xi^n\,u\,, \qquad (3.7)$$

and $D_\mu = \partial_\mu - i\alpha A_\mu$. In terms of A_μ, the previous expression for $<\mu\mu^*>$ becomes [7]:

$$<\mu(x)\,\mu^*(y)> = N^{-1}\int D\phi\,D\phi^*\,\exp-\int d^2z\left\{(D_\mu\phi)^*(D^\mu\phi) + V\right\} \qquad (3.8)$$

We see that path independence becomes now gauge invariance under the set of transformations

$$\phi(x) \longrightarrow e^{i\alpha\theta(S)}\,\phi(x) \qquad (3.9)$$
$$A_\mu(c) \to A_\mu(c) + \partial_\mu\theta(S) = A_\mu(c) + [A_\mu(c') - A_\mu(c)] = A_\mu(c')$$

From (3.8), we can draw an explicit operator realization for $\mu(x)$ in terms of ϕ in Minkowski space namely [7]

$$\mu(x) = \exp\left\{\alpha\int_{x,c}^\infty \varepsilon^{\mu\nu}\phi^*\,\overleftrightarrow{\partial}_\nu\,\phi\,d\xi_\mu\right\} \qquad (3.10)$$

This operator realizes our original dual algebra (3.1) and is the QFT generalization of the kink operator $\mu_3(n^*)$ of the Ising model, eq.(2.3).

125

The mixed function is given by

$$<\mu(x_1)\,\phi(y_1)\,\mu^*(x_2)\,\phi^*(y_2)> = N^{-1} \int D\phi D\phi^*\, e^{-S[\phi,A_\mu]} \phi(y_1)\,\phi^*(y_2) \quad (3.11)$$

Again, we find path independence except for an ambiguity $\exp \pm i\alpha$ which is nothing but a manifestation of the dual algebra in the functional integral.

For real fields one cannot couple the external field A_μ because the current identically vanishes. One would have to resort, in this case, to twisted boundary conditions, in order to simulate the presence of the external field.

Let us now apply our method to the computation of the kink correlation function in the case of φ^4-type theories.

A typical system would be described by the lagrangian density [8]

$$\mathcal{L} = \partial_\mu \phi^* \partial^\mu \phi - m^2 \phi^* \phi - \frac{\lambda}{4!}(\phi^4 + \phi^{*4}) - \frac{\mu}{(3!)^2}(\phi^*\phi)^3 \quad (3.12)$$

which possesses $Z(4)$ symmetry $\phi \to e^{i\frac{2\pi}{4}} \phi$. The system is shown to possess three phases, each of them associated with a different behavior of the soliton correlation functions:

i) symmetric phase

$$<\phi> = 0 \qquad <\mu(x)\,\mu^*(y)> \xrightarrow[|x-y|\to\infty]{} \text{const.}$$
$$<\mu> \neq 0 \qquad\qquad\qquad\qquad\qquad\qquad\qquad (3.13a)$$

that is μ is not a true kink creation operator,

ii) partially broken phase

$$<\phi> = 0 \qquad <\mu(x)\,\mu^*(y)> \xrightarrow[|x-y|\to\infty]{} |x-y|^{-\alpha}$$
$$<\mu> = 0 \qquad\qquad\qquad\qquad\qquad\qquad\qquad (3.13b)$$

and we see that in this case $\mu(x)$ creates massless kinks, in agreement with our general theorem [6],

iii) completely broken phase

$$<\phi> \neq 0 \qquad <\mu(x)\,\mu^*(y)> \xrightarrow[|x-y|\to\infty]{} \frac{e^{-|x-y|}}{|x-y|^\alpha}$$
$$<\mu> = 0 \qquad\qquad\qquad\qquad\qquad\qquad\qquad (3.13c)$$

that is, μ creates massive kinks with a mass $m \sim <\phi>^2$, in this case.

Let us apply now our method to a well known model, namely the sine-Gordon theory. We will see that it will reproduce the Coleman, Mandelstam [9] formulation of the sine-Gordon solitons. This was the first time, actually, that the Kadanoff-Ceva method was extended to continuum QFT [10].

The Lagrangian, given by

$$\mathcal{L} = \frac{1}{2} \partial_\mu \phi \partial_\mu - \alpha \cos \beta \phi \tag{3.14}$$

possesses the additive symmetry $\phi \to \phi + \frac{2\pi}{\beta}$. In order to apply our previous formalism, we define the following order variable

$$\sigma(x) = e^{i\frac{\beta}{2} \phi(x)} \tag{3.15}$$

The symmetry now becomes multiplicative: $\sigma(x) \to -\sigma(x)$. We introduce the disorder field as before, through the dual algebra (3.1), with $\sigma(x)$ instead of $\phi(x)$ and $g = -1$. We then write expression (3.2) for the disorder two point function. Imposing, then, path independence and going through the same steps as before, we find

$$\Psi_\mu = -\frac{2\pi}{\beta} \partial_\mu \phi$$

$$\delta \Psi_\mu = -\left(\frac{2\pi}{\beta}\right)^2 \partial_\mu \theta(S)$$

and

$$\mathcal{L}_c = \frac{1}{2} \left(\frac{2\pi}{\beta}\right)^2 \int_{x,c}^y d\xi^\mu \int_{x,c}^y d\eta_\mu \, \delta^2(z - \xi) \, \delta^2(z - \eta) \tag{3.16}$$

Observe that $\mathcal{L}_c$ is now field independent. From the expression of $<\mu\mu>$ and Ψ_μ, we can draw an explicit form for the soliton operator:

$$\mu(x,t) = \exp\left\{i \frac{2\pi}{\beta} \int_x^\infty \dot\phi(x,t) \, dz\right\}, \tag{3.17}$$

where we specialized to a curve belonging to the x^1 axis. Eq.(3.17) is the bosonic form of the soliton operator which realizes the original dual algebra.

If we define now

$$\Psi_1(x) = \sigma(x) \, \mu(x) \quad \text{and} \quad \Psi_2(x) = \sigma^*(x) \, \mu(x), \tag{3.18}$$

one easily finds that $\Psi(x) = \binom{\Psi_1}{\Psi_2}$ is a fermion field. Using the explicit forms of $\sigma(x)$ and $\mu(x)$, one gets

$$\Psi(x,t) = \exp i \left\{ \frac{\beta}{2} \gamma^5 \phi(x,t) + \frac{2\pi}{\beta} \int_x^\infty \dot\phi(z,t) \, dz \right\} \binom{1}{1}, \tag{3.19}$$

which is Mandelstam's [9] expression for the fermionic soliton operator. As was remarked in [11], the statistics of $1 + 1$ dimensional solitons is arbitrary.

We see, therefore, that through our method, we may reproduce the usual form of the sine-Gordon soliton operators. As it is well known, the dynamics of Ψ in (3.20) is determined by the massive Thirring model [9].

Let us discuss now a possible application of the above ideas to a realistic system of condensed matter. As a typical example, consider the linear ferromagnet $CsNiF_3$, in which the Ni magnetic atoms are aligned along linear weakly interacting chains. The magnetic hamiltonian may be written as

$$H = -J \sum_n \vec{S}_n \cdot \vec{S}_{n+1} + A \sum_n (S_n^z)^2 - g\mu_B B \sum_n S_n^x \qquad (3.20)$$

In this expressions, $\vec{S}_n$ is the spin of the nth Ni atom, B is a transverse magnetic field, μ_B is the Bohr magneton and g, the gyromagnetic facotr of Ni. The A term just reflects the anisotropy of the system: it forces the spins to remain orthogonal to the chain. A classical continuum theory for this system was developed in [12]. Choosing the z axis as coinciding with the chain and writing

$$\vec{S}_n = S(\cos\theta_n \cos\varphi_n, \cos\theta_n \sin\varphi_n, \sin\theta_n), \qquad (3.21)$$

where S is the spin modulus, φ is the azymuthal angle and θ, the angle between S and the xy plane, one may write H in the continuum limit ($\varphi_n \to \varphi(x,t)$, $\theta_n \to \theta(x',t)$) as

$$H = \int dx \left\{ \frac{1}{2}\left[(\partial_z \varphi)^2 + \frac{1}{C^2}\dot{\varphi}^2\right] + m^2(1-\cos\varphi) \right\} \qquad (3.22)$$

In the above expression ($\theta = \dot{\varphi}/2AS$)

$$m^2 = \frac{g\mu_B B}{JSa^2} \quad \text{and} \quad C = Sa(2JA)^{1/2}, \qquad (3.23)$$

where a is the lattice spacing.

As is well known, the sine-Gordon theory possesses classical soliton solutions. It was remarked in [12] that these will produce a peak in the neutron scattering cross section. This peak was claimed to be observed by Kjems and Steiner [13].

Another well know example of solitonic states in condensed matter is that of the defects in trans polyacetylene [14].

DUAL QUANTIZATION OF SOLITONS IN 2 + 1 DIMENSIONAL QUANTUM FIELD THEORY

Let us review now the method of soliton quantization based on the order disorder algebra, when applied to $2+1$ dimensional field theories. A prototype theory containing solitons in $2+1$ d is the relativistic Landau-Ginzburg model, described by

$$\mathcal{L} = -\frac{1}{4} F_{\mu\nu} F^{\mu\nu} + (D_\mu \phi)^* (D^\mu \phi) + m^2 \phi^* \phi - \frac{\lambda}{4}(\phi^*\phi)^2, \qquad (4.1)$$

where $D_\mu = \partial_\mu + ieA_\mu$.

This theory was shown [4] to contain classical static solutions of the form (vortices)

$$\phi(\vec{x}) = (\rho_0 + \rho(r))\, e^{i\arg(\vec{x})} \qquad \rho(r) \xrightarrow[r\to\infty]{} e^{-\alpha r}$$

$$\vec{A}(\vec{x}) = f(r)\, \hat{\theta} \qquad\qquad f(r) \xrightarrow[r\to\infty]{} \frac{1}{er} \qquad (4.2)$$

$$A_0(\vec{x}) = 0$$

where $r = |\vec{x}|$ and ρ_0 is the minimum of the potential $V(\phi)$ in (4.1). The asymptotic behavior of the vortex solution is

$$\phi(x) \xrightarrow[|\vec{x}| \to \infty]{} \rho_0 \, e^{-i \arg(\vec{x})} \qquad \vec{A}_i(x) \xrightarrow[|\vec{x}| \to \infty]{} \frac{1}{e} \partial_i \arg(\vec{x}) \qquad (4.3)$$

The topological current of this system is $J^\mu = \varepsilon^{\mu\alpha\beta} \partial_\alpha A_\beta$. The associated topological charge is

$$Q = \int d^2x \, J^0 = \oint \vec{A} \cdot d\vec{\ell} = \Phi_B \,, \qquad (4.4)$$

where Φ_B is the magnetic flux along the plane. It is easy to see that the vortex has $Q = \frac{2\pi}{e}$.

We are going to introduce now a fully quantized vortex creation operator through an algebra which is a generalization of (3.1). An important difference from the $1+1$ dimensional case now appears. Observe that the soliton operator in $1+1\,d$ was a local field and the algebra (3.1) is based on the concept of y being at the right or at the left of x. Since in $2+1$ dimensions these concepts no longer make sense, we arrive at the conclusion that an algebra which generalizes (3.1) should involve the local fields $\phi(x)$ and $\vec{A}(x)$ and a nonlocal vortex operator $\mu(c)$, defined on a certain curve C. The concept of being at the left or at the right is now exchanged by being inside on outside C. (investigations about local vortex operators were described in [15]. Related results on the lattice appeared in [16]).

In the same spirit of the algebra (3.1) and keeping in mind the asymptotic behavior (4.3), we introduce the vortex creation operator through the equal time commutation rules [17]

$$\mu(\vec{x},t;c)\,\phi(\vec{y},t) = \begin{cases} e^{i \arg(\vec{y}-\vec{x})} \, \phi(\vec{y},t)\,\mu(\vec{x},t;c) & \vec{y}-\vec{x} \notin T(c) \\ \phi(\vec{y},t)\,\mu(\vec{x},t;c) & \vec{y}-\vec{x} \in T(c) \end{cases} \qquad (4.5a)$$

and

$$\mu(\vec{x},t;c)\,A_i(\vec{y},t) = \begin{cases} [A_i(\vec{y},t) - \frac{1}{e}\partial_i^y \arg(\vec{y}-\vec{x})]\,\mu(\vec{x},t;c) & \vec{y}-\vec{x} \notin T(c) \\ A_i(\vec{y},t)\,\mu(\vec{x},t;c) & \vec{y}-\vec{x} \in T(c) \end{cases} \qquad (4.5b)$$

In these expressions C is a plane curve contained in the $t =$const. plane. $T(c)$ is the minimal surface bounded by C. $\vec{x}$ is a point belonging to $T(c)$ and characterizes the center of the vortex, i.e., the point in relation to which the angle $\arg(\vec{x}-\vec{y})$ is defined. This angle is measured with respect to an arbitrary direction characterized by a vector $\tilde{r}_0$. Without loss of generality we may choose $\tilde{r}_0$. pointing in the x^1 direction.

We now generalize for the present case our prescription for the computation of kink correlation functions in $1+1$ dimensions. Extending the arguments introduced in [7,10], we write in euclidean $3d$ space [17]

$$<\mu(x;c_1)\,\mu^*(y,c_2)> = Z^{-1}\int D\phi\,D\phi^*\,DA_\mu\,\exp\bigg\{-\int d^3z\Big[\mathcal{L}+\mathcal{L}(S,T,L)+$$
$$\int_{S(c_1c_2)} d^2\xi^\mu\,\Psi_\mu\,\delta^3(z-\xi)+\int_{x,L}^y d\eta^\mu\,\Phi_\mu(S)\,\delta^3(z-\eta)+\int_{T_1 U T_2} d^2\xi^\mu\,\chi_\mu\,\delta^3(z-\xi)\Big]\bigg\}$$
(4.6)

In this expression $S(C_1, C_2)$ is an arbitrary surface such that its boundary is $\partial S = C_1 U C_2$. T_1 and T_2 are the minimal surfaces bounded respectively by the plane curves c_1 and c_2. $d^2\xi^\mu$ are the surface elements along S, T_1 and T_2. L is an arbitrary curve connecting x and y and $d\eta^\mu$ is the line element along it. Ψ_μ, $\Phi_\mu(s)$ and χ_μ are functionals of the fields, to be determined (observe that we allow Φ_α to depend on S). $\mathcal{L}(S,T,L)$ is a renormalization counterterm, also to be determined, introduced in order to compensate for the eventual singularities coming from the line and surface terms. Z is the vacuum functional.

As in [7] and, we are going to determine Ψ_μ, Φ_μ, χ_μ and $\mathcal{L}(S,T,L)$, by imposing surface and path independence on (4.6). Let us take an arbitrary surface $S'(c_1,c_2)$ such that we also have $\partial S' = C_1 U C_2$ and call ΔV the volume bounded by $S U S'$. We will assume for simplicity that the surface S' is always exterior to S. Let us call Γ the closed surface made out off S and S' such that $\int_S = \int_{S'} - \oint_\Gamma (\partial(\Delta V) = \Gamma)$. Later on we will consider the most general case.

Let us perform now the following change in the functional integration variables inside ΔV:

$$\phi(z) \longrightarrow e^{-i\omega(z)\,\theta(\Delta V)} \qquad (4.7a)$$
$$A_\mu(z) \longrightarrow A_\mu(z)+\omega(z)\frac{1}{e}\partial_\mu\theta(\Delta V)+\theta(\Delta V)\frac{1}{e}\partial_\mu\omega(z) \qquad (4.7b)$$

($\theta(\Delta V)$ is the 3d Heaviside function with support on ΔV).

If we choose $\omega(z) = \alpha(z;x,y) \equiv \left[\theta(z^3-x^3)\arg(\vec{z}-\vec{x}) - \theta(z^3-y^3)\arg(\vec{z}-\vec{y})\right]$, corresponding to a vortex placed on $x = (\vec{x},x^3)$ and an anti-vortex placed on $y = (\vec{y},y^3)$, we see that the last term in (4.7b) corresponds to two Dirac strings going from $(\vec{x},x^3)$ to $(\vec{x},+\infty)$ and from $(\vec{y},+\infty)$ to $(\vec{y},y^3)$, respectively along the z^3 axis. Of course, this is equivalent by a $U(1)$ transformation to a configuration having a single dirac String going from x to y along an arbitrary curve. This configuration may be introduced by choosing $\omega(z) = \alpha_L(z;,x,y)$ in (4.7), where α_L is the obvious generalization of α defined by

$$F_{\mu\nu}^{\text{Dirac String}} = \frac{1}{e}\int_{x,L}^y \varepsilon_{\mu\nu\alpha}\,\delta^3(z-\xi)\,d\xi^\alpha = \frac{1}{2\pi e}[\partial_\mu,\partial_\nu]\,\alpha_L(z;x,y). \qquad (4.8)$$

In this expression, L is an arbitrary curve connecting x and y and $d\xi^\alpha$ is its line element. The $U(1)$ transformation which switches from one to another configuration of Dirac strings is given by (4.7) with $\omega = \alpha_L - \alpha$. Since L is arbitrary, we are going to choose it coinciding with the L in (4.6).

The only terms in (4.6) which are affected by the change of variable (4.7) (with

$\omega = \alpha_L$) are $F_{\mu\nu}$ and possibly Ψ_μ, Φ_μ and χ_μ. We assume $\mathcal{L}(S,T,C)$ to be invariant under it. An explicit computation taking in account the properties of the derivative of the Heaviside function shows that under (4.7) $F_{\mu\nu} \to F_{\mu\nu} + \tilde{F}_{\mu\nu}$, with

$$\tilde{F}_{\mu\nu} = -\frac{\alpha_L}{e}\left[\partial_\mu \oint_{\Gamma = S'-S} d^2\xi_\nu\, \delta^3(z-\xi) - (\mu \leftrightarrow \nu)\right] + \frac{1}{e}\theta(\Delta V)\left[\partial_\mu, \partial_\nu\right]\alpha_L \quad (4.9)$$

Introducing (4.9) in (4.6) (remenbering that $F_{\mu\nu} \to F_{\mu\nu} + \tilde{F}_{\mu\nu}$) and imposing the cancelation of all terms that spoil the surface invariance, we obtain [17]

$$\Psi^\mu = \frac{\alpha_L}{e}(z;x,y)\, F^{\mu\nu}\, \partial_\nu \quad (4.10a)$$

$$\Phi^\mu = \frac{1}{e}\theta(V_S)\, \epsilon^{\mu\nu\alpha}\, F_{\alpha\nu} \quad (4.10b)$$

$$\chi^\mu = -\frac{1}{e}\partial_\nu \alpha_L(z;x,y)\, F^{\mu\nu} \quad (4.10c)$$

$$\mathcal{L}(S,T,L) = \frac{1}{4}\tilde{F}_{\mu\nu}(S)\, \tilde{F}^{\mu\nu}(S) \quad (4.10d)$$

In (4.10d), $\tilde{F}_{\mu\nu}(S) = \partial_\mu \tilde{A}_\nu(S) - \partial_\nu \tilde{A}_\mu(S)$, where

$$\tilde{A}_\mu(S) = -\frac{\alpha_L}{e}(z;x,y)\int_{S(C_1,C_2)} d^2\xi_\mu\, \delta^3(z-\xi) + \frac{1}{e}\left[\theta(V_S)-1\right]\partial_\mu \alpha_L(z;x,y) \quad (4.11)$$

Observe that (4.9) may be written as $\tilde{F}_{\mu\nu} = \tilde{F}_{\mu\nu}(S') - \tilde{F}_{\mu\nu}(S)$.

Inserting (4.10) in the original expression (4.6), we may write, using (4.11)

$$<\mu(x;c_1)\mu^*(y,c_2)> = Z^{-1}\int D\phi\, D\phi^*\, DA_\mu$$
$$\times \exp\left\{-\int d^3z\left[\mathcal{L}(\tilde{F}_{\mu\nu} \to F_{\mu\nu} + \tilde{F}_{\mu\nu}(\tilde{A}_\mu(S)))\right]\right\} \quad (4.12)$$

It is clear now that surface invariance is just a consequence that under (4.7) $\tilde{A}_\mu(S) \to \tilde{A}_\mu(S')$ and $\tilde{F}_{\mu\nu}(S) \to \tilde{F}_{\mu\nu}(S')$.

Shifting the A_μ variable of functional integration as $A_\mu \to A_\mu - \tilde{A}_\mu(S)$, we get the equivalent expression

$$<\mu(x;c_1)\mu^*(y,c_2)> = Z^{-1}\int D\phi\, D\phi^*\, DA_\mu\, \exp\left\{-\int d^3z\left[\mathcal{L}(D_\mu\phi \to \tilde{D}_\mu\phi)\right]\right\}, \quad (4.13)$$

where $\tilde{D}_\mu = \partial_\mu + ie(A_\mu - \tilde{A}_\mu(S))$. Expressions (4.12) and (4.12) are our final result for the vortex two point function. Mixed correlation functions may be obtained by just introducing ϕ and A_μ fields in (4.12) and (4.13) in the usual way. Surface invariance for these functions is attained up to multiplicative factors $e^{\pm i\alpha_L(z;x,y)}$ for ϕ fields and additive factors $\frac{1}{e}\partial_\mu \alpha_L(z;x,y)$, for A_μ fields. These ambiguities are just a manifestation of the dual algebra (4.5) in the functional integral. By analytic continuation back to the Minkowski region, the different possibilities would correspond to the various operator orderings in the correlation function. Upper vortex correlation functions would be obtained by just introducing additional external fields $\widetilde{A}_\mu(S)$.

From (4.12) and (4.13), dropping the renormalization counterterms and analytically continuing to Minkowski space, we may extract two equivalent realizations for the vortex operator $\mu(x;c)$. We first use surface invariance and choose $S(c_1,c_2)$ in (4.12/13) as $S(c_1,c_2) = [\mathbb{R}_x^2 - T_1] U [\mathbb{R}_y^2 - T_2]$, where $\mathbb{R}_x^2$ is the plane at x^3=constant. We see, then, that V_S in (4.12/13) is the infinite slice between $\mathbb{R}_x^2$ and $\mathbb{R}_y^2$. Making the $U(1)$ gauge transformation which takes α_L back to α we see that the second term in (4.12/13) vanishes. With the choices above, we immediately see, from (4.12/13), that the vortex operator is given by (Minkowski space)

$$\mu(x;c) = \exp\left\{-i\int d^3z\, F^{\mu\nu}\partial_\mu \widetilde{A}_\nu(S_x)\right\} =$$
$$= \exp\left\{-\frac{i}{e}\int_{\mathbb{R}_x^2 - T_x} d^2z\, \arg(\vec{z}-\vec{x})\,\partial_i F^{io}(\vec{z},t)\right\}, \qquad (4.14a)$$

or

$$\mu(x;c) = \exp\left\{i\int d^3z\, j^\mu \widetilde{A}_\nu(S_x)\right\} =$$
$$= \exp\left\{\int_{\mathbb{R}_x^2 - T_x} d^2z\,[\phi^*(\vec{z},t)\pi^*(\vec{z},t) - \pi(\vec{z},t)\phi(\vec{z},t)]\arg(\vec{z}-\vec{x})\right\}(4.14b)$$

In the expressions above $j_\mu = ie[\phi^* D_\mu \phi - (D_\mu\phi)^*\phi]$, $\widetilde{A}_\mu(S_x) = -\frac{1}{e}\arg(\vec{z}-\vec{x})\int_{\mathbb{R}_x^2-T_x}\delta^3(z-\xi)d^2\xi_\mu$ ($\widetilde{A}_\mu(S) = \widetilde{A}_\mu(S_x) - \widetilde{A}_\mu(S_y)$) and $\pi = (D_0\phi)^* = \phi^* - ieA^\circ\phi^*$, is the momentum canonically conjugate to ϕ. That (4.14a) and (4.14b) are equivalent may be seen by the use of $\partial_\mu F^{\mu\nu} = j^\nu$ plus integration by parts.

Using the expansion for the operator product $e^A B e^{-A}$ it is straightforward to verify that $\mu(x;c)$ as given by (4.14) satisfies the dual algebra (4), from which we started.

Let us consider now the CP^1 model in $2+1$ dimensions, which is described by the action (euclidean)

$$S = \int d^3x\left[|D_\mu Z_1|^2 + |D_\mu Z_2|^2 + i\frac{2\theta}{\pi}J^\mu A_\mu\right], \qquad (4.15)$$

where Z_i, $i=1,2$ are complex scalar fields satisfying the constraint $|Z_1|^2 + |Z_2|^2 = 1$. $D_\mu = \partial_\mu + iA_\mu$, where A_μ is a $U(1)$ gauge field. A_μ acquires a kinetic term through radiative corrections to its self-energy. For $\theta = 0$, the A_μ field equation gives $A_\mu = iZ^+\partial_\mu Z$, where $Z = \binom{Z_1}{Z_2}$. The θ-term in (4.15) is the so called Chern-Simons term, in which $J^\mu = \frac{1}{2\pi}\varepsilon^{\mu\nu\alpha}\partial_\nu A_\alpha$ is the topological current. The CP^1 model possesses a $U(1)$ gauge symmetry.

The CP^1 model is equivalent to the $\mathcal{O}(3)$ Nonlinear σ-Model [18], given by

$$S = \int d^3x \frac{1}{2} \partial_\mu n^a \, \partial^\mu n^a + i\theta \, \widetilde{H} \; ; \qquad (4.16)$$

with $n^a n^a = 1$, $a = 1, 2, 3$, provided we make the identification $n^a = Z^+ \sigma^a Z$, σ^a being the Pauli matrices. In (4.16), $\widetilde{H}$ is the Hopf term, which was argued to be the Nonlinear σ-model version of the Chern-Simons term [19].

The above system contains static soliton solutions with $Q = \int d^2x \, J^0 \neq 0$, as was shown by Belavin and Polyakov [20]. In the Nonlienar σ-Model version, the soliton solution is given by

$$\vec{n}_s(\vec{x}, t) = \left(\sin f(r) \, \hat{x}, \, \cos f(r) \right) \; ; \; \vec{x} = r \, \hat{x} \, . \qquad (4.17)$$

In the CP^1 version, the solution reads

$$Z_S = \begin{pmatrix} \cos \frac{f(r)}{2} & e^{-(i/2) \arg(\vec{x})} \\ \sin \frac{f(r)}{2} & e^{(i/2) \arg(\vec{x})} \end{pmatrix} \qquad (4.18a)$$

$$A_i^S(\vec{x}, t) = \frac{1}{2} \cos f(r) \, \partial_i \arg(\vec{x}) \; ; \; A_0^S = 0 \qquad (4.18b)$$

For $\theta = 0$, an exact solution may be found for $f(r)$, namely

$$f(r) = 2 \arctan \frac{\lambda}{r} \; ; \; f(r) \xrightarrow[r \to 0]{r \to \infty} \begin{cases} 0 \\ \pi \end{cases} \qquad (4.19)$$

For $\theta \neq 0$ $f(r)$ has the same asymptotic behavior above.

Making a gauge transformation $Z_i \to e^{-i\Lambda} Z_i$; $A_\mu \to A_\mu + \partial_\mu \Lambda$, with $\Lambda = -\frac{1}{2} \arg(\vec{x})$, we may write

$$Z_S = \begin{pmatrix} \cos f/2 & e^{-i \arg(\vec{x})} \\ \sin f/2 & \end{pmatrix} \; ; \; A_i^S = \frac{1}{2} \left[\cos f(r) + 1 \right] \partial_i \arg(\vec{x}) \; ; \; A_0^S = 0 \quad (4.20)$$

We see that in the CP^1 version the soliton is vortex-like. These solitons were called skyrmions, in connection to the ones appearing in the Skyrme model [21].

Let us apply now the dual method of soliton quantization to the skyrmions of the CP^1/Nonlinear σ-Model [22]. We introduce, in analogy to the vortex case, a nonlocal skyrmion operator $\mu(x; c)$, through the dual algebra (equal times) [22]

$$\mu(x; c) \, Z_1(y) = \begin{cases} e^{-i/2 \, \arg(\vec{y} - \vec{x})} \, Z_1(y) \, \mu(x; c) & \vec{y} \notin T(c) \\ Z_1(y) \, \mu(x; c) & \vec{y} \in T(c) \end{cases} \qquad (4.21a)$$

$$\mu(x; c) \, Z_2(y) = \begin{cases} Z_2(y) \, \mu(x; c) & \vec{y} \notin T(c) \\ e^{i/2 \, \arg(\vec{y} - \vec{x})} \, Z_2(y) \, \mu(x; c) & \vec{y} \in T(c) \end{cases} \qquad (4.21b)$$

$$\mu(x; c) \, A_i(y) = \begin{cases} [A_i(y) + \frac{1}{2} \partial_i \arg(\vec{y} - \vec{x})] \, \mu(x; c) & \vec{y} \notin T(c) \\ [A_i(y) - \frac{1}{2} \partial_i \arg(\vec{y} - \vec{x})] \, \mu(x; c) & \vec{y} \in T(c) \end{cases} \qquad (4.21c)$$

In the above expressions C is a closed plane curve contained in the plane $t =$cosntant and $T(c)$ is the minimal surface bounded by it and containing $\vec{x}$.

We are going to determine the operator $\mu(x;c)$ satisfying the above algebra. In order to do that, we must in first place determine the basic commutators between the lagrangian variables Z_i and A_μ.

The system contains the following set of second class constraints

$$\varphi_1 = Z_i^+ Z_i - 1 \approx 0$$
$$\varphi_2 = \pi_i Z_i + \pi_i^* Z_i^* \approx 0$$
$$\varphi_3 = P_1 - \frac{\theta}{\pi^2} A_2 \approx 0$$
$$\varphi_4 = P_2 + \frac{\theta}{\pi^2} A_1 \approx 0$$
(4.22)

In the above equations, $\pi_i = \frac{\partial \mathcal{L}}{\partial \dot{Z}_i} = (D_0 Z_i)^*$ and $P_i = \frac{\partial \mathcal{L}}{\partial(\dot{A}_i)} = \frac{\theta}{\pi^2} \varepsilon^{ij} A_j$. The first class (gauge) constraints are implemented à la Gupta-Bleuler. Using the Method of Dirac, we obtain the following nonzero commutators

$$[Z_i, \pi_j] = i\hbar \left[\delta_{ij} - \frac{1}{2} Z_i Z_j^* \right] \delta^2(\vec{x} - \vec{y})$$
$$[Z_i^*, \pi_j] = \frac{i\hbar}{2} Z_i^* Z_j^* \, \delta^2(\vec{x} - \vec{y})$$
$$[\pi_i, \pi_j] = \frac{i\hbar}{2} [\pi_i Z_j^* - Z_i^* \pi_j] \delta^2(\vec{x} - \vec{y})$$
$$[\pi_i, \pi_j^*] = \frac{i\hbar}{2} [\pi_i Z_j - Z_i^* \pi_j^*] \delta^2(\vec{x} - \vec{y})$$
$$[A_i, A_j] = \frac{i\hbar}{2\pi^2 \theta} \varepsilon^{ij} \, \delta^2(\vec{x} - \vec{y})$$
(4.23)

Using these commutation rules and our previous experience with the vortex operator, we find the expression for $\mu(x;c)$ satisfying the dual algebra (4.21):

$$\mu(\vec{x},t;c) = \exp\left\{ -(1/2) \int_{\mathbb{R}^2 - T_x} d^2 x' \, \arg(\vec{x}' - \vec{x}) \left[Z_1^*(\vec{x}',t) \pi_1^*(\vec{x}',t) \right. \right.$$
$$\left. - \pi_1(\vec{x}',t) Z_1(\vec{x}',t) \right] + (1/2) \int_{T_x} d^2 x' \, \arg(\vec{x}' - \vec{x}) \left[Z_2^*(\vec{x}',t) \pi_2^*(\vec{x}',t) \right.$$
$$\left. - \pi_2(\vec{x}',t) Z_2(\vec{x}',t) \right] - (i\theta/\pi^2) \left[\int_{\mathbb{R}^2 - T_x} - \int_{T_x} \right] d^2 x'$$
$$\left. \arg(\vec{x}' - \vec{x}) \, \varepsilon^{0ij} \, \partial_i' A_j(\vec{x}',t) \right\} \equiv e^{B(\vec{x},t;c)}$$
(4.24)

We may write the skyrmion operator μ in the compact form

$$\mu(\vec{x},t;\,c) = \exp\left\{i\int d^3z\left[j_\mu^{(1)}\,\widetilde{A}^{(1)\mu} + j_\mu^{(2)}\,\widetilde{A}^{(2)\mu} - (4\theta/\pi)\,J^\mu\big(\widetilde{A}_\mu^{(1)} + \widetilde{A}_\mu^{(2)}\big)\right]\right\}, \quad (4.25)$$

by using the currents. $j_\mu^{(a)} = i\big[Z_a^+(D_\mu Z_a) - (D_\mu Z_a)^+ Z_a\big]$, $a = 1,2$, the topological current J^μ and defining the external fields

$$\widetilde{A}_\mu^{(a)}(z,x;T_x(c)) = (-1)^{a+1}(1/2)\,\arg(\vec{z}-\vec{x})\int_{\Omega_a} d^2\xi_\mu\,\delta^3(\xi-z)\,,\ a=1,2 \quad (4.26)$$

where $\Omega_1 = \mathbb{R}^2 - T_x$, $\Omega_2 = T_x$, $z = (\vec{z},z^0)$, $x = (\vec{x},t)$, $d^2\xi_\mu = \delta_{\mu 0}d^2\xi$ and $\xi^0 = t$.

Observe that in the dual algebra (4.21) as well as in our construction of μ, eqs. (4.24/26), only the asymptotic behavior of the classical configuration was used. This classical configuration, however, contains in addition the smeared out Heaviside functions $\cos f/2$, $\sin f/2$ and $\cos f$. Since we want to work with unsmeared fields, in the spirit of local field theory, we exchanged these smooth functions in equations (4.24/26) by the corresponding true Heaviside functions centered on C. This immediately allows us to interprete R, the radius of C, as a measure of the skyrmion size. The classical analog of R would be, then R_0, such that $f(R_0) = \pi/2$.

Let us evaluate now the equal time commutation relation of the skyrmion field. Taking eq.(4.24), using the equal time commutation relations (4.23) and the identity $\varepsilon^{ij}\partial_i\partial_j\arg(\vec{x}) = 2\pi\delta^2(\vec{x})$ which is a consequence of the Cauchy-Riemann equation for the function $\ell n\,x$, we find $[B(\vec{x},t),B(\vec{y},t)] = (i\theta/\pi)\big[\arg(\vec{x}-\vec{y}) - \arg(\vec{y}-\vec{x})\big]$. Using the Baker-Hausdorff formula and the fact that $\arg(x) - \arg(-x) = \pi$ we immediately see that $\mu(\vec{x},t;c_1)\mu(\vec{y},t;c_2) = \mu(\vec{y},t;c_2)\mu(\vec{x},t;c_1)e^{i\theta}$. It is clear that for $\theta = 0$ the skyrmion is a boson whereas for $\theta = \pi$ it is a fermion. For other values of θ it obeys a generalized statistics ($0 \leq \theta < 2\pi$) as was first observed in [19].

Let us consider now the correlation function $<\mu(\vec{x},x^0;c_1)\mu^*(\vec{y},y^0;c_2)>$. This is most conveniently expressed in the functional integral framework. Of course we expect the occurence of divergences associated with the time localized infinite surfaces in (4.25/26). These divergences also appear in the case of vortices and kinks and as in those cases, they may be eliminated by the introduction of counterterms whose explicit form is determined by the requirement of surface invariance of $<\mu\mu^*>$ [7,10,17]. Inserting (4.25) in the euclidean functional integral

$$<\mu\mu^*> = Z^{-1}\int\prod_{a=1}^{2}DZ_a\,DZ_a^*\,DA_\mu\,e^{-[S[Z_a,A_\mu]+S_{count}]}\mu\mu^*\delta\big[|Z|^2-1\big]\,, \quad (4.27)$$

and introducing the appropriate counterterms, we obtain [22]

$$<\mu(x;c)\,\mu^*(y;c)> = Z^{-1}\int DA_\mu\,DZ_a\,DZ_a^*\,\exp\Bigg\{-\int d^3z$$
$$\Big[|[\partial_\mu + i(A_\mu + \widetilde{A}_\mu^{(1)})]\,Z_1|^2 + |[\partial_\mu + i(A_\mu + \widetilde{A}_\mu^{(2)})]\,Z_2|^2$$
$$i\,\frac{2\theta}{\pi}\,J_\mu A^\mu + \frac{i2\theta}{\pi}\,J^\mu[\widetilde{A}_\mu^{(1)} + \widetilde{A}_\mu^{(2)}]\big) + \frac{i2\theta}{\pi}\,[\widetilde{J}_{(1)}^\mu + \widetilde{J}_{(2)}^\mu]A_\mu\Big]\Bigg\}\times$$

$$\delta\bigl[|Z|^2 - 1\bigr] \tag{4.28}$$

In this expression, $\widetilde{A}_\mu^{(a)} = \widetilde{A}_\mu^{(a)}(z;x) - \widetilde{A}_\mu^{(a)}(z;y)$; $a = 1,2$.

One may easily see that surface invariance is a consequence of gauge invariance. The θ-terms are surface invariant except for a term

$$I_\theta = -\int d^3z \left[i\, \frac{2\theta}{\pi}\, (\widetilde{J}_\mu^{(1)} + \widetilde{J}_\mu^{(2)})(\widetilde{A}^{(1)\mu} + \widetilde{A}^{(2)\mu})\right] \tag{4.29}$$

It happens that

$$I_\theta = \begin{cases} 0 \\ i\theta \end{cases} \tag{4.30}$$

depending on the surface chosen. This ambiguity is similar to the ones found previously in various systems. As we saw, it must reflect the commutation rule for the μ's. Indeed, taking the operator $\mu(x;c)$ and using the basic commutators (4.23), we find (equal times)

$$\mu(x;c)\, \mu(y;c) = e^{i\theta}\, \mu(y;c)\, \mu(x;c) \tag{4.31}$$

For $\theta \neq 0, \pi$, the skyrmions obey generalized statistics in agreement with the semiclassical analysis of Wilczek and Zee [19].

Using the relation $n^a = Z^+ \sigma^a Z$, we may evaluate the commutation rules of the skyrmion field μ with the nonlinear σ-field:

$$\mu(x;c)\, n^{\pm}(y) = e^{\pm i\, \arg(\vec{y}-\vec{x})}\, n^{\pm}(y)\, \mu(x;c) \tag{4.32}$$
$$[\mu(x;c),\, n^3(y)] = 0$$

where $n^{\pm} = n^1 \pm in^2$. We see that μ is dual to $n^{\pm}$.

Let us show now an interesting property the skyrmions have at $\theta = 0$. For this value fo θ, let us make the shift $A_\mu \to A'_\mu = A_\mu + \widetilde{A}_\mu^{(1)}$ in the functional integral (4.28). The effect of this (at $\theta = 0$) is that $\widetilde{A}_\mu^{(1)}$ decouples from Z_1 and the new external field coupled to Z_2 becomes $\widetilde{A}_\mu^{(2)} - \widetilde{A}_\mu^{(1)}$. It happens that $\widetilde{A}_\mu^{(2)} - \widetilde{A}_\mu^{(1)}$ is a pure gauge:

$$\widetilde{A}_\mu^{(2)} - \widetilde{A}_\mu^{(1)} = \partial_\mu \Lambda$$

$$\Lambda = \frac{1}{2}\Bigl[\arg(\vec{z}-\vec{x})\, \theta(z^3 - x^3) - \arg(\vec{z}-\vec{y})\, \theta(z^3 - y^3)\Bigr] \tag{4.33}$$

we may therefore completely eliminate the external field from (4.28) through a gauge transformation, obtaining the result

$$<\mu\mu^*>_{\theta=0}\, = 1 \tag{4.34}$$

The above argument, of course, may be extended to show that for an arbitrary correlation funciton of the μ's, we have

$$<\mu\ldots\mu^*\ldots>_{\theta=0} = 1 \qquad (4.35)$$

The skyrmions condense at $\theta = 0$. This fact, combined with the observation that the skyrmion is dual to the transversal components of $\vec{n}$, e.(4.32) leads us conclude that $<n^{\pm}>_{\theta=0} = 0$. This is still compatible with ordering in the 3-direction.

Let us discuss now the possible relevance of the above field theory models for realistic systems of condensed matter, namely High-T_c superconductors.

Let us take the typical compound La_2CuO_4, which becomes superconductor upon doping, say with Barium: $La_{2-\delta}Ba_\delta CuO_4$. There is evidence that superconductivity occurs in the CuO_2 planes, which contain C^{++} and O^{--} atoms, respectively in the configurations $3d^9$ and $2p^6$. Cu^{++} has spin $S = 1/2$ and one active electron and orbital. O^{--} has a perfect gas configuration, having therefore $S = 0$ and no active electron. The Cu^{++} are on the sites and the O^{--} on the links of a square lattice. The standard model for the system is the two dimensional Hubbard model, described by

$$H = t\sum_{<ij>,\sigma}(\psi_{i\sigma}^+\psi_{j\sigma} + H.C.) + U\sum_i(\psi_{i\uparrow}^+\psi_{i\uparrow})(\psi_{i\downarrow}^+\psi_{i\downarrow}) . \qquad (4.36)$$

In this expression $\psi_{i\alpha}^+$ is the creation operator for the active electron of Cu^{++}, with spin $\alpha = \pm 1/2$. Of course, there is one electron per site, corresponding a half filled band. It can be shown [23] that in the half-filling case and at strong coupling $U \gg t$, the above model is equivalent to the $\mathcal{O}(3)$ antiferromagnetic Heisenberg model, described by

$$H = J\sum_{<ij>}\vec{S}_i \cdot \vec{S}_j , \qquad (4.37)$$

where $J = 4t^2/U$ and $\vec{S}_i = \psi_{i\alpha}^+\vec{\sigma}_{\alpha\beta}\psi_{i\beta}$, $\vec{\sigma}$ being the Pauli matrices.

Recent Monte Carlo simulations indicate that the ground state of this system is an ordered (Néel) State [24]. This is in agreement with neutron scattering experiments [25].

The continuum limit of the model described by (4.37) has been studied in terms of the plaquette "staggered spin"

$$\vec{S}_{PL} = \frac{\vec{S}_a - \vec{S}_b + \vec{S}_c - \vec{S}_d}{4\sqrt{S(S+1)}} , \qquad (4.38)$$

where a, b, c, d are the vertices of a given lattice plaquette. In the continuum limit $\vec{S}_{PL} \to \vec{n}(\vec{x},t)$, with $|\vec{n}|^2 = 1$. It may be shown, then, that the continuum limit of (4.37) gives a field theory in $2+1$ dimensions whose action is [26]

$$S = \int d^3x \, \frac{1}{2} \partial_\mu n^a \, \partial^\mu n^a , \qquad (4.39)$$

that is, the Nonlinear σ-Model.

Last year it was suggested that a topological (Hopf) term could appear in (4.39) [27]. More recently, it was argued by many authors that $\theta = 0$ [28].

We see that the nonlinear σ-field can be thought of as the continuum limit of the staggered spin of the Cu^{++} atoms of the superconducting (upon doping) material La_2CuO_4.

In the this language, we see that our founding that $<\mu>_{\theta=0}=1$ implies that the transverse components of the staggered spin must be disordered

$$<S_x>=<S_y>=0 \qquad (4.40)$$

This is compatible with Z ordering. The priviledged role of the Z component may be understood as a consequence of the fact that the classical skyrmion solution (4.17/18), upon which we constructed our algebra was obtained by assuming a z-axis ordered ground state.

It would be interesting to investigate what would be the field theory associated with the dopped system and how could one describe the superconductivity mechanism in the continuum limit.

DUAL QUANTIZATION OF SOLITONS IN $3+1\, d$ QUANTUM FIELD THEORY

The prototype thoery containing solitons in this case is the $SO(3)$ Georgi-Glashow model [29], given by

$$\mathcal{L} = -\frac{1}{4} F^a_{\mu\nu} F^{\mu\nu a} + \frac{1}{2} (D_\mu \Phi)^T (D^\mu \Phi) + \frac{m^2}{2} \Phi^a \Phi^a - \frac{\lambda}{4!} (\Phi^a \Phi^a)^2 \qquad (5.1)$$

The fields A^a_μ, Φ^a, $a = 1, 2, 3$ are chosen to transform under the adjoint representation of $SO(3)$. The field streght tensor and covariant derivatives are given by

$$F^a_{\mu\nu} = \partial_\mu A^a_\nu - \partial_\nu A^a_\mu + e\,\varepsilon^{abc} A^b_\mu A^c_\mu \qquad (5.2a)$$

and

$$(D_\mu)^{ab} = \partial_\mu \delta^{ab} + \varepsilon^{abc} A^c_\mu \,. \qquad (5.2b)$$

The system possesses a topological current

$$J^\mu = \varepsilon^{\mu\nu\alpha\beta}\, \partial_\nu F^a_{\alpha\beta}\, \frac{\Phi^a}{|\vec{\Phi}|} \qquad (5.3)$$

Defining an electromagnetic field

$$F^{\mu\nu} \equiv F^{\mu\nu}_a\, \frac{\Phi^a}{|\vec{\Phi}|}\,, \qquad (5.4)$$

we find the topological charge

$$Q = \int d^3x J^0 = \oint \vec{B} \cdot d\vec{s}\,, \qquad (5.5)$$

where $\vec{B}$ is the magnetic field associated with $F_{\mu\nu}$. We see that the topological charge is nothing but a magnetic charge.

The lagrangian (5.1) possesses a local $SO(3)$ symmetry ($A_\mu \equiv A^a_\mu T^a$, T^a, $a=$

1, 2, 3 are the $SO(3)$ generators)

$$\Phi^a \xrightarrow[g]{} g^{ab}\Phi^b \equiv g \circ \Phi$$
$$A_\mu \xrightarrow[g]{} g A_\mu g^{-1} + \frac{i}{e} \partial_\mu g g^{-1} \equiv g \circ \Phi \quad (5.6)$$

with $g(x) \in SO(3)$.

In 74, 't Hooft and Polyakov [5] showed that this model possesses classical soliton solutions with $Q = \frac{2\pi}{e}$. Choosing the ground estate as

$$\Phi_v^a = \phi_0\, \delta^{a3}, \quad A_{\mu,v} = 0, \quad (5.7)$$

the soliton solution has the asymptotic behavior

$$\Phi^a(\vec{x},t) \xrightarrow[|\vec{x}|\to\infty]{} [\bar{g} \circ \Phi_v]^a = \phi_0 \frac{x^a}{|\vec{x}|}$$

$$A_\mu^a(\vec{x},t) \xrightarrow[|\vec{x}|\to\infty]{} [\bar{g} \circ A_\mu]^a = \begin{cases} \frac{1}{e} \varepsilon^{aib} \frac{x^b}{|\vec{x}|^2} & \mu = i = 1,2,3 \\ 0 & \mu = 0 \end{cases} \quad (5.8)$$

where $\bar{g}(\omega) = e^{-i\omega_0^a(x)\, T^a}$, $\omega_0^a = (-\theta \sin\varphi,\ \theta \cos\varphi,\ 0)$, θ, φ being the angles of the spherical coordinate system.

Since $Q \neq 0$, the 't Hooft-Polyakov soliton was called magnetic monopole.

A quantum theory of magnetic monopoles based on the order disorder duality ideas was recently establieshed [30]. Since the space is now three dimensional, we must have $\mu = \mu(x;s)$, s being a 2-surface. A dual algebra for the monopole may them be constructed [30], taking in account the asymptotic behavior (5.8) (equal times):

$$\mu(x;s)\, \Phi^a(y) = \begin{cases} [\bar{g} \circ \Phi]^a(y)\, \mu(x;s) & \vec{y} \notin T(s) \\ \Phi^a(y)\, \mu(x;s) & \vec{y} \in T(s) \end{cases} \quad (5.9a)$$

$$\mu(x;s)\, A_\mu^a(y) = \begin{cases} [\bar{g} \circ A_\mu]^a(y)\, \mu(x;s) & \vec{y} \notin T(s) \\ A_\mu^a(y)\, \mu(x;s) & \vec{y} \in T(s) \end{cases} \quad (5.9b)$$

In the above expressions, $T(S)$ is the minimal volume bounded by S, contained in the hiperplane $t =$ constant and containing $\vec{x}$.

As in the previous cases, we may find expressions for the correlation functions of $\mu(x;s)$ and in particular an operator realization, which reads [31]

$$\mu(x;s) = e^{R(x;S)}$$

with

$$R(x;s) = \frac{1}{ie} \int_{\mathrm{I\!R}^3 - T(s)} d^3\xi (D_i)^{ab}\, F^{iob}(\vec{\xi},t)\, \omega_{\vec{x}}^a(\vec{\xi}) \ . \quad (5.10)$$

In this expression, $\omega_{\vec{x}}^a(\vec{\xi})$ contains the angles θ,φ of $\vec{\xi}$ measured with respect to $\vec{x}$.

REFERENCES

1. H.A. Kramers and G.H. Wannier, *Phys. Rev.* **60** (1941) 252
2. L.P. Kadanoff and H. Ceva, *Phys. Rev.* **B3** (1978) 3918
3. E. Fradkin and L. Susskind, *Phys. Rev.* **D17** (1978) 2637; J.B. Kogut, *Rev. Mod. Phys.* **51** (1979) 659
4. II.B. Nielsen and P.Olesen, *Nucl. Phys.* **B61** (1973) 45
5. G.'t Hooft, *Nucl. Phys.* **B79** (1974) 276; A.M. Polyakov, *JETP Lett.* **20** (1975) 194
6. R. Köberle and E.C. Marino, *Phys. Lett.* **126B** (1983) 475
7. E.C. Marino, B. Schroer and J.A. Swieca, *Nucl. Phys.* **B200** [FS4] (1982) 499
8. E.C. Marino, *Nucl. Phys.* **B217** (1983) 413; *Nucl. Phys.* **B230** [FS10] (1984) 149; A.A.S. de Macedo and E.C. Marino, *Phys. Rev.* **D** (1989), in press
9. S. Coleman, *Phys. Rev.* **D11** (1975) 2088; S. Mandelstam, *Phys. Rev.* **D11** (1975) 3026
10. E.C. Marino and J.A. Swieca, *Nucl.Phys.* **B170** [FS1] (1980) 175
11. B. Schroer and J.A. Swieca, *Nucl. Phys.* **B170** [FS1] (1980) 175
12. H.J. Mikeska, *J. Phys.* **C11** (1988) L29
13. Kjems, Steiner, *Phys. Rev. Lett.* **41** (1978) 1137
14. W.P. Su, J.R. Schrieffer and A.J. Heeger, *Phys. Rev. Lett.* **42** (1979) 1698; *Phys. Rev.* **B22** (1980) 2099; H. Takayama, Y.R. Lin-Liu and K. Maki, *Phys. Rev.* **B21** (1980) 2388; D.K. Campbell and A.R. Bishop, *Nucl. Phys.* **B200** [FS4] (1982) 297
15. Z.F. Ezawa, *Phys. Rev.* **D18** (1978) 2091; *Phys. Lett.* **82B** (1979) 426; K. Bardakci and S. Samuel, *Phys. Rev.* **D18** (1978) 2849; *Phys. Rev.* **D19** (1979) 2357
16. J. Fröhlich and P.A. Marchetti, *Europhys. Lett.* **2** (1986) 933; *Communn. Math. Phys.* **112** (1978) 343
17. E.C. Marino, *Phys. Rev.* **D38** (1988) 3194
18. R. Rajaraman, "Solitons and Instantons", North Holland, Amsterdam, 1982
19. F. Wilczek and A. Zee, *Phys. Rev. Lett.* **51** (1983) 2250
20. A.A. Belavin and A.M. Polyakov, *JETP Lett.* **22** (1975) 245
21. T. Skyrme, *Proc. R. Soc. Lond* **262** (1961) 237
22. K. Furuya and E.C. Marino, PUC preprint (1989)
23. J. Hirsch, *Phys. Rev. Lett.* **54** (1985) 1317; S. Kivelson, D.S. Rokshar and J.P. Sethna, *Phys. Rev.* **B35** (1987) 8865; A.E. Ruckenstein, J.P. Hirschfeld and J. Appel, *Phys. Rev.* **B36** (1987) 857
24. E. Manousakis and R. Salvador, *Phys. Rev. Lett.* **62** (1989) 1310; *Phys. Rev.* **B39** (1989) 575
25. D. Vaknin et al, *Phys. Rev. Lett.* **58** (1987) 2802; G. Shirane et al, *Phys. Rev. Lett.* **59** (1987) 1613; Y. Endoh et al, *Phys. Rev. Lett.* **37** (1988) 7443
26. F.D.M. Haldane, *Phys. Lett.* **A93** (1983) 464; *Phys. Rev. Lett.* **50** (1983) 1153
27. I.E. Dzyaloshinskii, A.M. Polyakov, P.B. Wiegmann, *Phys. Lett.* **A127** (1988) 112; P.B. Wiegmann, *Phys. Lett.* **A127** (1988) 112; A.M. Polyakov, *Mod. Phys. Lett.* **A3** (1988) 325; P.B. Wiegmann, *Phys. Rev. Lett.* **60** (1988) 821
28. T. Dombre and N. Read, *Phys. Rev.* **B38** (1988) 7181; E. Fradkin and M. Stone, *Phys. Rev.* **B38** (1988) 7181; L.B. Ioffe and A.I. Larkin, *Int. J. Mod. Phys.* **B2** (1988) 203; X.G. Wen and A. Zee, *Phys. Rev.Lett.* **61** (1988) 1025; F.D.M. Haldane, *Phys. Rev. Lett.* **61** (1988) 1029
29. H. Georgi and S. Glashow, *Phys. Rev. Lett.* **32** (1974) 438
30. E.C. Marino and J. Stephany Ruiz, *Phys. Rev.* **D39** (1989), in press.

VARIATIONAL APPROACH

TO QUANTUM STATISTICAL MECHANICS

Riccardo Giachetti [a], Valerio Tognetti [b] and Ruggero Vaia [c]

[a] Dipartimento di Matematica, Università di Cagliari
Via Ospedale 72, I-09124 Cagliari, Italy

[b] Dipartimento di Fisica, Università di Firenze
Largo E. Fermi 2, I-50125 Firenze, Italy

[c] Istituto di Elettronica Quantistica - CNR
Via Panciatichi 56/30, I-50127 Firenze, Italy

INTRODUCTION AND SUMMARY

The path-integral formulation of quantum statistical mechanics has been accomplished several years ago by Feynman [1], who extended his treatment of quantum mechanics propagator to imaginary times, in order to give the expression for the density operator in the coordinate representation. This approach gives useful tools to reduce quantum statistical mechanics calculations to classical ones, in order to use again the configurational integral and eventually the phase-space integral for the evaluation of the partition function. This turns out to be useful expecially for numerical applications and can be alternative with the Wigner [2] expansion.

In order to get an effective potential, to be inserted in the configurational integral, Feynman introduced the variational approach based on the possibility to exactly calculate the path-integral of a free-particle. This approach was improved by two of us [3] taking into account all the quantum effects of the harmonic part of the potential, while the variational principle, in the first cumulant approximation, is used to account for the quantum corrections due to the anharmonic part. Applications to some one-dimensional non-linear fields were also given [3-4]. The results for the case of one particle in an anharmonic potential were later recovered also by Feynman and Kleinert [5]. Recently, we have generalized the method for Hamiltonian systems where the mixed "pq" terms are present [6] in order to construct an effective Hamiltonian using again the phase-space.

This paper is organized as follows. In sect.1 we review the path-integral formulation of the quantum statistical mechanics. In sect.2, an outline of the variational method is presented, confined to the case of one particle in an anharmonic potential. In sect.3 we present the effective potentials for non-linear Klein-Gordon fields and we evaluate them in sect.4, by a low-coupling expansion. Finally, in sect.5, we use the variational method for obtaining the effective Hamiltonian, starting from the Hamiltonian operator for the system of interest.

1. PARTITION FUNCTION AS PATH-INTEGRAL

The starting point is the path-integral formulation of the quantum mechanics introduced by Feynman [1]. The probability for a system to evolve in the state x at time t, being in the state x' at the initial time $t = 0$ is given by the following path-integral from $x(0) = x'$, $x(t) = x$:

$$K(x, x') = \int \mathcal{D}[x(\tau)] \, e^{i/\hbar \int_0^t d\tau L(x, dx/d\tau, \tau)} , \qquad (1.1)$$

where L is the Lagrangian density of the system. For the moment, in order to calculate the partition function, we limit ourselves to the case in which the kinetic energy is a quadratic form in $\dot{x}$ and is separated from the configurational energy. In this case the Lagrangian approach appears to be useful. Let us now remember the expression of the unnormalized density matrix of the system in thermal equilibrium at temperature $T \equiv \beta^{-1}$:

$$\rho(x, x') = < x | e^{-\beta H} | x' > . \qquad (1.2)$$

The partition function is $Z \equiv e^{-\beta F} = \int dx \rho(x, x)$, where F is the free-energy.

Performing the Wick rotation to the imaginary time $\tau = iu$, $t = i\beta\hbar$, we obtain the free energy F in terms of a path-integral calculated over all closed paths, i.e. $x(0) = x(\beta\hbar)$:

$$e^{-\beta F} = \int \mathcal{D}[x(u)] \, e^{-S[x(u)]} . \qquad (1.3)$$

In this expression $S[x(u)]$ is the euclidean action:

$$S[x(u)] = \frac{1}{\hbar} \int_0^{\beta\hbar} du \, L_E(x, \dot{x}) . \qquad (1.4)$$

In the following we prefer to use an equivalent expression. Introducing the average point of each path:

$$\bar{x}[(x(u)] = \frac{1}{\beta\hbar} \int_0^{\beta\hbar} du \, x(u) , \qquad (1.5)$$

we can first sum on classes of paths having the same average y, giving:

$$e^{-\beta F} = \int_{-\infty}^{+\infty} dy \int_{\bar{x}=y} \mathcal{D}[x(u)] \, e^{-S[x(u)]} . \qquad (1.6)$$

In order to outline the general method, we will describe in detail the case of a single particle in an anharmonic potential $V(x)$, whose euclidean action is:

$$S[x(u)] = \frac{1}{\hbar} \int_0^{\beta\hbar} du \left[\frac{1}{2} m \dot{x}^2(u) + V(x(u)) \right] . \qquad (1.7)$$

The classical limit can be easily obtained: for $\beta\hbar \to 0$ the relevant paths are very close to their average point, so that $V(x(u)) \simeq V(y)$ and the path-integral involves the only kinetic term, giving

$$e^{-\beta F} = \left(\frac{m}{2\pi\hbar^2 \beta} \right)^{1/2} \int_{-\infty}^{+\infty} dy \, e^{-\beta V(y)} , \qquad (1.8)$$

i.e. the usual classical configurational integral.

2. VARIATIONAL METHOD

The variational method starts from the Feynman-Jensen inequality [1]

$$F \leq F_0 + \frac{1}{\beta} < S - S_0 >_{S_0} , \qquad (2.1)$$

where S_0 is a trial action and F_0 the correspondent free energy. Since the path integral is *exactly computable for quadratic actions*, we choose as trial action:

$$S_0[x(u)] = \frac{1}{\hbar}\int_0^{\beta\hbar} du \left[\frac{m}{2}\dot{x}^2(u) + w(\bar{x}) + \frac{m}{2}\omega^2(\bar{x})(x(u)-\bar{x})^2\right] . \qquad (2.2)$$

Here $w(y)$ and $\omega^2(y)$ are variational parameters to be determined by minimizing the right hand side of (2.1). Feynman [1] does not introduce the last term, which is crucial in exactly accounting for the *quantum harmonic effects*.

The minimization gives:

$$<S - S_0>_{S_0} = 0 \qquad (2.3)$$

and hence

$$e^{-\beta F} \simeq e^{-\beta F_0} = \left(\frac{m}{2\pi\hbar^2\beta}\right)^{1/2} \int_{-\infty}^{+\infty} dy\, e^{-\beta V_{\text{eff}}(y)} , \qquad (2.4)$$

where

$$V_{\text{eff}}(y) = \int_{-\infty}^{+\infty} dz\, V(y+z)\frac{e^{-z^2/2a(y)}}{\sqrt{2\pi a(y)}} - \frac{ma(y)\omega^2(y)}{2} - \frac{1}{\beta}\ln\frac{f(y)}{\sinh f(y)} , \qquad (2.5)$$

$f(y) \equiv \beta\hbar\omega(y)/2$ is the dimensionless parameter related to the quantum character of the system; $\omega(y)$ is the solution of the self consistent equations:

$$m\omega^2(y) = \int_{-\infty}^{+\infty} dz\, V''(y+z)\frac{e^{-z^2/2a(y)}}{\sqrt{2\pi a(y)}} \qquad (2.6)$$

$$a(y) = \frac{\hbar}{2m\omega(y)}\left(\coth f(y) - \frac{1}{f(y)}\right) . \qquad (2.7)$$

Let us comment the results of eqs.(2.4-2.7):
i) $a(y)$ rules the quantum spread of the potential, and represents the difference between the total mean square fluctuation and the classical one, calculated in the one-loop approximation. Its behaviour varies from that of a free particle, as in the original Feynman approach, for $\beta \to 0$ [$a = \hbar^2\beta/12m$] to an harmonic oscillator in the ground state for $\beta \to \infty$ [$a = \hbar/(m\omega(y))$].
ii) The logarithmic term translates the free energy of a classical oscillator with frequency $\omega(y)$ to the corresponding quantum one, i.e. it accounts *exactly* for the quantum harmonic part.
iii) The *quantum nonlinear* effects are contained in the other terms of (2.5) and in the dependence on y of $\omega(y)$.

It is interesting to consider the following limiting cases.

Harmonic oscillator. $V(x) = m\Omega^2 x^2/2$ gives

$$V_{\text{eff}}(y) = \frac{1}{2}m\Omega^2 y^2 - \frac{1}{\beta}\ln\frac{\beta\hbar\Omega}{2\sinh(\beta\hbar\Omega/2)} \quad ; \quad e^{-\beta F_0} = \frac{1}{2\sinh(\beta\hbar\Omega/2)} , \qquad (2.8)$$

which is the well known exact result.

High temperature limit. We recover the first term of the Wigner [2] expansion

$$V_{\text{eff}}(y) = V(y) + \frac{1}{2}\frac{\hbar^2\beta}{12m}V''(y) + o(\beta) . \qquad (2.9)$$

Low temperature limit. Using the steepest descent method around the minimum y_m of $V_{\text{eff}}(y)$ we find the one-loop renormalization of the frequency:

$$F = \frac{1}{2}\hbar\omega(y_m) + \frac{1}{\beta}ln[1 - e^{-\beta\hbar\omega(y_m)}] + g(\beta^{-1}) \ . \qquad (2.10)$$

It is also interesting to evaluate the effective potential related to the double-well potential:

$$V(x) = \frac{\mu^2}{4\lambda}(y^2 - 1)^2 \ . \qquad (2.11)$$

Here we introduce the reduced temperature $t = (4\lambda)/(\mu^2\beta)$ and the coupling parameter $Q = (8\hbar\lambda)/(2m\mu^3)^{1/2}$. The latter rules the quantum character of the system. From Fig. 1, one can see that the quantum fluctuations makes softer the potential, so that the jumps across the barrier activated by the temperature are enhanced. It has been shown [5] that eqs.(2.4-2.7) give very accurate effective potentials and consequently accurate free-energies, for a large range of couplings Q and temperatures t.

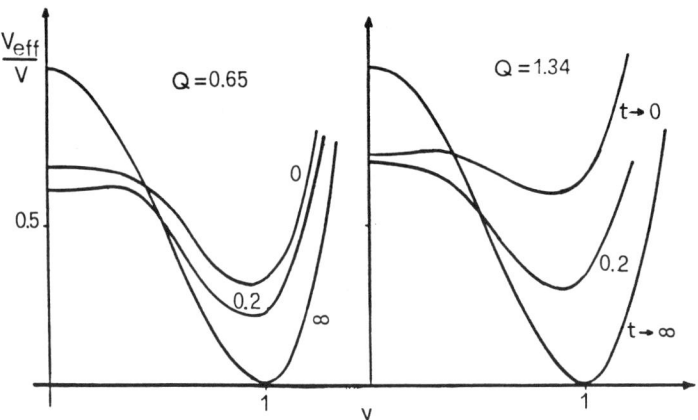

Figure 1 — *Effective potential for the double-well model [eq.(2.11)], for different values of coupling Q and temperature t.*

3. EFFECTIVE POTENTIAL FOR KLEIN-GORDON FIELDS

The discrete version of non-linear Klein-Gordon (K-G) fields, is described by the Lagrangian:

$$L = Aa \sum_{i=-N}^{N} \left[\frac{1}{2}\dot{\Phi}_i^2 - \frac{1}{2}\Omega_0^2(\Phi_i - \Phi_{i-1})^2 - \Omega_1^2 \mathcal{U}(\Phi_i)\right] \equiv \frac{Aa}{2}\sum_i \dot{\Phi}_i^2 - V(\Phi) \ , \qquad (3.1)$$

where $\Phi = \{\Phi_i\}$ are scalar variables on a periodic chain with $2N+1$ sites and spacing a, and $\mathcal{U}(\Phi_i)$ is a local potential with unitary second derivative in its absolute minima.

The Sine-Gordon (SG), Double Sine-Gordon (DSG) and φ^4 theories are obtained for:

$$\begin{aligned} \text{SG}: \quad & \mathcal{U}(\varphi) = 1 - \cos\varphi \\ \text{DSG}: \quad & \mathcal{U}(\varphi) = \frac{1}{\cosh^2\rho}(1 - \cos\varphi) + \frac{\tanh^2\rho}{4}(1 - \cos 2\varphi) \\ \varphi^4: \quad & \mathcal{U}(\varphi) = \frac{1}{8}(\varphi^2 - 1)^2 \end{aligned} \qquad (3.2)$$

The DSG potential contains the parameter $\rho \in [0, \infty]$, and for $\rho \to 0(\infty)$ it reduces to a $2\pi - (\pi-)$ SG model.

In the *continuum limit* ($a \to 0$, $\Omega_0 a = c_0 =$ constant) these fields admit classical kink-like excitations with rest energy $E_K = \nu A c_0 \Omega_1$ (SG: $\nu = 8$; DSG: $\nu = 4[1 + 2\rho/\sinh 2\rho]$; φ^4 : $\nu = 2/3$). The DSG kink resembles a pair of π-SG kinks at distance 2ρ.

The following dimensionless parameters are useful:
— $R \equiv \Omega_0/\Omega_1$, measures the kink length in lattice units, and $R \to \infty$ in the continuum limit.
— $Q \equiv \hbar\Omega_1/E_K = \hbar/(\nu A c_0)$, is the "coupling constant" (the usual field theoretic definition is $g^2 = \nu Q$).
— $t \equiv T/E_K$, is the reduced temperature.

The variational treatment applied to the above theories yields:

$$e^{-\beta F} \simeq e^{-\beta F_0} = \left(\frac{Aa}{2\pi\hbar^2\beta}\right)^{(2N+1)/2} \int d\varphi \, e^{-\beta V_{\text{eff}}(\varphi)} \, ,$$

$$V_{\text{eff}}(\varphi) = Aa \sum_{i=-N}^{N} \left[\frac{\Omega_0^2}{2}(\varphi_i - \varphi_{i-1})^2 + \Omega_1^2 \sum_{n=0}^{\infty} \frac{1-n}{n!}\left(\frac{D_i}{2}\right)^n \mathcal{U}^{(2n)}(\varphi_i)\right] - \frac{1}{\beta}\sum_k \ln\frac{f_k}{\sinh f_k} \, . \tag{3.3}$$

Here $f_k(\varphi) = \beta\hbar\omega_k(\varphi)/2$ and the frequencies $\omega_k(\varphi)$ are defined together with the orthogonal matrix $U_{ki}(\varphi)$ by the secular equation

$$\omega_k^2 \delta_{kl} = \sum_{i,j} U_{ki} U_{lj} \left[\Omega_0^2 B_{ij} + \delta_{ij} \Omega_1^2 \sum_{n=0}^{\infty} \frac{1}{n!}\left(\frac{D_i}{2}\right)^n \mathcal{U}^{(2n+2)}(\varphi_i)\right] \, , \tag{3.4}$$

where $B_{ij} = 2\delta_{ij} - \delta_{i,j+1} - \delta_{i,j-1}$. Eq. (4) has to be self consistently solved with

$$D_i(\varphi) = \frac{\hbar}{Aa} \sum_k U_{ki}^2 \frac{1}{\omega_k}\left(\coth f_k - \frac{1}{f_k}\right) \, . \tag{3.5}$$

$D_i(\varphi)$ is the *quantum renormalization parameter*, and generalizes the quantity α of the single particle. For zero temperature it is site- and configuration- independent, and coincides with that previously found in the semiclassical renormalization theory [7-8].

4. LOW-COUPLING EFFECTIVE POTENTIAL

The above equations are very difficult to solve, because of the complicated implicit dependence on the field configuration. However, in the low coupling limit, expansions can be performed which are very useful when we have to treat in fully quantum way the linear excitations, while the non-linear ones can be treated in one-loop approximation. In this low-Q limit we can split the frequencies given by (3.4) as

$$\omega_k^2(\varphi) = \Omega_k^2 + \delta\omega_k^2(\varphi) + O(Q^2) \, , \tag{4.1}$$

with $\Omega_k^2 = 4\Omega_0^2 \sin^2(ka/2) + \Omega_1^2$ and

$$\delta\omega_k^2(\varphi) = \Omega_1^2 \sum_i U_{ki}^2 \left[\sum_{n=0}^{\infty}\frac{1}{n!}\left(\frac{D}{2}\right)^n \mathcal{U}^{(2n+2)}(\varphi_i) - 1\right] = O(Q)$$

and in the expression for V_{eff} we can expand the logarithmic term obtaining

$$V_{\text{eff}}(\varphi) = Aa \sum_i \left[\frac{1}{2}\Omega_0^2(\varphi_i - \varphi_{i-1})^2 + \Omega_1^2 \, \mathcal{U}_{\text{eff}}(\varphi_i)\right] - \frac{1}{\beta}\sum_k \ln\frac{F_k}{\sinh F_k} + O(Q^2) \, , \tag{4.2}$$

where $F_k = \beta\hbar\Omega_k/2$ and

$$\mathcal{U}_{\text{eff}}(\varphi_i) = \sum_{n=0}^{\infty} \frac{1}{n!}\left(\frac{D}{2}\right)^n \mathcal{U}^{(2n)}(\varphi_i) - \frac{D}{2} \ . \tag{4.3}$$

This expansion for the effective potential is surely valid if the condition $\beta\hbar\delta\omega_k^2/(4\Omega_1) \ll 1$ is satisfied, which amounts to require

$$Q \ll \frac{16}{\nu\mathcal{U}^{(4)}(\varphi_{\min})} \quad ; \quad t \gg \frac{1}{8}\mathcal{U}^{(4)}(\varphi_{\min}) \, Q \, D(t) \ ; \tag{4.4}$$

where $\varphi_{\min}$ is an absolute minimum of $\mathcal{U}(\varphi)$ ($\nu\mathcal{U}^{(4)}(\varphi_{\min}) = 8$ for SG, $1/4$ for φ^4). The latter condition can be evaluated in the most unfavourable conditions replacing $D(t)$ with its maximum value $D(0) = \nu Q/(2\pi)[\ln 8R + O(R^{-2})]$, giving

$$t \gg \frac{\nu\mathcal{U}^{(4)}(\varphi_{\min})}{16\pi} Q^2 \ln 8R \ . \tag{4.5}$$

Note that the Wigner expansion is valid for $\beta\hbar\Omega_0/2 \ll 1$, or $t \gg QR/2$, so that our expansion is a very significant improvement.

In this limit U_{ki} is an ordinary real Fourier transformation (plus a contribution of order Q), so that $ka = 2\pi r/(2N+1)$, $r = -N, ..., N$. The quantum renormalization parameter $D = D(t)$ is calculated using Ω_k: it is site- and configuration- independent and can be easily computed numerically.

The advantage of eq.(4.2) is that the complicated implicit dependence on the field configuration is absent. We are left with an effective potential which usually resembles $V(\varphi)$, apart from the quantum renormalization of some parameters entering the single site potential. Therefore the configurational integral (3) can be calculated by any classical method, like temperature expansions [9] or numerical transfer matrix [10].

This method also allows for a straightforward passage to the continuum limit, according to the general pattern used in dealing with the renormalization constants of the theory. Indeed $D(t)$ diverges in the continuum limit, but we can replace Ω_k with its zero-T renormalized counterpart in (4.1) and yield an expression like (4.2) which only contains the convergent renormalization parameter $D'(t) \equiv D(t) - D(0)$. This has been shown in [4] for SG and φ^4.

The effective potentials for SG, DSG and φ^4 turn out to be:

$$\text{SG} : \mathcal{U}_{\text{eff}}(\varphi) = e^{-D/2}(1 - \cos\varphi) + D^2/8$$

$$\text{DSG} : \mathcal{U}_{\text{eff}}(\varphi) = \frac{1}{\cosh^2\rho} e^{-D/2}(1 - \cos\varphi)$$

$$+ \frac{\tanh^2\rho}{4} e^{-2D}(1 - \cos 2\varphi) + \frac{1 + 3\tanh^2\rho}{8} D^2$$

$$\varphi^4 : \mathcal{U}_{\text{eff}}(\varphi) = \frac{1}{8}(\varphi^2 - 1 + 3D)^2 - \frac{3}{4}D^2 \ . \tag{4.6}$$

The renormalization parameter is given by

$$D(t) = \frac{\nu Q \Omega_0}{2\pi} \int_0^\pi dk \, \frac{1}{\Omega_k}\left(\coth F_k - \frac{1}{F_k}\right) \ , \tag{4.7}$$

with $\Omega_k^2 = 4\Omega_0^2 \sin^2(k/2) + \Omega_1^2$ and $F_k = \beta\hbar\Omega_k/2$.

We have set up a numerical transfer matrix in order to evaluate the nonlinear contribution to the specific heat per site, namely its total value *minus* the corresponding harmonic contribution, which we report in Fig. 2; it can be seen that we are now involved only in classical calculations. In particular the $Q = 0$ results coincide with the classical ones (check with [9] and [10]), and are meaningful as long as from the classical point of view one can take $E_K = \hbar\Omega_1/Q$ finite, when letting $Q \to 0$, $\hbar \to 0$.

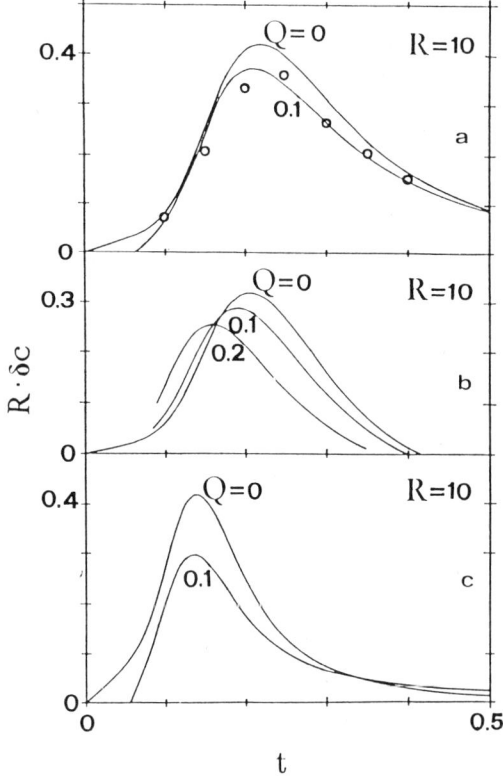

Figure 2 — *Non-linear specific heat per site: a-SG, b-φ^4, c-DSG for $\rho=1$ (full lines). Circles are Quantum Monte Carlo data for $R=10$ and $Q=0.1$, from [11].*

5. HAMILTONIAN PATH-INTEGRAL AND EFFECTIVE HAMILTONIAN

The above method does not apply to *nonstandard* Hamiltonians, where the dependence on the conjugated momenta is not a quadratic form, separated from the configurational part. Typical examples are the relativistic particle or interacting spins described by their action-angle variables.

In the following, for sake of simplicity, we put $\hbar=1$ except in the expression where the classical limit has to be evidenced. After having expressed the partition function of a general Hamiltonian, we introduce a variational method in order to derive an effective Hamiltonian function. The latter, in analogy with the effective potential, can be introduced in a classical like phase-space integral, which gives an approximation of the partition function and yields the exact quantum result in the case of quadratic systems.

Let us now give a outline of the Hamiltonian path-integral. For any operator $\hat{A} = \hat{A}(\hat{p},\hat{q})$

we define a function $A(p,q)$ in phase space by "p-left ordering":

$$\langle p|\hat{A}|q\rangle = (2\pi)^{-1/2} e^{-ipq} A(p,q) , \qquad (5.1)$$

where $\hat{q}|q\rangle = q|q\rangle$ and $\hat{p}|p\rangle = p|p\rangle$ and $\langle p|q\rangle = (2\pi)^{-1/2} e^{-ipq}$. Let $\hat{H}(\hat{p},\hat{q})$ be the Hamiltonian operator. We have

$$e^{-\beta\hat{H}} = \lim_{\substack{N\to\infty \\ N\epsilon=\beta}} (1-\epsilon\hat{H})^N \qquad (5.2)$$

$$\langle p|1-\epsilon\hat{H}|q\rangle = (2\pi)^{-1/2} e^{-ipq-\epsilon H(p,q)} + O(\epsilon^2) \qquad (5.3)$$

and by insertion of the appropriate completness relations we get the Hamiltonian form of the path-integral ($q_{N+1} \equiv q''$):

$$\langle q''|e^{-\beta\hat{H}}|q'\rangle = \int \mathcal{D}[p(u)] \int_{(q',0)}^{(q'',\beta)} \mathcal{D}[q(u)] e^{S[p(u),q(u)]} \qquad (5.4)$$

$$\equiv \prod_{i=1}^{N}\left(\int \frac{dp_i dq_i}{2\pi\hbar}\right) \delta(q_1-q') \exp\left\{\sum_{i=1}^{N}\left[\frac{i}{\hbar}p_i(q_{i+1}-q_i) - \epsilon H(p_i,q_i)\right]\right\}$$

$$S[p(u),q(u)] = \frac{1}{\hbar}\int_0^{\beta\hbar} du\ [ip(u)\dot{q}(u) - H(p(u),q(u))] . \qquad (5.5)$$

$S[p(u),q(u)]$ is the (euclidean) action functional. The partition function and the thermal averages are easily expressed in path-integral form using these formulas.

Classical limit $\hbar \to 0$. In (5.4) the p_i-integration contains a rapidly varying phase factor which yields vanishing contributions unless $q_{i+1} = q_i$. We can hence put $q_N = q_{N-1} = ... = q_1$ in the slowly varying factors $e^{-\epsilon H(p_i,q_i)}$. Then the last $N-1$ q-integrals give factors $2\pi\hbar\ \delta(p_{i-1}-p_i)$, and we are eventually lead to the classical phase-space integral $\mathcal{Z}_{cl} = (2\pi\hbar)^{-1}\int dp dq \exp[-\beta H(p,q)]$. Note that the terms which arise from the ordering prescription are at least of order $\hbar$ and are hence absent in $H(p,q)$, i.e. the classical Hamiltonian.

Let us consider a "trial" Hamiltonian $\hat{H}_0(\hat{p},\hat{q})$, with associated action S_0. The Bogoliubov inequality $F \leq F_0 + \langle\hat{H}-\hat{H}_0\rangle_0$ can be rewritten as in eq.(2.1): $F \leq F_0 + \beta^{-1}\langle S_0-S\rangle_{S_0}$, F and F_0 being the free energies corresponding to $\hat{H}$ and $\hat{H}_0$. This inequality can be used for providing an approximation from above to the free energy of a given system. The idea is to devise an approximate action S_0 involving a certain number of variational parameters.

In the standard case the validity of the Feynman-Jensen inequality has been established, while, in the general case, it is necessary to verify its validity for the particular form of any trial action. In the following we *assume* that the variational parameters of the trial quadratic Hamiltonian can be functionals. So we set

$$S_0 = \int_0^{\beta\hbar} \frac{du}{\hbar}\left[ip\dot{q} - \frac{(p-\bar{p})^2}{2m} - \sigma(p-\bar{p})(q-\bar{q}) - \frac{i\hbar\sigma}{2} - \frac{m(\omega^2+\sigma^2)}{2}(q-\bar{q})^2 - w\right] \qquad (5.6)$$

where the parameters m, σ, ω and w depend on the average point functional:

$$(\bar{p},\bar{q}) \equiv \beta^{-1}\int_0^\beta du\ (p(u),q(u)) . \qquad (5.7)$$

After minimizing with respect to w we find that this determination implies $\langle S_0-S\rangle_{S_0} = 0$, and

$$e^{-\beta F} \simeq e^{-\beta F_0} = \int \frac{dp dq}{2\pi\hbar} e^{-\beta H_{\text{eff}}(p,q)} , \qquad (5.8)$$

with the effective Hamiltonian

$$H_{\text{eff}}(p,q) = H_{\mathcal{S}}(p,q) - m\omega^2 \alpha - \beta^{-1} \ln(f/\sinh f) , \qquad (5.9)$$

where

$$f = \frac{\hbar\beta\omega}{2} , \qquad \alpha \equiv \frac{\hbar}{2m\omega}\left(\coth f - \frac{1}{f}\right) , \qquad (5.10)$$

and the "$\mathcal{S}$-smoothed" Hamiltonian is given by

$$H_{\mathcal{S}}(p,q) \equiv \langle\!\langle\, H(p+\eta, q+\xi)\,\rangle\!\rangle . \qquad (5.11)$$

Here the double brackets denote the average with a bivariate gaussian distribution in (η,ξ), which can be uniquely defined by its moments up to second order: $\langle\!\langle\, 1\,\rangle\!\rangle = 1$, $\langle\!\langle\,\eta\,\rangle\!\rangle = \langle\!\langle\,\xi\,\rangle\!\rangle = 0$ and

$$\langle\!\langle\,\eta^2\,\rangle\!\rangle = m^2(\omega^2+\sigma^2)\alpha , \quad \langle\!\langle\,\eta\xi\,\rangle\!\rangle = -(m\sigma\alpha + i\hbar/2) , \quad \langle\!\langle\,\xi^2\,\rangle\!\rangle = \alpha . \qquad (5.12)$$

The further minimization with respect to m,σ,ω gives (the subscripts denoting partial derivation):

$$m^{-1} = \{H_{pp}\}_{\mathcal{S}} , \qquad \sigma = \{H_{pq}\}_{\mathcal{S}} , \qquad m(\omega^2+\sigma^2) = \{H_{qq}\}_{\mathcal{S}} . \qquad (5.13)$$

In this way the effective Hamiltonian (5.9) is fully determined, and use of (5.8) gives the best approximation for the free energy $F(\beta)$.

Let us conclude this section on some general remarks about the physical meaning of the terms in eq.(5.9).
i) The smoothed $H_{\mathcal{S}}$ which appears in (5.9) reflects the fact that the particle is subjected to quantum fluctuations, which make it to feel only the average energy around a point in phase-space. Note that the classical part of these fluctuations is subtracted out: indeed (5.10) defines the pure quantum fluctuation in the one-loop approximation, whereas the classical contribution is exactly accounted for by the phase-space integral recipe.
ii) These quantum fluctuations are "effectively" less and less relevant for increasing temperature. For zero temperature ($f\to\infty$) (5.12) implies that $\Delta p \Delta q = \hbar\sqrt{1+\sigma^2/\omega^2}/2 \geq \hbar/2$, in agreement with the uncertainty principle.
iii) In spite of the intrinsically complex quantities involved in the determination of $H_{\mathcal{S}}$, it is possible to prove that $H_{\mathcal{S}}$ itself is a real function. The parameters determined by (5.13) turn also out to be real. It follows that the effective Hamiltonian H_{eff} is real.
iv) Application to a general linear problem yields the exact quantum free energy, as it can be easily verified, due to the logarithmic term in (5.9), which ultimately translates from classical to quantum the free energy of harmonic oscillators with frequency ω.
v) In the standard case the effective Hamiltonian gives rise to the previously introduced effective potential [3-4]
vi) We have also verified the agreement of the effective Hamiltonian (5.9) with the corresponding one defined by Wigner [2], up to order $\beta\hbar^2$.

REFERENCES

1. R.P. Feynman, "Statistical Mechanics", Benjamin, Reading MA (1972).
   R.P. Feynman and A.R. Hibbs, "Quantum Mechanics and Path-Integrals", Mc Graw Hill, New York (1965).
2. M. Hillery, R.F. O'Connell, M.O. Scully and E.P. Wigner, Phys. Rep. **106**, 122 (1984).
3. R. Giachetti and V. Tognetti, Phys. Rev. Lett. **55**, 912 (1985) and Phys. Rev. **B33**, 7647 (1986) and **B36**, 5512 (1987).
4. R. Giachetti, V. Tognetti and R. Vaia, Variational approach to quantum statistical mechanics, in "The Path Integral Method with Applications", eds. A. Ranfagni, V. Sa-Yakanit and L. Schulman, World Scientific, Singapore (1988); Phys. Rev. **A37**, 2165 (1988); Phys.Rev. **A38**, 1521 and 1638 (1988).
5. R.P. Feynman and H. Kleinert, Phys. Rev. **A34**, 5080 (1986). W. Janke, in "Path Integrals from meV to MeV", eds. V. Sa-Yakanit and L. Schulman, World Scientific, Singapore (1989).

6. R. Giachetti, V. Tognetti and R. Vaia, in "Path Integrals from meV to MeV", eds. V. Sa-Yakanit and L. Schulman, World Scientific, Singapore (1989).
7. K. Maki and H. Takayama, Phys. Rev. **B20**, 3223 and 5009 (1979).
8. R.F. Dashen, B. Hasslacher and A. Neveu, Phys. Rev. **D10**, 4114 and 4130 (1974) and **D11**, 3424 (1975).
9. K. Sasaki, Progr. Theor. Phys. **68**, 411 (1982) and **71**, 1169 (1984).
   K. Sasaki and T. Tsuzuki, J. Mag. Mag. Mat. **31-34**, 1283 (1983). Progr. Theor. Phys. (Suppl. N.94), (1988).
   T. Tsuzuki, Progr. Theor. Phys. **70**, 975 (1983)
10. T. Schneider and E. Stoll, Phys. Rev. **B22**, 5317 (1980).
11. S. Wouters and H. De Raedt, in "Magnetic Excitations and Fluctuations II", eds. U. Balucani, S.W. Lovesey, M. Rasetti and V. Tognetti. XXIII Springer Proceedings in Physics, Springer Verlag, Berlin (1987).

PATTERN CHANGES IN ELECTRODEPOSIT OF $CuSO_4$

V.M. Castillo, R.D. Pochy and L. Lam

Department of Physics, San Jose State University
San Jose, California 95192, USA

We have done experiments on the formation of patterns in electrodeposits of $CuSO_4$.[1,2] At low voltage, DLA-like fractal patterns are obtained. At relatively high voltage, there is a sudden change of pattern for the constant potential in a single run of the experiment (Fig. 1). Our results differs from those obtained by the Ann Arbor group.[3]

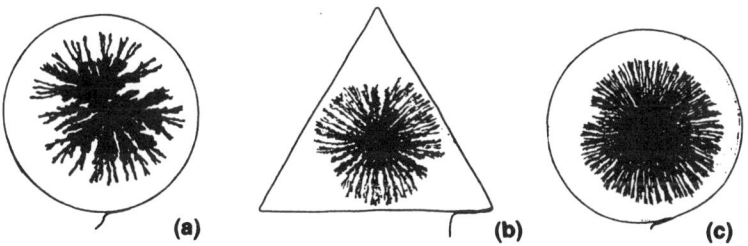

Fig. 1. Copper electrodes are used. Cell thickness is 0.6 mm. Concentration is 0.05 M. Voltage = 10, 20 and 25 V in (a), (b) and (c), respectively.

To understand and simulate these results, we wanted a model that will give filaments with controllable thickness. We developed a biased random walk model in which the particle can move sideward and downward. The control parameter, R (the ratio of the probability of the sideward to downward motion), is kept constant. The sticking probability when the random walker meets the growing cluster sideward and downward is S_H and S_V respectively. These models (Fig. 2) do generate filaments of controllable thickness but fail to produce abrupt changes in the patterns. Different approaches are in progress.

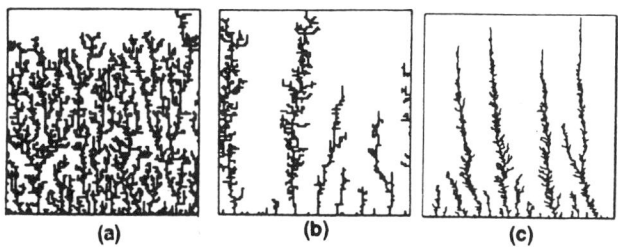

Fig. 2. Model A: $S_H = S_V = 1.0$ (a) R = 0.5. (b) R = 10. Model B: (c) R = 10. S_V varies linearly from 1/3 at the bottom to 0.8 at the top, $S_H = (1 - S_V)/2$.

References

1. K.C. Chan, V.M. Castillo, Y.J. Chang, K.L. Clague, M. Condron, J.K. Kim, H.S. Lakkaraju, L. Lam, W. Miner and R.D. Pochy, paper presented at the Conference on Nonlinear Science, Santa Cruz, California, March 25-26, 1988.
2. L. Lam, in "Wave Phenomena", L. Lam and H.C. Morris, eds., Springer, New York (1989).
3. L. Sander, in "Physics of Structure Formation", W. Güttinger and G. Dangelmayr, eds., Springer, New York (1987).

LINEAR DYNAMICS OF MODULATED SPIN MAGNETS

Christian J Lantwin

Department of Theoretical Physics, Oxford University

1 Keble Road, Oxford OX1 3NP, UK

Periodic systems with a superimposed (periodic) modulation are found in different manifestations, ranging from the magnetic phases of rare earth elements (1) to such recent developments as anyon systems(2).
  In the presented poster we focus on the description of the dynamic properties of sinusoidally modulated spin magnets and make an attempt to relate the calculatory results to experimental data for the element erbium(3).
  The formalism that we present (3,4) yields an analytic expression for the linear response function. The thus obtained results are identical to numerical studies on the same and similar problems(6), but give in addition to the mere functional dependence of the linear response function on the various parameters, the framework for a physical interpretation of the characteristics of the investigated systems. This is of particular importance in the study of incommensurate systems, where our formalism -- contrary to other approaches -- supplies the possibility of a continuation to the description of commensurate structures.
  In the case of erbium, the sinusoidally modulated phase occurs at a temperature near the Néel temperature of 84 K. We find a poor agreement of the experimental data with the calculation. This may be due to the influence of higher harmonics in the modulation, as the data are taken at a temperature of 70 K, where these are observed. A more comprehensive discussion is found in (1).

REFERENCES

1. C J Lantwin preprint of Oxford University, submitted to Z.Phys. B 1989
2. Y Hasegawa, P Lederer, T M Rice, P B Wiegmann preprint of the ETH Zurich ETH-TH089-17 1989
3. R M Nicklow, N Wakabayashi 1982 Phys.Rev. B 26 (3994)
4. S W Lovesey 1988 J.Phys.C:Solid State Physics 21 (2805)
5. S W Lovesey 1988 J.Phys.C:Solid State Physics 21 (4967)
6. T Ziman, P A Lindgaard 1986 Phys.Rev. B 33(3) (1976)

The stochastic ϕ^4 atomic chain

Angel Sánchez and Luis Vázquez

Departamento de Física Teórica I
Facultad de Ciencias Físicas
Universidad Complutense
28040 Madrid (Spain)

We study the ϕ^4 chain[1] with a stochastic distortion of the potential added as a perturbation. We deal mainly with its continuum limit,

$$\phi_{tt} - \phi_{xx} - \phi + \phi^3 + \left[-V_1(t)\phi + V_3(t)\phi^3\right] + \alpha\,\phi_t = 0,$$

where $V_i(t)$ are independent, gaussian white noises.

We have obtained analytical results using some approaches for weak noises[2], and carried out numerical simulations of the chain evolution in time. Our main preliminary findings are the following:

1. The stochastic generalization of the Strauss-Vázquez[3] numerical scheme is very accurate, and we reproduce some exact predictions[4] on the evolution of the mean value of the total energy.

2. Kinks are stable under noises of strength 0.1 or less in adimensional units. Around that order of magnitude radiation begins to appear and kinks seem to be sensitively distorted by energy spreading to its surroundings. Over dispersions greater than 1 they are quickly destroyed.

3. The velocity of slow kinks is unchanged by small perturbations. If $v > 0.1$, they suffer a slowing down via energy conversion into phonon-like radiation.

4. Kinks incoming the noise zone from the outside are rapidly trapped at the edge of that zone but energy is much less transformed into radiation.

We thank the use of the IBM 3090 of the C.I.E.M.A.T. (Spain) and a grant no. PB86-0005 from the C.I.C.Y.T. (Spain). A.S. thanks an Universidad Complutense fellowship.

---

[1] For a description of the ϕ^4 chain see J.F. Currie, J.A. Krumhansl, A.R. Bishop and S.E. Trullinger, *Phys. Rev.* **B22**, 477 (1980), and references therein.

[2] Details on the approaches can be found in ref. 1 and also in P.J. Pascual and L. Vázquez, *Phys. Rev.* **B32**, 8305 (1985).

[3] W. Strauss and L. Vázquez, *J. Comp. Phys.* **28**, 271 (1978).

[4] J.M.R. Parrondo, M. Mañas and F.J. de la Rubia, preprint (1989).

CHAOS AND TURBULENCE

Tomas Bohr

The Niels Bohr Institute
Blegdamsvej 17
DK-2100 Copenhagen

INTRODUCTION

The lecture notes presented here are an attempt at giving a coherent introduction to some concepts which are not usually presented together. "Low dimensional chaos" is, despite its recent absorption into the physics community, becoming a reasonably mature subject. At least one now knows practical ways of quantifying "chaos", one has an understanding of the underlying fractal geometry and for certain special cases one even has strong theoretical results. Turbulence, on the other hand, is an extremely ramified subject and, especially from the experimental side, rather well-studied, largely because of its many technical applications from refrigerators to airplanes. In these lectures the word "turbulence" will be used in a rather general sense, not limited to the motion of fluids. We shall use it to describe motion which is irregular both in space and time as distinguished from the word "chaos" which is usually used about the irregular temporal motion of a single (or a few) variables. This means that we can talk about turbulence in chemical reactions, dynamical interphases and many other non-hydrodynamical situations.

In recent years chaos-seminars have often begun with words about turbulence, maybe even pictures of clouds or oceans, and then, suddenly, the logistic map would appear on the blackboard or a picture of a strange attractor. The explanations offered was more in the nature of anology: both cases involve irregular, deterministic motion and fractal structures. But, in reality, it seems rather mind-boggling how the enormous number of degrees of freedom responsible for the turbulent motion of oceans or the atmosphere could boil down to the chaotic motion of a few variables, giving a low-dimensional strange attractor. The relation between chaos and turbulence is something like the relation between motion of a single particle in a given potential and a field theory describing a many-particle system. Clearly completely new concepts arise in the "field theory" since one now has a spatial manifold of states to work with; but, very probably, some of the concepts from the "single particle" chaos will be applicable and useful. And this represents an approach to turbulence differing in many respect from the classical one.

The subject of these notes is vast and I have had to restrict myself to a very limited subset. The choice has been made first of all from the subjects that I know anything about and, second, from those that will fit together into a reasonably coherent story. One omission, which anyone familiar with turbulence will notice immediately, is the scaling theory of fully develloped turbulence going back to Kolmogorov and others. This is a very interesting subject, connected with the fractal properties of turbulent flows, and I hope soon to understand it well enough to write something coherent about it!

If I have anything new to tell about any of the other subjects, it would certainly not have been possible without my colleagues, especially those with whom I have had the good fortune of working with. It is a pleasure to thank Geoff Grinstein, Mogens Jensen, C.Jayaprakash, Yu He, David Rand, Ole Bøssing Christensen, Rene Rasmussen and Anders Pedersen for collaboration on the subjects discussed below.

CHAOS

One of the powerful realizations of the last decade or so is that very complicated dynamics and very beautiful geometric structures can be obtained by extremely simple prescriptions. One of the most famous examples is Hénon's mapping [1],

$$x_{n+1} = 1 - a x_n^2 + y_n \tag{1.1a}$$

$$y_{n+1} = b x_n \tag{1.1b}$$

which captures many features of the dynamics of "real" systems described by differential equations. Hénon chose $b=0.3$ and $a=1.4$, he chose a starting point and he iterated (1.1) forward plotting the successive iterates. What he found was that, irrespective of starting point, the iterates seemed to converge, or be attracted to, the object shown in fig.1 called a *strange attractor*. On the attractor the dynamics is *ergodic* which means that every point is visited and revisited over and over again. More precisely: any given neighborhood, no matter how small, will always be revisited if we iterate far enough ahead. Closely connected to this is the "unstable" character of the dynamics on the attractor: two points initially very close will diverge exponentially away from each other. The rate of exponential separation (for sufficiently close points iterated sufficiently far) is a well-defined quantity independent of starting point. This quantity is called the Lyapunov exponent and since it is going to be a recurrent character (hero?) of this story we shall reserve the letter λ for it. Positive λ signifies chaos [2].

The simplest way to calculate the Lyapunov exponent is simply to choose two points close to each other and close to the attractor and iterate them both forward. The log of this separation will grow (on average) at the rate λ. It is hard to do this very accurately due to the limited precision available on the computer. For the Hénon attractor (fig.1) $\lambda \approx 0.42$ which means that the separation is scaled by a factor of ten for each $\log 10/0.42 \approx 6$ iterates and, therefore, that two points seprated by a distance of, say, 10^{-8} will be at random positions on the attractor after less than 50 iterates.

One can, in stead, take the mathematical limit of infinitesimal initial separation by working with the *tangent map*

$$dx_{n+1} = -2 a x_n dx_n + dy_n \tag{1.2a}$$

$$dy_{n+1} = b dx_n \tag{1.2b}$$

which is obtained by differentiating (1.1) and which is *linear* in (dx,dy). Imagine that we have computed a sequence $(x_0,y_0),(x_1,y_1),\cdots,(x_n,y_n)$ by iterating the map (say (1.1)) forward from (x_0,y_0). Now choose an arbitrary vector (dx_0,dy_0) of unit length. Then use (1.2) to calculate (dx_1,dy_1). Rescale it to obtain the unit vector $(dx_1,dy_1)/\chi_1$. Now insert this vector and x_1 into (1.2) to obtain (dx_2,dy_2). Rescale to a unit vector by $\chi_2 = |(dx_2,dy_2)|$ and reinsert in (1.2) with x_3 etc. We can keep this going as long as we please and the Lyapunov exponent is now hidden in the χ's giving the rate of exponential growth of the *length* of the iterated vector:

$$\lambda = \lim_{n \to \infty} \frac{1}{n} \sum_{i=1}^{n} \log \chi_i \tag{1.3}$$

Fig.2 shows an example of the use of this method for the Hénon attractor. Fig.2a shows the terms $\log \chi_i$ in (1.3) and fig.2b shows their sum. Although the terms in fig.2a fluctate strongly one can pick out a mean slope in fig.2b defining λ and the more iterates we use the more precisely we can pin down this number.

For the Hénon map (1.1) one can actually define more than one Lyapunov exponent. Since the map is two-dimensional, so is its tangent space. One can then ask not only: how does the *length* of a vector change under iteration, but also: how does the *area* spanned by two vectors change. By definition areas change by $e^{\lambda_1+\lambda_2}$, the sum of the two largest Lyapunov exponents, volumes by the first 3 etc. Again these can be found from the tangent map by very similar algorithms. Take for siplicity the case of the first two exponents. Start with two, orthogonal unit vectors $\vec{e}_1$, $\vec{e}_2$ as shown in fig.3. The tangent map takes them into $\vec{e}^*_1$ and $\vec{e}^*_2$. Now orthogonalize by projecting $\vec{e}^*_2$ orthogonally to $\vec{e}^*_1$ and normalize the resulting two vectors orthogonal vectors by factors $\chi_1(1)$ and $\chi_2(1)$. Now repeat the process always orthonormalizing in the same sequence, simply rescaling $\vec{e}_1$, but projecting $\vec{e}_2$ before rescaling. Then

$$\lambda_j = \lim_{n \to \infty} \frac{1}{n} \sum_{i=1}^{n} \log \chi_j(i) \qquad (1.4)$$

and since the area at each step scales by a factor $\chi_1 \chi_2$ its average scaling is given by $e^{\lambda_1+\lambda_2}$.

For the Hénon map the area scales by the same factor at each step, namely the absolute value of the determinant of the Jacobian (1.2) which is simply b. Thus $\lambda_2 = \log b - \lambda_1$ which means that the Hénon attractor with $\lambda_1 = 0.42$ has $\lambda_2 = -1.62$.

It is clear from looking at fig.1 that the Hénon attractor has a beautiful and complex geometric structure. It seems to be made out of threads or, really, one long folded thread. That is, the spatial density is very different in one direction (*along* the threads) where it looks smooth, from the *transverse* direction where it is extremely lumpy and has large gaps. In the latter direction it looks like a "Cantor set" and indeed the attractor is characterized by a fractal dimension larger than one but less than 2 - around 1.26 [3]. The standard way of defining a fractal dimension is to generalize the concept of length or area to noninteger dimensions. Assume that it takes $N(\varepsilon)$ small squares of size ε to cover the attractor (fig.1). Then the area is

$$A = \lim_{\varepsilon \to 0} N(\varepsilon) \varepsilon^2 \qquad (1.5)$$

For the Hénon attractor $A=0$ because of the large gaps in the transverse direction. One might then (with Haussdorff) try

$$A(D) = \lim_{\varepsilon \to 0} N(\varepsilon) \varepsilon^D \qquad (1.6)$$

for varying real D. If D is too large $A(D)=0$, but if D is too small it diverges (in particular $A_\varepsilon(0)=N(\varepsilon)$) so there might be a value $D=D_c$ seperating these two regions and giving some finite value of $A(D_c)$. This defines the "fractal dimension" and from (1.6) we can see that if $N(\varepsilon) \sim \varepsilon^{-D}$ for small ε then D is the fractal dimension. For the "standard" Cantor set shown in fig.4, which Cantor gave as an example in 1883, obtained by iteratively removing the middle third of a line segment, we have $N(3^{-n})=2^{-n}$ which means that $N(\varepsilon) \sim \varepsilon^{-D}$ where the fractal dimension is $D=\log 2/\log 3 \approx 0.63$ [4].

The thredded geometry and the chaotic dynamics are closely connected: the fact that even extremely close points will eventually diverge from each other is possible only because they sit on distinct windings of the threads which keep folding ad infinitum. The dynamics "magnifies" phase space so that more and more threads are seen. There is an explicit connection between the geometric properties (fractal dimension) and the dynamics ones (Lyapunov exponents) which was conjectured by

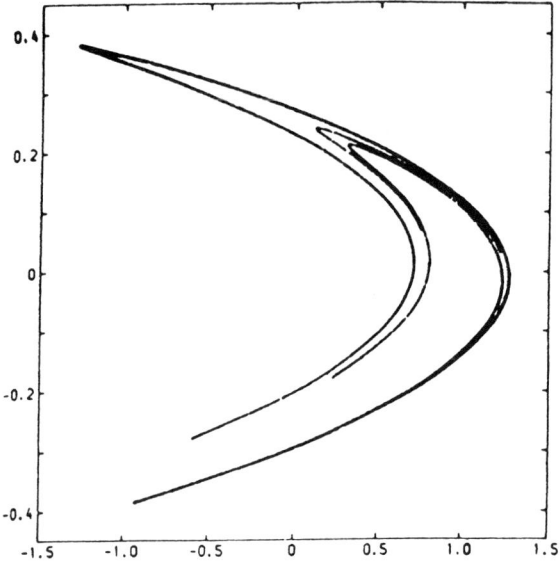

Fig. 1. The Hénon attractor. (From [1]).

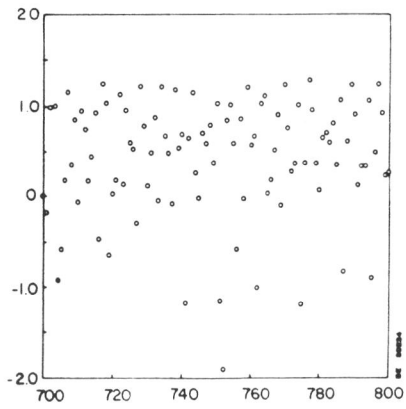

Fig. 2a.

Tangent vector rescaling factor for each timestep or the Hénon map. The plot is semilogarithmic and the Lyapunov exponent is the mean value.

Fig. 2b.

Accumulated length of iterated tangent vector on the Hénon map versus timestep. The plot is semilogarithmic and the mean slope defines the Lyapunov exponent.

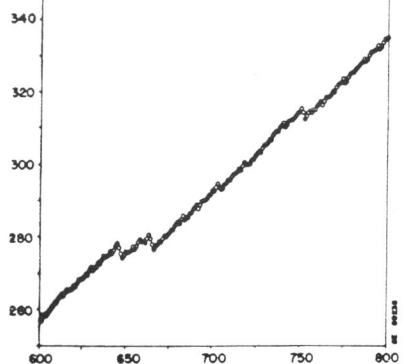

Kaplan and Yorke [5]. For a two-dimension system like the Hénon map it is given by

$$D = 1 + \frac{\lambda_1}{|\lambda_2|} \tag{1.7}$$

and it can be generalized to higher dimensions [5,2]. It is not known whether this relation is actually valid for the Hénon attractor, although numerically its very close: $1+0.42/1.62 = 1.26$. Let us give a simple argument for (1.7).

Imagine that we have covered our attractor with $N(\varepsilon)$ discs of radius ε. Since the attractor is *invariant* under the map (1.1) the iterate of our covering must be a new covering. But due to the expansion and contraction embodied by the Lyapunov exponents the covering is now made up of ellipses. Let's suppose, very roughly, that they're all equal which means that they all have major semiaxis $\varepsilon e^{\lambda_1}$ and minor semiaxis $\varepsilon e^{\lambda_2}$. If we use discs of radius = minor semiaxis, then approximately $e^{\lambda_1 - \lambda_2}$ discs are needed to cover each of the ellipses. Thus

$$N(\varepsilon e^{\lambda_2}) \approx N(\varepsilon) e^{\lambda_1 - \lambda_2} \tag{1.8}$$

Assuming now that $N(\varepsilon) \sim \varepsilon^{-D}$ we find $D = 1 - \lambda_1/\lambda_2 = 1 + \lambda_1/|\lambda_2|$ in accordance with (1.7).

The Lyapunov exponents and dimensions discussed above are computed by certain well-defined operations and numerically they often seem to give reasonable results. In a more rigorous sence, however, very little is known. It is not known whether the Hénon attractor of fig.1 really exists in the sense that it describes the asymptotic dynamics of the system. It is, on the contrary, well known that minute changes in the parameters can change the asymptotic behaviour drastically. As an example, if the parameter a in (1.1) is changed from 1.4 to 1.39945219 the attracting set is simply a 13-cycle [6], but it would take thousands of iterates to discover it.

For one-dimensional maps one knows a bit more [7,8]. For certain classes of maps one can prove that chaotic parameter values do exists and although the chaotic regions are dense with small windows of stable periodic motion they are "likely" in the measure-theoretic sense. In the limit $b \to 0$, (1.1) becomes a one-dimensional map. The subsitution $x = -4(x'-\frac{1}{2})/R$ with $R = 1+\sqrt{1+4a}$ takes it into the unit interval in the form

$$x_{n+1} = R x_n (1 - x_n) \tag{1.9}$$

the well-known and ubiquitous logistic map. The "bifurcation diagram" for this map is shown in fig.5. For a grid of R-values 300 subsequent iterates have been plottted after having discarded the first few hundreds as transients. The period doubling cascade on the left accumulates at $R = 3.5699456...$ and leads into a chaotic region. The chaotic region is dense with "windows" in which the motion is periodic so the Lyapunov exponent (fig.6) is a weird function of R. It's envelope, however, shows power-law behaviour

$$\lambda \sim (R - R_c)^\nu \tag{1.10}$$

with $\nu \approx 0.44$. The universal theory of period doubling [9,8] predicts the geometric accumulation of doubling points, R_n:

$$\lim_{n \to \infty} \frac{R_n - R_{n-1}}{R_{n+1} - R_n} = \delta = 4.66920... \tag{1.11}$$

and relates it to ν by

$$\nu = \frac{\log 2}{\log \delta} \tag{1.12}$$

which simply states that the Lyapunov exponent is doubled under the doubling transformation (since two iterates of the map are grouped together in one "new" map).

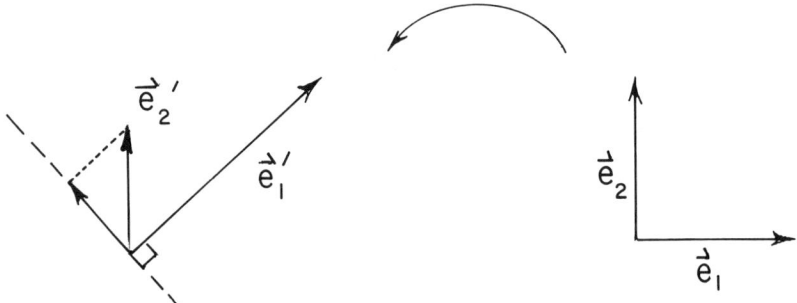

Fig. 3. Orthogonalization for two Lyapunov exponents as explained in the text.

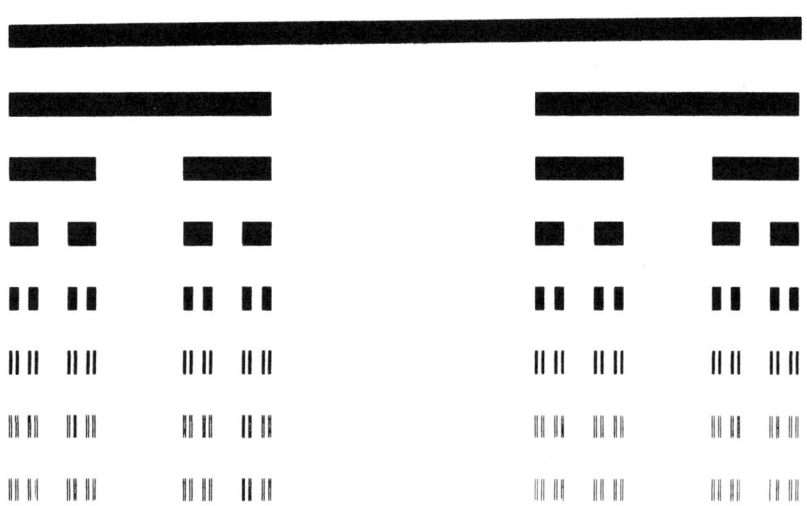

Fig. 4. Cantor's discontinuum. From [4].

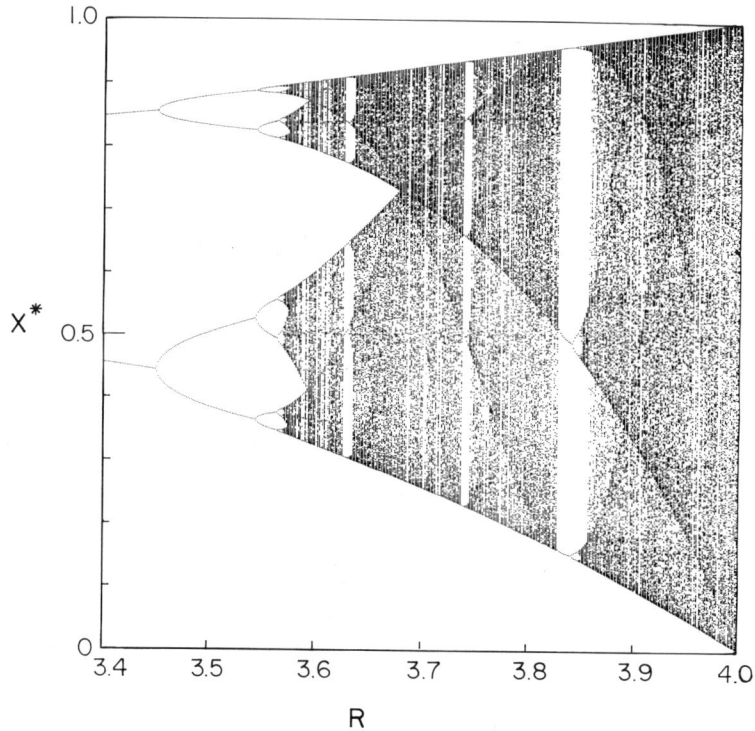

Fig. 5. Bifurcation diagram for logistic map. Picture curtesy of F.Christiansen.

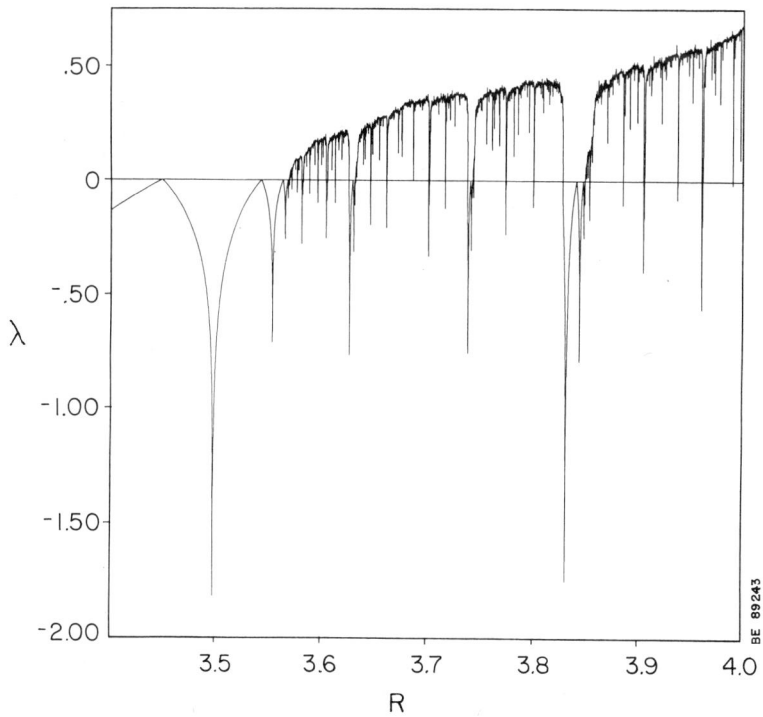

Fig. 6. Lyapunov exponent for logistic map. Picture curtesy of F.Christiansen.

TURBULENCE

The motivation for studying chaotic maps like (1.1) have been mainly two-fold. Firstly, questions related to the *integrability* of Hamiltonian systems, in particular bearing on the stability of the solar system (which is of some importance to all of us!), led Poincaré into the wilderness of chaotic motion and chaotic maps. The maps appearing in this context are *area preserving* expressing Liuvilles theorem on conservation of volumes in phase space. (For the map (1.1) this is the case when $b=1$.) The ideas of Poincaré, Birkhoff and others paved the way for the KAM (Kolmogorov, Arnold, Moser) theorem in which it is proved that a large class of invariant tori do survive even in nonintegrable (chaotic) dynamical systems [10].

The second motivation, which is the subject of these lectures, is to understand *turbulence* - especially its onset. The attempts to relate turbulence to chaotic motion were mainly spurred by a famous paper by Ruelle and Takens in 1971 [11]. The idea is that, even though a hydrodynamical system has lots of degrees of freedom most of them can be inactive leaving only a few active modes; and these modes can easily interact and generate a strange attractor.

The classical hydrodynamical experiments showing chaos [8] (e.g. period doublings) are very small in size; the limitations in system size is a crucial factor in damping away superfluos modes. On the other hand, these systems are not truly turbulent like the flow in the oceans or the atmosphere. To avoid confusion we shall therefore make a careful semantic distinction between "chaos" and "turbulence". A *chaotic* system has a "chaotic" time evolution - technically we can define this as meaning: "having a positive Lyapunov exponent". A *turbulent* system is "disordered" or "irregular" both spatially and temporally. A given point in space undergoes disorderly motion in time and a snapshot of the system shows disorder as well. Notice that I have been careful not to call the disorder "chaos". The reason is that it can be very hard to determine whether a time signal, e.g. the velocity field at a given point in space, has a positive Lyapunov exponent. The Lyapunov exponent relies on the notion of "distance" in phase space, so one has to know in which phase space to imbed the signal. As an example, consider the "disordered" time series of a quantity $f(t)$, e.g. a velocity component in a fluid, as function of time. Assume that at times $t_1, t_2, t_3, \cdots$ it has the same value so one might naively find a Lyapunov exponent by comparing the subsequent time evolution from these points. But, obviously, if the slopes at these different times are different, subsequent values of f will simply separate linearly from each other teaching us nothing whatsoever about chaos. To learn something relevant we must somehow extend the dimensionality of the space by using f.ex. the d-dimensional vector $f(t), f'(t), f''(t), \ldots, f^{(d-1)}$ or, simpler experimentally, $f(t), f(t+\tau), f(t+2\tau), \ldots, f(t+(d-1)\tau)$ where τ is a time delay. Such methods are widely used and described in the literature (see e.g.[2]), but not much is known about their applicability to fully turbulent systems where d can become very large.

Modelling turbulence on a computer one always has to resort to some sort of spatial discretization by which the PDEs are transformed to a, usually very large, system of ODEs. In the next chapter we shall go even further and use discrete mappings like (1.1) coupled together on the lattice. In any case it is quite straight forward to calculate Lyapunov exponents for the *entire* system. One has to work with phase space dimensions of several thousands but the algorithms for computing Lyapunov exponents are perfectly well-defined and functional. In the next chapter we shall see that the largest Lyapunov exponent, at least for a simple model system, has a well-defined limit in the turbulent state as the system size increases; it doesn't diverge or go to zero. Thus one can again be more precise and call a state turbulent only if it has a positive Lyapunov exponent.

Real turbulence has to do with spatial disorder and is therefore never truly low-dimensional. The spatial disorder implies that it might be sensible to define a *correlation length* beyond which the motion is basically uncorrelated. In a sense this is the spatial counterpart of the Lyapunov exponent since one can (very roughly) think of λ^{-1} as a "correlation time".

One can give a simple argument for the existence of a finite correlation length in a turbulent system [12]. Consider a system in which disturbances propagate with some finite speed, c. Imagine that we make a small, local perturbation of size δ_0 at time $t=0$ in a turbulent state at point A. Due to the positive Lyapunov exponent the disturbance will increase as $e^{\lambda t}$ and it will spread out with some speed bounded by c. At point B a distance R away nothing is changed at least up to time R/c at which the perturbation has grown to $\delta_0 e^{\lambda R/c}$. If this quantity is too large we cannot imagine that points A and B can remain correlated and this means that the largest distance, ξ, to remain correlated (the correlation length) should satisfy the upper bound

$$\xi \leq const. \frac{c}{\lambda} \qquad (2.1)$$

Upper bound because we might have overestimated the velocity with which the perturbation spreads. The existence of a correlation length means that distant parts of the system are effectively decoupled from each other. Thus the dimension of the "strange attractor" describing the motion should grow linearly with volume and be infinite in an infinite system. This means that turbulent motion in a large system is not describable in terms of low-dimensional strange attractors.

After these general remarks it might be appropriate to look at some pictures of "real" turbulence. Fig.7 shows the onset of turbulence in a unstable boundary layer - the appearance of an "Emmons spot" [13]. Somewhere downstream a little spot of turbulent fluid suddenly emerges. It grows, since its front and back move at different velocities, and takes on a characteristic boomerang shape. Inside, the motion is fully turbulent, outside it is laminar; a very impressive phenomenon. A similar thing happens in pipe flows. Here a "plug" (first described by Reynolds in 1883) appears and grows in size while being carried downstream. The growth velocity decreases with decreasing Reynolds number and vanishes at $R \approx 2300$ [14]. In these examples there are basically two things to explain: The "convective" properties i.e. growth and form of the spots which have to do with the turbulent-laminar interphase, and the properties of the fully developed turbulence inside, which appears to have many universal features.

Another classical example is thermal convection. Fig.8 shows a fluid which is heated strongly from below. Well above the critical Rayleigh number where convection sets in truly turbulent motion appears in which hot "plumes" are shot into the fluid from the bottom boundary layer.

The last example is chemical: the Belousov-Zhabotinsky reaction [15]. Here the local dynamics wants to go into a "limit cycle" where the concentration of Br (shading in the figure) varies periodically in time. But the period of the cycle varies with its "radius" (say, the maximal concentration) and spatial inhomogeneities therefore makes it hard for different parts of the system to keep in pace with each other. The result is that *vortices* appear where the phase can jump and the system becomes turbulent. We shall look at a model for this kind of turbulence in the last chapter. Unfortunately I only have a "computer picture" from the model studied there (fig.17) so the picture will be shown later.

In chapter 1 we saw how fractals appear in the phase space of chaotic systems. In turbulent systems I have tried to convince you that the analogous fractal will have very high dimension, namely something proportional to the system size. The reason is the finite correlation length which implies that each new "correlation volume" brings in new uncorrelated modes which just adds more dimensions. In *real space* however turbulent systems seem to generate low dimensional fractals. Fig.9 shows a jet of high velocity water injected into quiescent water [16]. The boundary seems to be fractal; and indeed pictures like this have been digitized and a fractal dimension has been estimated [17]. Similarly it is believed that the dissipation in a turbulent fluid takes place on a fractal set [18,17].

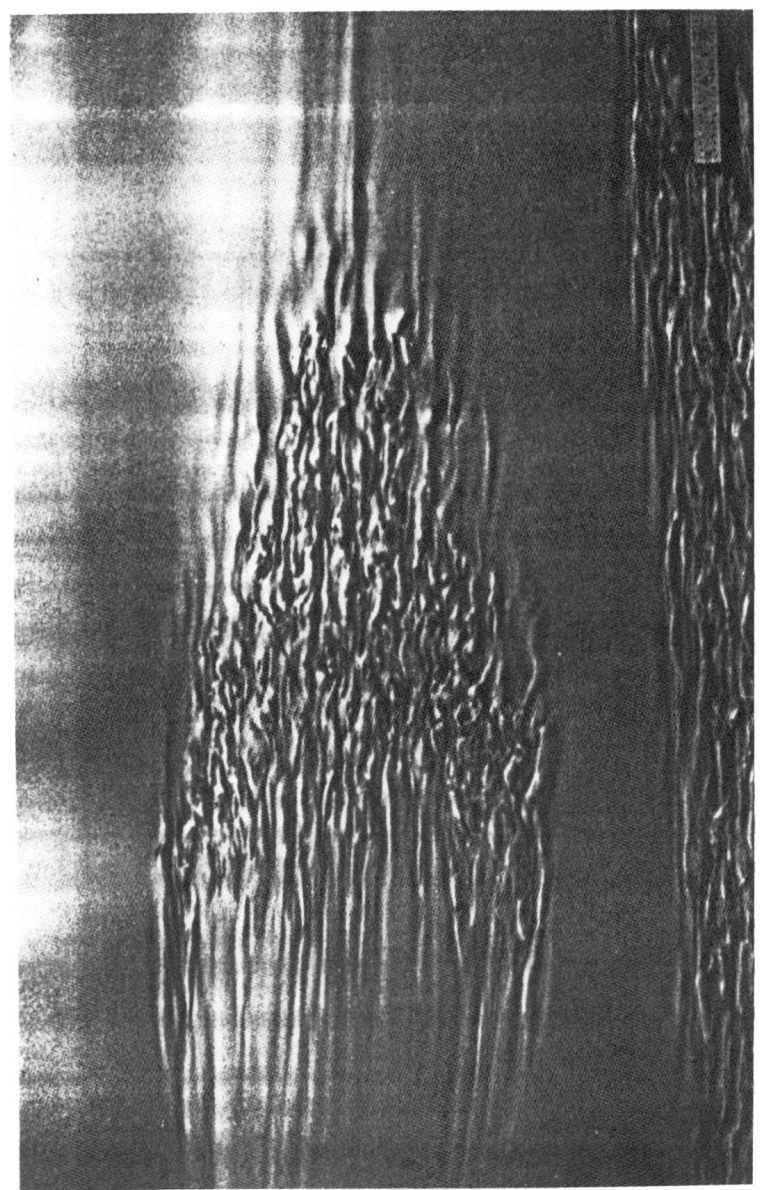

Fig. 7. Turbulent spot in unstable boundary layer. From [16], the picture is due to Cantwell, Coles and Dimotakis 1978.

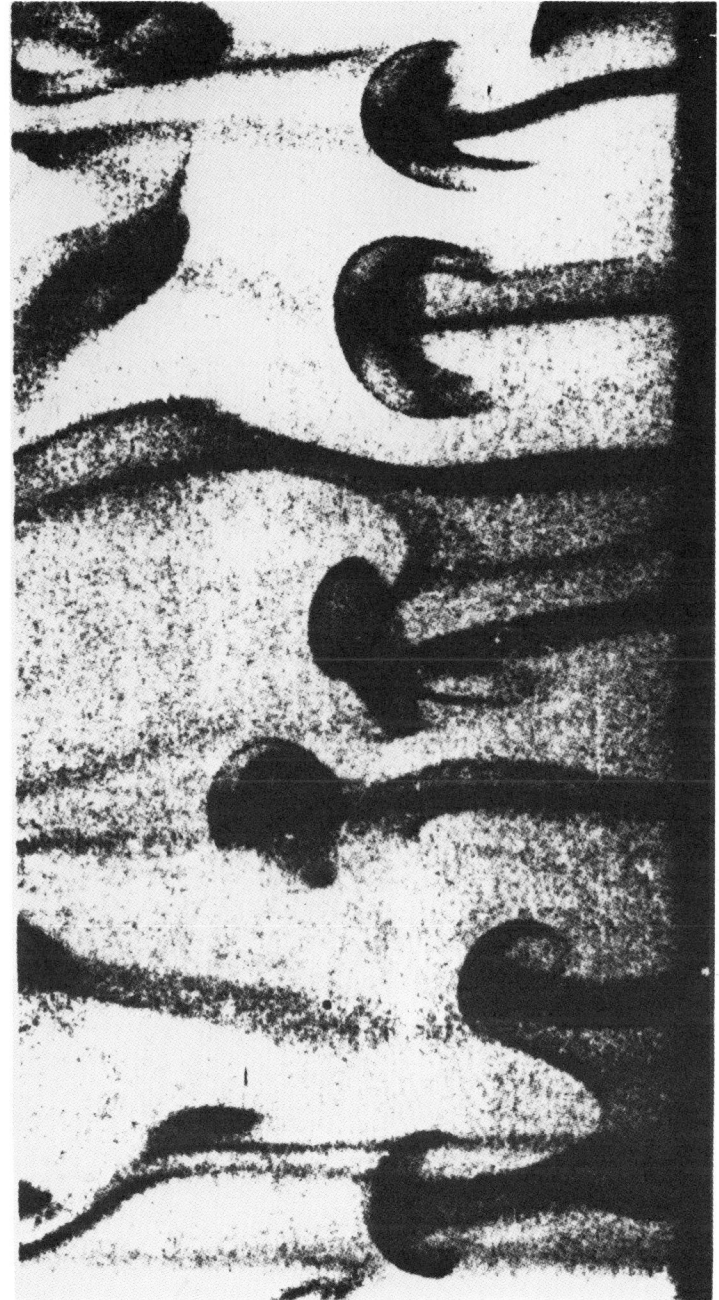

Fig. 8. Thermal convection. From [16], the picture is due to Sparrow, Husar and Goldstein 1970.

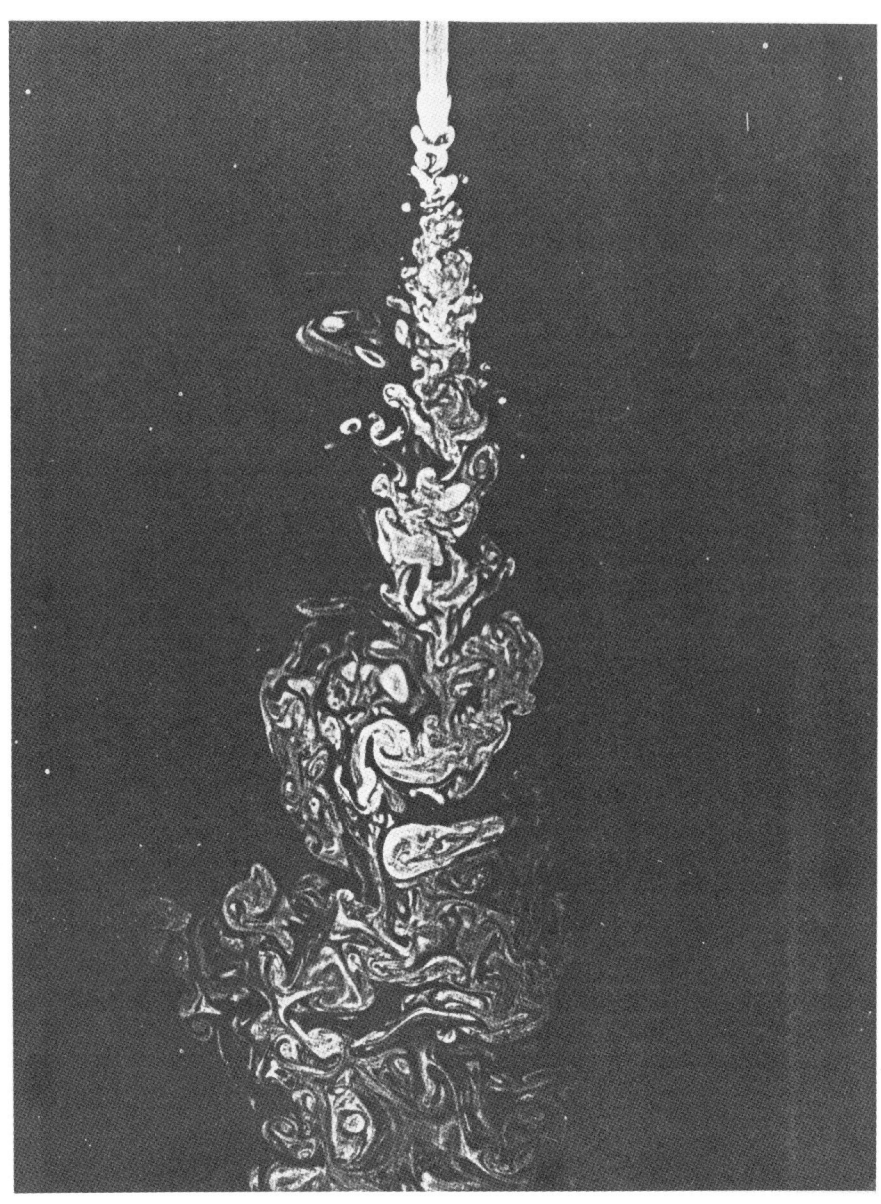

Fig. 9. Turbulent jet. From [16], the picture is due to Dimotakis, Lye and Papantoniou 1981.

COUPLED MAP LATTICES

A simple way of generating turbulent sytems is by coupling chaotic maps together on a lattice [19-22]. In this way one obviously gains no insight whatsoever into the *origin* of turbulence, one simply takes for granted that small regions in, say, a turbulent spot, seems to undergo chaotic motion. In stead we can hope to learn something about the correlations between different parts and the relations between quantities like dimensions, correlation lengths and Lyapunov exponents. In this way coupled map lattices have become a useful tool in building up intuition and understanding about spatially exteded systems.

Consider a two-dimensional, regular lattice indexed by (i,j), $i,j = 1,2,...,L$. On each site we have a variable $u_n(i,j)$ where n is the "discrete time" and we assume periodic boundary conditions in i and j. The dynamics is given by

$$u_{n+1}(i,j) = (1-\varepsilon) f(u_n(i,j)) + \frac{\varepsilon}{4} \sum_{n.n. \ i',j'} f(u_n(i',j')) \tag{3.1}$$

where (i',j') are nearest neighbors of (i,j). The function $f(x)$ is taken to be some non-linear map that can sustain chaotic motion, e.g. the logistic map $f(x) = Rx(1-x)$ discussed in chapter 1. The bifurcation diagram in fig.5 shows the period doubling cascade with accumulation point $R_c = 3.5699456...$ and for $R_c < R < 4$ the attractor is either chaotic or periodic.

Typical states, differing in R, obtained by iterating (3.1) with $L=30$ and noisy initial conditions is shown in fig.10. The figures show snapshots of the system at a given time. The next timestep looks different - the system is both spatially and temporally disordered - but certain statistical features are preserved. Small R ($> R_c$) and large ε (<1) gives smooth spatial variations whereas large R (<4) and small ε (>0) makes the picture rugged with rapid spatial variations. The variations with R is clearly illustrated by the figures and it is plausible that the correlation length of fig.10a is larger than fig.10b.

If we choose parameters corresponding to a periodic window in fig.5 we see something surprizing: although (3.1) certainly has a stable uniform cycle the system never finds it. The state again looks spatially disordered like fig.10.

To understand this one must realize that the periodic states of the logistic map are very different above and below R_c. Above R_c the stable cycle coexists with a *chaotic repellor* (see e.g. [23]). There are points in the unit interval that never settle down on the cycle, but they form a Cantor set of dimension less than 1 so they don't fill anything. If a starting point is not exactly on the Cantor set it will be repelled and finally end up on the cycle.

For the coupled system this is different. When the size of the system grows the typical transient time to get into the cycle becomes larger: since the repellor has a positive Lyapunov exponent, the motion on it will, as argued in last chapter, be characterized by a finite correlation length. Now to be caught by the attracting cycle the system has to get into its basin of attraction *coherently* (in order not to be knocked out again on the next step) and this becomes increasingly unlikely as the size, or more precisely L/ξ increases. For the infinite system the repellor has, in a sense, become an attractor and we have gotten rid of the weird dependence on parameters in the chaotic region of the single maps (fig.6) [12,24].

A lot of work has been done on coupled map systems in one dimesion i.e. coupled chains. In the present lectures, however, we shall always stick to two dimensional lattices because the equilibration properties are much better. In one dimension one often gets kink-like structures, which, although only "metastable", takes forever to decay. In two dimensions, analogosly to the equilibrium case, "surface tension" will speed up the equilibration process. Still, one has to keep in mind that the maps (3.1), like

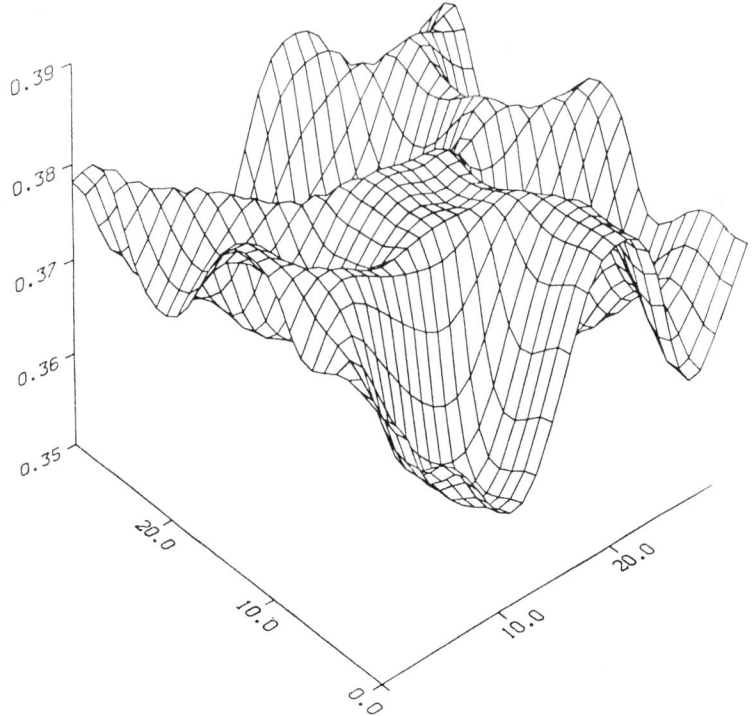

Fig. 10a. Typical state of the coupled logistic map system. $R = 3.5732$ and $\varepsilon = 0.4$. From [30].

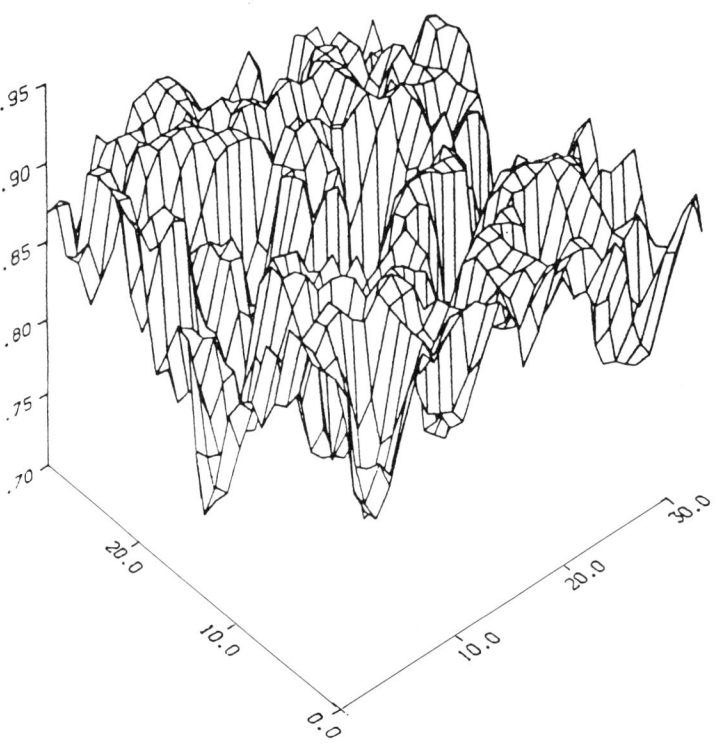

Fig. 10b. As fig.10a, but $R = 3.6786$ and $\varepsilon = 0.4$. From [30].

almost all other non-equilibrium dynamical systems, are "multistable". Which of the stable states is "most stable" can often be answered by starting e.g. one half of the system in one state and one in the other and looking at the motion of the "domain wall"[12]. Another way is to put on noise which ensures that the system is not trapped in some "unlikely" state.

One can define Lyapunov exponents for (3.1) just as for (1.1) by looking at differentials

$$du_{n+1}(i,j) = (1-\varepsilon)f'(u_n(i,j))du_n(i,j) + \frac{\varepsilon}{4}\sum_{n.n.}f'(u_n(i',j'))du_n(i',j') \qquad (3.2)$$

For a lattice of size L one can, in principle, define L^2 Lyapunov exponents, such that the sum of the first i characterize the exponential growth or decay of an i-dimensional hypervolume in phase space. For now we shall confine our attention to the *largest* one which describes the growth of distances analogosly to the single map case. Numerically this (as well as the lower exponents) can be found by the techniques described in chapter 1 and we might ask how this quantity depends on lattice size, whether it shows the characteristic lengthscales of the system. This question was addressed in [25] which we shall follow below.

A specific example is shown in fig.11, where the largest Lyapunov exponent of the map (1) with $R=3.5732..$ (i.e. $\log((R-R_c)/R_c) = -7$) and $\varepsilon=0.4$ is plotted against the size L of a quadratic lattice, varying from 1 to 50. For each lattice, noisy initial conditions were used. The first 5000 iterates were discarded and the next $N=10000$ iterates were used to evaluate λ. The uncertainity was estimated by comparing to the value of λ obtained half-way (i.e. with $N=5000$) - it is of the order of a few times the dot size in the figure.

For small lattices ($L\leq 8$ in the specific example) the Lyapunov exponent is close (within 5%) to the single map value 0.059. Between $L=8$ and $L=9$ it drops to a value close to zero where it remains up to $L=13$. Between $L=13$ and 15 it rises to its "large system" value -all the way up to $L=50$ it remains close (again within 5%) to $\lambda_\infty=0.030$. A closer look at the motion reveals that the lattice becomes absolutely flat for $L\leq 8$ i.e. the chaotic motion of each map is completely in phase (the differences in λ are numerical inaccuracy due to the finite waiting times).

Between $L=9$ and $L=12$ the state is modulated in one direction and flat in the other as shown in fig.12a. It is easy to see that this transition is simply determined by linear stability of the uniform chaotic state $u_n(\vec{x})=M_n$ where $M_{n+1}=f(M_n)$. In the chaotic state very long waves are unstable [26-28]. A state which is almost uniform, i.e. of the form $u_n(\vec{x}) = M_n+\delta_n(\vec{x})$ (where $\vec{x}=(i,j)$) will, to linear order in δ, change to $M_{n+1}+\delta_{n+1}(\vec{x})$ where

$$\delta_{n+1}(\vec{x}) = f'(M_n)\left[(1-\varepsilon)\delta_n(\vec{x}) + \frac{\varepsilon}{4}\sum_{\vec{x}'}\delta_n(\vec{x}')\right] \qquad (3.3)$$

and where the sum is over nearest neighbors $\vec{x}'$ of $\vec{x}$. By Fourier transformation, i.e. expanding δ as

$$\delta_n(\vec{x}) = \sum_{\vec{q}}\delta_n(\vec{q})e^{i\vec{q}\vec{x}} \qquad (3.4)$$

where $\vec{q} = (q_1,q_2)$ we get

$$\delta_{n+1}(\vec{q}) = f'(M_n)(1-\varepsilon+\frac{\varepsilon}{2}(\cos q_1+\cos q_2))\delta_n(\vec{q})) \qquad (3.5)$$

A perturbation $\delta_0 e^{i\vec{q}\vec{x}}$ of the flat state will therefore grow as $e^{\lambda_0+\delta\lambda_\varepsilon(\vec{q})n}$, where λ_0 is the Lyapunov exponent of the single map f (i.e. $\lim_{N\to\infty}\sum_{n=1}^{N}\log f'(M_n)$) and

171

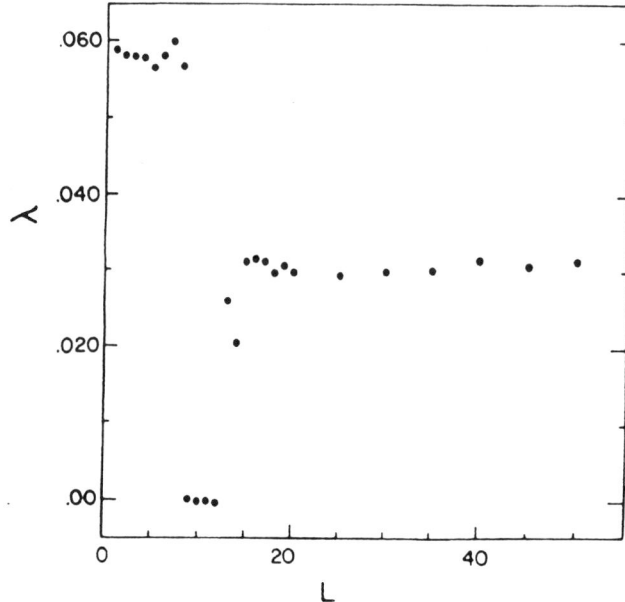

Fig. 11. Size dependence of the largest Lyapunov exponent. From [25]. $\log((R-R_c)/R_c)=-7.0$ and $\varepsilon=0.4$.

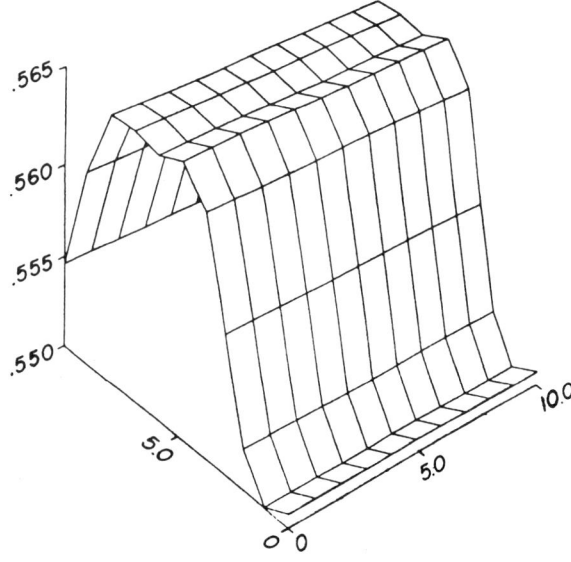

Fig. 12a. Snapshot of logistic map lattice with parameters as in fig. 11. Size $L=11$.

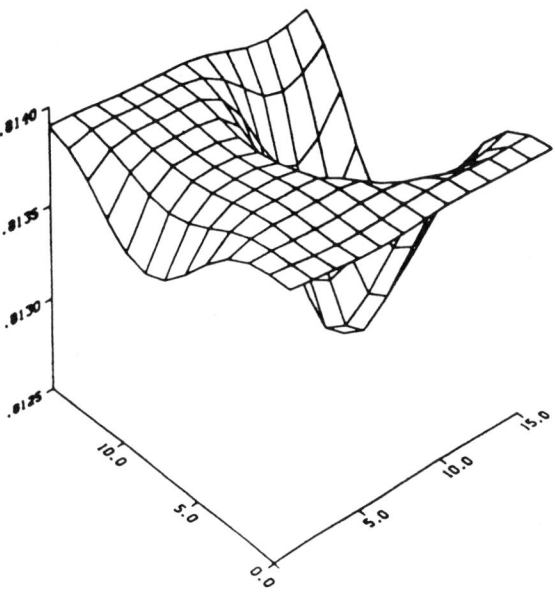

Fig. 12b. As fig.12a, but size L=13.

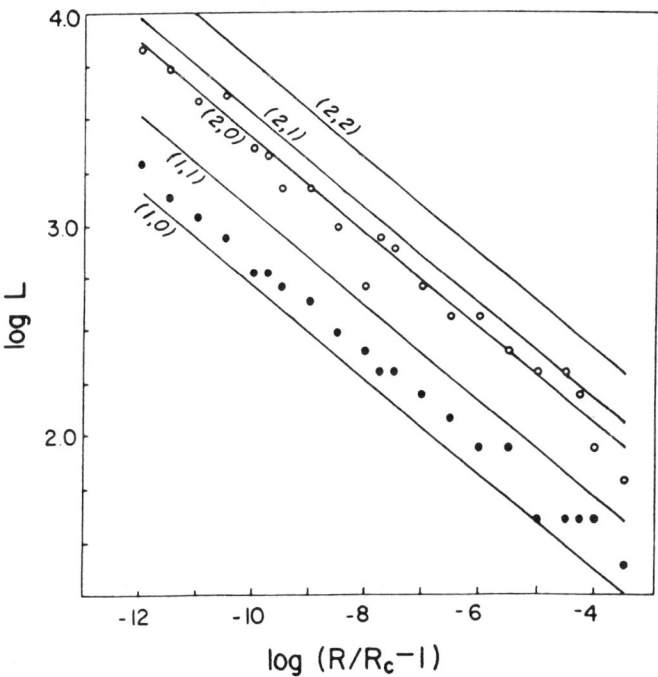

Fig. 13. Scaling of L_1 (black circles) and L_c (empty circles). Along the x-axis is $\log((R-R_c)/R_c)$ and ε is 0.4. The straight lines have slope $-\log2/2\log\delta \approx -0.2249$ and give the instability thresholds (in the uniform state) for wavevectors $\vec{q}=\frac{2\pi}{L}(n_1,n_2)$, where (n_1,n_2) are indicated on each line. From [25].

$$\delta\lambda_\varepsilon(\vec{q}) = \log|(1-\varepsilon(1-\frac{1}{2}(\cos q_1+\cos q_2)))| \qquad (3.6)$$

which, for small q, becomes

$$\delta\lambda_\varepsilon(\vec{q}) \approx -\frac{\varepsilon}{4}q^2 \qquad (3.7)$$

The allowed wavevectors are of the form $\vec{q}=\frac{2\pi}{L}(n_1,n_2)$, where n_1 and n_2 are integers. Thus the first wavevector $\vec{q}=\frac{2\pi}{L}(1,0)$ or $\vec{q}=\frac{2\pi}{L}(0,1)$ becomes unstable at system size L_1 determined by $\lambda_\varepsilon(\vec{q}=(2\pi/L_1,0))=0$ which for large L_1 means $L_1 \approx \pi\sqrt{\varepsilon/\lambda_0}$ [29].

For the parameters corresponding to fig.11 we get $L_1=8.02$ which means that $L=9$ is the first unstable system. Strangely enough at first sight the system responds to this instability by becoming *stable* - the vanishing Lyapunov exponent indicates (quasi)periodic behaviour. On further reflection this is perhaps not so strange. Above L_1 the system can suddenly use the spatial degrees of freedom (at least in one direction) and thus it has much more phase space to search for stable motion and can remain coherent by lowering the Lyapunov exponent.

At $L=13$ modulations in both directions appear as shown in fig.12b. This happens roughly at $L=\sqrt{2}L_1$ corresponding to $\lambda_\varepsilon(\vec{q}=\frac{2\pi}{L}(1,1))=0$ i.e. the instability of the second mode in the uniform state. Now the system begins to have problems keeping its different parts together and for increasing L the dynamics quickly approaches "large system" behaviour with a new Lyapunov exponent reflecting the fact that different parts are dephased. Already at $L=L_c=15$ the value of λ is indistinguishable from its value at $L=50$ indicating that the incoherent averaging responsible for its "large system" value is now effective. Fig.10a shows the system with $L=30$. The lowering of the Lyapunov exponent compared to the single map value is caused by the existence of both positive and negative slopes in $f(u)$. Continuity along the lattice (enforced by the coupling) implies that between regions with positive slopes and regions with negative ones there must be regions where the displacement is very small thus lowering the overall growth rate.

The number of positive Lyapunov exponents is a rough measure (lower bound if the measure is sufficiently smooth) of the dimension [2]. This number changes rapidly around L_c. For the above parameters it is zero between $L_1=9$ and $L=12$. From 3 at $L=13$ it changes to 6 at $L_c=15$ so indeed the system is quickly getting high-dimensional. This means that we can view L_c as a kind of *coherence length:* systems larger than L_c are effectively *large*. Naively one might have taken L_1 as a coherence length, but that would be misleading. At L_1 the *uniform* state is lost, but coherence is still maintained.

For other values of the parameter R one finds similar behaviour. There is a size L_1 at which the Lyapunov exponent jumps down to a value close to zero. At a later value L_2, λ starts moving up again and at L_c its value is indistinguishable from that of the infinite system (extrapolating from $L \approx 50$). On fig.13 L_1 and L_c is shown against $R-R_c$ in a log-log plot. There is a certain degree of subjectivity in the determination of these quantities since the drop at L_1 is not always as quick as in fig.11 and the rise around L_c can be more irregular. So the fluctuations (especially in L_c) are considerable, as is evident from the irregular dependence on R. For small L the staircase nature of the plot due to the discreteness of L is evident. The straight lines on fig.2 have slope $-v/2$, with v defined by (1.10), and are (approximately) the linear stability thresholds for the uniform chaotic state for different modes as marked on the lines. They are obtained from (3.6) using the scaling behaviour (1.10) It is seen that, very roughly, L_c is a factor of (slightly less than) 2 larger than L_1 and that they seem to scale in the same way when R approaches R_c:

$$L_c \sim L_1 \sim (R-R_c)^{-v/2} \qquad (3.8)$$

In order to check our interpretation of L_c and the scaling relation (3.8) we shall further compare our results with those obtained from looking at the spatial variations in systems of fixed length (L=50 or 100). Fig.14 shows the angular average $C(k=|\vec{k}|)$ of the (absolute value of the) Fourier transform $u_n(\vec{k})$ (averaged over several n) of (1) with R=3.570986 (i.e. $\log((R-R_c)/R_c)$=−8.14). The correlation length would be determined by the behaviour at the smallest values of k, but it is very hard to fit that part of the curve in a consistent way due to the limitations in system size. As shown in the figure, however, $C(k)$ shows exponential decay over a large range of intermediate k–values and this defines a lengthscale ρ through $C(k) \sim e^{-\rho k/2\pi}$. By this method we can determine $\rho(R)$ for a series of R-values approaching R_c^+ and the result is shown in fig.15 for a lattice of size 50 (L=100 for the points closest to R_c). Again the exponent is within a few percent of $-\nu$ very close to R_c, but an interesting deviation is seen from around $\log((R-R_c)/R_c)$=−7.5. To the right of this point ρ seems to be governed by a different, "non-critical", larger exponent. In fact it turns out to depend on the coupling ε varying from around -0.3 to -0.4. This behaviour is not seen in the "direct" coherence length L_c (fig.13), and I don't understand its origin. One should of course keep in mind that the exclusion of the small k-values means that ρ could characterize some "short range order" distinct from the largest relevant lengths. For further details of this method and results for logistic and Hénon type maps the reader is referred to ref.[30].

If we interpret L_c as a correlation length ξ the scaling relation (3.8) can be rewritten as

$$\xi \sim \lambda^{-\frac{1}{2}} \tag{3.9}$$

which is rather different from (2.1) (although (2.1) is still an upper bound). A similar result was found for a one-dimensional system of coupled linear discontinous maps [28]. The estimate (2.1) was, however, based on the notion that disturbances propagate with a fixed speed independent of λ.

Let us investigate this a little more carefully by going back to (3.2). As a crude model, let us assume that all f's that appear have the same value (>1). In that case fig.11 would show a constant $\lambda = \log f\,'$ independent of L and (3.2) would become

$$du_{n+1}(\vec{x}) = e^\lambda \left[(1-\varepsilon)du_n(\vec{x}) + \frac{\varepsilon}{4}\sum_{x'} du_n(\vec{x'})\right] \tag{3.9}$$

or

$$du_{n+1}(\vec{q}) = e^\lambda \left[(1-\varepsilon + \frac{\varepsilon}{2}(\cos q_1 + \cos q_2))du_n(\vec{q})\right] \tag{3.10}$$

Thus a local disturbance $du_0(\vec{x}) = \delta(\vec{x})$ evolves into

$$du_n(\vec{x}) = e^{n\lambda} \sum_{\vec{q}} e^{i\vec{q}\cdot\vec{x} + n\delta\lambda_\varepsilon(\vec{q})} \tag{3.11}$$

with $\delta\lambda_\varepsilon(\vec{q})$ given by (3.6). To check whether a propagating disturbance is amplified we insert $\vec{x}=\vec{v}\cdot t$ and look at the growth rate [31,32]. For a very large system the sum in (3.11) becomes an integral:

$$du_n(\vec{x}=\vec{v}n) = e^{n\lambda} \int_0^{2\pi}\int_0^{2\pi} e^{(i\vec{q}\cdot\vec{v}+\delta\lambda_\varepsilon(\vec{q}))n} d\vec{q} \tag{3.12}$$

and for large n the integral is dominated by its saddle point, found by differentiating by the components of $\vec{q}$, i.e. satisfying

$$iv_j = -\nabla_j \delta\lambda_\varepsilon(\vec{q}) \tag{3.13}$$

This determines a (generally complex) $\vec{q}^*$ and the exponential growth rate of (1.12) is then the real

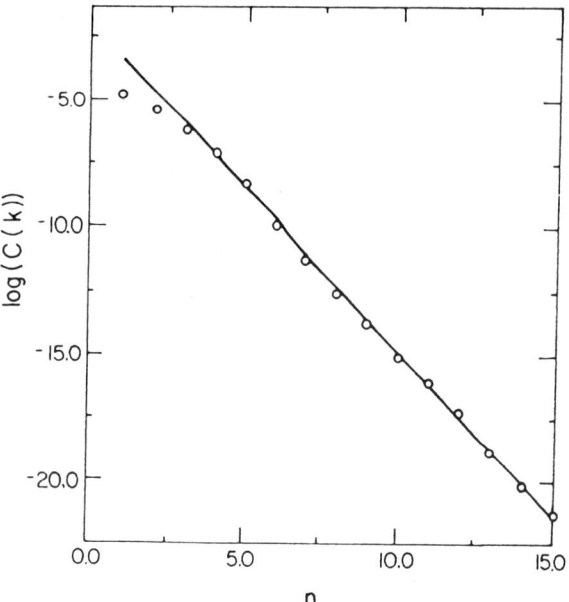

Fig. 14. Logarithm of the angle-averaged spatial power spectrum versus the wavenumber $n=kL/2\pi$. $L=30$, $\log((R-R_c)/R_c)=-8.14$ and $\varepsilon=0.4$. The slope of the straight line fit is $-\rho/L$. From [30].

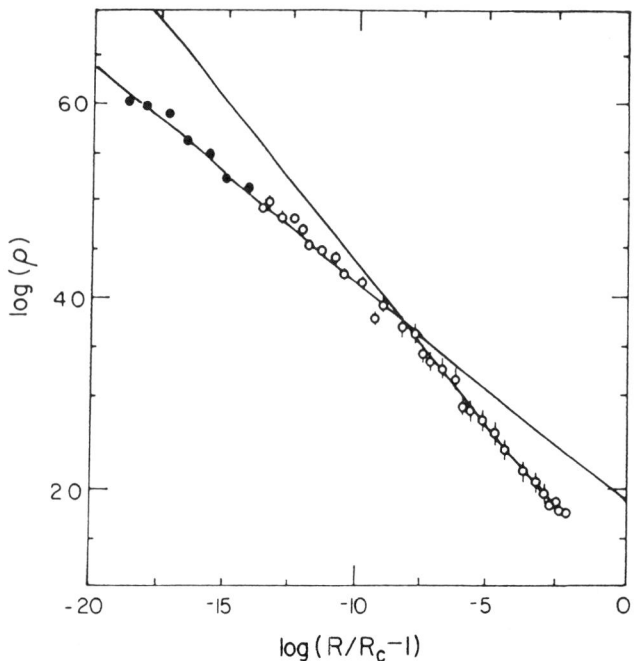

Fig. 15. Scaling of characteristic lenght scale ρ. L is 50 (open circles) and 100 (black circles). $\varepsilon=0.4$. The straight line fits have slopes -0.229 and -0.351. From [30].

part of

$$\lambda(\vec{v}) = \lambda + i\vec{q}^* \cdot \vec{v} + \delta\lambda_\varepsilon(\vec{q}^*) \tag{3.14}$$

For small $\vec{q}^*$ we get $iv_j = \frac{\varepsilon}{2}q_j^*$ and

$$\lambda(\vec{v}) = \lambda - \frac{v^2}{\varepsilon} \tag{3.15}$$

This shows that we'll see exponential growth in any frame of reference travelling with a constant speed less than

$$v_{max} = \sqrt{\varepsilon\lambda} \tag{3.16}$$

and if we insert this maximal speed into (2.1) we get

$$\xi \leq const.(\varepsilon/\lambda)^{\frac{1}{2}} \tag{3.17}$$

I agreement with (3.9) (Note that the small q expansion is valid for small λ). To prove something like (3.16) for the general case seems very hard, but at least we have made some progress in the right direction.

CHEMICAL TURBULENCE

Chemical reactions provide beautiful examples of non-linear dynamics - the non-linearities arise naturally from the interactions between different species involved in most chemical reactions [33,34]. There are famous, well studied examples, in particular the Belousov-Zhabotinsky reaction, where one sees limit cycles of many different periods and beautiful strange attractors [15], [35]. These phenomena are best studied in well stirred reactors where the concentrations are almost uniform over the entire volume. In the absence of stirring, inhomogeneities appear and play a crucial role. Since the period of the oscillations usually depends on the concentrations, different parts get out of step with each other and this gives rise to the nucleation of spiral waves and to turbulent states [36,37].

In this chapter we shall look at the turbulent states that can appear in such systems [37-41] following [40] and [41]. We shall assume that the local dynamics (i.e. the behaviour seen in the stirred reactor) simply is a stable limit cycle. More precisely we shall assume that a parameter μ changes the local dynamics from a stable fixed point for $\mu<0$ to a stable limitcycle (+ an unstable fixed pint) for $\mu>0$. At $\mu=0$ the system thus has a *Hopf bifurcation* and for small, positive μ the limit cycle has a small radius which scales like $\sqrt{\mu}$. The frequency of the limit cycle does *not* approach zero at the transition but some constant ω_0 (for $\mu<0$ this is approximately the frequency of the transients leading to the fixed point). Close to the transition we can approximate the chemical concentrations by their lowest harmonics

$$\rho(\vec{x}) = \rho_0 + Ae^{i\omega_0 t} + A^* e^{-i\omega_0 t} \tag{4.1}$$

where the order parameter, A is assumed to be slowly varying in space and time. It is important to note that A is necessarily *complex* due to the time dependence of ρ and it can be shown [42,37] that A, to lowest order in $|A|$, $|\dot{A}|$ and $|\nabla A|$ satisfies the so-called complex Ginzburg-Landau equation

$$\dot{A} = \mu A - (1+i\alpha)|A|^2 A + (1+i\beta)\nabla^2 A \tag{4.2}$$

where μ, α and β are real numbers. The parameter μ is the usual Landau coefficient: Negative μ implies a quiescent state ($A=0$) whereas positive μ gives nonzero values to the order parameter. In fact there is a homogeneous solution

$$A = \sqrt{\mu}e^{-i\alpha\mu t} \tag{4.3}$$

for positive μ and it is seen that the frequency of this periodic state is $\omega = \alpha\mu$ (so the frequency of the chemical state is $\omega_0-\alpha\mu$). The parameter β is related to differences betwee the diffusivities of the different species [37]. It is important to note that both α and β introduce a preferred sense of rotation in the complex A-plane so we anticipate that the relation between their signs will play a significant role. On the other hand, complex conjugation of (4.2) leads to an equation for A^* of the same form, but with opposite signs of α and β, and therefore the simultaneous change of sign of these two parameters doesn't affect the dynamics. In the following we can thus, without loss of generality, restrict our attention to the case $\beta \leq 0$. We further restrict ourselves to two-dimensional sytems i.e. chemical reactions in a shallow dish.

The linear stability of the homogeneous state (4.3) can be investigated by standard techniques [42]. If we look at weak perturbations of (4.3) in the form

$$A = (\sqrt{\mu}+\rho(x,y,t))e^{i(-\alpha\mu t+\phi(x,y,t))} \tag{4.4}$$

and treat (4.2) to linear order in ρ and ϕ we find that ρ and ϕ will decay exponentially as long as $1+\alpha\beta > 0$, whereas long wavelenth modes ($|\vec{k}|<k_c \propto |1+\alpha\beta|$) will be exponentially enhanced if $1+\alpha\beta < 0$. In particular, if α and β have the same sign, (4.3) is always linearly stable. For later convenience we shall denote the lowest unstable value of α for a given β as α_0 (thus $\alpha_0(\beta)=-1/\beta$).

Let us briefly show how this comes about. If we insert (4.4) into (4.2) and retain only *linear* terms in ρ and ϕ we get

$$\dot{\rho} \approx -2\mu\rho+\nabla^2\rho-\beta\sqrt{\mu}\nabla^2\phi$$

$$\dot{\phi} \approx \nabla^2\phi-2\alpha\sqrt{\mu}\rho+\frac{\beta}{\sqrt{\mu}}\nabla^2\rho$$

which in Fourier space becomes

$$(i\omega+2\mu+k^2)\rho = \beta\sqrt{\mu}k^2\phi$$

$$(i\omega+k^2)\phi = -(2\alpha\sqrt{\mu}+\frac{\beta}{\sqrt{\mu}}k^2)\rho$$

which implies the dispersion relation

$$(i\omega+2\mu+k^2)(i\omega+k^2) = -\beta k^2(2\alpha\mu+\beta k^2)$$

which, for small k, gives

$$\omega \approx i(\mu+k^2)\pm\sqrt{2k^2(1+\alpha\beta)-(\mu+k^2)^2}$$

Thus the imaginary part of ω is negative at small k when $1+\alpha\beta < 0$.

The turbulent states seem closely connected to the appearence and dynamics of *topological defects* in the form of *vortices* like the ones that drive the phase transitions in the planar XY-model and in superfluid Helium films [43]. In fact, taking $\alpha=\beta=0$ in (4.2), leads to a *potential* equation (i.e. one in which $\dot{A}$ can be written as the (functional) derivative of some functional $F[A]$) which, in the presence of noise describes the XY-model at finite temperature.

Vortices are defined in this model, like in the XY-model, as singularities of the angle field. The total variation of the angle over a closed loop, i.e. $\Delta\phi = \oint d\phi$, doesn't necessarily vanish if the loop encloses vortex centers. Instead $\Delta\phi = 2\pi n$, where the integer n is the total vorticity of the region enclosed. In the vortex center the angle is not defined so for the order parameter itself ($A = re^{i\phi}$) to remain well-defined requires that r vanishes in the center. Typical vortex ($\Delta\phi=2\pi$) and antivortex ($\Delta\phi=-2\pi$) configurations are shown in fig.16.

In order to simplify our understanding of the system and to save computer time one can replace the PDE (4.2) by a coupled map lattice [40]. It can be viewed as a rough approximation to the complex Ginzburg-Landau equation (which can be made exact by taking certain limits of the parameters) or as an interesting dynamical system in its own right. We hope of course that properties related to the onset of turbulence will be universal and thus not affected by the particular model chosen.

We split the coupled map lattice into two parts: a *local* map $A' = F(A)$ representing the two first terms of (4.2) and a *nonlocal* part representing the complex heat equation which results from omitting the local terms. The latter part has the solution

$$A(t+\tau_0) = e^{\tau_0(1+i\beta)\nabla^2} A(t) \qquad (4.5)$$

On the lattice we approximate the Laplacian by an average ΔA over neighbors. Thus on a two-dimensional hexagonal lattice $(x,y)=(i+j/2,\sqrt{3}/2\,j)$ with $i,j = 1,...,N$ we take

$$\Delta A(i,j) = \frac{2}{3} \sum_{i',j'} A(i',j') - A(i,j) \qquad (4.6)$$

where the sum is over the six nearest neighbors (i',j'). The nonlocal map is then given by $\bar{A} = (1 + \frac{\tau_0}{M}(1+i\beta)\Delta)^M A$. Here M is an integer that determines the range of the effective interaction. The limit $M \to \infty$ reproduces the exponential above (except, of course, that Δ and ∇^2 are not the same). We take M around 5, large enough to ensure that short wavelength instabilities do not occur.

The properties of the local map F are very simple. In contrast to most of the literature on coupled map systems they are completely *non-chaotic*. If we look at (4.2) without the last term it can be written as

$$\dot{r} = \mu r - r^3 \qquad (4.7a)$$

$$\dot{\phi} = -\alpha r^2 \qquad (4.7b)$$

The general structure of this can be easily reproduced by maps

$$r_{n+1} = f(r_n) \qquad (4.8a)$$

and

$$\phi_{n+1} = \phi_n - \tau\alpha r_n^2 \qquad (4.8b)$$

where the map f has an unstable fixed point in 0 and a stable one in $r=\sqrt{\mu}$. Specifically, one can integrate (4.7a) as

$$r(t+\tau) = \frac{\sqrt{\mu}r(t)}{\sqrt{\lambda\mu+(1-\lambda)r(t)^2}} \qquad (4.9)$$

where $\lambda = e^{-2\mu\tau}$, which fixes the map f. The full map lattice can now be written

$$A_{n+1}(\vec{r}) = F(\bar{A}_n(\vec{r})) \qquad (4.10)$$

and its properties closely resemble equation (4.2). We have mostly worked with periodic boundary conditions, but in special cases like a single vortex (see below), we have chosen "free boundary" conditions meaning that terms in $\Delta A(i,j)$ extending outside the boundary are omitted. Most of our

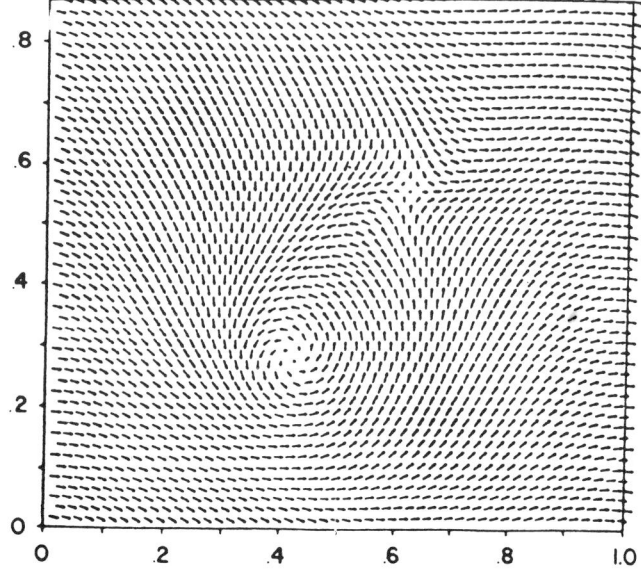

Fig. 16. Vortex and antivortex (periodic boundary conditions). From [40].

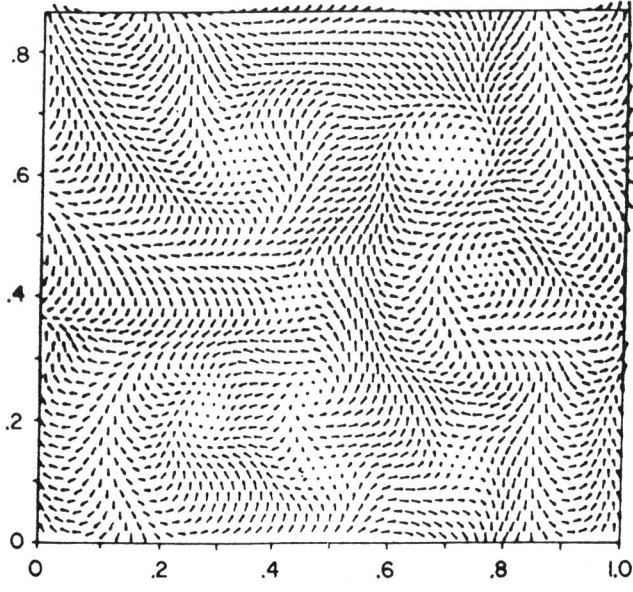

Fig. 17. Turbulent state obtained by iterating the slightly perturbed harmonic state with $\vec{k}=(4,4)$ and $\alpha=0.7$.

simulations have been done with lattice size $N=50$ with occasional excursions up to $N=100$.

One can again ask for linear stability of the homogeneously rotating state $A_n = \sqrt{\mu}e^{-i\alpha\mu\tau n}$ and the resulting criterion (replacing $1+\alpha\beta>0$) turns out to be

$$1+2\frac{\mu}{1-s}\tau\alpha\beta > 0 \tag{4.11}$$

where s is the stability parameter for the stable fixed point of f, i.e. $s=f'(\sqrt{\mu})$. For the map (4.9) $s=e^{-2\mu\tau}$, and it is seen that the new stability criterion approaches the old one as $\tau\to 0$. Again, instability occurs at α_0, where now $\alpha_0 = \frac{1-s}{2\mu\tau|\beta|}$.

The linear stability at $\alpha<\alpha_0$ means only that small perturbations of the uniform state (4.3) will damp out. It tells us nothing about large fluctuations or the fate of *random* initial conditions. The PDE (4.2), as well as the coupled maps (4.10), admits e.g. a family of harmonic solutions

$$A = Re^{i\vec{k}\cdot\vec{x}-i\omega t}$$

with R and ω determined by the wavevector $\vec{k}$. These states have lower linear stability threshold such that the threshold in α diminishes with $|\vec{k}|$ - even at $\alpha=0$ some set of small $\vec{k}$ vectors are unstable. Fig.17 shows the result of iterating the state with $\vec{k}=(4,4)$ around 1000 times at a value of α well below α_0. It is indeed unstable and seems to go into a turbulent state.

In order to quantify this we have to compute - well what do you think? Yes, the Lyapunov exponent! By our standard recipy we must differentiate (4.10) and iterate forward. The result is shown in fig.18a. As in fig.2 we plot the sum of the logs of the rescaling factors for each time step and the average slope should determine the Lyapunov exponent. We get a clearly positive exponent over the 7000 iterates shown and its value is around 0.02. The parameters are $\alpha=0.74$ and $\beta=0.2$ and this should be compared to the linear instability threshold $\alpha_0=0.82$. Going further down in α something interesting appears as illustrated in fig.18b: we seem to get a positive exponent only for a finite time T - then it goes to zero. For α around 0.75, T becomes so large that we have not been able to determine it, presumably it diverges at some α around this value. Lowering α lowers T as well and the value of λ measured (up to T). These "finite time" Lyapunov exponents approach zero at $\alpha=\alpha_c \approx 0.5$ where [41]

$$\lambda \sim \sqrt{\alpha-\alpha_c}$$

In a sense the system has two transitions: one at α_c where the finite time Lyapunov exponent becomes positive and one later where T diverges and the turbulent motion becomes "stable". Now, what happens to the turbulent states at $t=T$ when it decays and becomes laminar? This seems to be due to the *entanglement* described in [40]. Opposite vortices basically attract, but if α is not small enough, or they are too far away, they never quite make it all the way to each other. Somewhere along the way they get caught in each others' spiral arms and are stuck as shown in fig.19. The picture isn't frozen since the vortex pair keeps sending out periodic wavetrains, but the centers don't move. At present we don't know whether this effect is "real" i.e. exists for the Ginzburg-Landau equation (4.2) or it is introduced by the lattice which we use for the coupled map system (4.10). If not the turbulent states might persist further down in α and T might always be infinite. Our belief is that the effect would be present also in the continuum equation and that the reason is that the vortices screen each other with their spiral arms an effect completely missing in the XY-model where the vortices don't spiral.

In ref.[41] it is argued that the infinite system might have a *continuos* transition where the real, infinite time Lyapunov exponent starts becoming nonzero. The reason is that the transient time T seems to grow with system size, whereas the finite time Lyapunov exponent is basically independent of system size. If these trends persist we would expect any finite T to grow to infinity in an infinite system and thus all systems above α_c would be turbulent. The mechanism determining the transient

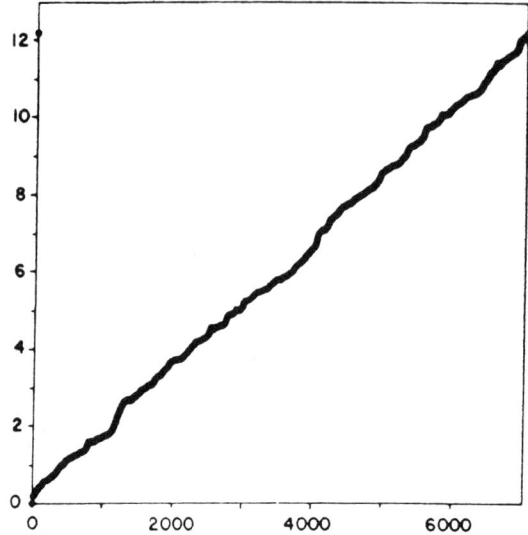

Fig. 18a. Growth rate of tangent vector. The Lyapunov exponent is the mean slope. $\alpha=0.74$. From [41].

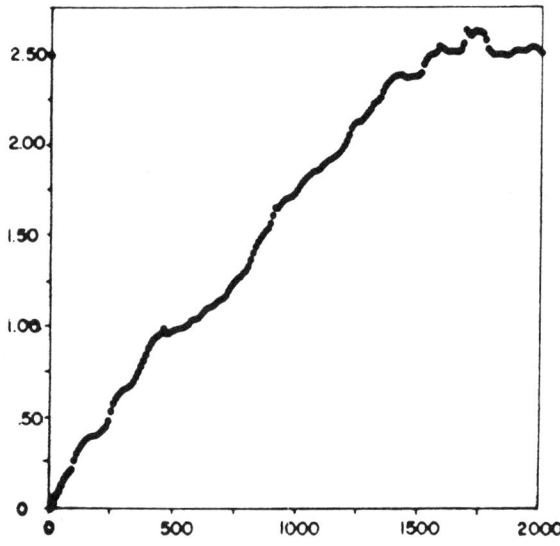

Fig. 18b. The same as fig.18a but with $\alpha=0.71$. It is seen that the positive slope lasts only for a finite number of iterates (around 1700). From [41].

time and its growth with system size is not well understood. In the transient turbulent state, which can persist for thousands of timesteps, vortices are created and annihilated and interact violently. Then suddenly, and this moment strongly depends on the initial, random state, one (or maybe two) vortices start to outgrow the others; in a few hundred iterates it has basically eaten them up and we end with a few vortices entangled in each others' arms, a state which, although not static, has zero Lyapunov exponent.

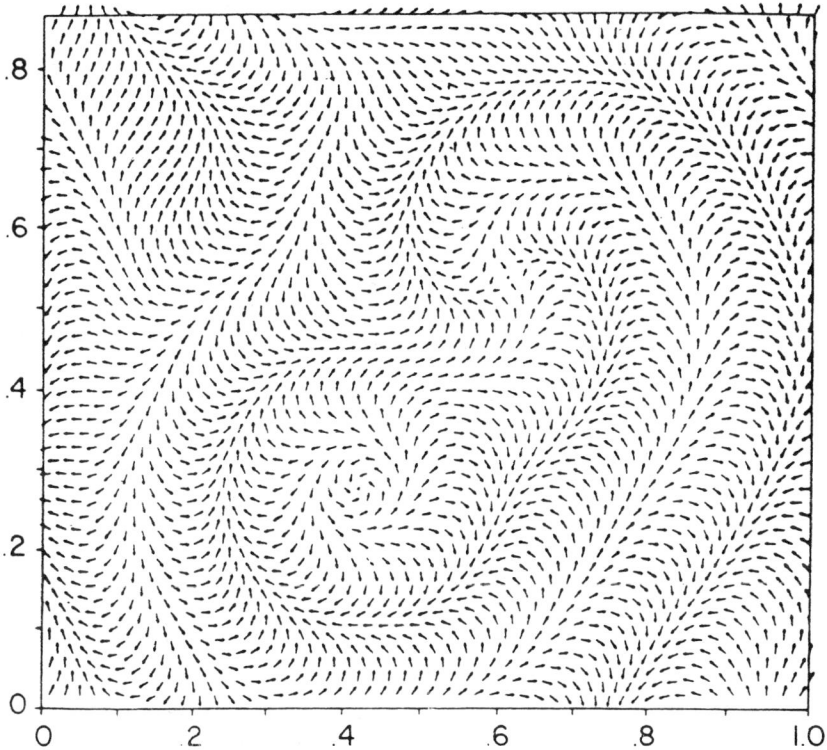

Fig. 19. Entangled (stuck) pair of vortices (periodic boundary conditions). Parameters are $\alpha=0.5$, $\beta=-0.25$, $\mu=0.2$, $\tau_0=0.2$, $\tau=1$. We show the 400'th iterate of the state $A=(z-z_1)(z*-z_2*)/(|z-z_1||z*-z_2*|)$. From [40].

This describes our observations so far, and, as you can see, lots of questions are left unanswered. To understand the mechanism of turbulence and entanglement in these systems in a clearer and more quantitative way seems an important challenge. Maybe one of you will take it up?

REFERENCES

1. M.Hénon, Comm.Math.Phys. $\underline{50}$, 69 (1976)
2. J.-P.Eckmann and D.Ruelle, Rev.Mod.Phys. $\underline{67}$, 617 (1985).
3. D.A.Russel, J.D.Hanson and E.Ott, Phys.Rev.Lett. $\underline{45}$, 1175 (1980).
4. B.B.Mandelbrot, "The Fractal Geometry of Nature" (Freeman 1982).
5. J.L.Kaplan and J.A.Yorke, Comm.Math.Phys. $\underline{67}$, 93 (1979).
6. P.Cvitanović, Phys.Rev.Lett. $\underline{61}$, 2729 (1988).
7. P.Collet and J.-P.Eckmann, "Iterated Maps on the Interval as Dynamical Systems" (Birkhauser 1980).
8. P.Cvitanović, ed. "Universality in Chaos" (Adam Hilger 1984).
9. M.J.Feigenbaum, J.Stat.Phys. $\underline{19}$, 25 (1978) and $\underline{21}$, 669 (1979).
10. McKay, Meiss and Percival.
11. D.Ruelle and F.Takens, Comm.Math.Phys. $\underline{20}$, 167 and $\underline{21}$, 21 (1971).
12. T.Bohr, G.Grinstein, Yu He and C.Jayaprakash, Phys.Rev.Lett. $\underline{58}$, 2155 (1987).
13. J.W.Elder, J.Fluid Mech. $\underline{9}$, 235 (1960).
14. D.J.Tritton, "Physical Fluid Dynamics" (Van Nostrand Reinhold 1977).
15. A.N.Zaikin and A.M.Zhabotinsky, Nature $\underline{225}$, 535 (1970).
16. M.van Dyke, "An Album of Fluid Mechanics" (The Parabolic Press 1982).
17. K.R.Sreenivasan and C.Meneveau, J.Fluid Mech. $\underline{173}$, 357 (1986). C.Meneveau and K.R.Sreenivasan in "CHAOS '87" ed. M.Duong-van (Nucl.Phys. B $\underline{2}$, 49 (1987).
18. G.Paladin and A.Vulpiani, Phys.Rep. $\underline{156}$, 147 (1987).
19. J.P.Crutchfield and K.Kaneko, "Phenomenology of Spatio-Temporal Chaos" in "Directions in Chaos" vol.1, ed. Hao Bai-Lin (World Scientific 1987).
20. H.Chaté and P.Manneville, "Spatio-temporal Intermittency in Coupled Map Lattices", Physica D $\underline{32}$, 409 (1988).
21. K.Kaneko, Physica D to appear.
22. K.Kaneko, Prog.Theor.Phys. $\underline{74}$, 1033 (1985).
23. T.Bohr and D.Rand, Physica $\underline{25D}$, 387 (1987).
24. K.Kaneko and J.P.Crutchfield, Phys.Rev.Lett. $\underline{60}$, 2715 (1988).
25. T.Bohr and O.B.Christensen, Phys.Rev.Lett., to be published.
26. T.Yamada and H.Fujisaka, Prog.Theor.Phys. $\underline{72}$(1984)
27. S.P.Kuznetsov and A.S.Pikovsky, Physica $\underline{19}$ D, 384 (1986). A.S.Pikovsky, Z. Phys. B $\underline{55}$, 149 (1984).
28. F.Kaspar and H.G.Schuster, Phys.Lett. A $\underline{113}$, 451 (1986).
29. Note that small wavelength instabilities can occur if ε becomes too large. It is easy to check that $|1-\varepsilon+\frac{\varepsilon}{2}(\cos q_1+\cos q_2)| \leq 1$ for all q_1, q_2 as long as $0\leq\varepsilon\leq 1$. For $\varepsilon>1$ the "staggered" state with $q_1=q_2=\pi$ becomes unstable.
30. O.B.Christensen, "Spatial Correlations in Coupled Map Lattices", Thesis, University of Copenhagen (1987). Unpublished.
31. R.Deissler and K.Kaneko, Phys.Lett. A $\underline{119}$(1987).
32. T.Bohr and D.Rand, sumitted to Physica D.
33. G.Nicolis and I.Prigogine, "Self-organization in Non-equilibrium systems" (Wiley 1977).

34. H.Haken, "Synergetics" (Springer 1977).
35. J.-C.Roux, R.H.Simonyi and H.L.Swinney, Physica D $\underline{8}$, 257 (1983).
36. A.T.Winfree, "The Geometry of Biological Time" (Springer 1980).
37. Y.Kuramoto, *Chemical Oscillations, Waves and Turbulence* Springer, Berlin (1980)
38. A.V.Gaponov-Grekhov and M.I.Rabinovich, Sov.Phys.Usp. $\underline{30}$, 433 (1987). A.V.Gaponov-Grekhov, A.S.Lomov, G.V.Osipov and M.I.Rabinovich, "Pattern formation and dynamics of two-dimensional structures in nonequilibrium dissipative media". Gorky preprint (1988).
39. P.Coullet, L.Gil and J.Lega, Phys.Rev.Lett. $\underline{62}$, 1619 (1989).
40. T.Bohr, M.H.Jensen, A.W. Pedersen and D.Rand, to appear in "New Trends in Nonlinear Dynamics and Pattern Forming Phenomena" ed. P.Coullet and P.Huerre (Plenum 1989).
41. T.Bohr, M.H.Jensen and A.W. Pedersen:"Transition to turbulence in a discrete complex Ginzburg-Landau model", preprint (1989).
42. A.C.Newell and J.A.Whitehead, J.Fluid Mechanics $\underline{38}$, 279 (1969); A.C.Newell in *Lectures in Applied Mathematics,* vol. 15, Am. Math. Society, Providence (1974).
43. See e.g. J.M.Kosterlitz in "Nonlinear Phenomena at Phase Transitions and Instabilities" ed. T.Riste (Plenum 1982) p.397, or D.R.Nelson in "Phase Transitions and Critical Phenomena" vol. 7 ed. C.Domb and J.L.Lebowitz (Academic Press 1983) p.1.

ELECTRON LOCALIZATION IN DISORDERED SYSTEMS

K. B. Efetov

Max-Planck-Institut für Festkörperforschung
Heisenbergstrasse 1, 7000 Stuttgart 80
Federal Republic of Germany

ABSTRACT

These lectures contain a review of results obtained by the supersymmetry method in the theory of disordered metals. It is shown, how such problems as level statistics in a limited volume, localization in wires, films and granulated materials, Anderson transition, can be reduced to the study of supermatrix σ-models in spaces of different dimensionality. The solution of the σ-models in these cases is presented. The group of the symmetry of Q-matrices entering the σ-model under consideration is noncompact. The role of the noncompactness is discussed. The phenomenon of localization is a formal consequence of this property. The noncompactness results in a very unusual critical behavior near the Anderson transition which is inconsistent with the hypothesis of one parameter scaling.

1 INTRODUCTION

Many interesting phenomena in disordered systems are described by a Hamiltonian with a random potential U(r)

$$\hat{H} = \hat{H}_0 + U(r) \tag{1.1}$$

The first term in eq. (1.1) is the electron kinetic energy. It can include also a magnetic field B. The random potential U(r) describes interaction with impurities. This interaction can be also magnetic and spin orbital. In order to calculate physical quantities one should solve the Schrödinger equation

$$H\Phi = E\Phi \tag{1.2}$$

to calculate the physical quantities for an arbitrary potential U(r) and to average over the random potential. If the potential U(r) varies at distances $a \gg \lambda$, where λ is the electron wavelength, the problem becomes classical. This case reduces to the problem of percolation. The electron can move only in regions where $E > U(r)$. A macroscopic conductivity exists only if $E > E_c$, where E_c is the critical energy which depends on the distribution of the random potential. For $E < E_c$ the macroscopic conductivity is zero.

In the opposite limit $a \ll \lambda$ the electron motion becomes quantum mechanical and hence more complicated. However, as first proposed by Anderson [1], the critical energy E_c exists in this case too. If the electron energy E is large enough $E > E_c$, the electron wavefunction is extended. For $E < E_c$ the electron wavefunction decreases exponentially at distances ℓ. In this case the electron is localized at a point r_0 and ℓ is the localization length. In metals only electrons near the Fermi energy E_F contribute to the kinetic energy. If the disorder is weak $E_c < E_F$, wavefunctions of the electrons near the Fermi energy are extended and the macroscopic conductivity is finite. For strong disorder $E_c > E_F$ the system does not conduct because electrons are localized and cannot move.

The existence of the critical energy E_c is closely related to the dimensionality of the space. For example, electrons in one dimensional chains are always localized for an arbitrary weak disorder [2]. The conductivity can be zero for any disorder not only in chains but also in wires [3]. Later it was proposed [4] that even in two dimensions all states are localized.

The Hamiltonian (1.1) can be used to describe the level statics in a limited volume. The problem of the level statistics originates from the nuclear physics [5], where some statistical hypotheses were used to describe properties of complex nuclei. Starting from the Hamiltonian (1.1) one can check these statistical hypotheses because when proposing these hypotheses it was not specified to what systems they were applied. The level statistics determines absorption and reflection properties of the system with small volumes and can be studied experimentally.

All the problems described by the one particle Hamiltonian (1.1) can be studied using the super symmetry method based on the integration over both commuting and anticommuting variables. The main scheme of this method and its application to the disordered system is presented below.

2. MATHEMATICAL FORMULATION OF THE PROBLEMS: PERTURBATION THEORY

Let us first write down basic formulae for physical quantities.

The simplest quantity which determines all thermodynamic quantities is the density of states $\rho(\epsilon)$

$$\rho(\epsilon) = < \sum_k \psi_k \psi_k^*(r) \delta(\epsilon - \epsilon_k) > \qquad (2.1)$$

where $\psi_k(r)$ and ϵ_k are eigenfunctions and eigenenergies of the Hamiltonian (1.1). The angular brackets in eq. (2.1) stand for the averaging over impurities.

Kinetic quantities can be calculated provided one knows the density-density correlation function $\chi(r,t)$

$$\chi(r,t) = i\theta(t) < \sum_{k,k'} exp(i(\epsilon_k - \epsilon_{k'})t)(n(\epsilon_k) - n(\epsilon_{k'})) \cdot$$
$$\cdot \psi_k(r)\psi_k^*(0)\psi_{k'}(0)\psi_{k'}^*(r) \qquad (2.2)$$

where $\theta(t) = \begin{Bmatrix} 1 \ t > 0 \\ 0 \ t < 0 \end{Bmatrix}$, $n(\epsilon)$ is the Fermi function. For studying the level statistics it is very convenient to introduce the level correlation function $R(\omega)$.

$$R(\omega) = \frac{1}{4\omega\nu^2} < \sum_{k,m}(n(\epsilon_k) - n(\epsilon_m))\delta(\omega - \epsilon_m + \epsilon_k) > \qquad (2.3)$$

where ν is the density of states at the Fermi energy.

All these formulae can be rewritten in terms of the retarded G^R and advanced G^A Green functions

$$G^{R,A}(r,r',\epsilon) = \sum_k \psi_k(r)\psi_k^*(r')G_k^{R,A},$$
$$G_k^{R,A} = (\epsilon - \epsilon_k \pm i\delta)^{-1} \qquad (2.4)$$

For example eq. (2.1) takes the form

$$\rho(\epsilon) = \frac{1}{\pi} <ImG^R(r,r)>$$

The equation (2.2) for the density-density correlation function X in the frequency representation can be rewritten as follows

$$X(r,\omega) = \frac{1}{2\pi i} \int (n(\epsilon) - n(\epsilon - \omega)) K(r, \epsilon.\omega) d\epsilon +$$
$$+ X_1(r,\omega)$$
$$K(r,\epsilon,\omega) = <G^R_\epsilon(o,r) G^A_{\epsilon,\omega}(r,0)>$$
$$X_1(r,\omega) = -\frac{1}{2\pi i} \int n(\epsilon)(<G^R_{\epsilon+\omega}(o,r) G^R_\epsilon(r,o)> -$$
$$- <G^A_\epsilon(o,r) G^A_{\epsilon-\omega}(r,o)>)d\epsilon \qquad (2.5)$$

In the same way one can obtain the following formula for the level correlation function

$$R(\omega) = \frac{1}{8\pi^2 \nu^2 \omega} \ Re \int_{-\infty}^{\infty} <\sum_{m,k}(n(\epsilon) - n(\epsilon - \omega)) G^A_m(\epsilon - \omega)$$
$$(G^A_k(\epsilon) - G^R_k(\epsilon)) > d\epsilon \qquad (2.6)$$

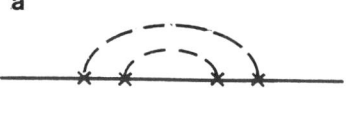

Fig. 1

In order to calculate the quantities given by eqs.(2.4-2.6) one should first solve the Schrödinger equation for an arbitrary potential U(r), to substitute the found Green functions into (2.4- 2.6) and then to average. However, it is not possible to find a solution of the Schrödinger equation for an arbitrary potential in an explicit form. The usual way to overcome this difficulty is to use a perturbation theory. For simplicity let us assume that the distribution of the random potential U(r) is Gaussian and that it is described by the following expressions

$$<U(r)> = 0$$
$$<U(r)U(r')> = \frac{1}{2\pi\nu\tau}\delta(r-r') \qquad (2.7)$$

The procedure of the expansion over the potential U(r) and of the averaging is described in the book [6]. If one calculates the averaged Green function $<G>$, diagrams without intersections of impurity lines are the most important. A typical diagram of this sort is drawn in Fig. 1a.

Summation of the diagrams without intersections gives for the averaged Green function

$$G^{R,A}_{\epsilon,p} = (\epsilon - \epsilon(p) \pm \frac{i}{2\tau})^{-1} \qquad (2.8)$$

Of course, this technique can be used for the calculation of more complicated quantities such as the conductivity. Just as in the case of the calculation of the density-density correlation function (2.5) one calculates the average of products of two Green functions. A typical diagram in this case is drawn in Fig. 1b. Using this technique one can obtain many useful results. However, this technique works only in the limit of a large mean free path τ, such that the condition $\tau\epsilon_0 \gg 1$, where ϵ_0 is the Fermi energy is fulfilled. The Anderson transition occurs in the region $\tau\epsilon_0 \sim 1$ and cannot be studied by a perturbation theory. Besides, in low dimensionality the perturbation theory does not work even in the limit $\tau\epsilon_0 \gg 1$, because certain sequences of diagrams give diverging contributions to the conductivity. The simplest sequence has the form [7].

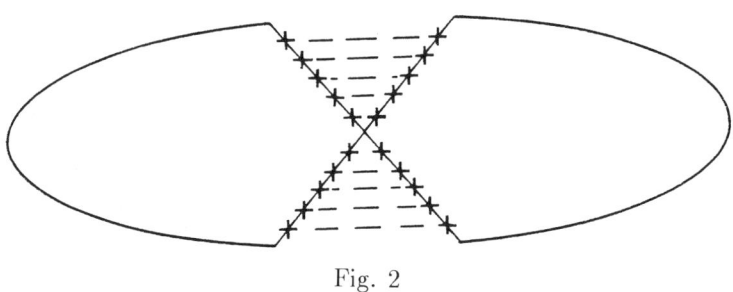

Fig. 2

After summing the ladder in Fig. 2 one can obtain

$$\sigma(\omega) = \sigma_0(1 - \frac{1}{\pi\nu} \int \frac{d^d k}{(2\pi)^d} \frac{1}{-i\omega + Dk^2}) \qquad (2.9)$$

where D is the classical diffusion coefficient, σ_0 is the classical conductivity. The integral in eq. (2.9) diverges in the limit $\omega \to 0$ provided the dimensionality d is low ($d \leq 2$). The correction to the classical conductivity in eq.(2.9) can be interpreted as the contribution of a "diffusion mode" which is massless in the limit $\omega \to 0$. The dimensionality in eq. (2.9) is determined by the geometry of the sample. For example, a system with a small volume corresponds to the zero dimensional case.

The diffusion modes can be considered as "Goldstone modes" which appear as a consequence of a spontaneous breaking the symmetry. Below, a method based on integration over both commuting and anticommuting Grassmann variables is considered. This method gives the possibility to speak about the spontaneous breaking the symmetry. Using this method one can solve very difficult problems which can not be solved by other methods.

3. SUPERMATHEMATICS

Let us recall first the basic concept of supermathematics, which deals with objects consisting of both commuting and anticommuting variables. By definition, the classical anticommuting variables χ_i, i = 1,2 ... n satisfy the following relations [8]

$$\{\chi_i, \chi_j\} \equiv \chi_i\chi_j + \chi_j\chi_i = 0 \qquad (3.1)$$

for any $1 \leq i,j \leq n$
From the property (3.1) it follows that the square of an arbitrary variable χ_i is zero

$$\chi_i^2 = 0 \qquad (3.2)$$

Owing to eq. (3.2) any function of the anticommuting variables is a finite polinomial.

It is very convenient to define for any χ the complex conjugate χ^*. The variables χ_i^* anticommute with each other and with all the variables χ_i. By the complex conjugate of a product of variables $X_i, ... X_n$ we mean the product of the complex conjugates

$$(\chi_1\chi_2\cdots\chi_n)^* = \chi_1^*\chi_2^*\cdots\chi_n^*$$

Let as define the inverse operation as

$$(\chi_i^*)^* = -\chi_i$$

It follows from these definitions that the quantity $\chi^*\chi$ is "real".

Let us call a sequence of anticommuting χ and commuting S variables a supervector Φ

$$\Phi = \begin{pmatrix} X \\ S \end{pmatrix}, \chi = \begin{pmatrix} \chi_1 \\ \chi_2 \\ \vdots \\ \chi_n \end{pmatrix} \quad S = \begin{pmatrix} S_1 \\ S_2 \\ \vdots \\ S_n \end{pmatrix} \quad (3.3)$$

In a usual way one can introduce the transposed supervector Φ^T and the supervector $\Phi^+ = (\Phi^T)^*$. Then one can introduce the scalar product and all other constructions of the linear algebra. The simplest supermatrix F has the form

$$F = \begin{pmatrix} a & \delta \\ \rho & b \end{pmatrix} \quad (3.4)$$

where a and b are matrices consisting of commuting variables, δ and ρ contain only anticommuting ones. By definition, the transposed supermatrix has the form

$$F^T = \begin{pmatrix} a^T & \rho^T \\ \sigma^T & b^T \end{pmatrix}$$

With this definition one has

$$(F_1 F_2)^T = F_2^T F_1^T$$

The supertrace STr of the supermatrix F is defined as follows

$$STrF = Tra - Trb \quad (3.5)$$

It follows from (3.5) that

$$STr(F_1 F_2 ... F_n) = STr(F_n F_1 ... F_{n-1})$$

Now, let us consider integrals over the anticommuting variables. These integrals were first introduced by Berezin [8]. They are defined as follows

$$\int d\chi_i = \int d\chi_i^* = 0, \int \chi_i d\chi_i = \int \chi_i^* d\chi_i^* = 1 \quad (3.6)$$

It follows immediately from the definitions (3.6) that for an arbitrary matrix A one has

$$\int exp(-\vec{\chi}^* A \vec{\chi}) d\chi_1^* d\chi_1 ... d\chi_n^* d\chi_n = det A, \quad \vec{\chi} = (\chi_1, \chi_2 \cdots \chi_n) \quad (3.7)$$

The result of the integration (3.7) differs from the corresponding result for the integration over commuting variables by the presence of detA in R.H.S instead of $(detA)^{-1}$.

This difference is the basis of the super symmetry method. For more details about the supermathematics one can use the review[9].

4. THE σ-MODEL

Using the definition of supervectors (3.3) and eq. (3.7) one can write down the Green functions $G^{R,A}$ in the following form

$$G_\epsilon^{R,A}(r,r') = \pm i \int \Phi^1(r)\Phi^1(r') e^{i\Phi^+(\pm(\epsilon-H\pm i\delta))\Phi} \, D\Phi \qquad (4.1)$$

Let us note the absence of the weight denominator in eq.(4.1), which follows from the eq. (3.7). It enables us to average the Green function (4.1) over the impurities before all other calculations. With the Gaussian distribution (2.7) one can obtain immediately

$$< G_\epsilon^{R,A}(r,r') > = \mp \int \Phi^1(r)\Phi^1(r') e^{i\Phi^+(\pm(\epsilon-H\pm i\delta))\Phi - \frac{1}{4\pi\nu\tau}(\Phi^+\Phi)^2} \, D\Phi \qquad (4.2)$$

The representation for the averaged Green function (4.2) was used for the first time in [10].

In order to calculate a product of two Green functions one should double the number of the components of the supervectors. For the correlation function $K(r,\epsilon,\omega)$ (2.5) one can obtain

$$K(r,\epsilon,\omega) = \int \psi_\alpha^1(r)\bar\psi_\alpha^1(0)\psi_\beta^2(0)\bar\psi_\beta^2(r) e^{-L}$$

$$L = \int \left[-i\bar\psi H_0 \psi - \frac{i(\omega+i\delta)}{2} \bar\psi \wedge \psi + \frac{1}{4\pi\nu\tau}(\bar\psi\psi)^2 \right] dr \qquad (4.3)$$

where

$$\psi^a = \frac{1}{\sqrt{2}} \begin{pmatrix} \chi^a \\ \chi^{a*} \\ S^a \\ S^{a*} \end{pmatrix}, a = 1,2$$

$\bar\psi$ is the vector conjugate with respect to ψ. The definition of the conjugation can be found in the review [9]. The matrix $\wedge$ is a 8 x 8 matrix which can be written in the form

$$\wedge = \begin{pmatrix} 1 & 0 \\ 0 & -1 \end{pmatrix}$$

The eq. (4.3) has the form of a field theory without any disorder. It enables us to use very well developed methods.

The most natural way to study the Langrangian (4.3) is to use a mean field theory. In this approximation one substitutes the integration term in eq. (4.3) by the following quadratic expressions

$$\begin{aligned}(\bar\psi\psi)^2 &\equiv \bar\psi_\alpha\psi_\alpha\bar\psi_\beta\psi_\beta \to <\bar\psi_\alpha\psi_\alpha>\bar\psi_\beta\psi_\beta + \\ &+ \bar\psi_\alpha<\psi_\alpha\bar\psi_\beta>\psi_\beta + \underbrace{\bar\psi_\alpha\psi_\alpha\bar\psi_\beta\psi_\beta}\end{aligned} \qquad (4.4)$$

The averages in eq. (4.4) in the mean field scheme do not depend on coordinates and can be found from self consistency equations. These averages can play a role of an order parameter. In problems under consideration fluctuations of this order parameter are very important, because they give divergent contributions analogous to the one given by eq. (2.9). The region of small momenta is the most important. One can generalize the mean field scheme (4.4) to take into account slow variations of the order parameter. First, let us approximate the interaction term L_{int} in eq. (4.3) by the following expression

$$L_{int} = \frac{1}{4\pi\nu\tau} \sum_{p_1+p_2=p_3+p_4} (\bar{\psi}_{p_1}\psi_{p_2})(\bar{\psi}_{p_3}\psi_{p_4}) =$$

$$= \sum_{\substack{p_1 p_2 \\ q<q_0}} [(\bar{\psi}_{p_1}\psi_{p_2})(\bar{\psi}_{p_1+q}\psi_{-p_2-q}) + (\bar{\psi}_{p_1}\psi_{p_2})(\bar{\psi}_{-p_2-q}\psi_{-p_1+q}) +$$

$$+ (\bar{\psi}_{p_1}\psi_{-p_1+q})(\bar{\psi}_{p_2}\psi_{-p_2-q})] \quad (4.5)$$

where q_0 is a cutoff. By the order of magnitude q_0 is equal to the inverse mean free path.

The term L_{int} can be decoupled with the help of integration over additional slowly varying fields (Hubbard- Stratonovich transformation). Each term is decoupled separately. The third term is decoupled by integration over a scalar. This term only renormalizes the Fermi energy. The first and the second terms are equal to each other and can be decoupled by integration over a supermatrix $Q_{\alpha,\beta}$. As a result one obtains

$$exp(-L_{int}) = \int [exp(-\frac{1}{2\tau}\int(\bar{\psi}Q\psi + \frac{\pi\nu}{4}STrQ^2)dr)]D\psi \quad (4.6)$$

Using eqs.(4.3, 4.6) one can integrate over ψ. As the result one reduces the calculation of the integrals over ψ to the calculation of integrals over Q with the effective "free energy" F[Q]

$$F[Q] = \int dr[-\frac{1}{2}STrln(iH_0 + \frac{i\omega\Lambda}{2} + \frac{Q}{\tau}) + \frac{\pi\nu}{4\tau}STrQ^2] \quad (4.7)$$

where Q is a 8x8 supermatrix with certain symmetry properties.

The next step is to find the saddle point of the free energy functional (4.7). Assuming that the saddle point solution does not depend on coordinates one obtains

$$Q = \frac{1}{\pi\nu}\int[i(\epsilon-\epsilon(p)) - \frac{1}{2}\omega\Lambda) + \frac{Q}{2\tau}]^{-1}dp \quad (4.8)$$

The imaginary part of Q renormalizes the energy ϵ. The real part is more interesting and below only this quantity is considered. At $\omega \neq 0$ only $Q = \Lambda$ is the solution. However, at $\omega = 0$ the solution is degenerate. One can see easily that any supermatrix Q of the form

$$Q = V \Lambda \bar{V} \quad (4.9)$$

for an arbitrary V, satisfying the condition $V\bar{V} = 1$, is the solution.

The situation is completely analogous to that in the theory of superconductivity where one can determine using the BCS equation the modulus of the superconducting order parameter but the phase remains arbitrary. The saddle point approximation used for the calculation of the eigenvalues of the matrix Q is accurate provided $\epsilon_0\tau \gg 1$ and $d > 1$, where ϵ_0 is the Fermi energy, and d is the microscopic dimensionality. In order to describe fluctuations related to the degeneracy of the solution (4.9) one should expand the free energy functional F[Q] (4.7) in gradients of V. The result can be written in the form of a non linear σ-model

$$F = \frac{\pi\nu}{8}\int STr[D(\nabla Q)^2 + 2i\omega \wedge Q]dr \quad (4.10)$$

where $D = \frac{v_0^2 \tau}{d}$ is the classical diffusion coefficient, the supermatrix Q has the form (4.9). Yet, nothing has been said about magnetic and spin-orbit interactions. One can show [9] that they result in the presence of external fields in F[Q] (4.10). This lowers the symmetry of the free energy (4.10). However, in the limit of small moments one can obtain the functional (4.10) again, provided the symmetry of the matrix Q is changed. Depending on the presence of magnetic and spin-orbit interactions three possible symmetries exist. According to the classification [5] the case without magnetic and spin-orbit interactions corresponds to the orthogonal ensemble, a system with magnetic interactions - to the unitary ensemble, and a system which has no magnetic interactions but which has spin-orbit ones, corresponds to the symplectic ensemble.

The supermatrix Q (4.9) can be written in a form which is more convenient for calculations

$$Q = U \begin{pmatrix} \cos \hat{\theta} & i \sin \hat{\theta} \\ -i \sin \hat{\theta} & -\cos \hat{\theta} \end{pmatrix} \bar{U} ,$$

$$\hat{\theta} = \begin{pmatrix} \theta_{11} & 0 \\ 0 & \theta_{22} \end{pmatrix}, U = \begin{pmatrix} u & 0 \\ 0 & v \end{pmatrix}, U\bar{U} = 1 \quad (4.11)$$

The form of the matrices $u, v, \hat{\theta}$ depends on the symmetry of the ensemble under consideration. The most simple is the unitary ensemble. In this case

$$\theta_{11} = \theta, \theta_{22} = i\theta_1, 0 < \theta < \pi, \theta_1 > 0 \quad (4.12)$$

It is seen from (4.11, 4.12) that the structure of the matrix Q is unusual, because it contains not only conventional sines and cosines but also hyperbolic ones. In other words a group of the order parameter Q is noncompact. This property is extremely important and gives rise to many interesting results.

5. LEVEL STATISTICS

Let us first calculate the level correlation function $R(\omega)$(2.6). This quantity, just as the density-density correlation function, contains the products $G^R G^A$. Due to the restricted volume of the sample the corresponding σ-model is effectively zero dimensional. Of course, the frequency ω must be small

$$\omega \ll min\{D/a^2, \tau^{-1}\},$$

where a is the size of the sample. After the calculations described in the previous chapter one obtains

$$R(\omega) = \frac{1}{2} + \frac{1}{128} Re \int Tr((1+\wedge)Q) Tr((1-\wedge)Q) exp(-F_0[Q]) dQ$$

$$F_0[Q] = \frac{i\pi}{\Delta}(\omega + i\delta) STr(\wedge Q) \quad (5.1)$$

where $\Delta = (2\nu V)^{-1}$ is the average level spacing, V is the volume.

The eq. (5.1) shows that the level correlation function $R(\omega)$ can be expressed in terms of a definite integral over the elements of the supermatrix Q. The calculation of the integral can be simplified considerably due to the symmetry of $F_0[Q]$. Using the parametrization (4.11) one can see that $F_0[Q]$ contains the elements of the matrix $\hat{\theta}$ only. The elements of the matrix u enter only the preexponential in eq. (5.1). After the integration over the elements of the matrix u one obtains an integral over the elements of the matrix $\hat{\theta}$. We write down this integral in the explicit form for the unitary case

$$R_{unit}(x) = 1 + \frac{1}{2} Re \int_{-1}^{1} \int_{1}^{\infty} exp(i(x-i\delta)(\lambda - \lambda_1)) d\lambda d\lambda_1 \quad (5.2)$$

where
$$x = \frac{\pi\omega}{\Delta}, \lambda = \cos\theta, \lambda_1 = \cosh\theta_1$$

The corresponding integrals for the orthogonal and symplectic cases contain three variables and are much more complicated. However, all these integrals can be calculated exactly [9]. As the results one has

$$R_{ort}(x) = 1 - \frac{\sin^2 x}{x^2} - \frac{d}{dx}\left(\frac{\sin x}{x}\right)\int_1^\infty \frac{\sin xt \, dt}{t}$$

$$R_{init}(x) = 1 - \frac{\sin^2 x}{x^2}$$

$$R_{symp}(x) = 1 - \frac{\sin^2 x}{x^2} + \frac{d}{dx}\left(\frac{\sin x}{x}\right)\int_0^1 \frac{\sin xt}{t}dt \qquad (5.3)$$

The formulae (5.3) coincide exactly with the famous results obtained from the statistical hypotheses [11]. Let us emphasize that eqs.(5.3) were obtained from the first principles. Besides, the calculations presented above are much more simple than those used in [11]. The method of the study of the level statistics with the help of the supermatrix σ-model is successfully applied for studying problems of complex nuclei [12].

The knowledge of the level correlation function gives the possibility to determine the response to an electromagnetic field. The calculation of the response was done in the work [13]. The expression (5.3) were obtained in the limit $\epsilon_0\tau \gg 1$ when the sample is not very dirty. Another important condition is $\tau^{-1} \gg \Delta$. Only in this case one can speak about a random level distribution. Apparently eqs. (5.3) are not valid for dirty samples when $\epsilon_0\tau \simeq 1$. The maximal size of samples for which eqs. (5.3) can be used is determined by inelastic processes. At low temperatures the size can be very large.

6. WIRES

The next problem which can be solved exactly is the problem of localization in wires. Using scaling arguments Thouless [3] predicted the localization for any weak disorder. For his arguments it was not important whether a chain or a thick wire was considered. At the same time from the formal point of view this models are quite different. For the one-dimensional chain very well-developed methods exist [2] which allow one to calculate the frequency-dependence of the conductivity. The N-conducting-channel model was considered using the Landauer method [14]. These authors also made the assertion about localization.

Using the supermatrix σ-model one can solve this problem for arbitrary frequencies. We consider the case of a microscopically three dimensional system. Only the geometry of the sample is assumed to be one-dimensional. The system is microscopically three-dimensional provided the condition

$$\tau^{-1} \gg (sm)^{-1} \qquad (6.1)$$

is fulfilled, where s is the cross section of the wire.

In this limit the classical diffusion exists at small distances. It is supposed as before that the mean free path is much larger than atomic lengths. Provided the frequency is low enough $\omega \ll \{D/s, \tau^{-1}\}$, all non-zero transverse harmonics can be neglected and the σ-model becomes one-dimensional. The density- density correlation function (2.5, 4.3) takes the form

$$K(0,r) = -2\pi^2\nu^2 \int Q_{13}^{12}(0)Q_{31}^{21}(r)exp(-F_1[Q])DQ$$

$$F_1[Q] = \frac{\pi\nu^2 s}{8}STr\int[D\left(\frac{dQ}{dx}\right)^2 + 2i\omega\Lambda Q]dx \qquad (6.2)$$

As usual, when considering one-dimensional models one can use the transfer matrix technique. Using this method one can reduce the calculation of the functional integral (6.2) to the study of solutions of an effective Schrödinger equation corresponding to the Lagrangian $\tilde{F}_1[Q]$ (6.2). The main scheme of the calculations is as follows.

First one introduces a function $\Psi(Q)$

$$\Psi(Q) = \int_{x<x_1} e^{F_1'[Q']} DQ' \qquad (6.3)$$

In eq. (6.3) the integration is performed over the supermatrices Q' in all sites $x < x_1$. It is assumed that at the point x_1 the matrix Q is fixed and therefore the integral depends on this Q. Provided the wire is infinite the function $\Psi(Q)$ does not depend on the coordinate.

Comparing neighboring sites one can obtain the equation for $\Psi(Q)$

$$\Psi(Q) = \lim_{\delta \to 0} \int exp[-\delta STr\left(\frac{D(Q-Q')^2}{\delta^2}\right) + 2i\omega\Lambda Q')]\Psi(Q')dQ' \qquad (6.4)$$

In the limit $\delta \to 0$ eq. (6.4) is a partial differential equation with the variables $\hat{\theta}$. Eq. (6.4) can be rewritten in the form

$$H\Psi(Q) = 0 \qquad (6.5)$$

where H is an effective Hamiltonian, which has a symbolic form

$$H = -\frac{1}{D}\frac{\partial^2}{\partial Q^2} + 2i\omega \wedge Q$$

The explicit form of the Hamiltonian H can be found in the review[9].

The function $\psi(Q)$ enables to calculate correlation functions in one point. For calculation of two point correlation functions one needs the function Γ

$$\Gamma(0,r;Q_0,Q_r) \int_{0<x<r} e^{-F''[Q]} dQ \qquad (6.6)$$

In eq.(6.6) one should integrate over all Q in sites between 0 and r. For the function Γ one obtains

$$\left(-\frac{\partial}{\partial r} + H(Q)\right)\Gamma(0,r;Q_0,Q_r) = \delta(r)\delta(Q_0 - Q_r) \qquad (6.7)$$

Then the correlation function K(0,r) (6.2) takes the form

$$K(r) = -2\pi^2\nu^2 \int (Q_0)_{13}^{12}(Q_r)_{31}^{21}\Psi(Q_0)\Psi(Q_r)\Gamma(0,r;Q_0,Q_r)dQ_0 dQ_r \qquad (6.8)$$

Eqs. (6.5, 6.6, 6.7, 6.8) give the solution of the problem. From these equations one can derive for arbitrary ω partial differential equations which depend only on the variables $\hat{\theta}$. These equations can be solved in the explicit form in the most interesting limit $\omega \to 0$. As the result one obtains for the correlation function K in the time representation at large distances $x \gg L_c$

$$K(t \to \infty, r) \equiv p_\infty(r) = \frac{1}{4\sqrt{\pi}L_c}\left(\frac{\pi^2}{8}\right)^2\left(\frac{4L_c}{r}\right)^{3/2} exp\left(-\frac{r}{4L_c}\right) \qquad (6.9)$$

where $L_c = \pi\nu s D$ is the localization length.

The eq. (6.9) shows that all states are localized though the localization length L_c can be very large. It is interesting to note that the eq. (6.9) coincides exactly with the corresponding expression obtained for chains [2]. Of course this coincidence occurs only in the limit $\omega \to 0$. The eq. (6.9) describes the behavior of all three ensembles. Only the localization L_c length depends on the presence of magnetic and spin-orbit impurities. Using the method presented above one can also calculate the dielectric permeability ϵ

$$\epsilon = 32\zeta(3)e^2\nu L_c^2 \qquad (6.10)$$

where $\zeta(x)$ is the Riemann ζ-function.

In the limit $\omega \gg D/L_c^2$ classical formulae are applicable and one has for the conductivity σ the usual Einstein relation

$$\sigma = 2\nu e^2 D \qquad (6.11)$$

7. WEAK LOCALIZATION. RENORMALIZATION GROUP

The σ-model (4.10) describes the quantum motion of a particle. At large frequencies the motion becomes classical. In this region quantum corrections are small in any dimensionality and can be studied using a perturbation theory. A large number of works on this subject exists. Usually the diagrammatic technique suggested in [6] is used. All these calculations can be considerably simplified with the help of the σ-model (4.10). It is not the aim of these lectures to present all interesting effects arising when taking into account the quantum corrections. Therefore we only show how these corrections can be calculated using the σ-model.

It is very convenient to parametrize the matrix Q. We take the following parametrization

$$Q = \Lambda(1+iP)(1-iP)^{-1}, P = \begin{pmatrix} 0 & B \\ \bar{B} & 0 \end{pmatrix} \qquad (7.1)$$

At high frequencies the main contribution comes from Q which are close to Λ. It corresponds to small P. Therefore, in order to calculate corrections to the classical diffusion one should insert eq.(7.1) into eq.(4.10) and to make an expansion in P. First terms of this expansion in the free energy F can be written as follows

$$F = F^{(0)} + F^{(1)}$$
$$F^{(0)} = \pi\nu \int STr[D\nabla B \nabla \bar{B} - i\omega B\bar{B}]dr$$
$$F^{(1)} = \pi\nu \int STr[D(\nabla B \nabla \bar{B} B\bar{B} + \nabla \bar{B}\nabla B \bar{B}B) - i\omega(B\bar{B})^2]dr \qquad (7.2)$$

The eq.(6.2) for the density-density correlation function K can be used in any dimensionality. Neglecting $F^{(1)}$ in eq.(7.2) one can obtain

$$K(k,\omega) = \frac{2\pi\nu}{Dk^2 - i\omega} \qquad (7.3)$$

Taking into account non quadratic terms in eq.(7.2) one can see that eq.(7.3) is valid in all orders, provided one substitutes the classical diffusion coefficient D by an effective diffusion coefficient $\tilde{D}$ In the lowest order the coefficient $\tilde{D}$ is equal to

$$\tilde{D} = D\left(1 - \frac{1}{\pi\nu}\int \frac{d^dk}{(2\pi)^d} \cdot \frac{1}{-i\omega + Dk^2}\right) \qquad (7.4)$$

Eq. (7.4) coincides exactly with the result (2.9) of the summation of the sequence of the diagrams in Fig. 2. All the diffusion modes are described by fluctuations of the supermatrix Q. The result (7.4) is written for the orthogonal ensemble. Magnetic and spin- orbit interactions change the quantum correction (7.4). Magnetic fields or magnetic impurities destroy the quantum correction, spin-orbit interactions change the sign in eq.(7.4)

In principle, one can calculate next terms in the expansion for any dimensionality. Of course this expansion is meaningful only if the disorder is weak. In low dimensionality the additional restriction on the frequency must be also satisfied. In one and zero dimensionality all terms of the expansion can be written using exact solutions obtained in the previous chapters. In other dimensionalities exact solutions are not available and a calculation of terms of the expansion is not easy. In the two dimensional case the corrections to the classical diffusion coefficient are logarithmic. This fact enables us to use for the calculations the renormalization group scheme. This scheme was first proposed to study the classical Heisenberg model [15] and then applied to many other nonlinear σ-models [16]. It was also used to study localization using the replica σ-models [17].

According to this approach one starts with the free energy F

$$F = \frac{1}{t} \int STr[(\nabla Q)^2 + 2i\tilde{\omega}\Lambda Q]dr \qquad (7.5)$$

where $t = \frac{8}{\pi\nu D}, \tilde{\omega} = \omega/D$

In the two dimensional case all logarithmic integrals must be cut at large momenta by the inverse mean free path l^{-1}. The main idea of the renormalization group approach is to integrate when calculating the partition function Z

$$Z = exp(-F[Q])D \, Q \qquad (7.6)$$

over matrices Q changing in the space with momenta in the interval $(b\lambda, \lambda)$, where $b < 1$. After the integration one must obtain the same expression (7.5) because eq. (7.5) is the only form which can be written for small gradients and frequencies. However, the coefficients t and $\tilde{\omega}$ in this new form differ from the initial ones. After the scaling $\lambda \to \frac{\lambda}{b}$ one obtains the same model (7.5) with the same cutoff but with different coefficients t, $\tilde{\omega}$. It gives the possibility to write equations for t and $\tilde{\omega}$. It turns out that in the supermatrix model under consideration the coefficient $\tilde{\omega}/t$ does not change after the integration. Corrections to the coefficient t are logarithmic in the two dimensional space. In the formal case of small but finite $\epsilon = d - 2$ the procedure can be also made, provided one makes the expansion not only in t but in ϵ too. The result of these calculations depends on the symmetry of the matrices Q. Taking into account the first two orders of the perturbation theory one can obtain

$$\beta(\tilde{t}) \equiv \frac{d\tilde{t}}{d \ln \lambda} = (d-2)\tilde{t} + \alpha \tilde{t}^2 + \frac{1}{2}\tilde{t}^3(1-\alpha^2),$$

$$\tilde{t} = \frac{t}{4 \cdot 2^d \pi (d/2) \Gamma(d/2)} \qquad (7.7)$$

The coefficient $\alpha = 0$ for the unitary case, $\alpha = -1$ for the orthogonal ensemble and $\alpha = 1$ for the symplectic one. Eq. (7.7) was written for the first time by Wegner[18] using the replica σ-model. Let us notice that now four loop contributions have also been calculated [19].

Let us first discuss the two dimensional case. In order to solve eq.(7.7) one needs a boundary condition. It is easy to see that the procedure presented above is valid for momenta λ_0 larger than $\tilde{\omega}^{1/2}$. This quantity serves as a cutoff for small momenta. At large momenta the natural cutoff is $(v_0\tau)^{-1}$, because the σ-model (7.5) is derived for

$\lambda \ll (v_0\tau)^{-1}$. Using these cutoffs one can write the solutions of eq. (7.7). For the orthogonal and symplectic ensembles the solution takes the form

$$\tilde{t}(\omega) = \frac{\tilde{t}_0}{1 + \alpha \tilde{t}_0 \ln 1/\omega\tau} \tag{7.8}$$

where $\tilde{t}_0$ is proportional to the classical resistivity.

For the orthogonal ensemble $\tilde{t}(\omega)$ which is proportional to the resistivity grows when the frequency ω decreases.

Eq. (7.8) is applicable only in the region $\tilde{t}(\omega) \ll 1$. The question about the resistivity in the limit $\omega \to 0$ can not be solved using the renormalization group scheme presented above though sometimes people claim that the eq. (7.8) proves the localization of all states in two dimensions. For the unitary ensemble one can also obtain the growth of the resistivity. However, in this case again one cannot obtain any formulae for small ω.

A completely different situation occurs for the symplectic ensemble. In this case the resistivity decreases when lowering the frequency. In this case eq.(7.8) becomes exact in the limit $\omega \to 0$ because all high order terms in eq. (7.7) can be neglected if $\tilde{t} \to 0$. Therefore one comes to a very surprising result that the static resistivity for a disordered system with spin-orbit impurities vanishes [20]. Let us emphasize that this result is the direct consequence of the existence of the one parameter renomalization group. If everything is correct in this scheme one must accept the result about the zero resistivity.

In $2 + \epsilon$ dimensions the fixed point t_c, where the function $\beta(t)$ turns to zero, exists

$$\beta(t_c) = 0 \tag{7.9}$$

The point t_c corresponds to the metal-insulator transition. In the vicinity of this point one can obtain [21] for the conductivity σ

$$\delta \sim \left(\frac{t_c - t}{t_c}\right)^s, s = -\frac{\epsilon}{\beta'(t_c)} \tag{7.10}$$

The result (7.10) which is obtained with the help of the renormalization group procedure shows that the metal-insulator transition has usual properties of a second order phase transition.

The same result was obtained from the scaling hypothesis [4].
According to this hypothesis the only quantity determining the behavior of the system is the total resistivity of the system. Using this conjecture the behavior of a disordered metal in an arbitrary dimension d was considered. The conductance g of the system with the size bL is related to the conductance of the system with the size L by the following equation

$$g(bL) = f(b, g(L)) \tag{7.11}$$

Taking the limit $b \to 1$ one can rewrite eq. (7.11) in a differential form

$$\frac{d \ln g(L)}{d \ln L} = \beta(g(L)) \tag{7.12}$$

Eq (7.12) has the same form as eq.(7.7) provided one identifies g^{-1} with $\tilde{t}$. It means that the renormalization group study of the σ-model in $2 + \epsilon$ dimensions confirm the scaling hypothesis.

The self-consistent approach developed in [22,23] also gives the power low behavior (7.10). All these results are consistent with each other and seemed to give a very good description of the metal-insulator transition. However, now a clear evidence exists that everything is not so simple and the character of the transition has not still been understood. To the author's opinion the most clear evidence of that comes from the study of the supermatrix σ-model using methods different from the $2 + \epsilon$ expansion. In the next chapter such a study is carried out using an approximation which is analogous to the Bethe-Peierls approximation in the theory of phase transitions. This approximation becomes exact in high dimensionality.

8. EFFECTIVE MEDIUM APPROXIMATION

Eq. (7.5) does not define completely the σ-model for $d \geq 2$ because one needs a cutoff at short distances. In order to have a completely defined model let us consider a lattice version of the σ-model. It can be written in the same way as the Heisenberg model for spin systems.

$$F = -\sum_{i,j} J_{ij} STr\, Q_i Q_j - \beta \sum_i STr \Lambda Q_i \tag{8.1}$$

where β is proportional to $i\omega$.

The matrix Q is specified as before by eq. (4.11). This form of the matrix Q implies that the density of states is constant. One can come to this conclusion expressing the Green function G(4.2) in terms of the matrix Q. Then the density of states is proportional to the following integral

$$\int Q_{11}^{11} e^{-F[Q]}\, DQ \tag{8.2}$$

Calculating the integral in eq.(8.2) one can see that it is equal exactly to unity. The expression (8.1) corresponds to a disordered granulated material. Provided each granule is not very dirty one can introduce the matrix Q. The first term in eq. (8.1) describes the hopping of electrons from one granule to another. It is analogous to the Josephson coupling in granulated superconductors. In spite the fact that each granule is not dirty the macroscopic conductivity can be zero if the coupling constant J_{ij} is small enough. Therefore the metal-insulator transition is in the region of the applicability of eq. (8.1). The free energy (8.1) can be also derived from the N-orbital model [18] in the limit of large N. In the lattice σ-model described by eq. (8.1) the density of states is an irrelevant quantity. However, the point of view that this simplification does not change the critical behavior near the transition is quite reasonable because the density of states cannot have any singularity at the transition point.

In order to obtain kinetic quantities one needs to calculate the density-density correlation function K which takes the form

$$K(o,r) = -2\pi^2 \nu^2 \int Q_{13}^{12}(0) Q_{31}^{21}(r) exp(-F[Q]) DQ \tag{8.3}$$

with the free energy F (8.1)

The form of the free energy F[Q] (8.1) looks like the free energy of the classical Heisenberg model of a ferromagnet. The frequency β plays the role of an external magnetic field. The most natural way to study the model given by eqs. (8.1,8.3) is to use very well developed methods of statistical physics. However, it is not so simple. A "high temperature" expansion does not work because the symmetry group of matrices Q is not compact. If one expands exp(-F[Q]) in J_{ij} assuming that J_{ij} are small one obtains inevitably integrals of the type

$$J^n \int (cosh\theta_1)^n\, e^{-\beta\, cosh\, \theta_1} sinh\, \theta_1 d\, \theta_1 \tag{8.4}$$

In the limit of small frequencies $\beta \to 0$ this integral grows and the high temperature expansion is not valid.

The mean field theory cannot say anything about the kinetics too. According to this method one substitutes one of the matrices Q_i in the first term in eq.(8.1) by the expectation value $<Q>$

$$\sum_{ij} J_{ij} STr \, Q_i Q_j \to 2 STr \sum_j J_{ij} Q_j <Q> \tag{8.5}$$

Then the quantity $<Q>$ is calculated in a self consistent way using this approximate free energy functional. However, in the model under consideration $<Q>$ is equal to the matrix Λ for arbitrary J_{ij}. This result is quite natural because $<Q>$ is proportional to the density of states which is a constant in this model. Therefore the mean field theory cannot help to study the model.

Let us try now to use a "low temperature" expansion which can be valid for large J_{ij}. The expansions presented in the previous chapter were written for this case. In the calculations of the previous chapter one had to make an artificial cutoff at large momenta. In that expansion the noncompactness of the model could not be felt. However, one can make a low temperature expansion without using any artificial cutoff. Following this procedure [24] one rewrites eq. (8.1) in the form

$$F = F_0 + F_1$$
$$F_0 = -2\tilde{J} \sum_i STr\Lambda Q_i \, , \, \tilde{J} = \sum_j J_{ij} - \frac{\beta}{2}$$
$$F_1 = -\sum_{i,j} J_{ij} STr(Q_i - \Lambda)(Q_j - \Lambda) \tag{8.6}$$

Substituting eq. (8.6) in eq. (8.3) and expanding $\exp(-F[Q])$ in F_1 one can write down all terms of the series. It is convenient to represent J_{ij} by a thin line connecting the sites i and j.

When calculating the density-density correlation function K (8.3), only continuous lines consisting of segments J_{ij} contribute. Taking into account one line without any intersections (Fig. 3a)

Fig. 3

one can obtain for K in the momentum representation

$$K(k) = \frac{2\pi^2 \nu^2}{J - J(k)} \tag{8.7}$$

In the limit of small k and ω eq. (8.7) reduces to the usual diffusion pole (7.3). The next step is to take into account graphs, containing loops. For small k and ω all these contributions coincide with the contribution given by the expansion procedure presented in the previous chapter. At first glance such a perturbation theory works well in the case of large J. However, it is not so due to the existence of a divergent subseries arising when expanding $exp(-F_1)$. In order to see this divergence let us estimate the contribution given by the graph.3b which consists of n lines connecting neighboring points i and j. The result depends on the ensemble but the mechanism of the growth of the terms of the perturbation theory is the same for all ensembles. Therefore we present the result only for the unitary ensemble.

Provided the number n of lines connecting two neighboring sites i and j is large the main contribution L_n when integrating over Q_i and Q_j comes from the region $\theta_1 \gg 1$ (we use the parametrization (4.11, 4.12)) and can be written as follows

$$L_n \approx (4J_{ij})^n(-1)^n cosh^2 8\tilde{J}[\int_0^\infty exp(n\theta_1 - 8\tilde{J}cosh\theta_1)d\theta_1]^2 \tag{8.8}$$

In the limit $n \gg \tilde{J}$ one can use the saddle point method. The saddle point θ_{1s} is equal to

$$\theta_{is} \simeq ln(n/4\tilde{J}) \tag{8.9}$$

Integrating near the saddle point in eq. (8.8) one can find

$$L_n = (-1)^n(n-1)!(J_{ij}/\tilde{J})^n(4\tilde{J})^{-2n}cosh^2 8\tilde{J} \tag{8.10}$$

We see from eq. (8.10) that L_n is large for large n even if J is large. Of course, this growth of the terms of the perturbation theory occurs on lattices of any geometry and dimensionality. This growth is not seen in the continual σ-model discussed in the previous chapter.

In order to overcome this difficulty let us perform first a partial summation of the repeated lines. After this all graphs will contain effective lines obtained by the summation of the single lines. No segments of the effective lines can coincide. Representing the effective lines by thick lines Fig. 3c one can draw several typical diagrams.

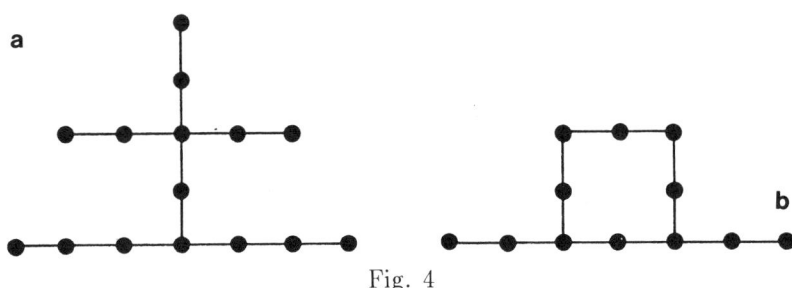

Fig. 4

Of course, it is not possible to sum all the diagrams exactly and one should select the most important ones. In the real three dimensional case all approximations are not controlled. Such a situation is typical in the theory of phase transitions. The problem becomes more simple in the case of high dimensionality $d \gg 1$. In this limit diagrams without loops are most important. All such diagrams have a tree like structure. For example one should take into account the diagram 4a and neglect the diagram 4b. In the main approximation trees and branches do not touch each other and so they are independent. The structure of the tree diagrams enables us to write down an integral equation for the partition function $\psi(Q)$ of an infinite tree with one fixed Q.

$$\Psi(Q) = \int N(Q,Q')Z(Q')\Psi(Q')dQ' \tag{8.11}$$

where

$$N(Q,Q') = exp\left(\frac{\alpha}{4}STrQQ'\right)$$

$$Z(Q) = \Psi^{m-1}(Q)exp\left(\frac{1}{4}\beta STr\Lambda Q)\right)$$

$$\alpha = 8J_{12}, m = 2d - 1$$

The supermatrix σ-model has a very important property. The thermodynamic free energy F

$$F = -ln \int e^{-F[Q]} DQ \tag{8.12}$$

is equal to zero identically. This fact gives the possibility to write the recurrence equation (8.1) for the infinite tree. In this model the question about boundary conditions is not important. Let us notice that the equation (8.11) coincides with the corresponding equation for the Bethe lattice where it is exact. All details concerning the study of the σ-model on the Bethe lattice can be found in the works [25]. More detailed study in a high dimensional cubic lattice was presented in [26] In this lecture we present only the main ideas of these works.

The density-density correlation function K(r) (8.3) can also be written in the considered approximation in a closed form

$$K(r) = -2\pi^2\nu^2 \int Q_{13}^{12} P_{31}(r,Q)Z(Q)\Psi(Q)dQ \tag{8.13}$$

where

$$P(r,Q) - \sum_{r'} w(r-r') \int N(Q,Q')P(r',Q')Z(Q')dQ' +$$

$$+ m \int N_2(Q,Q')P(r,Q')Z(Q')dQ' = \delta(r)Q^{21}\Psi(Q)$$

$$N_2(Q,Q') = \int N(Q,Q)N(Q,Q')Z(Q)dQ$$

$$w(r-r') = \begin{cases} 1, /\bar{r}-\bar{r}'/ = 1 \\ 0 \ /\bar{r}-\bar{r}'/ \neq 1 \end{cases} \tag{8.14}$$

Eq. (8.14) is written for the case of a simple cubic lattice. The form of this equation differs from the form of the corresponding equation for the Bethe lattice [25]. This is due to the difference between the geometry of the Bethe lattice and the geometry of a real d-dimensional lattice.

Eq. (8.11) can also be derived from eq. (8.1) using the following self consistent procedure. Let us single out two neighboring sites 1 and 2. Let the interaction between these two sites be described as before by the first term in (8.1). Let us substitute the interaction of these two sites with all other sites by an interaction with an effective medium. If a matrix Q' corresponds to the site 2, then the interaction Y of the site 2 with the effective medium depends only on Q' and is proportional to m. The condition of the self-consistency can be written in the form

$$expY(Q) = \int N(Q,Q')exp[m\ Y(Q') + \frac{1}{4}\beta STr\Lambda Q']dQ' \tag{8.15}$$

Denoting $\Psi(Q) = expY(Q)$ one can come to eq. (8.1).

In order to study kinetic quantities one should write self-consistency conditions for quantities characterizing hopping from one granule to another. That is why it is necessary to take into account at least two sites. One site quantities can describe only the density of states. It is natural that the conventional mean field theory can not describe the metal-insulator transition.

The self-consistent approximation under consideration is similar to the effective medium approximation used in the percolation theory [27]. It is quite reasonable to call it also the effective medium approximation. This approximation is analogous to the Bethe-Peierls approximation used in the theory of phase transitions. The Bethe-Peierls approximation becomes exact on the Bethe lattice and therefore it is not surprising that eq. (8.11) coincides with the corresponding equation obtained for the Bethe lattice.

Now let us present the main scheme of the calculation of the density-density correlation function K(o,r) with the help of eqs. (8.11, 8.13, 8.14). Eq. (8.14) is linear and can be solved using the expansion in terms of eigenfunctions $\Phi_E(Q)$ of the following integral equation

$$\int N(Q,Q')[Z(Q)Z(q')]^{1/2}\Phi_E(Q')dQ' = E\Phi_E(Q) \tag{8.16}$$

Using this expansion one can rewrite eq. (8.13) in the form

$$K(k) = 2\pi^2 \nu^2 V \sum_E B_E[1 - Ew(k) + mE^2] \tag{8.17}$$

The coefficients B_E can be written in terms of some integrals over Q [26].

It was shown [25] that eq. (8.11) has a singularity at a critical point α_c. For $\alpha < \alpha_c$ a solution $\Psi(Q) = 1$ exists in the limit $\beta \to 0$. At $\alpha > \alpha_c$ a non trivial solution $\Psi(Q)$ appears which decreases in the limit $\theta_1 \to \infty$. The value α_c separates the metal $\alpha > \alpha_c$ and the dielectric $\alpha < \alpha_c$ regions. For the unitary ensemble in the limit $m \gg 1$ the transition point α_c is determined by the equation

$$m\left(\frac{\alpha_c}{2\pi}\right)^{1/2} \ln\frac{2}{\alpha_c} = 1 \tag{8.18}$$

In the metal region the spectrum of eigenvalues E of the integral equation (8.16) is discrete. In the limit of low frequencies it is sufficient to keep in the sum (8.17) the term with the largest eigenvalue E_0 only. In the limit of low frequencies and small momenta one can obtain exactly

$$K(k) = \frac{4\pi\nu}{Dk^2 - i\omega} \tag{8.19}$$

The correlation function K(k) has the usual diffusion form. The diffusion coefficient D can be written as an integral over Q. Due to the rotational symmetry of eq. (8.11) the solution Ψ depends only on variables $\hat{\theta}$ (4.14, 4.15). In the case of the unitary ensemble the explicit expression for D takes the form

$$D = \frac{m}{2\pi\nu} \int_0^\infty \int_0^\pi \left[\left(\frac{\partial \Psi_0}{\partial \theta}\right)^2 + \left(\frac{\partial \Psi_0}{\partial \theta_1}\right)^2\right] \Psi_0^{m-1} \frac{d\theta d\theta}{(\cosh\theta_1 - \cos\theta)^2} \qquad (8.20)$$

Eq. (8.20) determines the diffusion coefficient for arbitrary α in the metal region. For large α eq. (8.20) gives the classical diffusion coefficient. The diffusion coefficient determined by eq. (8.20) has a very interesting critical behavior [25]

$$D = p(\alpha - \alpha_c)^{-3/2} exp(-q(\alpha - \alpha_c)^{-1/2}) \qquad (8.21)$$

Eq. (8.21) shows that the metal-insulator transition is continuous but it is not similar to conventional second order phase transitions, where the critical behavior obeys power lows. An analysis shows that the exponential decrease (8.21) of the diffusion coefficient is due to the non compact group of the matrices Q. The approximation used to obtain eq. (8.21) is quite usual for the theory of phase transitions and any compact model would give in this approximation the power low behavior.

In the dielectric region $\alpha < \alpha_c$ the spectrum of eigenfunctions Φ_E (8.16) is continuous.

In the limit of low frequencies the coefficients B_E (8.17) take the form [25]

$$B_E = -\frac{\epsilon^2 a(\epsilon)}{i\omega\pi\nu}$$

where $a(\epsilon)$ is a function of a variable ϵ. This function can be expressed in terms of integrals of a combination of functions Ψ and Φ_E [25]. After some transformations one can obtain

$$K(k) = \frac{2\pi\nu}{-i\omega} \int_0^\infty \frac{a^2(\epsilon)\epsilon^2 d\epsilon}{1 - w(k)E(\epsilon) + mE^2(\epsilon)} \qquad (8.22)$$

where $E(\epsilon) = \Gamma_\epsilon(\alpha)$.

The form of the function $\Gamma_\epsilon(\alpha)$ can be found in [25].

A behavior of wave functions of the disordered systems can be described by the function $p_\infty(r, \epsilon)$

$$p_\infty(r, \epsilon) = < \sum_{diser} \delta(\epsilon - \epsilon_n)(u_n(0)u_n^\star(r))^2 > \qquad (8.23)$$

where u_n and ϵ_n are eigenfunctions and eigenenergies of the original Schrödinger equation. In eq. (8.22) the brackets stand for the averaging over impurities, the sum is taken over discrete levels. In the considered model the dependence on the energy ϵ reduces to a dependence on α which is a function of ϵ. Knowing the density-density correlation function K one can obtain the function $p_\infty(r, \alpha)$

$$p_\infty(r, \alpha) = \tilde{K}(r, t \to \infty) \qquad (8.24)$$

In the limit of large distances $r \gg 1$ the calculation of $p_\infty(r, \alpha)$ becomes simpler because the main contribution comes from the region of small ϵ. In principle, an expression for $p_\infty(r, \alpha)$ can be written for any $r \gg 1$.

In the limits $1 \ll r \ll l$ and $r \gg l$, where l is the localization length, the asymtotics take the form

$$p_\infty(r, \alpha) = \begin{cases} const.r^{\frac{d+2}{2}} l^{-\frac{d}{2}} exp(-r/4l) &, r \ll l, \\ \frac{const}{r^{d+1}}, r \ll l \end{cases} \qquad (8.25)$$

The localization length near the critical point α_c is large and diverges as $\alpha \to \alpha_c$

$$l = const(\alpha_c - \alpha)^{-1} \qquad (8.26)$$

Eq. (8.25) shows that at large distances $r \gg l$ the decrease of wave functions is exponential. In the region $r \ll l$ it obeys the power low. At the transition point l diverges and the wave function decreases at any distances in the power low.

Using the function $p_\infty(r, \alpha)$ one can calculate the polarizability κ

$$\kappa \delta_{\alpha\beta} = e^2 \int r_\alpha r_\beta P_\infty(r, \alpha) dV \qquad (8.27)$$

Simple calculations show that the polarizability is proportional to the localization length l

$$\kappa = const \cdot l \qquad (8.28)$$

The result (8.27) contradicts to the hypothesis about the existence of a one parameter scaling. If this hypothesis were true the polarizability would be proportional to the square of the localization length. This disagreement is again a consequence of the noncompactness of the group of matrices Q. Not only the localization length but also the intersite distance (which is taken as unity) enters eqs. (8.25, 8.27). As we have seen in the previous chapter the $2 + \epsilon$ expansion confirms the one parameter scaling. However, this method takes into account only perturbative terms. The question about the noncompactness of the group does not arise in this approach. Apparently, one should consider properly the contribution of short distances which is neglected in continuum σ-models.

9. CONCLUSIONS

It is seen from the results of the previous chapters that the supersymmetry approach enables us to study many interesting problems of disordered metals using methods of phase transitions and field theory. The supersymmetry method is more efficient than the replica one. Of all results presented above only the perturbation theory and $2 + \epsilon$ expansion can be obtained using the replica σ-models [17]. These models fail already to give results for the problem of level statistic which correspond to the zero dimensional σ-models. The results obtained for the level statistics coincide with the corresponding results derived by other methods starting from statistical hypotheses. The results for wires in the limit $\omega \to 0$ coincide with the results for chains [2]. Such a coincidence cannot be occasional and therefore confirms the correctness of the approximations made when deriving the supermatrix σ-model. The main property of the σ-model is the noncompactness of the group of the symmetry of matrices Q. In fact this property determines the form of the density-density correlation function K(r)

$$K(r) = \frac{p_\infty(r)}{-i\omega} \qquad (9.1)$$

in the dielectric region.

As we have seen in the previous chapters the quantity ω in the σ-model is analogous to an external field in spin models. For compact models any small external field cannot be in the denominator of a spin-spin correlation function. In the model under consideration integration over large $\theta_1 \sim ln\ 1/\omega$ gives ω in the denominator in eq.(9.1). So, the noncompactness is not a result of an approximation but is an intrinsic property for the disordered systems.

A very strong disagreement between the results obtained by the $2 + \epsilon$ expansion and the results obtained by the effective medium approximation exists in the supermatrix σ- model. Of course, one can argue that the results obtained in the dimensionality $2 + \epsilon$ for small ϵ need not agree with the results obtained with the help of an approximation which becomes exact in high dimensionality. However, this disagreement seems to be more deep. As we discussed in the previous chapter the critical behavior is determined by the whole group of matrices Q. At the same time $2 + \epsilon$ expansion is in fact a perturbation theory near $Q = \lambda$. This perturbation theory does not feel the whole group. Therefore it is quite probable that something is lost in this approach. Unfortunately this question is still not clear.

The results of the $2 + \epsilon$ expansion fit very well on the scheme of the one parameter scaling [4]. The main quantity entering this hypothesis is the conductance g of the system. However, it is not so clear what the conductance of a disordered system is. According to calculations [28] this quantity fluctuates very much when changing the configuration of impurities

$$<g^2> - <g>^2 \sim e^2/h \qquad (9.2)$$

Far from the critical point in the region of a good metal the conductance is large and the fluctuations can be neglected. In this region the one-parameter scaling works very well. However, near the transition and in the localized region fluctuations can be of the order of the mean value and it is meaningless to speak about scaling the conductance. It is quite reasonable to suggest that a not trivial distribution function scales with changing the volume [29]. The property of the noncompactness can be related to the existence of the distribution function. Unfortunately this question is far from to be clear.

References

1. P.W. Anderson, Phys. Rev. **109**, 1492, 1958

2. N.F. Mott, W.D. Twose, Adv. Phys. **10**, 107, 1961;
   V.L. Berenzinsky, Zh. eksp. teor. Fiz. **65**, 1251, 1973
   (Sov. Phys. JETP, **38**, 620, 1974)

3. D.J. Thouless, J. Phys. C **8**, 1803, 1975;
   Phys. Rev. Lett. **39**, 1167, 1977

4. E. Abrahams, P.W. Anderson, D.C. Licciardello, T.V. Ramakrishnan, Phys. Rev. Lett. **42**, 673, 1979

5. E.P. Wigner, Ann. Math. **53** 36, 1951

6. A.A. Abrikosov, L.P. Gor'kov, I.E. Dzyaloshinsky: Methods of Quantum Field Theory in Statistical Physics (New York: Prentice Hall).

7. L.P. Gor'kov, A.I. Larkin, D.E. Khmel'nitsky, Pis'ma Zh. eksp. theor. Fiz. **30**, 248, 1979 (Sov. JETP Lett. **30**, 228, 1979)

8. F.A. Berezin, Dokl. Akad. Nauk SSSR, **137**, 31, 1961; Method vtorichnogo Kvantovaniya "Nauka", 1965 (The Method of Second Quantization). (English translation published by Academic Press, N.Y: Math. Z. **1**, 3, 1967

9. K.B. Efetov, Adv. in Physics **32**, 53, 1983

10. A.J. McKane, Phys. Lett. A **76**, 33, 1980.

11. F.J. Dyson, J. Math. Phys. **3**, 140,157,166, 1962;
    M.L. Mehta, F.J. Dyson, J. Math. Phys. **4**, 713, 1963

12. J.J.M. Verbaarschot, H.A. Weidenmüller, M. R. Zirnbauer, Phys. Rep. **129**, 367, 1985

13. L.P. Gor'kov, G.M. Eliashberg, Zh. eksp. teor. Fiz. **48**, 1407, 1965 (Soviet Phys. JETP, **21**, 940)

14. P.W. Anderson, D.J. Thouless, E. Abrahams, D.S. Fisher, Phys. Rev. B**22**, 3519, 1980

15. A.M. Polyakov, Phys. Lett. B**59**, 79, 1975

16. E. Brezin, J. Zinn-Justin, Phys. Rev. B**14**, 3110, 1976
    E. Brezin, J. Zinn-Justin, J.C. Le Guillou, Phys. Rev. D**14**, 2615, 1976
    E. Brezin, S. Hikami, J. Zinn-Justin, Nucl. Phys. B**165**, 528, 1980

17. L. Schäfer, F. Wegner, Z. Phys. B**38**, 113, 1980
    K.B. Efetov, A.I. Larkin, D.E. Khmel'nitskii, Zh. eksp. teor. Fiz. **79**, 1120, 1980 (Sov. Phys. JETP, **52**, 568, 1980)

18. F. Wegner, Z. Phys. B**44**, 9, 1976

19. F. Wegner, Nucl. Phys. B, **316**, 663, 1989

20. S. Hikami, A.I. Larkin, Y. Nagaoka, Progr. Theor. Phys. **62**, 707, 1980

21. F. Wegner, Z. Phys. **25**, 327, 1976

22. W. Götze, Sol. St. Commun. **27**, 1393, 1978; Phil. Mag.B**43**, 219, 1981; J. Phys. C**12**, 1279, 1979

23. P. Wölfle, D. Vollhardt, Phys. Rev. Lett. **45**, 842, 1980; Phys. Rev. B**22**, 4666, 1980

24. V.G. Vaks, A.I. Larkin, S.A. Pikin, Zh. eksp. teor. Fiz. **52**, 1089, 1967

25. K.B. Efetov, Pis'ma v Zh. eksp. teor. Fiz. **40**, 17, 1984 (Sov. Phys. JETP Lett. **40**, 738, 1984); Zh. eksp. teor. Fiz. **88**, 1032, 1984 (Sov. Phys. JETP **61**, 606, 1984), Zh. eksp. teor. Fiz. **92**, 638, 1987 (Sov. Phys. JETP **65**, 360, 1987); Zh. eksp. teor. Fiz. **93**, 1125, 1987 (Sov. Phys. JETP **66**, 634, 1987); M.R. Zirnbauer, Nucl. Phys. B, 265, 375, 1986; Phys. Rev. **34**, 6394, 1986

26. K.B. Efetov, Zh. eksp. teor. Fiz. **94**, 357, 1988 (Sov. Phys. JETP, **67**, 199, 1988)

27. S. Kirkpatrick, Rev. Mod. Phys. **45**, 574, 1973

28. B.L. Altshuler, Pis'ma Zh. eksp. teor. Fiz. **41**, 530, 1985 (Sov. Phys. JETP Lett. **41**, 648,1985); P.A. Lee, A.D. Stone, Phys. Rev. Lett. **55**, 1622, (1985)

29. B. Shapiro, Phys. Rev. B**34**, 4394, 1986

AN INTRODUCTION TO THE DYNAMICS OF QUENCH-DISORDERED SPIN SYSTEMS

R. Bruinsma

Physics Department & Solid-Sate Science Center
University of California, Los Angeles
Los Angeles, CA, 90024

INTRODUCTION

The interest in quench-disordered spin systems derives from two sources. On the one hand, there are many problems both in experimental physics and in technology where we are interested in the effect of impurities on the material properties of magnets. Impurity effects are, for instance, important for the operation of commercial magnets (coercive fields) as well as for magnetic recording devices. Frequently, non-magnetic systems can also be represented by some appropriate spin model and again impurities may play an important role. An example is the phase separation of binary fluids in a porous medium which can be represented as a random-field Ising model. Spin-models with disorder are also useful outside condensed-matter physics. Spin-glass type models are, for instance, used as models for neural networks.

A second source of interest derives from the fact that quenched disorder produces extraordinary technical difficulties for the conventional methods of statistical mechanics. In particular, the powerful field-theoretic methods which were so succesfull for pure systems have not been very productive for quenched disordered spin-systems. In certain cases they even led to explicitly erroneous results. Developing more appropriate methods is a major challenge for condensed-matter theory.

In this lecture, I will discuss simple theoretical methods (domain-wall arguments and linear response theory) which have proven themselves to be helpful tools for understanding systems with quenched disorder. I will concentrate on random field effects and not address the spin-glass problem although these methods have been applied as well to spin-glasses. The lecture will be introductory and I will try to emphasize the underlying physical assumptions of the arguments. In addition to my own work - which was in collaboration with G.Aeplli - the arguments discussed below represent the work of many colleagues. However, because the literature is so mountainous, no references are given. You can find references in recent reviews by J.Villain and by T.Natterman.

Incidentally, we use the name "quenched disorder" to indicate the presence of immobile impurities. If the impurities are mobile and in thermal equilibrium, we call the disorder "annealed". Both terms originate

in metallurgy. "Quenching" is the process where red-hot steel is immersed
in cold water. As a consequence, impurities are frozen in. "Annealing" is
the inverse process: thermal cycling to get rid of impurities and defects.

LIFSHITZ-SHLYOZOV LAW

Before we address ourselves to the problem of the dynamics of quenched
disorder, it is important to first go over the dynamics of pure spin
systems. This will act as a "base-line" against which we can gauge the
effect of impurities. Let

$$H = -J \sum_{i,j} S_i S_j - h \sum_i S_i \qquad (1)$$

be the Hamiltonian of the Ising model with J the nearest-neighbour
exchange constant and with h an applied magnetic field. Suppose we first
introduce dynamics via the Monte-Carlo method, as one would for a computer
calculation. Heat up the spin system to a temperature far above the Curie
temperature T_c so we are deep into the paramagnetic phase. Next, rapidly
cool (quench) to a temperature far below T_c. The equilibrium phase has
long-range ferromagnetic order.

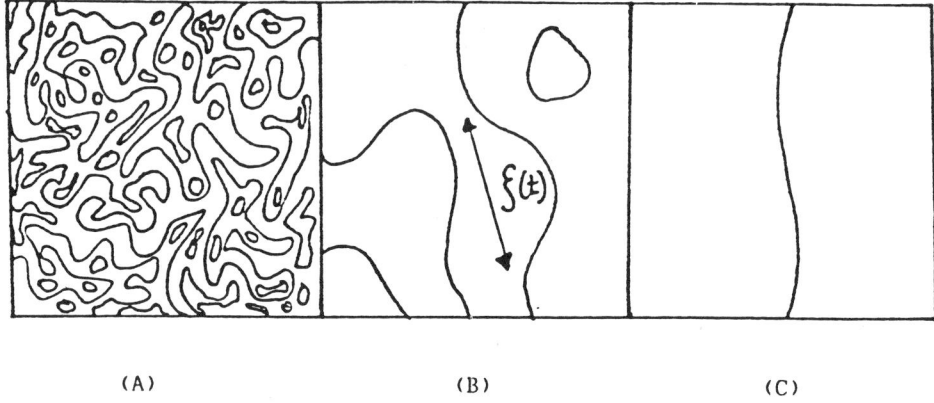

(A)　　　　　　　　　　(B)　　　　　　　　　　(C)

Fig.1. Coarsening of a quench-cooled Ising model

But we are not in equilibrium. We are starting from a highly
disordered spin configuration and it is going to take a while until we can
reach the ordered groundstate. How long? Let's take a look at the result
of a computer simulation for h=0 (Fig.1 A-C).

Initially, you have a spin configuration with disorder on atomic
length scales. After a short time (of order the spin-flip relaxation
time), you see an intricate pattern of small magnetic domains. Each domain
has either up or down spins. The domains are separated by a net of highly
curved domain-wall boundaries (Fig.1A). With time, the number of domains

decreases and the domain walls straighten out (Fig.1B). This process is called "coarsening". We can assign a characteristic length-scale $\xi(t)$ to the domain pattern which is both the typical curvature radius of the boundary as well as the typical domain size. With time $\xi(t)$, increases until eventually it becomes of the size of the system (Fig.1C). The growth of ξ obeys a power-law:

$$\xi(t) \propto t^p \qquad (2)$$

with p close to 1/2.

To describe coarsening theoretically, we first go to the short-time regime. We will not try to tell what individual spins are doing, but instead group them together into "cells" whose size is large compared to the lattice constant but small compared to the system size. Denote by $\vec{r}$ the location of a cell and by $M(\vec{r})$ the average of S_i over a cell. The quantity $M(\vec{r})$ is continuously varying for large enough cell size while it varies smoothly from cell to cell. This coarse-graining procedure considerably simplifies the book-keeping. Change from S_i to the new variable $M(\vec{r})$. After a standard calculation, the effective Hamiltonian for $M(\vec{r})$ is found to be the well-known Ginzburg-Landau free energy :

$$F = \int d^3r \, f(\vec{r}) \qquad (3)$$

with

$$f(\vec{r}) \cong 1/2 \left(\vec{\nabla}M\right)^2 + 1/2 \, r(T) \, M^2 + 1/4 \, u \, M^4 - h \, M \qquad (4)$$

the free-energy density. The temperature dependent parameter r(T) changes sign at T_c :

$$r(T) \propto (T-T_c) \qquad (5)$$

The free-energy density f has, as a function of M, a double-well shape (see Fig.2).

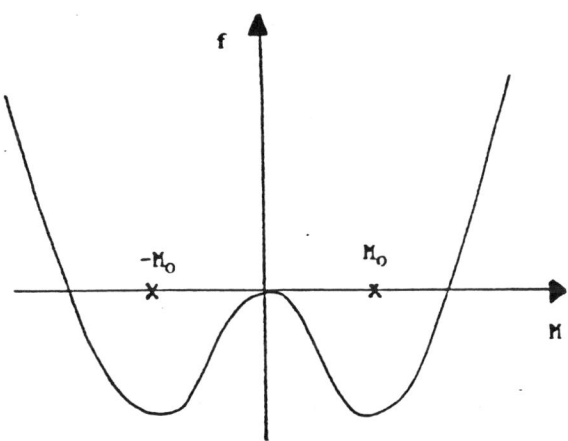

Fig.2. Ginzburg-Landau free energy

The two minima of f are located at $M = \mp M_0$, where $M_0 = \sqrt{(-r/u)}$. These two free energy minima correspond to states with up or down uniform magnetization. In the groundstate $M(\vec{r})$ is M_0 (or $-M_0$). The short-range order seen in our computer experiment is achieved by having $M=M_0$ or $M=-M_0$ nearly everywhere so the spins are in nearly everywhere in local thermal equilibrium. The boundaries between the regions with $M=M_0$ and $M=-M_0$ are the domain-walls. We will not concern ourselves with the very early formation history of domain-walls but assume that somehow a random network of walls has been established.

It is not difficult to calculate the internal structure of as domain wall. If as we assumed, the spin system is in local thermal equilibrium then the magnetization can be obtained by minimizing the free energy density:

$$\delta f/\delta M = 0 \qquad (6)$$

or, with h=0,

$$d^2M/dz^2 - rM - uM^3 = 0 \qquad (7)$$

with z the direction along which M varies, i.e, along the local normal of the domain-wall. Domain-walls correspond to solutions of Eq.7 with boundary conditions $M(z=\infty)=M_0$ and $M(z=-\infty)=-M_0$. It is straightforward to construct such solutions but we are here only interested in the characteristic internal length-scale of the domain-wall: the width w. By redefining $z'=z\sqrt{(-r)}$ and $M'=M/M_0$ in Eq.7, we arrive at a scale-free equation for $M'(z')$. This means that $1/\sqrt{(-r)}$ is the only length-scale of the problem so the domain-wall width must be proportional to

$$w = \sqrt{(-1/r)} \qquad (8)$$

As we shall see, this is also the characteristic decay length for the spin-spin correlation function (in the mean-field approximation). It diverges at T_c since r vanishes at T_c.

Having found the domain-wall width, we can compute the domain wall energy. We will define the domain-wall energy by

$$E\{\text{domain wall}\} = F - F\{\text{no domain-wall}\} \qquad (9a)$$

$$= (\text{domain-wall area}) * \int_{-\infty}^{\infty} dz\, (f(z)-f(\infty)) \qquad (9b)$$

If Δf is a typical value for $f(z)-f(\infty)$, then we can estimate the domain-wall energy as

$$E\{\text{domain-wall}\} \cong \sigma * (\text{domain-wall area}) \qquad (10)$$

with $\sigma = w*\Delta f$. The quantity Δf can be estimated from Fig.2: it must be of order the barrier between the two minima of f. This barrier in turn is the the depth of the free-energy minima so, using Eq.4, $\Delta f \sim rM_0^2$.

The proportionality factor σ between domain wall energy and area is called the surface tension σ. This quantity will play an important role in what follows. From our estimates for w and Δf, it follows that

$$\sigma \cong r^{3/2}/u \qquad (11)$$

The surface tension vanishes, according to Eq.11, at the Curie temperature as $(T_c-T)^{3/2}$ while the domain-wall width diverges as $1/(T_c-T)^{1/2}$. The domain-wall thus becomes fat and soft at T_c. It

indicates that a domain-wall description really is only suitable as
long as we are not too close to T_c. For the present purposes that is of
course just fine since we quench-cooled to a temperature far below T_c but
we should be careful not to extrapolate our results to T_c. (Actually,
Eq.11 is only a mean-field result; the real exponent of r is not 3/2).

Assume we are now in a late stage of the relaxation process towards
equilibrium. We have a dilute network of gently curving domain-walls. The
typical domain size $\xi(t)$ is large compared to the width of the wall so
Eq.10 should be valid. We now must find a convenient parametrization for
the different possible configurations of a domain-wall. Let $d(\vec{r})$ be the
elevation of the domain-wall over some two-dimensional flat surface (see
Fig.3).

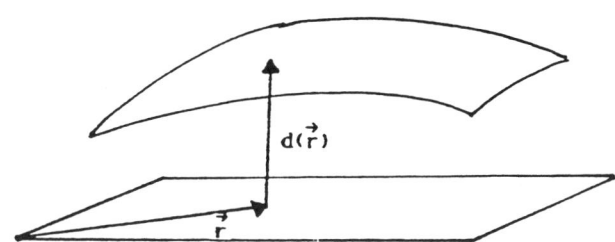

Fig.3. Solid-on-Solid parametrization of a domain wall

Using analytical geometry, we can express the area of a surface with no
overhangs in terms of $d(\vec{r})$. The domain-wall energy in this "Solid-on-
Solid" parametrization is

$$E = E\{\text{Domain wall}\} + E\{\text{Magnetic field energy}\} \qquad (12a)$$

or

$$E = \int d^2r \left[\sigma\left(1 + (\vec{\nabla}d)^2\right)^{1/2} - 2hd(\vec{r}) \right] \qquad (12b)$$

The last term is the energy cost of flipping the spins in a field.

We have studiously avoided specifying the actual equation of motion.
That is because results arrived at from pure statistical mechanics
arguments alone without using a particular dynamics scheme have a broader
generality. To predict how domain walls smooth out their curvature, we
must construct such an equation of motion. This we will do as follows.
Slowly moving macroscopic objects, such as domain walls, generate energy
dissipation. The simplest dissipative kinetics assumes that the velocity
is proportional to the driving force, just like a marble sinking in a
pond. In a quantitative form,

$$\partial d/\partial t = -\Gamma \; \delta E\{d\}/\delta d \qquad (13)$$

with Γ the "Onsager constant". (On occasion, you may see the use of
Hamiltonian dynamics for extended defects. In a condensed matter context

this is always incorrect). Using Eq.12b

$$\partial d/\partial t = \Gamma \{ \sigma \nabla^2 d + hd \} \qquad (14)$$

As an example how Eq.14 works, consider a spherical domain of radius R(t) and watch it collaps (for h=0) under its own surface tension. Since $-\nabla^2 d$ is the radius of curvature, Eq.14 reduces to

$$dR/dt = - \Gamma\sigma/R \qquad (15)$$

with solution

$$R(t) = \left(R^2(0) - 2\Gamma\sigma t \right)^{1/2} \qquad (16)$$

According to Eq.16, after a time

$$t^*(R(0) = R^2(0)/2\Gamma\sigma \qquad (17)$$

the droplet has collapsed. Now return to our original problem: how does a quench-disordered state evolve in time ? After a time t, all droplets whose initial radius was less then R(0) with $t=R(0)^2/2\Gamma\sigma$, (and all interfacial roughness on length-scales less then R(0)) should have vanished. The characteristic length-scale $\xi(t)$ of the domain wall pattern thus should exceed R(0). However, on length-scales large compared to R(0), we expect the spin-configuration to be disordered. This suggests that we should identify R(0) with $\xi(t)$, or

$$\xi(t) \cong (\Gamma\sigma t)^{1/2} \qquad (18)$$

This is a well known result, called the Lifshitz-Shlyozov law. The typical domain-size grows, according to this law, as a power-law. For a sample of size L, we expect equilibration after a time $t(L) = L^2/\Gamma\sigma$. The Lifshitz-Shlyozov law is actually valid only if the order-parameter M is non-conserved. For conserved order-parameters, the total magnetization $\int d^3r \, \vec{M(r)}$ is constant in time. In that case, the exponent in Eq.18 is 1/3 instead of 1/2. Magnetic systems, at long enough times, are non-conserved but if we would use an Ising model to describe, say, phase-separation of an alloy then we do have a conserved order parameter.

LINEAR RESPONSE THEORY

Our next step is to consider the effect of introducing a low concentration of impurities in our magnetic system. Such impurities may couple linearly or bilinearly to the order-parameter, depending on the physical system of interest. Linear-response analysis is particularly useful away from critical points. We will apply it to the paramagnetic phase only.

Introducing foreign atoms or vacancies into a ferromagnet system normally produces bilinear coupling of the impurities to the order parameter. However, introducing quenched disorder into an anti-ferromagnet in an applied magnetic field leads to a linear coupling to the order-parameter. Also quenched disorder in phase separation problem leads to linear coupling. In the following, we will restrict ourselves to linear coupling. Experimental results which are quoted refer to diluted Ising anti-ferromagnets in a uniform field.

We first introduce a single impurity at the origin of the cooordinate system and represent this impurity by a local magnetic field h. The Hamiltonian is then

$$H = -J \sum_{i,j} S_i S_j - h S_0 \quad (19)$$

with S_0 the spin at the origin. For a pure model, the groundstate is normally translationally invariant but impurities destroy translation invariance so the groundstate should have a position-dependent magnetization $\langle S_i \rangle$. To find $\langle S_i \rangle$, we will use linear-response theory. First, we define the correlation function $\chi_{i,j}$ of the pure Ising model:

$$\chi_{i,j} = \langle S_i S_j \rangle - \langle S \rangle^2 \quad (20)$$

where $\langle ... \rangle$ stands for a thermal average. The Fourier transform of the correlation function is

$$\chi(\vec{q}) = \int d^3r \, e^{i\vec{q}\cdot\vec{r}_{i,j}} \chi_{i,j} \quad (21)$$

Within Ginzburg-Landau theory it is given by

$$\chi(\vec{q}) = \chi/(1+(q\bar{\xi})^2) \quad (22)$$

You can prove Eq.22 from Eq.4 with u=h=0. First do a Fourier transform on $f(\vec{r}$, apply the equipartition theorem to $|M(\vec{q})|^2$ and finally use Eq.20. We call Eq.22 a "Lorentzian". It turns up frequently in the scattering cross-section for X-ray or neutron scattering studies of near-critical systems.

The quantity $\bar{\xi}$ is the equilibrium correlation length while is χ the uniform magnetic susceptibility. We will from now one always use the symbol ξ for a domain size and the symbol $\bar{\xi}$ for the equilibrium correlation length. Except close to T_c, $\xi \gg \bar{\xi}$. Within Ginzburg-Landau theory:

$$\bar{\xi} \propto (-1/r)^{1/2} \quad (23.a)$$
$$\chi \propto 1/(-r) \quad (23.b)$$

The correlation length $\bar{\xi}$ is thus proportional to the domain wall width w (Eq.8). Linear response theory asserts that we can also write $\chi_{i,j}$ as

$$\chi_{i,j} = \partial \langle S_i \rangle / \partial h_j \Big|_{h_j=0} \quad (24)$$

i.e., as the perturbation in the magnetization on site i due to an infinitesimal magnetic field on site j. Now think of our impurity at the origin as just such a perturbation. The perturbed magnetization at site i is then, according to Eq.24,

$$\langle S_i \rangle = \langle S \rangle + h \chi_{i,0} \quad (25)$$

By performing an inverse Fourier transform on $\chi(\vec{q})$, you can compute $\chi_{i,0}$ from Eq.22. Inserting the result in Eq.25 gives

$$\langle S_i \rangle - \langle S \rangle \propto (h/r)\exp(-r/\bar{\xi}) \quad (26)$$

with r the distance of site i from the origin. The perturbation thus decays exponentially as you move away from the impurity and the decay length is the ubiquitous correlation length.

Now suppose you have a collection of impurities instead of just one. The magnetization will look like a bunch of Arab tents in the desert. If the impurities get really dense, then they may even destroy the magnetization, but more about that later. The Hamiltonian is now

$$H = -J \sum_{i,j} S_i S_j - \sum_i h_i S_i \qquad (27)$$

Again applying Eq.24 gives, after a Fourier-transform

$$\langle S(\vec{q}) \rangle \cong h(\vec{q}) \chi(\vec{q}) \qquad (28)$$

What would the susceptibility be in the presence of impurities? The groundstate is not anymore uniform so you cannot use Eq.20. In fact, we must be careful because there is more then one relevant susceptibility. The generalization of Eq.20 to non translation-invariant systems is

$$\chi_{i,j} = \langle S_i S_j \rangle - \langle S \rangle_i \langle S_j \rangle \qquad (29)$$

In the low-temperature regime, thermal fluctuations are become rare so you can replace S_i by its thermal average $\langle S_i \rangle$ in which case $\chi_{i,j} \cong 0$. There is a second interesting correlation function, $\langle S_i S_j \rangle$, which remains finite in the low-temperature regime. It is related to the neutron-scattering form-factor $\mathcal{S}(\vec{q})$ by

$$\mathcal{S}(\vec{q}) \propto \sum_{i,j} \exp^{i\vec{q}\cdot\vec{r}_{i,j}} \langle S_i S_j \rangle \qquad (30)$$

At low temperatures, we may replace S_i by $\langle S_i \rangle$ in Eq.30. If we use Eq.28 in Eq.30, we get an answer which is explicitly dependent on the particular impurity configuration h_i. To arrive at a definite result, we perform an average of $\mathcal{S}(\vec{q})$ over all allowed impurity configurations h_i - a so-called "quenched average". It is assumed that for a "typical" experiment, the result of a measurement of a quantity coincides with the calculated quenched average of the quantity in question. This assumption is reasonable if the statistical fluctuations around the average value of the quantity are small ("self-averaging"). Quantities such as the free-energy are normally self-averaging. We will denote quenched averages by $\overline{\ldots}$.

Assuming that there are no correlation between impurities on different sites, the quenched-averaged correlation function of the random fields is

$$\overline{h_i h_j} = h^2 \delta_{i,j} \qquad (31)$$

After performing the quenched average of $\mathcal{S}(\vec{q})$, you should find (using Eq.31)

$$\overline{\mathcal{S}(\vec{q})} \propto h^2 / \left(1+(q\vec{\xi})^2\right)^2 \qquad (32)$$

This structure factor is called a Lorentzian squared and it indeed is observed during neutron scattering experiments on magnetic realizations of the random-field Ising model.

In a similar way, we can estimate the effect of impurities on the thermal behaviour of magnetic systems. The interaction energy between the spins and the random fields contributed a term Δf to the free energy of order

$$\Delta F \propto \sum_i h_i \langle S_i \rangle \qquad (33.a)$$

216

or
$$\Delta F \propto \int d^3q \, h(\vec{q}) \langle S(\vec{q}) \rangle \tag{33.b}$$

Using Eq.28 in Eq.33b and performing a quenched average gives for the correction to the free-energy density Δf:

$$\overline{\Delta f} \propto h^2 \chi / \overline{\xi}^3 \tag{34}$$

This result is valid beyond Ginzburg-Landau theory. If you put in the true critical behaviour for the susceptibility and correlation length:

$$\chi \propto 1/(T_c-T)^\gamma \tag{35.a}$$

$$\overline{\xi} \propto 1/(T_c-T)^\nu \tag{35.b}$$

you find

$$\overline{\Delta f} \propto (T_c-T)^{3\nu-\gamma} \tag{36}$$

The specific heat/unit volume $c(T)$ is proportional to $\partial^2 f/\partial T^2$. The correction term Eq.36 gives a correction Δc to the specific heat

$$\Delta c \propto 1/(T_c-T)^{2-3\nu+\gamma} \tag{37}$$

The so-called "hyper-scaling" relation asserts that the specific heat critical exponent α (i.e., $c(T) \propto 1/(T_c-T)^\alpha$) should be related to the correlation-length exponent ν by

$$\alpha = 2 - 3\nu \quad (?) \tag{38}$$

This relation is obeyed by most critical problems known to mankind, but, apparently, not in our case because of the γ term in Eq.37. The breakdown of hyperscaling will be important for the dynamics.

Hyperscaling really means that $\Delta f \propto 1/\overline{\xi}^3$. This relation seems nearly self-evident for a free-energy density (it is called the Pippard relation). Nevertheless, if there are impurities present which couple linearly to the order-parameter then it is wrong. In general, we can write

$$\Delta f \cong \overline{\xi}^{\theta-3} \tag{39}$$

with θ an index for the breakdown of hyperscaling. For our simple linear-response theory, $\theta=\gamma/\nu$ (see Eq.34). Experimentally, θ is found to be of order one.

IMRY-MA ARGUMENT : STATICS

The linear response theory, outlined above, has in practice proven to be useful for analyzing random-field systems above (and not too close to) the critical temperature. It is possible to include non-linear corrections, using the time-honoured Wilson-Fisher renormalization group method. The results have not been good and we will not discuss it. Instead, we will turn our attention to the low-temperature phase and return to our domain-walls. The following line of reasoning on the nature of the low-temperature phase was devised by Imry and Ma. It has proven to be extremely succesfull for random field systems and it is believed to be useful for spin-glasses as well.

Assume that we have a uniformly magnetized groundstate in which we introduce a domain of flipped spins of size R. The energy cost E(R) of creating this "drop" consists of two parts (see Eq.12): a surface tension term and a field energy. The surface energy term is, as before, the product of surface tension and interfacial area. The field energy is the magnetic energy change associated with flipping the spins inside the droplet.

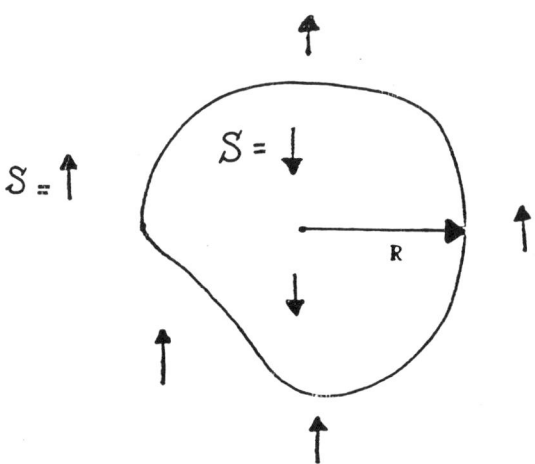

Fig.4. Imry-Ma Argument

Adding the two terms gives, in d dimensions,

$$E(R) \cong \sigma R^{d-1} - \sum_{\substack{i \text{ inside drop}}} h_i \qquad (40)$$

Now use the central limit theorem which states that the sum of N uncorrelated random variables is a Gaussian random variable of variance proportional to $N^{1/2}$. In our case, we have of order R^d uncorrelated random variables under the summation sign so the field energy is proportional to $R^{d/2}$. If the field energy is positive then obviously no mileage can be gained by creating the drop, so let's assume it to be negative:

$$E(R) \cong \sigma R^{d-1} - h R^{d/2} \qquad (41)$$

Minimize E(R) with respect to R. This should be the typical equilibrium domain-size $\bar{\xi}$:

$$\bar{\xi} \simeq \begin{cases} (\sigma/h)^{2/(d-2)} & d > 2 \\ \infty & d < 2 \end{cases} \qquad (42)$$

Below d=2 you produce unlimited amounts of domains since the field-energy exceeds the surface energy for large R. The uniform groundstate is unstable. Above d=2, the domain are finite in size and the uniform groundstate is stable. We call d=2 the "lower critical dimension". It turns out that in d=2 itself there is no long-range order.

HYSTERESIS & DOMAIN-WALL STATE

The Imry-Ma argument thus predicts long-range order in d=3. In computer experiments you do not find long-range order. Experiments initially led to contradictory results. The specific heat indicated a true phase transition, as expected from the Imry-Ma argument (for d=3). On the other hand, neutron-scattering experiments showed that there was no true long-range order. The measured $S(q)$ agreed with the line-shape Eq.32 expected for a disordered (equilibrium) system. As a result, the Imry-Ma argument was scrutinized and re-examined for over 10 years. It is indeed right.

To see what is really going on we will repeat the same computer experiment we discussed when deriving the Lifshitz-Shlyozov law but now in the presence of random fields. Quench-cool a random-field Ising model and observe what happens. Initially, the scenario is the same as for a pure Ising model: domains are formed and, with time, they increase in size. However, the growth, after a while, practically stops. Let ξ be the late-time typical domain size. We call this length the "frozen" correlation length while the associated spin-configuration is called the "domain-wall state". Since there is no long-range order, we are not in the equilibrium phase described by Imry and Ma. Suppose you now lower the random-field strength h and watch what happens to ξ. Since for h=0, we have a pure Ising model you should expect ξ to diverge as h goes to zero. Indeed it does, as

$$\xi(h) \propto 1/h^2 \qquad (43)$$

Suppose you now turn h back up. What happens ? Nothing! The frozen correlation length remains frozen at $(1/h_{min})^2$ with h_{min} the lowest value of the random field strength encountered. The spin configuration thus depends on the preparation history of you sample. For instance, you may cool you system with h=0 ("zero-field cooling") and then turn it up. Since $h_{min}=0$, $\xi=\infty$ and we have long-range order. But, if you cool in a field ("field cooling"), then $h_{min}=h$ and we have short range order! Neutron-scattering experiments using the field-cooling method indeed show short-range order while the zero-field cooling method produces long-range order (Fig.4).

This is not supposed to happen for well behaved statistical-mechanics models. In fact, it would appear to violate Boltzmann's ergodicity principle. This principle roughly states that a statistical mechanics system performs with time an average over all allowed states of the system. If the system could perform this time average, then it should have come across the true groundstate. At low temperatures this state should dominate the partition function and the free-energy. Apparently, this does not happen so ergodicity seems to be violated. Equilibrium statistical mechanics is simply not relevant for the low-temperature phase. Hysteresis is an intrinsic feature of the physics of random-field systems.

Back to more mundane issues. How come experimentalists found a Lorentzian squared line-shape for low-temperature field-cooled neutron-scattering experiments ? Eq.32 was derived under the assumption that we were dealing with an equilibrium phase. To see why, ask yourself the question what the $\mathcal{S}(\vec{q})$ would be for a domain-wall state. A really crude guess would be that $\mathcal{S}(\vec{q})$ is proportional to the form factor of a sphere of size $R=\xi$ so

$$\mathcal{S}(\vec{q}) \cong \left| \int_{r<R} d^3r \, \exp^{i\vec{q}\cdot\vec{r}} \right|^2 \tag{44a}$$

$$\cong \begin{cases} 1/q^4 & q \gg 1/R \\ R^4 & q \ll 1/R \end{cases} \tag{44b}$$

This structure-factor pops up all the time when you study random media. Whenever you have an array of reasonably flat surfaces with some characteristic length R, this is what you will see in a scatterring experiment. It is called "Porod's Law".

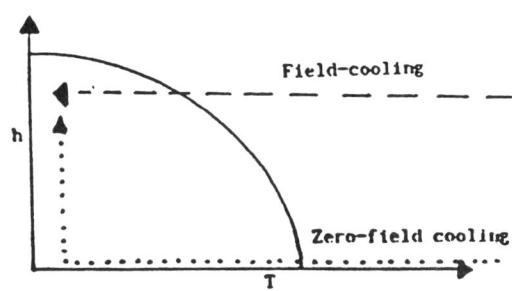

Fig.4. Field-cooling and Zero-field cooling sample preparation

Now compare Eq.44 with Eq.32. They look very similar for both small and large q and in practice they are hard to distinguish. That is a funny result: the neutron-scattering line-shape of a disordered low-temperature random-field magnet in equilibrium (Eq.32) is about the same as that of a non-equilibrium frozen domain-wall state (Eq.44). There is no good reason for this. It is just a coincidence and in two dimensions one could distinguish them easily. This unhappy accident considerably complicated the analysis of the scattering experiments.

How about the specific heat ? Why did experiments measuring c(T) find a phase transition ? Well, typical frozen correlation-lengths are of order 500 - 2000 A so that will also be the average separation between domain-walls. Specific heat measurements are insensitive to such dilute structures. Only the short-range order inside the domains contributes to c(T). That is why you sometimes hear the statement that specific heat measurements got the right answer (there is long range order in the groundstate in d=3) for the wrong reason (they had a lower sensitivity then the neutron-scattering experiments). As you may have guessed, this led to some lively discussions.

IMRY-MA ARGUMENT : DYNAMICS AND COERCIVE FIELD

Where is all this hysteresis is coming from ? We will repeat the Imry-Ma argument but now apply it to the dynamics. Recall our derivation of the Lifshitz-Shlyozov law. We computed how long it took a droplet of size R to collaps under its own surface tension. The droplet felt a pressure $P=\sigma/R$ (Eq.15). If the radius R shrinks by an amount δ, then the gain in surface energy is $P\Delta V$ with $\Delta V \cong \delta R^2$. If we add random fields then we must add to $P\Delta V$ the change in magnetic field energy. The spins inside the volume ΔV are all reversed so this change in magnetic field energy is proportional to $h(\Delta V)^{1/2}$ by the central-limit theorem. Assume this to be a positive number. The total energy gain, $\Delta E(\delta)$, is then

$$\Delta E(\delta) \cong -\sigma R\delta + hR\delta^{1/2} \qquad (45)$$

As a function of δ, ΔE first increases, reaches a maximum and finally decreases for large δ. A domain of size R trying to shrink has to overcome the energy maximum. Let the maximum of $\Delta E(\delta)$ be ΔE^*. You can determine it by setting $d\Delta E(\delta)/d\delta=0$ with the result

$$\Delta E^* \cong Rh^2/\sigma \qquad (46)$$

The time scale $\tau(R)$ required to overcome the barrier once more is given by the Arrhenius law:

$$\tau(R) \cong \tau_0 \exp(\Delta E^*(R)/k_bT) \qquad (47)$$

with τ_0 a microscopic time-scale (e.g. the spin-flip time). You may notice the similarity between this argument and the classical theory of nucleation and growth during a first-order phase-transition. Eq.47 is the random field analog of Eq.17, the law governing the collaps of a spherical domain. Following exactly the same line of arguments we used below Eq.17 gives for the time-dependent correlation length $\xi(t)$:

$$\xi(t) \propto (\sigma/h^2)\ln(t/\tau_0) \qquad (48)$$

If you compare Eq.48 with the Lifshitz-Shlyozov law (Eq.18), you see that the introduction of random fields has led to a terrible slowing down. If you want an increase in ξ by a factor two, you have to wait an order of magnitude longer in time. It is immediately clear that we have no hope of ever seeing the groundstate of the random-field Ising model. The origin of the "extra" slowing down lies in the fact that we applied the central-limit theorem here to a (thin) spherical shell of thickness δ. For the "statics" Imry-Ma argument you apply it to the whole volume of the domain. Statistical fluctuations in small samples are always bigger then in large samples, so the random term is relatively bigger in the dynamics case then in the statics.

Worse is to come. The logarithmic growth law, Eq.48, is indeed seen experimentally but only very close to the critical temperature. We define here, incidentally, the critical temperature as the temperature where we see a spike in the specific heat. Close to this temperature, one also sees (in d=3) the onset of hysteresis. Below the critical temperature, you see no growth at all in the frozen correlation length. This is due to the fact that the interfaces are not smooth, as we implicitly assumed. By stretching and deforming, they can wrap themselves around regions where the pinning action of the random field is particularly effective. With time, the domain-wall slows down and eventually it stops (Fig.5).

The fact that moving domain walls eventually are pinned down by impurities has long been known in the study of hysteresis in magnetism. This effect is responsible for the coercive field in the hysteresis loop (the M(H) vs. H curve) of a magnet. The coercive field H_c is the minimal external field you need to apply to start a domain-wall moving. It is possible to compute the coercive field of a random field magnet, again using the Imry-Ma argument, with the result:

$$H_c \propto h^2/\sigma \qquad (49)$$

Remember that σ/R is the pressure driving the collaps of a domain. From Eq.14 (with $-\nabla^2 d = 1/R$) you can see that we may think of this pressure as an applied magnetic field. A domain of size R can thus only

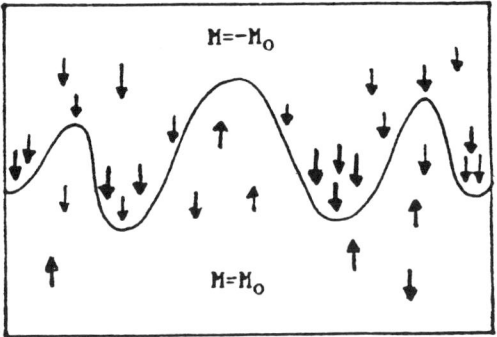

Fig.5. Impurity pinned domain-wall

shrink if σ/R exceeds H_c. This means that small domains collaps. But, if if σ/R is less then H_c then the the domains does not collaps so large domains survive. The borderline between these two cases are domains of size

$$\xi(h) \propto (\sigma/h)^2 \qquad (50)$$

Assume that, during the preparation history of the sample, the random field changed with time. Let h_{min} be the smallest random-field strength encountered during the preparation. All the domains whose initial size was less then $(\sigma/h_{min})^2$ must have collapsed, if the random field was changed sufficiently slowly. The characteristic domain size is then given by Eq.50 with h replaced by h_{min}. This conforms very well with our description of the hysteresis in the domain-wall state (Eq.43). We can thus explain both the h dependence of the hysteresis and the neutron-scattering lineshape (Eq.44) with very simple arguments.

DYNAMICAL SCALING

Suppose we were foolish enough to cool really, really slowly, starting in the paramagnetic phase, in the hope of seeing the "true" phase transition and the onset of long-range order. What would happen? We will apply dynamical scaling arguments to this situation. Dynamical scaling asserts that near the critical tempreture T_c of a second-order transition, there is only a single length-scale (the correlation length $\bar{\xi}$) and a single time-scale (the order-parameter relaxation time τ). For conventional second-order transitions, these length and time scales are related by

$$\tau \propto \bar{\xi}^z \quad (?) \tag{51}$$

with z the dynamical scaling exponent. Since $\bar{\xi}$ diverges near T_c as a powerlaw (Eq.35.b), so does τ. Close to a critical point we thus have slow relaxation ("critical slowing down"). How about the random-field case? We saw that in d=3, T_c is the temperature where the energy-barriers begin to appear which lead to hysteresis. One can use scaling arguments to estimate this barrier height near T_c. According to scaling, close to T_c, there can be only a single diverging energy scale ΔE. If Δf is the singular contribution to the free-energy density, then this energy scale ΔE must be $\bar{\xi}^3 \Delta f$. Using Eq.39 gives

$$\Delta E \cong \bar{\xi}^\theta \tag{52}$$

The characteristic energy-scale ΔE diverges at T_c with the exponent θ which measures the breakdown of hyperscaling. Since relaxation must involve activation over energy barriers of size ΔE, we can estimate the order-parameter relaxation time using once more the Arrhenius law:

$$\tau \cong \tau_0 \exp(\Delta E/k_b T) \tag{53a}$$

or

$$\tau \propto \exp(\bar{\xi}^\theta) \tag{53b}$$

The order-parameter relaxation time thus diverges exponentially fast instead of as the powerlaw expected from conventional critical slowing down (Eq.51). This exponential divergence is really a by-product of the breakdown of hyperscaling since for $\theta=0$, none of this would have happened. Because of the rapid growth of τ near T_c, it is inevitable that the relaxation time eventually will exceed any reasonable measurement time as you approach T_c (and not even all that close to T_c). After you leave equilibrium, the correlation length will stop diverging and we will enter the domain-wall state discussed previously. Note that the relation between Eqs.51 and 53 for critical slowing down is analogous to the relation between Eqs.18 and 48 for coarsening below T_c. In both cases, there is a tremendous slow-down due to the introduction of random-fields. The exponential divergence also dashes all hopes of measuring accurate critical exponents for random-field systems. Measuring exponents requires approaching T_c closely where, as we just saw, you unavoidably leave thermal equilibrium.

In summary, we have good (albeit heuristic) arguments for the relaxation processes just above T_c (Eq.53), just below T_c (Eq.48) and at lower temperatures (Eq.50). Further above T_c, we are in equilibrium and conventional methods of statistical mechanics should work. The breakdown of ergodicity below T_c is of fundamental interest. A similar breakdown of ergodicity was demonstrated (using more rigorous methods) for the infinite-range spin-glass. It appears that conventional statistical mechanics simply is not the proper tool to study highly disordered systems. The strong history dependence directly conflicts with the foundations of equilibrium statistical mechanics. Of course, history dependence is the rule rather then the exception in nature and we should rather be amazed how long we were able to postpone having to face this difficult and fascinating problem.

EXPERIMENTAL STUDIES OF SPIN GLASSES AND HEAVY FERMIONS:

THEIR MAGNETISM AND SUPERCONDUCTIVITY

J.A. Mydosh

Kamerlingh Onnes Laboratorium
der Rijksuniversiteit Leiden
2300 RA Leiden, The Netherlands

INTRODUCTION

In the most interesting and important areas of condensed-matter physics we must include the topics of spin glasses and heavy fermions. It is remarkable the influence these two subjects have exerted on certainly today's most topical area of all: high-temperature oxide superconductors. Not only is there believed to be a coexistence of a spin-glass-like state with the onset of superconductivity in $La_{2-x}Sr_xCuO_4$ and $YBa_2Cu_3O_{7-x}$, but also the concepts of "flux flow" and "flux creep" are related to the relaxation processes of spin glasses. In addition the symmetries and mechanisms of the high-T_c superconductors are constantly being compared to what occurs at very low temperatures in the heavy-fermions. Thus the concepts and models involved in the spin glasses and heavy fermions are of fundamental significance and extend by analogy to many other areas of physics.

The purpose of this article is to consider three topics from the phenomenological and experimental points of view, namely, spin-glasses, heavy-fermions and magnetic superconductors. The latter involves the former via a delicate interplay between the magnetic and superconducting tendencies of intermetallic compounds and alloys. Indeed, there is here an intrinsic relationship between these two, no longer mutually exclusive, ground states.

Firstly, after a brief introduction to the basic ingredients of a spin-glass emphasis will be placed on ideal spin glasses and their dynamical behavior. Then some of the "archetypal" spin-glass alloys will be treated regarding their relaxation and aging properties. Secondly, I will introduce the meaning of "heavy fermion" and list the most interesting compounds. Some relations among the unusual normal-state properties (C, χ and ρ) are given. Of particular generic importance is URu_2Si_2, which exemplifies the salient experimental features of heavy-fermion materials. Finally, superconductivity in magnetic systems is considered. As the foremost examples I succinctly treat: paramagnetic pairbreaking, Kondo superconductors, spin-glass freezing and superconductivity, magnetic superconductors (Chevrel and $RERh_4B_4$), and heavy-fermion superconductivity. Such a review should nicely set the stage for the subsequent lectures on the high-T_c oxide superconductor.

SPIN GLASSES: AN EXPERIMENTAL INTRODUCTION

A number of review articles [1, 2] and monographs [3, 4] have appeared recently to survey the spin-glass field. Of special pedagogical interest are the detailed reviews by Binder and Young [1] and Fischer [2]. More experimental overviews are given in the Proceedings of the (two) Heidelberg Colloquia on spin glasses [5, 6]. Since the field remains active and new results are continually being reported, up-to-date collections of the latest findings can be found in the proceedings of every major conference on magnetism [7].

A spin glass may be defined as a random, mixed-interacting, magnetic system characterized by a random, yet cooperative, freezing at a well-defined temperature T_f below which a highly irreversible, metastable frozen state occurs without the usual long-range spatial order. The randomness intrinsic to the material and the frustration generated by the competing exchange interactions are the basic ingredients. A highly-unusual frozen state results with many interesting properties [7]. For the past twenty years the canonical spin glasses have been taken to be CuMn and AuFe, i.e. randomly substituted metallic alloys. Very recently the "idealization" of these materials has been called into question [8, 9] so that their usefullness as model systems for comparison with theory seems limited and uncertain.

Three simple experimental properties should be measured to determine if a given material is a "good" spin glass. These are the ac-susceptibility, the magnetic specific heat and the low-temperature magnetization versus field. For a complete description of the experimental criterion, see Ref. 10.

During the past ten or more years the theory of spin glasses has advanced quite far using simple models and elegant solutions [11]. Yet the experimental situation has concentrated upon finding new and exotic spin glasses which greatly differ from the idealized models. We now know and accept that the spin-glass phenomenon is a general and fundamental type of magnetism (over 500 systems have been called spin glasses). Hence the experimentalist should now return to simple, ideal materials which closely mimic the theoretical models. Only then can experiment confront the theories and thereby establish a basic understanding of the physics.

One such ideal spin glass is achieved in $Rb_2Cu_{1-x}Co_xF_4$. This material for x between 0.18 and 0.40 is a "good" spin glass [12]. More important, it is an Ising system characterized by short-range, competing bond randomness via the mixed superexchange ($\pm J$) between the Cu and Co. Both magnetic ions have a spin 1/2 and the space dimensionality is 2. Thus we have a model compound for a D = 2, Edwards - Anderson, $\pm J$, Ising spin glass. All theoretical predictions agree that the equilibrium freezing temperature T_c = 0 for an Ising, D = 2 spin glass. Extensive experimental works by Dekker et al. [13] have shown this to be true and with the help of non-linear susceptibility measurements the critical exponents (γ, β, Δ) can be determined and favorably compared with theory [14].

More interesting now that the T_c = 0 question has been settled, is the dynamical behavior . Recently a new theory of activated dynamical scaling [15] has made definite predictions concerning the divergence of the characteristic relaxation time for a 2D spin glass, viz.

$$\ln\left(\frac{\tau}{\tau_o}\right) \propto \left(\frac{1}{T^{1+\psi\nu}}\right).$$

By performing a series of ac-susceptibility measurements over a wide frequency range $\tau(T)$ may be evaluated using various extraction procedures [16]. The result is independent of the specific procedure and is shown in Fig. 1.

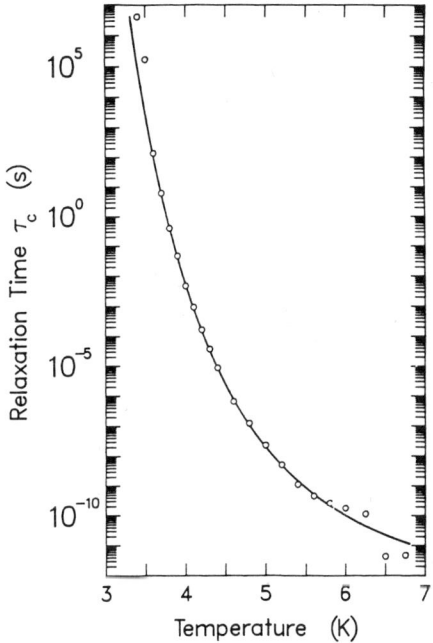

Fig. 1. Median relaxation time τ_c as a function of temperature. The solid line represents the fit which determines the exponents $\psi\nu$. After Dekker et al. Ref. 16.

Here $\tau_o = 2 \times 10^{-13}$ s and $\psi\nu = 2.2$ so that the divergence is even stronger than exponential. Note that albeit $T_c = 0$, τ leaves the experimentally accessible window ($\approx 10^5$ s) already around 3.5 K. This means that the dynamics are playing such an important role that the 2D spin glass is already out of equilibrium far above T_c. Consequently, it is impossible to study the equilibrium properties near T_c. For a 3D spin glass a similar, but less severe, divergence of τ should occur as $T \to T_c \neq 0$. Now the critical divergence is of the conventional form where $\tau \propto (T - T_c)^{-z\nu}$ with an unusually large $z\nu$ of order 7-8. Even here the characteristic relaxation times become too long for equilibrium to be maintained close to T_c. This in turn prohibits an exact determination of T_c which further rules out an accurate evaluation of the critical exponents.

At present no ideal 3D spin exists for the confrontation of experiment with theory. The best available system is $Fe_{0.5}Mn_{0.5}TiO_3$ [18] which unfortunately has different values of spin on the Fe and Mn ions. So the experimental challenge for the future is to create a truly ideal 3D system and study it systematically and carefully with a variety of techniques. This should then answer the key questions is there a phase transition in 3D and is it attainable (resolvable) in experiment.

For more than five years there has been considerable debate concerning the relaxation of the spin-glass magnetization M within the frozen state [19]. In order to resolve the different experimental results for M(t), a new variable, the waiting time t_w, was needed. This has to do with the aging of the spin-glass state after cooling to $T < T_f$. In a typical experiment CuMn (5 to 10 at.% Mn) is zero-field cooled (ZFC) and one waits a particular t_w before turning on a small, external field H. Once the field is applied the time dependence of the magnetization is tracked over many decades of time. Fig. 2 illustrates the increase of M(t) after the field is switched on.

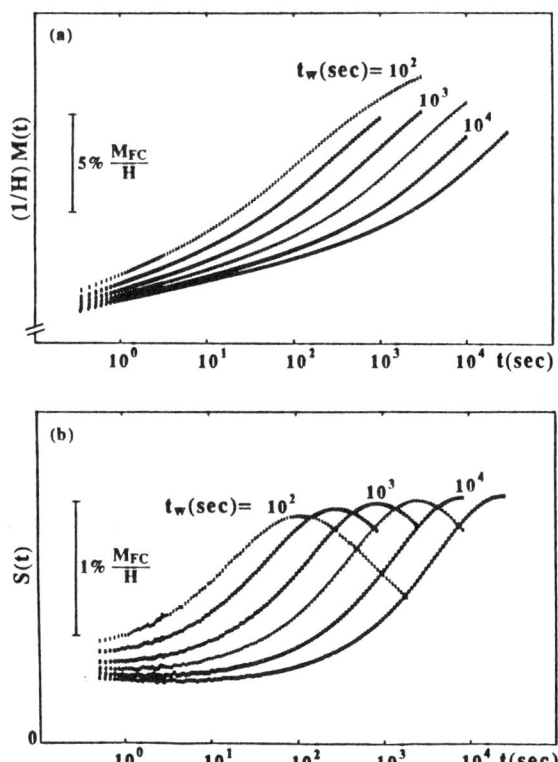

Fig. 2. ZFC susceptibility M(t)/H and the corresponding relaxation rate $S(t) \equiv (1/H)(\partial M/\partial \ln t)$ for CuMn (10 at.%) at $T = 0.9\ T_f$. After Sandlund et al. Ref. 20.

Note the strong influence of the waiting time which is especially evident in the derivative (see Fig. 2).

$$S(t) \equiv \frac{1}{H} \frac{\partial M(t)}{\partial \ln t}.$$

This illustrates the expanding evolution of the ZFC spin glass through its phase space, the extent of which governs the relaxation when the field is finally turned on.

There are many ways to manifest such effects in an experiment, e.g. field cooling or field quenching [21]. Two general properties emerge from such experiments performed in sufficiently small fields: linear response and linear superposition [22]. The expected decay of the saturated remanent magnetization is sketched in Fig. 3 over more decades than the available experimental "time-window" (compare with Fig. 2).

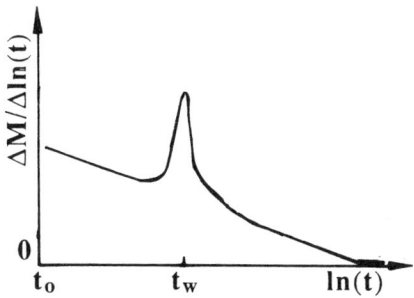

Fig. 3. Schematic representation of the relaxation rate S(t) over many decades of time. After Lundgren. Ref. 22.

The idealized sketch discloses that for $t \approx t_w$ a streched-exponental behavior is superimposed up a slowly decaying $\ln(t)$. Finally, at very long time scales the remanent decays to zero. Here we must distinguish non-equilibrium dynamics for the streched-exponential regime, quasi-equilibrium dynamics for the logarithmic decay and finally true equilibrium, if a phase transition occurs, when the remanent reaches zero.

From a theoretical point of view these processes (aging and non-equilibrium dynamics) can be associated with the growth of domains [23] within which equilibrium spin-glass order exists. The domain size, R, grows with the age of the system $t_a = t + t_w$, while the experimental probing length scale, L, increases with the observation time t. For R >> L there is quasi - equilibrium dynamics. And at a real phase transition $R \to \infty$, so equilibrium is attained. In contrast for $R \approx L$ [or $\ln(t) \approx \ln(t_w)$] the non-equilibrium dynamics is playing the major role and creating the streched-exponential behavior.

Despite its long history the spin-glass problem remains one of topical interest. Nevertheless, the present-day activity should focus upon ideal or model spin glasses and their experimental - theoretical comparisons, especially regarding the dynamics. It is only through a good understanding of the dynamical processes that the behavior of spin glasses (or even the real glasses) and the many other random analogs will be clarified.

HEAVY FERMIONS: AN EXPERIMENTAL INTRODUCTION

Although its history is much briefer (only 10 years old) the area of heavy fermions or highly-correlated electron systems does have its review articles [24 - 27]. Before the discovery of the high-T_c superconductors, the heavy-fermions were the focal point of intense interest and great activity. Even though the activity has diminished, the interest remains high since the superconductivity of heavy fermions is unconventional (other symmetry and non-BCS mechanism) and is often used for comparison and contrast with the high-T_c superconductors [28]. The latest results on the heavy fermions may be found in the proceedings of two recent conferences [29, 30].

The most simple definition of a heavy-fermion material concerns its specific heat. Here the electronic coefficient γ is exceptional large, ≥ 400 mJ/mole-K^2. Since $\gamma = (\pi^2/3)k_B^2 N(E_F) = k_B^2 m^* k_F/(3\hbar^2)$ the direct proportionality between γ and m^* [where $k_F = (3\pi^2 n)^{1/3}$]

thus means a very large effective mass. Accordingly, the 4f or 5f electrons become "heavy" through strong intra-electron correlations which create a many-body hybrization of the various electronic bands near E_F. Correspondingly there is a large density of states at E_F along with the many-body enhancement.

The most popular heavy-fermion systems are $CeCu_2Si_2$, UBe_{13} and UPt_3. More recently URu_2Si_2 has joined this category although its γ-value (180 mJ/mole-K^2) is somewhat smaller. These four intermetallic compounds exhibit low-temperature (T ≤ 1 K) superconductivity which is exceptional because magnetism evolving from the enhanced specific heat and susceptibility is expected as the ground state. The latter material, URu_2Si_2, does indeed show definite magnetic order of the itinerant [spin density wave(SDW)] type, but at a higher temperature (T_N = 17.5 K) and the magnetism coexists with the superconductivity as T → 0.

Another group of heavy fermions, namely, $NpBe_{13}$, U_2Zn_{17}, UCd_{11} and $CePd_3$, display only weak-moment magnetic ordering at low temperatures. Finally, a third class remains that becomes and stays heavy-Fermi liquids down to the mK temperature regime, e.g. $CeAl_3$, $CeCu_6$, $CeRu_2Si_2$ etc. So there are four different ground states possible; and the particular one cannot be predicted by either theoretical or emperical reasoning. The number of such "large γ" systems are constantly increasing, yet none of the new materials is superconducting.

The normal-state properties of the heavy fermions are very different than ordinary metals. We would, using the free-electron model with an effective (quasi-particle) mass, describe a metal in terms of its specific heat (γT), susceptibility [$\mu_B^2 N(E_F)$], and resistivity (AT^2). Tables of these "metallic" properties may be found in any text book on solid state physics [31]. In distinction to the normal metals the heavy fermions have greatly enhanced γ, χ and A values. These may be inter-related by Fermi-liquid theory: γ and $\chi \propto (1/T_F)$, and A $\propto (1/T_F)^2$, where $k_B T_F = E_F = \hbar^2 k_F^2 / 2m^*$. Hence the Fermi-temperature is greatly reduced from $\approx 10^5$ to $\approx 10^2 K$ and $\gamma \propto \chi \propto \sqrt{A}$. This is borne out experimentally where $\chi/\gamma = (3\mu_B^2)/\pi k_B^2)$, i.e. the free-electron value, and in Fig. 4 where the slope of A versus γ is two. The latter is important since it is a quite general connection of the thermodynamics with the transport properties of the normal state.

As mentioned above the system URu_2Si_2 shows such a variety of generic heavy-fermion behavior that it may be used to illustrate the full gamut of experimental effects. Although a recent review collects these various properties [32], let us now briefly focus upon one highly illustrative behavior, namely, the resistivity's temperature dependence as shown in Fig. 5. At high temperatures there is a strong crystalline anisotropy and negative temperature coefficient ($d\rho/dT < 0$). The latter signifies the Kondo effect which due to interactions and coherency, is broken up as the Fermi-liquid state is reached below 50 K. Then at 17.5 K the maximum/minimum effect in ρ represents the Néel temperature which is for a SDW-type of antiferromagnetic order. Here an additional T^2 dependence is found in $\rho(T)$ corresponding to the remaining Fermi-liquid behavior in the antiferromagnetic state. At yet lower temperatures the anisotropy disappears and the sample goes superconducting just under 1 K. Such a plot of the resistivity gives the overall picture for a typical heavy-fermion material although some special diversities are found from system to system. The same heavy electrons seem to participate in both SDW and superconducting transitions using different portions of the Fermi surface to form the necessary energy gaps.

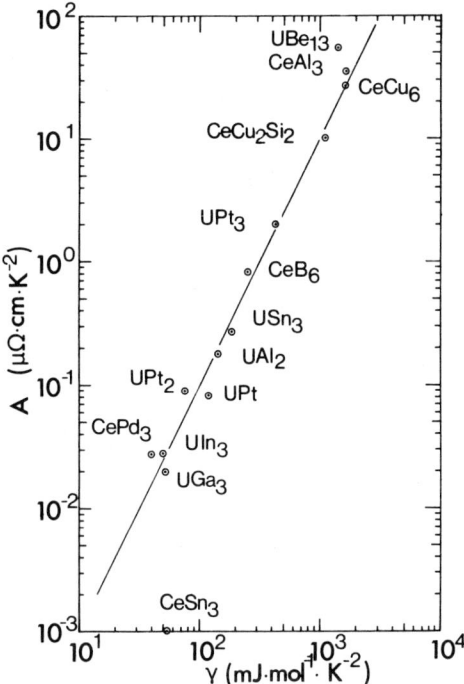

Fig. 4. T^2-coefficient of the resistivity, A, plotted again T-coefficient of the specific heat, γ, for various heavy (and "semi-heavy") fermion compounds. After Kadowaki and Woods. Ref. 32.

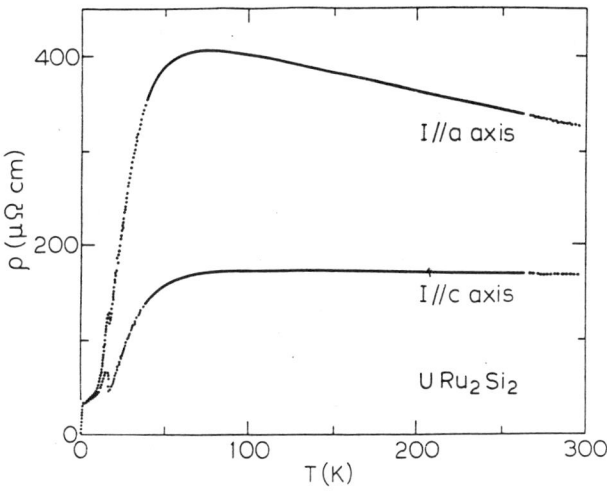

Fig. 5. Overview of the temperature dependence of the resistivity for URu_2Si_2. After Palstra et al. Ref. 34.

The heavy-fermion problem is not solved. Many new and different materials have been recently discovered with γ-values a few hundred or more. Magnetic ordering always antiferromagnetic, is often encounted, yet the superconductivity remains rare. Only the above-mentioned four compounds exhibit it. The microscopic origins of the heavy-fermion state continue to be a mystery along with the exact "unconventional" nature of the superconductivity (see following section). A further unresolved question concerns the reduced moment magnetism and its ordering. What is responsible for the small surviving moment? And do crystal-electric-field effects play a major (or even a minor) role? The heavy-fermion area will remain with us in the coming years as it becomes a general and fundamental phenomenon in need of a full microscopic description.

SUPERCONDUCTIVITY IN MAGNETIC SYSTEMS

The delicate balance between superconductivity and magnetism in a metal has been of constant importance almost since the beginning of modern low-temperature physics [35]. Various facets of this subject become highly fashionable, generate intense interest and activity, and, once sufficient progress is made, disappear only to be replaced with another ramification. Presently there is great commotion over the interplay and possible coexistence of magnetic effects, i.e. spin correlations or excitations, in the formation of superconductivity in the high-T_c oxides. This latter topic with its many intriguing possibilities will not be considered in this article, instead a brief historical outline will be sketched for superconductivity in magnetic systems up to the discovery of the high-T_c materials.

First back in the late 50's and 60's there was paramagnetic painbreaking. The Abrikosov - Gor'kov theory [36] predicted the depression of T_c for a superconducting host by diluting it with paramagnetic impurities. A simple relation was derived for the initial decrease of T_c per concentration of magnetic impurity

$$\frac{dT_c}{dx} = \frac{-4.93}{k_B} N(E_F) J_{sd}^2 S(S+1)$$

Here J_{sd} or J_{sf} is the exchange coupling between the conduction electrons and the local moments formed via d or f electrons; S is their spin value. This formular became so powerful that, in combination with ESR measurements, it was used to determine the host's density of states at the Fermi level $N(E_F)$. A nice review of the field was given by Maple [37] and a novel extension to a hydride superconductor (PdH) doped with Cr, Fe, Mn impurities is discussed in Ref. [38]. There exists good experimental-theoretical agreement, except at high impurity concentrations where inter-impurity interactions become important (see below).

At the beginning of the 70's the Kondo effect was still very much in vogue [39]. Here a single magnetic impurity in a metal interacts with its cloud of conduction electrons and thereby loses its spin. This occurs slowly around a temperature, the Kondo temperature T_K, given by $T_K = (T_F/k_B)\exp(-1/N(E_F)J_{sd})$. Now what happens when the host's superconducting T_{co} lies in the same range? The answer proposed by Müller-Hartmann and Zittart [40] is called temperature-dependent pair breaking. In other words, there is a maximum in the pair-breaking function which in turn leads to a maximum depression in T_c when $T_K \approx$

T_{co}. The two other limits ($T_K \ll T_{co}$ and $T_K \gg T_{co}$) and their transitions are also of interest and experimentally accessible. A recent contribution to exactly solving the $T_K \approx T_{co}$ case has been presented by Jarrell [41] and over the years Winzer [42] has attempted an experimental confirmation of the theoretical predictions with reasonable success.

The conventional wisdom [43] of the 60's and 70's was that magnetic ordering, expecially ferromagnetism, destroys the superconductivity. Simply put, the internal fields produced are larger than $H_c(T)$. But what about spin-glasses? Now there is no long-range order, so the random fields should cancel and paramagnetic pair breaking is forbidden by the frozen moments. To experimentally test this situation a proper system is needed. Finally after various compounds were tried a suitable material was found [44], namely $Th_{1-x}Nd_xRu_2$. In Fig. 6 I show the phase diagram determined mainly by ac-susceptibility measurements.

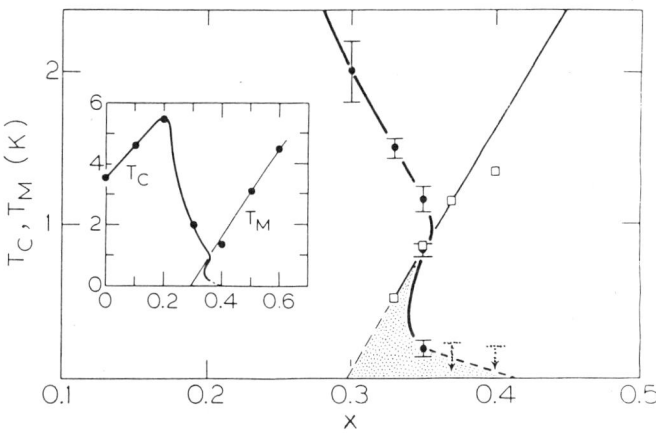

Fig. 6. Coexistence phase diagram for the superconducting spin-glass $Th_{1-x}Nd_xRu_2$. T_c is the superconducting transition temperature (●), and T_M the magnetic one (□). After Hüser et al. Ref. 44.

There is a definite region of coexistence: spin-glass freezing can occur far below T_c without destroying the superconductivity! If $T_f \leq T_c$ quenching of the superconductivity is found which then recovers at very low temperatures. Here there are three transitions: normal to super-conducting, a return to the normal state at the spin glass freezing temperature, and finally a reentry superconductivity at the lowest temperatures, see Fig. 6 at $x = 0.35$. A further property of the spin-glass state can be used to destroy the reentry superconductivity and this is the field-memory effect associated with the isothermal remanent magnetization. In all cases the critical fields are reduced by the onset of feezing at T_f.

In the late 70's and early 80's the discovery of two new ternary intermetallic compounds led to the "magnetic superconductors" [45, 46]. The Chevrel phases [46], molybdenum chalcogenides of the stoichiometry

PbMo$_6$Se$_8$ or PbMo$_6$S$_8$ exhibited T$_c$ ≈ 15 K and H$_{c2}$(0) ≈ 60 T. A whole series of magnetic rare earth elements could be substituted for the Pb. For example, Ho creates ferromagnetism and Dy antiferromagnetism both with T$_c$ > T$_M$. At the same time the ternary rhodium borides [47] (LuRh$_4$B$_4$ or YRh$_4$B$_4$ with T$_c$ ≈ 12 K, H$_{c2}$(0) ≈ 2.5 T) were displaying similar superconducting/magnetic behavior again with T$_c$ > T$_M$. The results of many experiments [45, 46] when summarized showed: (i) Coexistence of superconductivity and long-range antiferro- magnetic order. (ii) Destruction of superconductivity due to long-range ferromagnetism. However, preceding the ferromagnetic transition there occurs a sinusodially modulated magnetic state which coexists with the superconductivity in a narrow temperature region above the reentry transition. (iii) Always a depression of H$_{c2}$ at the magnetic transition (SmRh$_4$B$_4$ is an exception due to its fluctuating moment). (iv) Superconductivity is due mainly to the d-electrons of the non-rare-earth elements (Mo and Rh). Magnetism is carried by the f-electrons of the rare-earth. Between these two distinct (non-hybridized) electronic systems there is only a very weak RKKY interaction, so that the electrons are effectively decoupled.

The opposite occurs in the superconductivity of the heavy fermions. The large jump at T$_c$ in the specific heat [49] means that the heavy electrons themselves are participating in the superconducting transition. Other anomalous feature of heavy-fermion superconductivity include the large slopes of the critical fields H$_{c2}$(T) and their anisotropy. For T < T$_c$ power-law behavior is found in the specific heat, inverse spin-lattice relaxation and ultrasonic attenuation [27]. Such behavior contrasts with the BCS exponential laws of the conventional superconductors. Finally there is a most unusual effect upon alloying with non-magnetic elements. For U$_{1-x}$Th$_x$Be$_{13}$ a second lower transtion is observed below the upper superconducting one [50], see Fig. 7 for the phase diagram.

At present great controversy exists on how to describe the lower transition [51]. Is it a new superconducting phase with different order parameter symmetry or a subtle magnetic one? Much experimental and theoretical effort is going into solving this problem. In conclusion the heavy-fermions possess an unconventional type of superconductivity. The necessary attractive interaction seems not solely to be caused by phonons, Cooper pairs are more probably formed by antiferromagnetic spin fluctuations. When the coexistence of superconductivity and magnetism occurs it is mediated by the same 4f or 5f heavy electrons which are strongly hybridized with other electron bands near E$_F$.

CONCLUSIONS

This article, based upon three lectures, has attempted to treat three quite vast and very important areas of modern condensed-matter physics. These are the spin-glasses, the heavy fermions and superconductivity in magnetic systems. My approach has been to introduce the underlying physics in a phenomenological way and then survey the salient experimental properties. Since all three areas are extremely large each comprising many thousands of publications, the reader is referred to the bibliography, which although shamefully incomplete, does contain the key review articles for each area. Hopefully through these reviews the interested researcher will rapidly become familiar with his or her particular engrossment. I wish to acknowledge the long-standing support of the Nederlandse Stichting voor Fundamenteel Onderzoek der Materie (FOM).

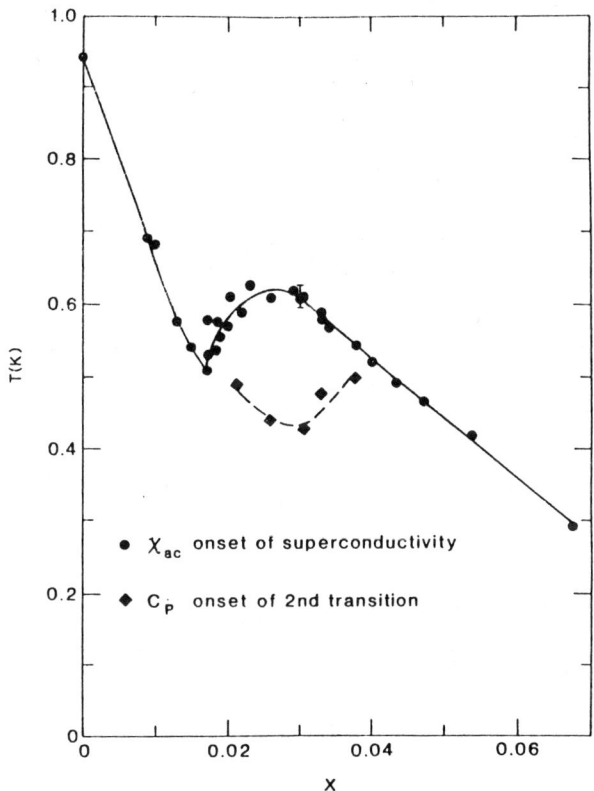

Fig. 7. Phase diagram for the heavy-fermion superconductor $Th_xU_{1-x}Be_{13}$. Whether the lower of the two transitions is a superconducting or a magnetic one remains an open question. After Smith et al. Ref. 50.

REFERENCES

1. K. Binder and A.P. Young, ReV.Mod.Phys. **58**, 801 (1986).
2. K.H. Fischer, Phys.Stat.Sol.(b) **116**, 357 (1983) and **130**, 13 (1985).
3. M. Mezard, G. Parisi and M. Virasoro, Spin Glass Theory and Beyond (World Scientific, Singapore, 1987).
4. D. Chaudhury, Spin Glasses and Other Frustrated Systems (Princeton University, Princeton, 1988).
5. Heidelberg Colloquium on Spin Glasses, edited by J.L. van Hemmen and I. Morgenstern, Lecture Notes in Physics **192** (Springer, Berlin, 1983).
6. Heidelberg Colloquium on Glassy Dynamics, edited by J.L. van Hemmen and I. Morgenstern, Lecture Notes in Physics **275** (Springer, Berlin, 1987).
7. See, for example, Proceedings of the International Conference on Magnetism - Paris '88, J. Physique (Paris) **49** C-8, 995-1172 (1988).
8. J.A. Mydosh, J.Magn.Magn.Mater. **73**, 247 (1988).
9. S.A. Werner, to be published in Comments Cond. Mat. Phys.
10. J.A. Mydosh in Ref. 5, pg. 38 and Ref. 6, pg. 24.
11. The starting point was the Edwards - Anderson model [J.Phys. F**5**, 965 (1975)], and the Sherrington - Kirkpatrick solution [Phys.Rev. Lett. **35**, 1972 (1975)].
12. C. Dekker, A.F.M. Arts and H.W. de Wijn, Phys.Rev. B**38**, 11512 (1988).
13. C. Dekker, A.F.M. Arts and H.W. de Wijn, J.Appl.Phys. **63**, 4334 (1988) and Phys.Rev. B**38**, 8985 (1988).
14. C. Dekker, Ph.D. Thesis, University of Utrecht (1988).
15. D.S. Fisher and D.A. Huse, Phys.Rev. B**36**, 8937 (1987). For more theoretical background, see A.J. Bray, Comments Cond.Mat.Phys. **14**, 21 (1988).
16. C. Dekker, A.F.M. Arts, H.W. de Wijn, A.J. van Duyneveldt and J.A. Mydosh, Phys.Rev.Lett. **61**, 1780 (1988) and Phys.Rev. B to be published.
17. A.T. Ogielski, Phys.Rev. B**32**, 7384 (1985).
18. K. Gunnarson, P. Svedlindh, P. Nordblad, L. Lundgren, H. Aruga and A. Ito, Phys.Rev. Lett. **61**, 754 (1988).
19. P. Nordblad, L. Lundgren, P. Svedlindh, L. Sandlund and P. Granberg, Phys.Rev. B**35**, 7181 (1987) and R. Hoogerbeets, Wei-Li Luo and R. Orbach, Phys.Rev. B**35**, 7185 (1987).
20. L. Sandlund, P. Svedlindh, P. Granberg, P. Nordblad and L. Lundgren, J.Appl.Phys. **64**, 5616 (1988).
21. L. Sandlund, Ph.D. Thesis, Uppsala University (1989).
22. For a recent review see L. Lundgren, J.Physique (Paris) **49** C-8, 1001 (1988).
23. D.S. Fisher and D.A. Huse, Phys.Rev. B**38**, 373 (1988).
24. G.R. Steward, Rev.Mod.Phys. **56**, 755 (1984).
25. P.A. Lee, T.M. Rice, J.W. Serene, L.J. Sham and J.W. Wilkins, Comments Cond. Mat. Phys. **12**, 99 (1986).
26. P. Fulde, J. Keller and G. Zwicknagel in Solid State Physics Vol. 41 edited by H. Ehrenreich and D. Turnbull (Academic, San Diego, 1988).
27. Z. Fisk, H.R. Ott, T.M. Rice and J.L. Smith, Nature **320**, 124 (1986).
28. See, for example, R. Tournier, A. Sulpice, P. Lejay, O. Laborde and J. Beille, J.Magn.Magn.Mater. **76 + 77**, 522 (1988).
29. Proceedings of the 6th International Conference on Crystal Field Effects and Heavy Fermion Physics, J.Magn.Magn.Mater. **76 + 77** (1988).

30. *Proceedings of the International Conference on Magnetism - Paris '88*, J. Physique (Paris) **49** C-8, 681-799 (1988).
31. N.W. Ashcroft and N.D. Mermin, *Solid State Physics* (Holt, Rinehart and Winston, New York, 1976).
32. K. Kadowaki and S.B. Woods, Solid State Commun. **58**, 507 (1986).
33. J.A. Mydosh, Physica Scripta **T19**, 260 (1987).
34. T.T.M. Palstra, A.A. Menovsky and J.A. Mydosh, Phys.Rev. B**33**, 6527 (1986).
35. See, for example, B.T. Matthias, H. Suhl and E. Corenzwit, Phys.Rev.Lett. **1**, 449 (1958).
36. A.A. Abrikosov and L.P. Gor'kov, Sov.Phys. JETP **12**, 1243 (1961).
37. M.B. Maple, Appl.Phys. **9**, 179 (1976) and also in *Magnetism V*, edited by H. Suhl (Academic, New York, 1973) Chap. 10.
38. J.C.M. van Dongen, D. van Dijk and J.A. Mydosh, Phys.Rev. B**24**, 5110 (1981).
39. J. Kondo in *Solid State Physics* edited by F. Seitz, D. Turnbull and H. Ehrenreich (Academic, New York, 1969) Vol. 23.
40. E. Müller-Hartmann and J. Zittartz, Phys.Rev.Lett. **26**, 428 (1971) and Z.Phys. **234**, 58 (1970). See also E. Müller-Hartmann in *Magnetism V* edited by H. Suhl (Academic, New York, 1973) Chap. 12.
41. M. Jarrell, Phys.Rev.Lett. **61**, 2612 (1988).
42. K. Winzer, Solid State Commun. **24**, 551 (1977).
43. V.L. Ginzburg, Sov.Phys. JETP **4**, 153 (1957).
44. D. Hüser, M.J.F.M. Reviersma, J.A. Mydosh and G.J. Nieuwenhuys, Phys.Rev.Lett. **51**, 1290 (1983); D. Hüser, Ph.D. Thesis, University of Leiden (1985).
45. For an excellent review see *Superconductivity in Ternary Compounds I and II*, edited by Ø. Fischer and M.B. Maple (Springer, Berlin, 1982).
46. *Ternary Superconductors*, edited by G.K. Shenoy, B.D. Dunlop and F.Y. Frodin (North Holland, Amsterdam, 1981).
47. R. Chevrel, M. Sergent and J. Prigent, J.Solid State Chem. **3**, 515 (1971).
48. B.T. Matthias, E. Corenzwit, J.M. Vanderberg and H. Barz, Proc.Nat.Acad.Sci. U.S.A. **74**, 1334 (1977).
49. See, for example, J.P. Brison, A. Ravex, J. Flouquet, Z. Fisk and J.L. Smith, J.Magn.Magn.Mater. **76 + 77**, 525 (1988).
50. J.L. Smith, Z. Fisk and H.R. Ott in *Theoretical and Experimental Aspects of Valence Fluctuations*, edited by L.C. Gupta and S.K. Malik (Plenum, New York, 1987).
51. M. Sigrist and T.M. Rice, J.Magn.Magn.Mater. **76 + 77**, 487 (1988) where additional references may be found.

HEAVY FERMIONS: THEORETICAL ASPECTS

Herbert Capellmann

Institut Laue Langevin, F-38042 Grenoble Cedex, France
and
Institut für Theoretische Physik C
Technische Hochschule Aachen, D-5100 Aachen
Federal Republic of Germany

INTRODUCTION

"Heavy fermion" systems belong to the class of "intermediate valent systems", in which the occupancy of the interior f-shell ($4f$- in rare earth, $5f$ in actinide compounds) is close to an instability. For simplicity the discussion will be restricted to Ce compounds (e. g. $CeAl_3$), where the concept of "valence fluctuations" presents itself in its simplest way: In the ground state the occupancy of the f-shell may fluctuate between ^1f (one f electron present, the dominant configuration, having a probability of order 0.9), and ^0f (no f electron). A large variety of anomalies in thermal, lattice, magnetic and electrical behaviour are observed in these systems. The experimentally determined entropy at low temperatures [1], which is of order $k_B \ln 2$ per Ce around 10K (representing the entropy of a magnetic doublet) for $CeAl_3$, similar to $CeAl_2$ and Ce_3Al_{11}, drops to zero when the temperature is lowered in a continuous and smooth way, whereas in the companion compounds $CeAl_2$ and Ce_3Al_{11} most of the entropy is lost rapidly due to magnetic phase transitions. Whereas in the latter the magnetic ("spin") degrees of freedom are frozen out in phase transitions, in the heavy fermion compound $CeAl_3$ these degrees of freedom are frozen out in a continuous way, certainly an unusual and unexpected behaviour when first discovered. Recent reports about magnetic order with an extremely small ordered moment ($0.01\mu_B$) at low T does not "explain" this behaviour, the main question remains why the dominant part of the entropy is not lost in a phase transition, the extremely small fraction, corresponding to the $0.01\mu_B$ ordered moment, which actually might participate in long range order being of secondary importance only.

Due to the slow decrease of entropy towards zero at zero T the slope dS/dT becomes very large at low T, accordingly the γ coefficient in the specific heat C_p is unusually large compared to normal metallic systems. Taking $CeAl_3$ as an example, if C_p is forced into a form

$$C_p = \gamma T + \beta T^3 \tag{1}$$

the coefficient γ rises sharply with decreasing T from a value of order 0.2 J/MolK^2 at 10K towards a maximum of 1.8 J/MolK^2 at 0.4K, then decreasing again below 0.4K, extrapolating towards 1.2 J/MolK^2 at zero temperature. This behaviour indicates a high density of states for excitations at low energy, several orders of magnitude larger than in normal metals, where the coefficient γ, proportional to the electronic effective mass $m*$, is of order 1mJ/MolK^2. To describe the much higher density of states in $CeAl_3$ (and other "heavy fermion" systems) an "effective mass" several orders of magnitude higher than the bare mass m_e has been defined. It is important to keep in mind that this is only a parametrization of the observed entropy and specific heat behaviour: γ is not constant in the region between zero and several degrees Kelvin, the entropy becomes quickly a sizable fraction of $k_B \ln 2$ per Ce, which is certainly a magnetic entropy to a large extent. The concept of an "effective mass" to describe this is a parametrization at best. At extremely low temperatures, probably in the mK region, a Fermiliquid description will be applicable and useful, but at higher T ($\sim 0.5K$ and higher) a Fermi liquid parametrization does not seem appropriate.

Another important observation in direct relation to the specific heat anomaly at 0.4K in CeAl$_3$, is an anomaly in the thermal expansion coefficient α [2].

Defining

$$\alpha' = \frac{d\alpha}{dT} \qquad (2)$$

α' is negative below 0.4K, with an extremely large magnitude $|\alpha'|$, changing sign at 0.4K (where the peak in γ occurs) becoming positive above, again with an extremely large magnitude. If we use the usual form to relate α' to the specific heat coefficient γ

$$\alpha' = \Gamma \kappa \frac{C}{T}. \qquad (3)$$

where κ is the compressibility and Γ the Grüneisen parameter (of order 2-3 in normal metals) a "normal" behaviour of α' should scale with $\gamma = C/T$ (which itself is already unusually large by 3 orders of magnitude). Instead α' has a magnitude **5** orders of magnitude larger than usual (besides changing sign from negative to positive at 0.4K) requiring Grüneisen parameters Γ of order 10^2. These extremely large Γ values are characteristic for all heavy fermion systems, indicating that the lattice is not simply responding to whatever happens in the electronic system (the "response coefficient" being the compressibility κ) but plays an active role in the low T behaviour.

A further anomaly, to be discussed below, is that of the electrical resistivity ρ. Again taking CeAl$_3$ as an example [3] the resistivity ρ is very large at high temperature ($T > 40K$) increasing with decreasing temperature below 300K, going through a maximum around 35K, then decreasing sharply below 35K towards very low values. Remark that the temperature scale of the maximum ($\sim 35K$) is two orders of magnitude higher than the temperature T_0 ($\sim 0.4K$) where C/T has its sharp maximum and where α' changes sign. Below it will be argued that these different anomalies occurring on different temperature scales are indirectly related.

THE ELECTRICAL RESISTIVITY, "KONDO"-BEHAVIOUR?

The observation of resistivity maxima at temperatures of several tens K in heavy fermion systems (e. g. CeAl$_3$, CeCu$_6$, CeCu$_2$Si$_2$) has often been taken as the signature of the Kondo effect. This "Kondo" type explanation, based on magnetic couplings of conduction electron spin density to magnetic f-electron moments is highly questionable. A Kondo coupling

$$H_\mathbf{K} = J \sum_i \boldsymbol{\sigma}(R_i) \cdot \mathbf{S}_i, \qquad (4)$$

where $\boldsymbol{\sigma}(R_i)$ is the conduction electron electron spin density, coupling to the (f-electron) moments $\mathbf{S}_i$ at sites $\mathbf{R}_i$, cannot explain the experimentally observed behaviour.

Wohlleben and Wittershagen [4] pointed out that the high temperature resistivity (where perturbation theory should be applicable) is far too high to be due to magnetic interactions. Estimates of magnetic coupling constants by comparison with other (stable valent) rare earths systems revealed that any magnetic contribution to the resistivity is *three orders of magnitude* (a factor 1000!) too small to explain the high temperature resistivity in Ce systems. Wohlleben and Wittershagen concluded, that *charge* couplings due to the valence fluctuations and different charge distributions of ^0f and ^1f states are of the correct order of magnitude to explain the effects.

Nozières [5], furthermore, pointed out that the "Kondo-effect", resulting from the coupling (4) in the case of *dilute magnetic impurities* cannot occur if the moments S_i are too numerous (the concentrated regime): In a simplified way the single impurity Kondo effect can be viewed as the developement of a bound state singlet between the impurity spin S_i and a spin density induced by the impurity in the conduction electron system. This singlet bound state develops at a low temperature scale of order

$$T_K \sim \frac{1}{N(0)} e^{-1/\tilde{J}}; \tilde{J} = J.N(0) \tag{5}$$

(where $N(0)$ is the conduction electron density of states at the Fermi energy E_F) when the weak (logarithmic) infrared divergencies occurring in the conduction electron t-matrix due to scattering off the magnetic impurity become important. The singlet eventually formed at low T consists (as far as the conduction electron spin density is concerned) of electronic states within an energy shell of width kT_K around E_F. For finite but small concentration of impurities this picture is still valid as long as the number of impurities is small compared to the number of conduction electrons within the shell of width kT_K. For a larger concentration

$$N_{imp} \stackrel{\sim}{>} N(0) k T_K \tag{6}$$

the conduction electrons within this shell are exhausted (= not sufficient in number) to form the Kondo singlets. "Nozières' counting argument", if applied to CeAl$_3$ for example, states that the number of conduction electrons within a shell $k_B T_0$, where T_0 is the experimentally observed temperature for the low T anomalies, is several orders of magnitude smaller than the number of Ce-ions, and consequently the number of Ce moments S_i is too large by at least 3 orders of magnitude (again a factor of 1000) to be compatible with a Kondo-effect. Therefore a theory, the basic ingredients of which are Kondo type behaviour, even if corrected at low T for coherence effects due to finite "impurity" concentration, is ruled out by Nozières' counting argument, according to Nozières the basic ingredients of a consistent theory must be different.

A further theoretical argument excluding the Kondo effect as playing an important role in the case that local spins S_i are *distributed on a dense lattice* coupling to conduction electrons via an interaction H_K is the following: the contact interaction (4) gives, in second order perturbation theory, an induced magnetic interaction $K_{ij}^{eff} S_i S_j$ between different moments, the RKKY interaction. Indeed the form of H_K (eq. (4)) was exactly that used by Ruderman and Kittel [6] to derive the interaction between nuclear moments (the S_i in the Ruderman-Kittel problem were those of nuclei). For a regular lattice arrangement, the moments S_i being separated by distances of order k_F^{-1} (k_F being the conduction electron Fermi momentum), the nearest neighbour interaction K_{nn}^{eff} is of order J^2

$$K_{nn}^{eff} \sim N(0) J^2 = \frac{1}{N(0)} \cdot \tilde{J}^2 \tag{7}$$

(falling off with increasing distance R as $R^{-3} \cos R k_F$). The energy scale $\sim \tilde{J}^2$ will result in *magnetic order* at a temperature $T_m \sim N(0) J^2 = N(0)^{-1} \tilde{J}$. For small $\tilde{J}$ any Kondo-type singularities, occurring on an energy and tempreature scale which is exponentially small (smaller than any power in $\tilde{J}$),

$$\tilde{J}^2 >>> e^{-1/\tilde{J}} \quad \text{for small} \quad \tilde{J}, \tag{8}$$

are irrelevant in the concentrated regime, where neighbouring moments feel a magnetic interaction K^{eff} much stronger than any coupling of order T_K. This interaction K^{eff} will induce magnetic order at temperatures of order T_m and will prevent any Kondo type singularities to occur at exponentially small temperature scales T_K. We have to keep in mind that the Kondo type infrared singularity is very weak (logarithmic) and that any interaction or energy scale larger than the exponentially small scale T_K will cut off the occurrence of the singularity (e. g. applied magnetic fields, couplings to other moments, temperature, etc.). The singularities come into play only if no other physical coupling lifts the internal local spin degeneracy of the magnetic impurity: The exponentially small enrgy scale T_K is relevant only for exponentially small impurity concentrations where the residual interaction between different moments is smaller than T_K.

For concentrated systems interactions between different moments dominate, inducing magnetic order at temperatures much higher than T_K. Once magnetic order has occurred the local magnetic degeneracies are lifted of course, the corresponding infrared singularities are therefore eliminated.

In many theoretical treatments an exchange interaction J_{ij}^{exch} is added (by hand) to the Hamiltonian H_K (4)

$$\hat{H} = H_K + \sum J_{ij}^{exch} \mathbf{S}_i \cdot \mathbf{S}_j \tag{9}$$

the "Kondo coupling" J and the interaction J_{ij}^{exch} being treated as independent parameters. This is valid only if J_{ij}^{exch} is to represent some effects different from the RKKY interaction (second order effects in J). It is invalid to use a Hamiltonian of type $\hat{H}$ (9) and extract exponentially small effects only from H_K ($\sim$ of order T_K) while neglecting second order effects, which lead to interactions similar to J_{ij}^{exch}. The assumption (often made to justify such treatments)

"T_K large compared to J_{ij}^{exch}"

is of no use since the "Kondo coupling" J will generate further relevant couplings similar in form to J^{exch} (J_{RKKY} in second order in J). If the scale T_K were large compared to the direct exchange J^{exch}, the latter might be neglected altogether, since the induced coupling J_{RKKY} will be of the same form and much larger than J^{exch}. Of course the assumption that T_K is large compared to the interaction J_{RKKY} induced by the coupling J itself cannot be justified for the concentrated case, since $\tilde{J}^2$ is large compared to $\exp\left\{-1/\tilde{J}\right\}$ for small $\tilde{J}$.

We conclude that the "Kondo lattice problem" is not really a problem since it was solved more than 30 years ago by Ruderman and Kittel: A contact interaction of type H_K (4) leads to indirect magnetic couplings between different moments, which induce magnetic ordering phenomena. The "single impurity Kondo problem" occurs only if these couplings between different moments are negligible on the scale T_K, a condition which can be met only for exponentially small impurity concentrations.

Although the usual *magnetic* Kondo effect is excluded by the arguments above, the physical explanation proposed by Wohlleben and Wittershagen [4] to explain the experimentally observed resistivity maximum as being due to slow charge fluctuations and charge couplings lead to a mathematical description [7] with *formal* similarities to those of the Kondo effect. The mathematical analogies should, however, not distract from the physically different mechanisms (magnetic couplings versus charge couplings).

Different time and energy scales are important in intermediate valency. The expression "valence fluctuations" implies the existence of a characteristic *slow* time scale in the correlation function

$$\Gamma_{n_f n_f}(t) \sim \langle n_f^{(i)}(t) n_f^{(i)}(0) \rangle, \qquad (10)$$

where $n_f^{(i)}$ is the f-electron number operator for site i. In our discussion we have in mind Ce systems with fluctuations between ^0f (no f-electron) and ^1f, one electron occupying the f-shell. To clarify the meaning of "slow" and fast we specify: the characteristic time and energy for valence fluctuations is of order

$$\varepsilon_{vf} \simeq 5 meV \simeq 50 k_B K; \qquad (11a)$$

$$t_{vf} = \frac{\hbar}{\varepsilon_{vf}} \simeq 1.5 \cdot 10^{-13} s \qquad (11b)$$

Another characteristic time and energy scale ("fast" in our terminology) is set by the conduction electron bandwidth

$$W \sim 5 eV \qquad (12a)$$

$$t_W = \frac{\hbar}{W} \sim 1.5 \cdot 10^{-16} s \qquad (12b)$$

A third important time scale is due to the interaction between f-electrons and conduction electrons: this Coulomb interaction is typically of the order of several eV, comparable to the conduction electron bandwidth. This Coulomb interaction is responsible for screening the f-hole, after a fluctuation from ^1f to ^0f. Although the details of the screening process are difficult to describe quantitatively, a rough estimate yields that the characteristic screening time t_s approaches t_W, it is much faster than the valence fluctuation time t_{vf}

$$t_s \ll t_{vf} \qquad (13)$$

The corresponding energy scale for the screening process $\varepsilon_s = h/t_s$ is of order eV much larger than ε_{vf} and also much larger than thermal energies kT. This implies that for the discussion of temperature dependent properties (the electrical resistivity ρ for example) and low energy excitations we can approximate the screening process to be instantaneous: on a coarsegrained (slow) timescale the valence fluctuation takes place (we chose Ce as an example) between the two states:

$$|a\rangle = |^1f\,^0X\rangle, \quad \text{and } |b\rangle = |^0f\,^1X\rangle \qquad (14)$$

Here X symbolizes a "screening orbital": $|^0f\,^1X\rangle$ is a state with no f-electron, the f-hole being screened by the occupation of the "screening orbital X". In reality X is not a strictly localized orbital but a resonance, for the low energy processes, however, the approximation chosen above is adequate, the resonance being centered at energies far away from the Fermi energy.

To understand the temperature dependence of the resistivity we have to discuss the scattering of the low energy conduction electrons by the valence fluctuations between states $|a\rangle$ and $|b\rangle$. Two processes occur: a conduction electron is scattered from state k to k' either with or without inducing a valence transition from $|a\rangle$ to $|b\rangle$ or vice versa:

$$V_{k'k}^{ab} c_{k'}^+ c_k |a\rangle\langle b| + V_{k'k}^{ba} c_{k'}^+ c_k |b\rangle\langle a|; \qquad (15a)$$

$$V_{k'k}^{aa} c_{k'}^+ c_k |a\rangle\langle a| + V_{k'k}^{bb} c_{k'}^+ c_k |b\rangle\langle b|. \qquad (15b)$$

The difference $V^{aa} - V^{bb}$ indicates that $|a\rangle$ and $|b\rangle$ have different charge distributions and exercise different potentials on the low energy conductions electrons. The average potential $1/2(V^{aa} + V^{bb})$ is only important for spatially inhomogeneous systems (e. g. La Ce ... mixtures). For regular lattices it is part of the translationally invariant crystal potential. To stress that magnetic couplings are not involved in the scattering processes discussed here, the conduction electron spin indices have been omitted in (15).

The model presented above is formally identical to the one used to describe the scattering of conduction electrons off *two level systems* (TLS), which was studied extensively to discuss resistivity anomalies in metallic glasses. Some of the results can be taken over directly: Vladar and Zawadowski (VZ) [8] showed that the effective scattering amplitudes V increase strongly with decreasing temperature (= energy cutoff), as indicated by the divergence of perturbation theory for the scattering t-matrix. In the notation used by VZ

$$V^x = \frac{1}{2}(V^{ab} + V^{ba}), \tag{16a}$$

$$V^z = \frac{1}{2}(V^{aa} - V^{bb}), \tag{16b}$$

the second order perturbation diverges due to the noncommutativity of V^x and V^z:

$$t_{kk'}^{(2)} \sim \sum_{k_1} K_{kk'}^{xz} \frac{1 - 2f(\varepsilon_1)}{\varepsilon_1 - \varepsilon} + R; \tag{17}$$

$f(\varepsilon)$ is the Fermifunction, R is the regular (non divergent) contribution and K^{xz} is the commutator

$$K_{kk'}^{xz} = (V_{kk_1}^x V_{k_1 k'}^z - V_{kk_1}^z V_{k_1 k'}^x). \tag{18}$$

The integral on the right hand side of (17) diverges logarithmically for ε and T tending towards zero.

Finite K^{xz} indicates that he sequence of the scattering processes matters ($V^x V^z \neq V^z V^x$), and as a consequence the divergent parts of direct and exchange processes, do not cancel anymore (as they do for simple potential scattering). Physically this is due to the different charge distributions in the configurations $|a\rangle$ and $|b\rangle$, which are altered both in range and (probably more important) in angular distribution.

Formally there are similarities and differences to the divergence of perturbation theory in the Kondo effect. The scattering process of conduction electrons by magnetic impurities couples the internal (= magnetic) degree of freedom of the impurity to the internal degree of freedom (= spin) of the conduction electron. This process is neccessarily "noncommutative" (VZ) because the conduction electron spin operators S^x and S^z do not commute. In the case of scattering of conduction electrons by a valence (= charge) fluctuation (a TLS) the internal spin degree of freedom of the conduction electron is unaffected, and the noncommutativity of the matrices V^x and V^z is essential for the occurrence of the logarithmic singularities (which are absent in the "commutative model" of VZ).

From the above (eq. (17)) it follows that the electrical resistivity will have a contribution due to scattering off valence (= charge) fluctuations which *increases* when the temperature is lowered from high values. At high T the different rare earth sites can be treated as incoherent scattering centers, perturbation theory (applicable at high T) yields the result (17). When lowering T the scattering matrix starts to increase, raising the resistivity ρ. Eventually the divergence of the scattering t matrix will be cut off, three effects might in principle contribute to this crossover: in the TLS as discussed by VZ a Kondo-type resonance in the conduction electron system might develop, binding with the TLS to form a singlet. We recall that in distinction to the Kondo effect (where a singlet involving different *spin* configurations develops) the singlet here invokes different *charge* configurations. This effect is probably of minor importance in intermediate valent Ce systems ("Nozières' counting problem " also applies to this effect in concentrated systems.

The overall *average* Ce-valence is stabilized by the balance between the average conduction electron-f electron Coulomb interaction and the conduction electron kinetic energy. This average valence for Ce systems is in the vicinity of 90% $|^1 f\rangle$ (configuration $|a\rangle$) and in a simplified TLS-Hamiltonian can be imposed by a molecular field term favouring state $|a\rangle$ over $|b\rangle$:

$$-\Delta^{(z)}(|a\rangle\langle a| - |b\rangle\langle b|) \qquad (19)$$

(the notation is that of VZ). This term (comparable to the action of a magnetic field in the Kondo problem) results in a cutoff of the divergence in the scattering matrix for $kT \sim \Delta^{(z)}$. Physically this is probably the dominant term imposing the valence fluctuation frequency.

A third reason for a crossover behaviour is the development of coherence between different Ce sites. (So far our discussion was applicable to the "impurity" = single site processes). Whereas for the dilute case (e. g. Ce impurities in LaAl$_3$) a finite $T = 0$ contribution to the resistivity remains, ρ will decrease to zero for $T \to 0$ in pure systems (e. g. CeAl$_3$) due to coherence between the periodically arranged Ce sites (and have the usual T^2 behaviour of Fermi liquids at low T).

But these effects need further study and are beyond the scope of the present discussion.

THE LOW TEMPERATURE MAGNETIC PROPERTIES

The main problem for the understanding of the low T (up to 10K) behaviour is the description of the magnetic properties. Let us consider the simplest case again: a single Ce ^1f ion. The magnetic f-electron degree of freedom is described by a Hamiltonian

$$H_f = \frac{p^2}{2m} + V(r) + \zeta \boldsymbol{\ell} \cdot \mathbf{S} \qquad (20)$$

where $V(r)$ represents the lattice potential and the third term on the right hand side is the spin orbit coupling (strong compared to the energy scale of 10K) resulting in a total angular momentum of $J = 5/2$ in our case. The average crystalline potential will (for low enough symmetry) split the 6 fold degeneracy ($J = 5/2$) into doublets $|X_\pm\rangle$ (due to Kramer's theorem the degeneracy must be even). We chose phases such that under time reversal operation Θ we have

$$\Theta |X_\sigma\rangle = \sigma |X_{-\sigma}\rangle; \quad \sigma = \pm 1. \qquad (21)$$

Typically the splitting between ground state $|\alpha_\pm\rangle$ and first excited multiplet $|\beta_\pm\rangle$ are of order 10 to 20 meV (e. g. for CeAl$_3$, CeAl$_2$, CeCu$_2$Si$_2$). The ground state doublet $|\alpha_\pm\rangle$ might be called an "effective spin" (we shall use the word "isospin") doublet. In the following we shall discuss the influence of slow charge fluctuations on magnetic transitions. Charge and lattice fluctuations, at zero T the zero point motion, will through spin orbit coupling induce "spin"-flip terms in the f-electron system [9]. In the introduction it was already pointed out that the low T specific heat anomaly coincides with a large anomaly in the lattice expansion coefficient. There is also direct evidence from neutron scattering experiments [10], that lattice vibrations couple strongly to crystalline electric field transitions (the observed splittings in CeAl$_2$ due to these couplings are of order 10 meV!). Charge and lattice fluctuations constitute fluctuations in the potential $V(r)$ felt by the magnetic ions and in the following we explore the consequence of these fluctuations on a single Ce ^1f ion [9]. We shall proceed in two steps: In the first part we discuss how magnetic transitions ("spin"-flips) are induced by an external time dependent electric field. At this first stage the origin of the time dependence is not treated, only its consequences concerning magnetic transitions are explored. In the second step we discuss the origin of the time dependent electric fields: charge and lattice fluctuations are treated as dynamic variables and their coupling to crystalline electric field excitations is formulated. An effective Hamiltonian will then be derived coupling the lattice and charge dynamics to magnetic transitions. Based on its exact diagonalization results for eigenstates and correlation functions are given.

The *average* potential $\overline{V}$ results in an energy level scheme with even degeneracy due to time reversal symmetry

$$(h_0 + \overline{V}) \mid X_\sigma\rangle = \varepsilon^{(x)} \mid X_\sigma\rangle. \tag{22}$$

For low enough symmetry the level scheme will consist of (Kramer's degenerate) doublets. For the following discussion we only retain a ground state doublet $\mid \alpha_\sigma\rangle$ and one excited doublet $\mid \beta_\sigma\rangle$. This is sufficient to demonstrate the principle of the coupling between charge fluctuations and magnetic transition.

Now consider a disturbance in the crystal potential ΔV, which at first we take to be static (= time independent). We chose the basis $\mid \alpha_\sigma\rangle, \mid \beta_\sigma\rangle$ such that the matrix elements of ΔV are real. The allowed matrix elements, compatible with time reversal invariance, are

$$\Delta V = \begin{array}{c|cccc} & \alpha_+ & \alpha_- & \beta_+ & \beta_- \\ \hline \alpha_+ & \Delta_1 & 0 & a & -b \\ \alpha_- & 0 & \Delta_1 & b & a \\ \beta_+ & a & b & \Delta_2 & 0 \\ \beta_- & -b & a & 0 & \Delta_2 \end{array} \tag{23}$$

We now examine the consequence of a *time dependent* perturbation $\delta V(t)$: the time dependence considered is expressed through time dependent matrix elements

$$a(t); \quad b(t); \quad \Delta_1(t); \quad \Delta_2(t)$$

The main question to be studied is: Under which circumstances can a flucutation $\delta V(t)$ induce transitions between states which in the absence of δV are time reversed, e. g. between $\mid \alpha_+\rangle$ and $\mid \alpha_-\rangle$. This process is what can be called "spin-flip" in a reduced model.

Let us suppose the system is in the state $\mid \alpha_+\rangle$ at time $t = 0$. In second order perturbation theory the transition amplitude to $\mid \alpha_-\rangle$ at time t becomes (we use units such that $\hbar = 1$):

$$C_{\alpha_-\alpha_+}(t) = -\sum_\nu \int_0^t dt' e^{-i\omega \cdot t'} V_{\alpha_-\nu}(t') \int_0^{t'} dt'' e^{i\omega \cdot t''} V_{\nu\alpha_+}(t''), \tag{24}$$

where the sum is over $\nu = \beta_+, \beta_-,$

$$\omega = \varepsilon_\beta - \varepsilon_\alpha \tag{25}$$

and the $V_{\alpha\beta}$ are the matraix elements of δV.

Two processes are possible, involving a transition from $\mid \alpha_+\rangle$ to $\mid \beta_+\rangle$ and back to $\mid \alpha_-\rangle$, or the similar process via $\mid \beta_-\rangle$:

$$C_{\alpha_-\alpha_+}(t) = \int_0^t dt' \int_0^t dt'' e^{i\omega(t''-t')} \{-a(t')b(t'') + b(t')a(t'')\} \tag{26}$$

For a time independent potential the transition amplitude vanishes, as required by time reversal. A *finite* amplitude is induced by the time dependence of the matrix elements $a(t)$ and $b(t)$, provided that their functional form differs (equ. (26) gives a vanishing amplitude if $a(t) = a_0 g(t)$ and $b(t) = b_0 g(t)$). This difference is a natural consequence of the interplay between charge and lattice fluctuations: the matrix elements a and b result from different symmetry operators, and since δV is a superposition of several components (rearrangement of electronic charges e. g. due to a fluctuation $^1f \leftrightarrow {}^0f$ on a Ce site in the vicinity of the origin and the slowly responding lattice rearrangement triggered by such a valence fluctuation) differing time dependences for $a(t)$ and $b(t)$ are obtained.

Up to now we treated the fluctuating field $\delta V(t)$ as being imposed from the outside. As explained before the origin of $\delta V(t)$ lies in the dynamical behaviour of the lattice and charge degrees of freedom. We shall now proceed to discuss these lattice and charge dynamics and their interplay with the crystalline electric field transition explicitely by formulating a Hamiltonian for the coupled system.

To work out the principle we want to study a Hamiltonian containing only the bare essentials for the effects to be studied. Like in chapter 2 we concentrate on the properties of a single rare earth ion in its crystalline environment. The dynamics of the environment (charge and lattice degrees of freedom) are represented by *harmonic oscillators*. From equ. (26) we have learned that "spin"-flips can be coupled to charge fluctuations only if there are at least two different frequencies in the charge fluctuation spectrum (leading to different time characteristics for the two transition amplitudes a and b possible). As a bare minimum we therefore ratain two harmonic oscillators of frequencies ω_a and ω_b. The noninteracting Hamiltonian (without coupling of charge and lattice degrees of freedom to crystalline electric field transitions) then takes the form:

$$H_0 = \omega_a(d_a^+ d_a + 1/2) + \omega_b(d_b^+ d_b + 1/2) \\ + \varepsilon_\alpha \sum_\sigma |\alpha_\sigma\rangle\langle\alpha_\sigma| + \varepsilon_\beta \sum_\sigma |\beta_\sigma\rangle\langle\beta_\sigma| \quad (27)$$

The $d_a^+(d_a)$, $d_b^+(d_b)$ are Bose creation (annihilation) operators. The ε_α, ε_β are energies of the two crystalline electric field doublets (Kramer's degenerate) due to the average crystal potential $\bar{V}$.

Transitions within the crystalline electric field manifold are induced, if the harmonic oscillators are displaced out of their equilibrium positions. To lowest order the matrix elements are proportional to the displacement operators.

$$q_a = \frac{1}{\gamma_a}(d_a^+ + d_a), \quad (28a)$$

$$q_b = \frac{1}{\gamma_b}(d_b^+ + d_b), \quad (28b)$$

where

$$\gamma_{a,b} = (2m_{a,b}\omega_{a,b})^{1/2} \quad (29)$$

The $m_{a,b}$ are the "masses" of the oscillators. Later we shall also need the momentum operators

$$p_{a,b} = \frac{i\gamma_{a,b}}{2}(d_{a,b}^+ - d_{a,b}) \quad (30)$$

The coupling between oscillator variables and rare earth variables in its simplest possible form then becomes

$$H_1 = aq_a\{|\beta_+\rangle\langle\alpha_+| + |\alpha_+\rangle\langle\beta_+| + |\beta_-\rangle\langle\alpha_-| + |\alpha_-\rangle\langle\beta_-|\} \\ + bq_b\{|\alpha_-\rangle\langle\beta_+| + |\beta_+\rangle\langle\alpha_-| - |\alpha_+\rangle\langle\beta_-| - |\beta_-\rangle\langle\alpha_+|\}. \quad (31)$$

The coupling constants a and b are real. The special form of H_1 is dictated by the behaviour of the $|X_\sigma\rangle$ under time reversal (21). Hermiticity and time reversal invariance are thus guaranteed for the full Hamiltonian

$$H = H_0 + H_1. \quad (32)$$

No explicit time dependence appears any more in H. The role of this time dependence $V(t)$, appearing explicitly in equs. (24) and (26), is now represented by the fact that the displacements q_a and q_b appearing in H_1 are dynamical variables

$$[q_a, H] \neq 0; \qquad [q_b, H] \neq 0. \tag{33}$$

We shall comment on the typical strength of the coupling constants, whose scale is set by $|a\sqrt{\langle q_a^2 \rangle}|\,|b\sqrt{\langle q_b^2 \rangle}|$, in the discussion.

To explore the consequences of the couplings H_1 for "spin"-flip transitions we use a canonical transformation:

$$\tilde{H} = e^{-s} H e^{s}. \tag{34}$$

S is determined to eliminate terms linear in a, b from $\tilde{H}$.

To explore the low T properties we then project back the transformed Hamiltonian $\tilde{H}$ onto the subspace spanned by $|\alpha_\pm\rangle$. Within the ground state doublet $|\alpha_\sigma\rangle$ we define isospin operators I_x, I_y, I_z,

$$I^{\pm} = I_x \pm i I_y \tag{35}$$

$$I^+ |\alpha_-\rangle = |\alpha_+\rangle;\; I^- |\alpha_+\rangle = |\alpha_-\rangle, \text{etc.} \tag{36}$$

This allows us to write the effective Hamiltonian acting in the subspace spanned by the $|\alpha_\sigma\rangle (d_a^+)^n (d_b^+)^m |0\rangle$ in the following compact way ($|0\rangle$ is the vacuum for oscillator excitations):

$$\tilde{H} = \frac{1}{2m_a} p_a^2 + \frac{1}{2m_b} p_b^2 + A_q q_a^2 + B_q q_b^2 + I_y \left(\frac{K_1}{m_a} q_b p_a + \frac{K_2}{m_b} q_a p_b \right) \tag{37}$$

The p, q are momentum and position operators and the constants A_q, B_q, K_1, K_2 are given by

$$A_q = \frac{1}{2} m_a \omega_a^2 + a^2 \frac{2(\varepsilon_\beta - \varepsilon_\alpha)}{\omega_a^2 - (\varepsilon_\alpha - \varepsilon_\beta)^2}, \tag{38a}$$

$$B_q = \frac{1}{2} m_b \omega_b^2 + b^2 \frac{2(\varepsilon_\beta - \varepsilon_\alpha)}{\omega_b^2 - (\varepsilon_\alpha - \varepsilon_\beta)^2}. \tag{38b}$$

$$K_1 = a \cdot b \frac{2}{\omega_a^2 - (\varepsilon_\alpha - \varepsilon_\beta)^2}, \tag{39a}$$

$$K_2 = -a \cdot b \frac{2}{\omega_b^2 - (\varepsilon_\alpha - \varepsilon_\beta)^2}. \tag{39b}$$

The new effective Hamiltonian satisfies all the "proper" symmetry rquirements:

$\tilde{H}$ is even under time reversal: the operator I_y (odd under time reversal) is coupled to operators $q.p$ with real constants, the odd power of p guaranteeing time reversal invariance. Also the behaviour under spatial inversion is the required one: I_y is even (similar to an angular momentum operator), as is the combination $q.p$ appearing in (34), again having similarity to an angular momentum.

$\tilde{H}$ can be exactly diagonalized, because as far as operators acting on "spin" variables are concerned, only I_y appears explicitly apart from the unit amtrix. A constant term (ε_α) has actually been omitted from $\tilde{H}$. Eigenstates to $\tilde{H}$ can therefore be chosen as eigenstates to I_y with eigenvalues ± 1. Using eigenstates to I_y

$$I_y \mid \eta \rangle = \eta \mid \eta \rangle; \quad \eta = \pm 1 ,\tag{40}$$

$\tilde{H}$ is diagonalized by

$$\begin{pmatrix} \tilde{p}_{a\eta} \\ \tilde{p}_{b\eta} \\ \tilde{q}_{a\eta} \\ \tilde{q}_{a\eta} \end{pmatrix} = \frac{1}{\sqrt{1-k\varepsilon}} \begin{pmatrix} 1 & 0 & 0 & \eta k \\ 0 & 1 & \eta k & 0 \\ 0 & \eta\varepsilon & 1 & 0 \\ \eta\varepsilon & 0 & 0 & 1 \end{pmatrix} \begin{pmatrix} p_a \\ p_b \\ q_a \\ q_b \end{pmatrix} \tag{41}$$

We obtain two oscillators with renormalized masses and frequencies not depending on η: All eigenvalues of course retain even degeneracy.

$$\frac{1}{2\tilde{m}_a} = \frac{1}{1+k\varepsilon}\left(\frac{1}{2m_a} - \varepsilon^2 B_q\right) \tag{42a}$$

$$\frac{1}{2}\tilde{m}_a\tilde{\omega}_a^2 = \frac{1}{1+k\varepsilon}\left(A_q - \frac{k^2}{2m_b}\right) \tag{42b}$$

For $\tilde{m}_b$, $\tilde{\omega}_b$ the constants A_q, B_q are interchanged. The ε and k are solutions of

$$\frac{\varepsilon}{1+k\varepsilon} = \frac{K_1 - K_2}{2m_a B_q - 2m_b A_q} \tag{43a}$$

$$\frac{k}{1+k\varepsilon} = \frac{2m_a B_q K_2 - 2m_b A_q K_1}{2m_a B_q - 2m_b A_q} \tag{43b}$$

The ground state doublet $\mid 00\eta \rangle$ takes the form of a "coupled two mode squeezed state" [11]:

$$\mid 00\eta \rangle = g \exp\left\{-R_a q_a^2 - R_b q_b^2 - i\eta R_c q_a q_b\right\} \mid \eta \rangle \equiv \mid \tilde{\Theta}_\eta \rangle \mid \eta \rangle .\tag{44}$$

The R_a, R_b, R_c are constants (see [9]), g is a normalization factor.

We want to derive several ground state correlation functions defined as

$$\Gamma_{xy}^{(\eta)}(t) \equiv \langle \eta, 0, 0 \mid e^{i\tilde{H}t} x e^{-i\tilde{H}t} y \mid 0, 0, \eta \rangle. \tag{45}$$

We obtain the following displacement - displacement correlation functions:

$$\Gamma_{q_a q_a} = \frac{1}{1-k\varepsilon}\left\{\frac{1}{\tilde{\gamma}_a^2} e^{-i\tilde{\omega}_a t} + \varepsilon^2 \frac{\tilde{\gamma}_b^2}{4} e^{-i\tilde{\omega}_b t}\right\}, \tag{46a}$$

$$\Gamma_{q_b q_b} = \frac{1}{1-k\varepsilon}\left\{\frac{1}{\tilde{\gamma}_b^2} e^{-i\tilde{\omega}_b t} + \varepsilon^2 \frac{\tilde{\gamma}_a^2}{4} e^{-i\tilde{\omega}_a t}\right\}, \tag{46b}$$

$$\Gamma_{q_a q_b}^{(\eta)} = \frac{i\eta}{1-k\varepsilon}\left\{\frac{\varepsilon}{2} e^{-i\tilde{\omega}_a t} + \frac{\varepsilon}{2} e^{-i\tilde{\omega}_b t}\right\}. \tag{46c}$$

Both bare oscillators q_a, q_b participate in each of the zero point oscillations of frequencies $\tilde{\omega}_a$, $\tilde{\omega}_b$ with a relative phase difference of $\pi/2$. From (46c) we see that this phase difffference is the only change in the displacement - displacement correlation functions between the two different states η of the ground state doublet, changing from $\pi/2$ to $-\pi/2$.

Contrary to the correlation functions above, which only contain simple harmonics, the magnetic correlation functions contain contributions from higher harmonics.

$$\begin{aligned}\Gamma_{I_z I_z}^{(\eta)} &= \sum_{n,m} \langle \eta, 0, 0 \mid e^{i\tilde{H}t} I_z e^{-i\tilde{H}t} \mid n, m, -\eta \rangle \langle -\eta, m, n \mid I_z \mid 0, 0, \eta \rangle \\ &= \sum_{n,m} e^{-i(n\tilde{\omega}_a + m\tilde{\omega}_b)t} \mid P_{mn} \mid^2 \quad .\end{aligned} \qquad (47)$$

P_{nm} is the projection

$$P_{nm} = \frac{1}{\sqrt{n!m!}} \langle \tilde{\Theta}_\eta \mid e^{2i\eta R_c q_a q_b} (\alpha_\eta^+)^n (\beta_\eta^+)^m \mid \tilde{\Theta}_\eta \rangle, \qquad (48)$$

The $\mid nm\eta \rangle = \frac{1}{\sqrt{n!}} \frac{1}{\sqrt{m!}} (\alpha_\eta^+)^n (\beta_\eta^+)^m \mid 00\eta \rangle$ are the excited states (see [9]). Due to the coupling of the "spin"-variables to the oscillators the spectral function of the "spin"-"spin" correlation function now is non-zero for *finite* frequencies. This is in contrast to the situation without such couplings ($K_1, K_2 = 0$), in this latter case all the spectral weight is concentrated at $\omega = 0$. The shift of the weight to finite frequencies due to the couplings K_1, K_2 reflects the fact that the (time reversal invariant) Hamiltonian induces "spin flips" in the f-electron system in transitions between states of the sort

$$\mid \phi \rangle = \mid \varphi(q_a, q_b) \rangle \mid \alpha_+ \rangle$$

and

$$\mid \Psi \rangle = \mid \psi(q_a, q_b) \rangle \mid \alpha_- \rangle,$$

where the $\mid \varphi(q_a, q_b) \rangle$ and $\mid \psi(q_a, q_b) \rangle$ are oscillator states. The appropriate transition amplitudes are nonzero in general (vanishing due to time reversal symmetry only for special cases like $\mid \psi \rangle = \Theta \mid \varphi \rangle$).

One of the conseqences of the shift to *finite* frequencies in the spectral function of the "spin"-"spin" correlation function is a reduction of the single-ion low temperature Curie constant.

The real part of the susceptibility χ' is related via the Kramer's-Kronig relation (to χ'') and the fluctuation-dissipation theorem (χ'' to Γ) to the correlation function

$$\chi'(0) = \int \frac{\chi''(\omega)}{\omega} d\omega \sim \int \Gamma(\omega) \frac{1 - e^{-\hbar\omega/kT}}{\omega} d\omega, \qquad (49)$$

yielding for low T

$$\chi' \sim \frac{1}{T} A_M^0 + K, \qquad (50)$$

where the Curieconstant A_M^0 contains the $\omega = 0$ contribution of $\Gamma(\omega)$, whereas the finite frequency parts contribute to the term K, which is constant at low T. For the model discussed above the reduction in Curieconstant due to the coupling of "spin" variables to the oscillators is given by a factor R:

$$R = \mid \langle \tilde{\Theta}_\eta \mid e^{2i\eta R_c q_a q_b} \mid \tilde{\Theta}_\eta \rangle \mid^2 \qquad (51)$$

The model discussed above was the simplest (non trivial) one for the coupling of the "spin"-degrees of freedom to dynamic charge and lattice fluctuations. A more general (but still single ion) coupling to more than two oscillators will in general take the form

$$H = \sum_i h_i^{osc} + \sum_{a=x,y,z} I_a K_{ij}^{(a)} q_i p_j \qquad (52)$$

where i distinguishes the different oscillators, the $K_{ij}^{(a)}$ being the "spin"-boson coupling constants. If more than one isospinoperator (I_x, I_y, I_z) and more than two oscillators are coupled, the Hamiltonian cannot be diagonalized exactly in general. For some special cases of high symmetry diagonalizations are still possible and it was demonstrated [12] that the "spin boson" coupling may in prinicple lead to a vanishing of the Curieconstant, even though the exact eigenvalues still have even (Kramer's) degeneracy. A further consequence is the possibility of lattice instabilities.

Already in the simple model (37) discussed above lattice anomalies develop at low T due to the establishment of the "squeezed states" (44) at low T: the zero point lattice vibrations are changed in frequency and amplitude due to the "spin-boson" coupling.

An estimate of the strength of the coupling constants based on the experimental neutron scattering results for CeAl$_2$ (where the spin-boson coupling has been studied most extensively [10]) yields values of order 1 meV [12], in accord with the temperature scale at which the low T anomalies occur.

CONCLUSIONS

Heavy fermion and intermediate valent behaviour are closely related to instabilities of the f-electron occupancy.

Whereas the usual (magnetic) Kondo effect does not play a role for "concentrated" systems [5] (where f-electron moments do not constitute very dilute impurities) slow charge fluctuations [4] lead to resistivity anomalies similar in experimental signature and mathematical treatment to the Kondo effect [7].

The magnetic properties are strongly influenced by the coupling of the magnetic degrees of freedom to charge and lattice fluctuations. The source for this coupling lies in the very strong f-electron spin orbit coupling. At low T the zero point charge fluctuations (also lattice vibrations constitute electric charge fluctuations) influence the orbital and (due to spin orbit coupling) spin degrees of freedom. So far only single ion effects of this coupling have been studied [9]. Already at this stage very rich and unusual behaviour results from the "spin-boson" coupling: Coupled modes involving different lattice and charge degrees of freedom (oscillators) and crystalline electric field transitions develop at low temperature. A reduction in the single ion Curie constant (in some models even to zero) results, together with a change in lattice frequencies and zero point amplitudes. Thus the relation between magnetic and lattice anomalies is established.

The single ion models used so far to treat the "spin-boson" coupling constitute a first step only: they demonstrate that qualitatively new effects result from these couplings and have to be considered for the explanation of the low T anomalies in heavy fermion systems. The extension to interaction effects between different rare earth ions remains to be carried out, only then we might expect a consistent understanding of heavy fermions.

References

[1] J. Flouquet, J. C. Lasjaunias, J. Peyrard, M. Ribault; J. Appl. Phys. **53**, 2127 (1982).

[2] M. Ribault, A. Benoit, J. Flouquet, J. Palleau, J. de Phys. Lett. **40**, L413 (1979).

[3] H. R. Ott, O. Marti, F. Hulliger, Sol. St. Comm. **49**, 1129 (1984).

[4] D. Wohlleben, B. Wittershagen, Advances in Physics **34**, 403 (1985).

[5] P. Nozières, Ann. de Physique (Fr.) **10**, 19 (1985).

[6] M. A. Ruderman, C. Kittel, Phys. Rev. **96**, 99 (1954).

[7] H. Capellmann, Sol. St. Comm. **65**, 797 (1988).

[8] K. Vladar, A. Zawadowski, Phys. Rev. **B28**, 1564, 1582, 1596 (1983) and references therein.

[9] H. Capellmann, S. Lipinski, K. U. Neumann, Z. Phys. **B75**, 323 (1989).

[10] M. Loewenhaupt, W. Reichardt, R. Pynn, E. Lindley, J. Magn. Magn. Mat. **63-64**, 75 (1987).

[11] B.L. Shumaker, Phys. Rep. **135**, 318 (1986).

[12] H. Capellmann, S. Lipinski (to be published).

GENERAL MANY-BODY SYSTEMS

S. A. Trugman

Theoretical Division
Los Alamos National Laboratory
Los Alamos, NM 87545

INTRODUCTION

The problem of how to visualize and sometimes solve a general many-body system is considered. The ideas are established in the context of very simple small systems, a Hubbard model and a coupled electron-phonon model. These models are also solved to good approximation in the thermodynamic limit, although the Hubbard model is restricted to a small number of holes added to the Mott insulating state. Response functions are also considered.

A fairly general many-body Hamiltonian is

$$H = H_{t,el} + H_{el-el} + H_{el-ph} + H_{ph} , \qquad (1)$$

consisting of an electron or other fermion kinetic energy and electron-electron interactions, which may be coupled to a bose field such as a phonon. The phonons themselves may be nonlinear (have self-interactions). The system may be strongly coupled (H_{el-el} and H_{el-ph} may be large). One may also add coupling to an external driving field, such as an AC electric field. The methods discussed are nonperturbative, and so differ from the standard methods of diagrammatic perturbation theory.[1] A comparison is made with diagrammatic methods in the context of the random phase approximation.

HUBBARD MODEL (SMALL SYSTEM)

The first example is the Hubbard model, which describes interacting electrons and contains only the first two terms of Eq. (1):

$$H_H = -t \sum_{<j,k>,s} (c^\dagger_{j,s} c_{k,s} + h.c.) + U \sum_j c^\dagger_{j\uparrow} c_{j\uparrow} c^\dagger_{j\downarrow} c_{j\downarrow} . \qquad (2)$$

The operator $c^\dagger_{j,s}$ creates an electron of spin s on a Wannier orbital on lattice site j. The first term (electron kinetic energy) causes electrons to hop to nearest neighbor sites without changing their spin. The last term is a repulsive on site electron-electron interaction. To illustrate the exact solution of Eq. (2) for a small system, consider the problem with two sites, two electrons, and the z-component of the spin S_z, which is conserved, equal to zero. (Infinite systems will be considered later.)

The Hilbert space in coordinate representation is given by

$$|1\rangle = \uparrow \quad \downarrow \qquad (3)$$

$$|2\rangle = \uparrow\downarrow$$

$$|3\rangle = \qquad \uparrow\downarrow$$

$$|4\rangle = \downarrow \quad \uparrow \ ,$$

where the first site is on the left and the second on the right. The Hamiltonian operating on $|1\rangle$ connects (has nonzero matrix elements) to states $|2\rangle$ and $|3\rangle$. State $|4\rangle$ also connects to $|2\rangle$ and $|3\rangle$, see Fig. (1). The diagonal energy of states $|2\rangle$ and $|3\rangle$ is U, and that of $|1\rangle$ and $|4\rangle$ is 0.

Note that an interacting many-body problem (containing the product of four fermion operators) has been mapped onto a non-interacting one particle tight-binding problem. If the operator $b_j^\dagger$ creates many-body state $|j\rangle$ in Eq. (3), the new Hamiltonian is

$$\tilde{H} = \sum_{j,k} \tilde{t}_{jk} (b_j^\dagger b_k + h.c.) + \sum_j \tilde{\varepsilon}_j b_j^\dagger b_j \ , \qquad (4)$$

with no interactions (four-fermion operators). The sites in the tight-binding problem, however,

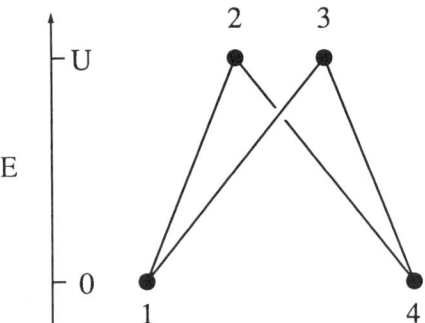

Figure 1 The tight-binding model represents the two-site Hubbard model with two electrons, $S_z = 0$. The bonds are off-diagonal matrix elements of amplitude $-t$.

represent many-body states, not the usual atomic or Wannier orbitals. It is a general result that one can always exactly map the ground and excited states of an interacting many-body problem onto those of a noninteracting one-body problem in this way.

In matrix notation, $\tilde{H}$ is

$$\tilde{H} = \begin{bmatrix} 0 & -t & -t & 0 \\ -t & U & 0 & -t \\ -t & 0 & U & -t \\ 0 & -t & -t & 0 \end{bmatrix} . \qquad (5)$$

This matrix is simple to diagonalize exactly. The four eigenvalues are

$$(E_1, E_2, E_3, E_4) = ((U - \sqrt{U^2 + 16t^2})/2, \ 0, \ U, \ (U + \sqrt{U^2 + 16t^2})/2) \ .$$

The two lowest energy (unnormalized) eigenvectors are ·

$$|\psi_1\rangle = |1\rangle + |4\rangle + a(|2\rangle + |3\rangle)$$

$$|\psi_2\rangle = |1\rangle - |4\rangle \ ,$$

where $a = -E_1/2t$. $|\psi_1\rangle$ is a singlet state and $|\psi_2\rangle$ is the $S_z = 0$ triplet state. For $U \gg t$, the low energy part of the Hilbert space is described by

$$H = const + J \vec{\sigma}_1 \cdot \vec{\sigma}_2 \ ,$$

with $J = t^2/U$.

It appears that the singlet and triplet states are written with incorrect signs. The signs are in fact correct, which brings up an issue that was glossed over: the ordering of anticommuting fermion operators. For the problem above, the reference ordering is to have the up spins operate first,

$$c^\dagger_{2\downarrow} c^\dagger_{1\downarrow} c^\dagger_{2\uparrow} c^\dagger_{1\uparrow} |0\rangle, \qquad (6)$$

where $|0\rangle$ is the vacuum. A state with a positive sign is given by operators in the above order, for example $|2\rangle = c^\dagger_{1\downarrow} c^\dagger_{1\uparrow} |0\rangle$. Suppose one had chosen a different ordering convention, such as putting site 1 operators first:

$$c^\dagger_{2\downarrow} c^\dagger_{2\uparrow} c^\dagger_{1\downarrow} c^\dagger_{1\uparrow} |0\rangle. \qquad (7)$$

The new tight binding model is shown in Fig. (2). The 2-4 bond sign, for example, is obtained with the convention of Eq. (7) as follows:

$$H_t |2\rangle = -t (c^\dagger_{2\uparrow} c_{1\uparrow}) c^\dagger_{1\downarrow} c^\dagger_{1\uparrow} |0\rangle = +t \, c^\dagger_{2\uparrow} c^\dagger_{1\downarrow} |0\rangle = +t |4\rangle.$$

The eigenvalues of the new problem (Fig. 2) are the same as those of Fig. 1, and the wavefunctions are "covariant": $(\psi_1, \psi_2, \psi_3, \psi_4)_1 \to (\psi_1, \psi_2, \psi_3, -\psi_4)_2$. Now a singlet is written in the conventional way.

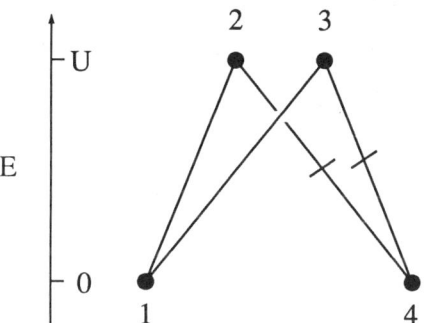

Figure 2 The modified tight-binding model for the two-site Hubbard model, using the ordering convention in Eq. (7). The slashed bonds are off-diagonal matrix elements of amplitude $+t$, and the unslashed bonds of amplitude $-t$.

This is in fact a type of gauge transformation. An example of a general tight-binding model is given in Figure (3a). If one changes the definition of a basis state (e.g. $|\phi_1\rangle \to -|\phi_1\rangle$), all of the bonds coming from $|\phi_1\rangle$ change sign, as shown in Figure (3b). In general any loop with an even number of $+t$ bonds may be transformed into a loop with no $+t$ bonds by a suitable choice of gauge (sign of basis functions). All $+t$ bonds may also be removed from bonds that are not part of a loop, such as 8-9. However, loops with an odd number of $+t$ bonds are frustrated (the $+t$ bonds may not be gauged away). The gauge transformation generalizes to $|\phi_1\rangle \to e^{i\theta} |\phi_1\rangle$, where θ was taken equal to π above.

Similar issues arize in the quantum Hall effect when a magnetic field penetrates a lattice. In that case a flux through a loop that is an integer times the flux quantum ϕ_0 is the same as zero flux under a gauge transformation.

One can write down an approximate ground state of an unfrustrated tight binding model almost by inspection. First gauge transform away all $+t$ bonds. Then all the ψ_j have the same sign in the groundstate, with ψ_j larger on sites that have a lower diagonal energy, and more or larger connected t_{ij} bonds.

One can solve larger Hubbard models exactly. For example, with 2N sites and 2N electrons, half of which are spin up, one must diagonalize a matrix of size $\left[\begin{array}{c} 2N \\ N \end{array}\right]^2$ on a side. Six sites yield a 400 x 400 matrix, or equivalently a tight-binding model with 400 sites, which can be diagonalized completely on a computer. Ten sites yield a 63,504 x 63,504 matrix, which can be solved for the ground and low lying excited states by the Lanczos method.[2] There is no exact solution for the infinite Hubbard model.

POLARONS (SMALL SYSTEM)

The second example is a coupled electron-phonon system, described by

$$H = -t \sum_{<j,k>,s} (c_{j,s}^\dagger c_{k,s} + h.c.) + U \sum_j n_{j\uparrow} n_{j\downarrow} + V \sum_{<j,k>} n_j n_k \qquad (8)$$

$$- \lambda \sum_j (n_{j\uparrow}+n_{j\downarrow})(a_j+a_j^\dagger) + \Omega \sum_j a_j^\dagger a_j ,$$

where $n_{j\uparrow}=c_{j\uparrow}^\dagger c_{j\uparrow}$ and $n_j = n_{j\uparrow}+n_{j\downarrow}$. In the 1-d version, electrons run along a chain, possibly interacting with each other on site and on nearest neighbor sites (U and V terms respectively).

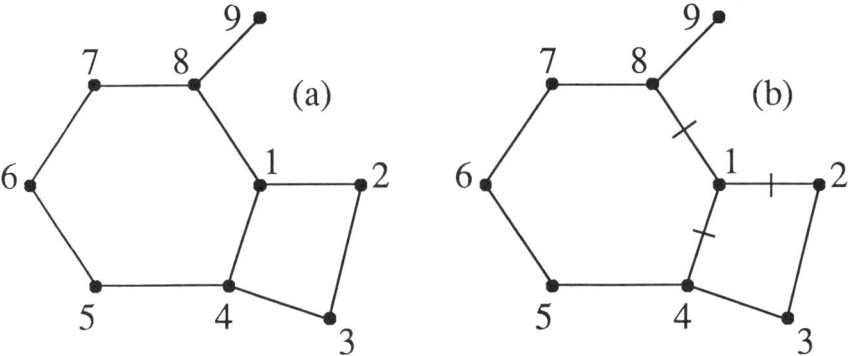

Figure 3 (a) A general tight-binding model may contain loops, dead ends, and sites with different coordination numbers. (b) A gauge transformation accomplished by $|\phi_1\rangle \rightarrow -|\phi_1\rangle$. The slashed bonds are matrix elements $+t$.

Each site is coupled to a harmonic oscillator, so that the oscillator feels an extra force when an electron is on that site, $\delta H = -\lambda_1 \hat{x}$, where $\hat{x}$ is the phonon coordinate. In terms of the creation operator $a^\dagger$ for the oscillator,

$$\hat{x} = \sqrt{\frac{\hbar}{2m\omega}} (a+a^\dagger).$$

The last term in Eq. (8) is the energy of the oscillators, with $\Omega=\hbar\omega$ and the zero point energy subtracted off. This model is for a phonon energy that is independent of $\vec{k}$, or optical phonons. (If $a_j^\dagger a_k$ terms were added to the Hamiltonian, the phonon energy would have a nonzero dispersion.) The Hamiltonian in Eq. (8) describes the system shown in Figure (4).

For simplicity, consider first a two site problem with 1 electron and two phonons. With only one electron present, the U and V terms do not operate. The basis functions can be taken either in position or momentum space. For variety, and to compare with the random phase approximation (RPA), the calculation will be done in momentum space. For a two site lattice, only $\vec{k}=0,\pi$ are allowed.

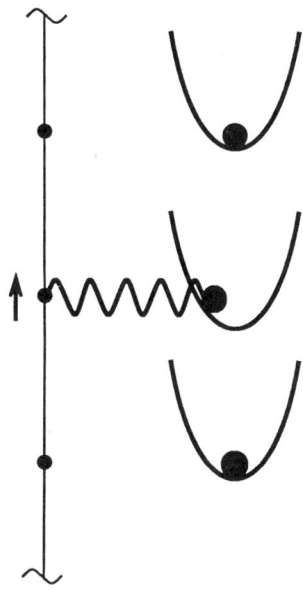

Figure 4 A polaron system in which electrons hop along a chain. Each site on the chain is associated with a harmonic oscillator. If an electron, represented by an arrow, is present on a site, an additional force is applied to the oscillator on that site.

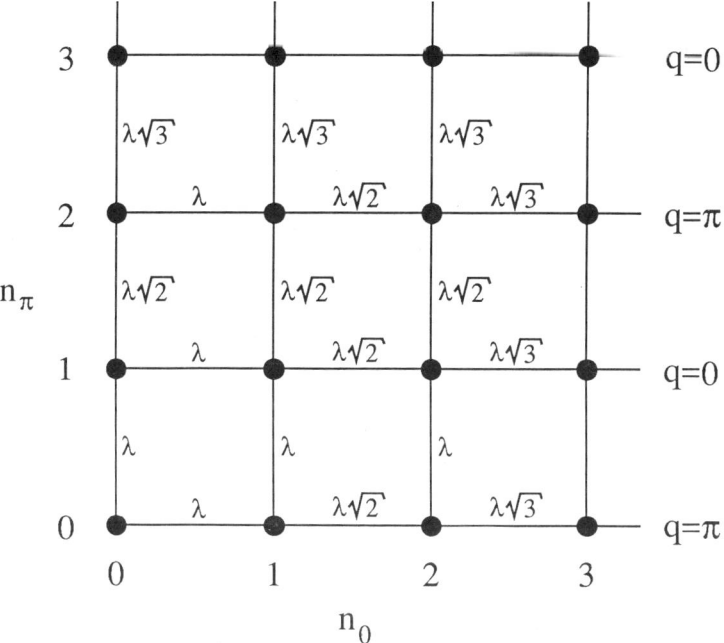

Figure 5 A portion of the infinite tight-binding model representing a coupled electron-phonon system. The sector pictured has total momentum $K = \pi$. A state (site) is labeled by (n_0, n_π), where n_0 is the number of momentum zero phonons and n_π is the number of momentum π phonons. The electron momentum q is shown on the right for each row. The bonds are off-diagonal matrix elements of amplitude $-\lambda$ times a numerical constant.

There are 2 electron basis states,

$$|0_{el}\rangle = \frac{1}{\sqrt{2}}(|1_{el}\rangle + |2_{el}\rangle)$$

$$|\pi_{el}\rangle = \frac{1}{\sqrt{2}}(|1_{el}\rangle - |2_{el}\rangle) .$$

The energy of the first is $-t$ and of the second is $+t$. There are also two phonon creation operators

$$a_0^\dagger = \frac{1}{\sqrt{2}}(a_1^\dagger + a_2^\dagger)$$

$$a_\pi^\dagger = \frac{1}{\sqrt{2}}(a_1^\dagger - a_2^\dagger)$$

Each $a^\dagger$ can create arbitrarily many quanta. A many-body state is specified by

$$|q_{el}\rangle |n_0\rangle |n_\pi\rangle ,$$

where $q_{el} = 0$ or π, and the phonon occupation numbers are $n_0 = 0,1,2,...$, $n_\pi = 0,1,2,...$. The electron phonon interaction conserves momentum. Its strength is momentum independent in this model.

The equivalent 1-body tight binding model consists of two disconnected pieces for this model, one for each total momentum. The total momentum $K = \pi$ piece of the Hilbert space is shown in Figure (5). The q_{el} can be deduced from the total K, $q = (K - \pi n_\pi) \mod 2\pi$, so by specifying the phonon state, one also specifies the electron state. The lowest row of vertices has $q = \pi$, etc.

The diagonal energy of a site is

$$E(n_0, n_\pi) = \Omega(n_0 + n_\pi) + (-1)^{n_\pi} \quad (K = \pi) .$$

The numerical factors in the off-diagonal matrix elements can be obtained using $a^\dagger |n\rangle = \sqrt{n+1} |n+1\rangle$. The energy in Fig. (5) therefore increases linearly to the upper right, with an additional corrugation as a function of y. In this basis, t appears in a diagonal (site) energy,

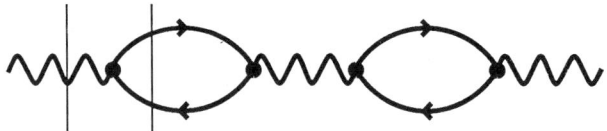

Figure 6 A diagram retained in the RPA.

in contrast to the real-space basis in the previous example, where t is an off-diagonal bond strength. There is an identical lattice for the $K = 0$ sector, except that the y corrugations are opposite,

$$E(n_0, n_\pi) = \Omega(n_0 + n_\pi) - (-1)^{n_\pi} \quad (K = 0) .$$

To find the ground state and low lying excited states numerically, one truncates the lattice (keeping states to the lower left, with low diagonal energies). The remaining problem is solved numerically. One should check that the truncation does not effect the physics, by verifying that the wavefunctions and energies of the low lying eigenstates have converged.

The electron interaction with the phonons is said to be retarded or frequency-dependent. In this formulation, however, one need not include an explicitly frequency-dependent interaction, but merely couple in phonon states of various energies on an equal footing with all other states.

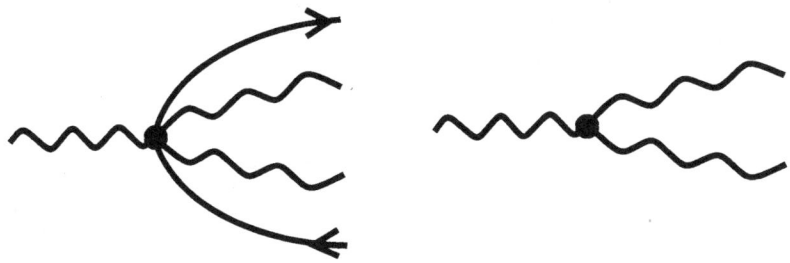

Figure 7 Some vertices that are neglected in the RPA. The matrix element connecting state (0,1) to state (0,2) is shown on the left, and the one connecting (0,1) to (1,1) on the right.

How do standard diagrammatic methods, such as the random phase approximation (RPA), compare with solving the equivalent tight-binding lattice? Consider the question of how the $k = \pi$ phonon is changed by the electron-phonon interaction. If λ were zero, the bare $k = \pi$ phonon would be an eigenstate of energy $-t + \Omega$. This is the state $(n_0, n_\pi) = (0,1)$ in Fig. (5). The RPA sums all diagrams of the form shown in Figure (6), with any number of bubbles. The vertical line on the left cuts through the state with one $k = \pi$ phonon and no electron-hole pairs, which is state (0,1) in Fig. (5). The vertical line on the right cuts through the state with an electron-hole pair and no phonons, which is the state (0,0). In the exact problem, there are also matrix elements from state (0,1) to (0,2) and to (1,1). These matrix elements are the vertices shown in Figure (7), which are neglected by the RPA.

The RPA thus keeps only the two states (0,1) and (0,0) and the bond between them, and throws away the rest of the lattice. It solves this tiny "two site" problem exactly. When is this a good approximation? One requires $\lambda \ll t$ for there to be no significant admixtures of the neglected state (0,2) in the ground state. Furthermore, $\lambda \ll \Omega$ is required to prevent significant admixtures of state (1,1), which was also neglected. In this limit however, the bare phonon state (0,1) is essentially exact, so that one need not have bothered with more than one state. The RPA is thus not very useful for this case.

Various response functions (Green's functions) can be calculated directly from the eigenstates. For example the optical absorption is

$$\alpha(\omega) = \frac{1}{\omega} \sum_n |\langle n | \hat{J} | 0 \rangle|^2 \delta(\omega - (\varepsilon_n - \varepsilon_0)), \tag{9}$$

where $\hat{J}$ is the current operator. If the Hilbert space is truncated, $\alpha(\omega)$ becomes unreliable for very large ω.

LARGE SYSTEMS

A general large many-body problem cannot be solved by any method, including this one. Consider the Hubbard model with 10^{23} sites at arbitrary filling in the momentum space basis. In this treatment, the first state (noninteracting fermi ground state) connects via U to an enormous number of states $O(10^{69})$. The problem at this level, which is a very large "star" tight binding model (Fig. 8a), is still straightforward to solve exactly. However, there is no justification for stopping at this level. Each perimeter state connects to a large number of other states, sometimes forming loops, and each of them connects to many new states, etc., so the problem finally becomes intractable. This is illustrated schematically in Fig. (8b). The many-body problem has still been mapped exactly onto a one-body tight-binding problem, but one that is too large to solve.

There is, however, a class of problems that can be solved essentially exactly or to good approximation on an *infinite* lattice. These problems describe the quantum dynamics of one or several "defects" in a well-understood background. Two examples are: (1) The problem in which a small number of electrons interact with optical phonons to form polarons, bipolarons, etc. on an infinite lattice. (2) The problem of a hole and the interaction between pairs of holes in the Mott insulating state of the Hubbard model on an infinite lattice in two or more dimensions.

The polaron problem is the same electron-optical phonon problem described by Eq. (8), but calculated on an infinite lattice in real space rather than k-space. The many-body basis states are

$$|j\rangle \; |n_j\rangle \; |n_{j+1}\rangle \; |n_{j-1}\rangle \; |n_{j+2}\rangle \; |n_{j-2}\rangle \ldots$$

The first ket is the electron location, followed by the number of phonons on the same site, on nearest neighbor sites, etc. Again one constructs an arbitrarily large variational space, and then checks that the space is big enough.

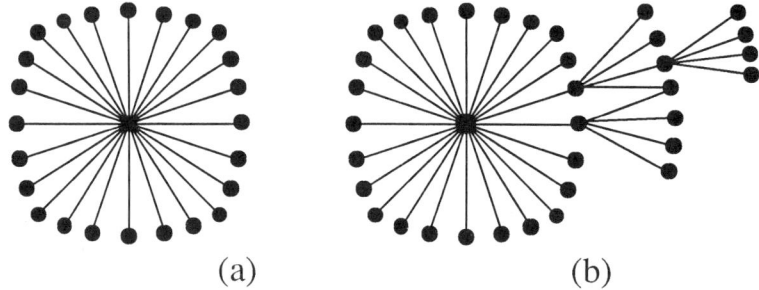

Figure 8 (a) The large "star" obtained as the first approximation to the Hubbard model with many sites. The off-diagonal matrix elements operate only once. (b) If the off-diagonal matrix elements act repeatedly, each of the sites at the edge connects to many other sites, and each of those connects to many others, etc., sometimes forming loops. (Shown schematically.)

A small variational space might allow for zero or one phonon on the site that the electron is on or on a nearest neighbor site. (A much larger space is used for accurate calculations.) The small variational space can be written

state	-1	0	1	(10)
1	0	0	0	
2	0	1	0	
3	0	0	1	
4	1	0	0	
5	0	1	1	
6	1	1	0	
7	1	0	1	
8	1	1	1	

The columns show the number of phonons on the site to the left of the electron (-1), on the same site as the electron (0), and to the right of the electron (1). All translations of these states are also included in the Hilbert space. The tight-binding lattice is shown in Fig. (9). The vertical bonds have strength $-\lambda$, and the others strength $-t$. The diagonal energy is zero for state 1, Ω for states 2, 3, and 4, and 2Ω for states 5 and 6. Different states with the same number represent translations of a state. For example, the leftmost state 3 represents the state with an electron on site 0 and a phonon on site 1. The middle state 3 represents the state with an electron on site 1 and a phonon on site 2. States 7 and 8 form a disconnected part of the Hilbert

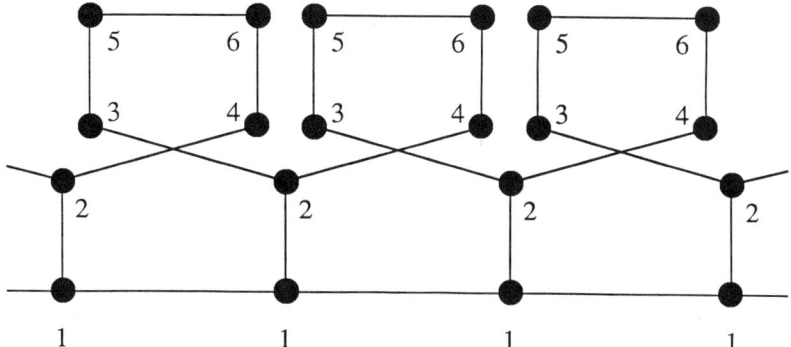

Figure 9 The tight-binding model for a polaron on an infinite lattice with the small basis set of Eq. (10). The tight-binding lattice extends to infinity and is periodic.

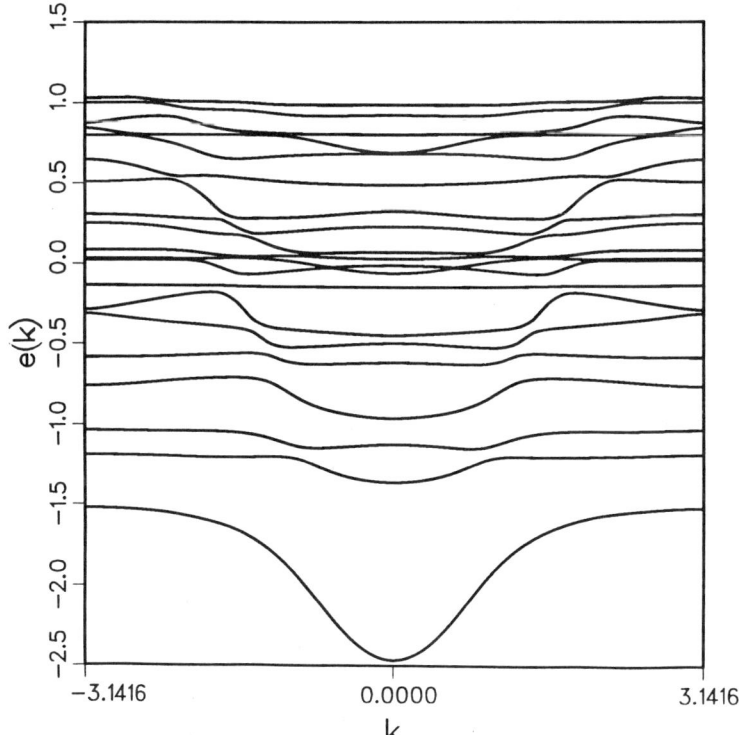

Figure 10 The twenty lowest eigenvalues of the polaron problem plotted as a function of wavevector k.

space and are not shown. There are 6N states in the Hilbert space, where the number of sites N is infinite. The Hilbert space is translation invariant, with 6 states per unit cell. The translation invariance implies that the exact eigenstates obey Bloch's theorem. For any crystal momentum $\vec{k}$, one need diagonalize only a 6 x 6 hermitian matrix, rather than a 6N x 6N matrix.

The ground state eigenfunction of the tight-binding model with fixed $\vec{k}$ describes a polaron with momentum $\vec{k}$. Note that even with this small Hilbert space, there is already more than one way to propagate through the lattice, either directly along the baseline or through a high loop. The loop route shows that the polaron can move by making virtual internal excitations and then getting rid of them. Different propagation routes interfere in the physical polaron. The polaron is never "self trapped," but rather is delocalized in a Bloch state of wavevector $\vec{k}$, possibly with a large effective mass.

Plotting the tight-binding eigenvalues as a function of $\vec{k}$ gives a graph that looks like a band structure, although it describes many-body physics. The lowest band is the polaron quasiparticle energy $\varepsilon(\vec{k})$ for the electron dressed with phonons. The higher bands are either excited states of the quasiparticle or unbound electron-phonon states, which suffer phase shifts. One can examine how the quasiparticle energy, wavefunction, and residue Z vary with k. Figure (10) shows these bands for a larger variational Hilbert space that includes 86 variational states per lattice site, with up to three phonons. Other calculations with several thousand variational states, involving many phonons in a large neighborhood of the electron, have been performed. Calculations have also been done for bipolarons, nonlinear phonons, and for the AC conductivity of a polaron.[3]

The same method can be applied to other many-body problems on an infinite lattice. The Hubbard model with one electron per site on a square lattice forms a Mott insulating state with antiferromagnetic long-range order.[4] The problem of one and two holes in the Mott insulating state and the interaction between the holes has been studied.[5] In this case the variational space consists of the location of the hole(s), and a number of spin-flips relative to the Néel state in the vicinity of the holes. The Green's function for hole propagation, which contains excited state information, has also been obtained.[6] These studies have used a variational space as large as 609 states per real space lattice site.

SUMMARY

For a small many-body system, one can solve essentially exactly for the low-lying eigenvalues, eigenfunctions, and linear response to an external probe. These systems may include electron-electron, electron-phonon, and nonlinear phonon interactions. Standard diagrammatic techniques, like the RPA, may be inadequate for these problems. Some infinite systems can be solved exactly or to good approximation by the same technique, which exactly maps a many-body problem onto a one-body tight-binding model.

I would like to thank I. Batistic and S. Marianer for their substantial contributions to the polaron sections. This work was supported by the US DOE. I would also like to thank the University of Evora for their hospitality, and the ITP at Santa Barbara for their additional support under NSF Grant PHY82-17853, supplemented by funds from NASA.

REFERENCES

(1) See, e.g., A. A. Abrikosov, L. P. Gorkov, and I. E. Dzyaloshinski, *Methods of Quantum Field Theory in Statistical Physics* (Pergamon, Elmsford, NY, 1965), S. Doniach and E. H. Sondheimer, *Green's Functions for Solid State Physicists* (Benjamin, Reading, MA, 1974), A. L. Fetter and J. D. Walecka, *Quantum Theory of Many Particle Systems* (McGraw-Hill, NY, 1971), G. D. Mahan, *Many-Particle Physics* (Plenum, NY, 1981).

(2) S. Pissanetzky, *Sparse Matrix Technology* (Academic, NY, 1984).
(3) I. Batistic, S. Marianer, and S. A. Trugman, unpublished.
(4) J. D. Reger and A. P. Young, Phys Rev. B **37**, 5978 (1988).
(5) S. A. Trugman, Phys Rev. B **37**, 1597 (1988).
(6) S. A. Trugman, LANL preprint 89-2230.

NUMERICAL METHODS FOR MANY BODY PROBLEMS
A) EXACT DIAGONALIZATION OF SMALL SYSTEMS

X. Zotos

Institut für Theorie der Kondensierten Materie

Universität Karlsruhe, Karlsruhe, FRG

INTRODUCTION

In recent years numerical simulations have been increasingly used in the study of models describing strongly interacting systems. In particular after the discovery of high temperature superconductivity and the realization that it involves strongly interacting fermions a great deal of work appeared in the literature either on exact diagonalization studies or Quantum Monte Carlo studies on systems like the $t - J$ model[1], Hubbard model[2] or more complicated models involving several bands[3]. In these two lectures we will attempt to give an impression about the systems that can be studied using exact diagonalization techniques and which using Monte Carlo methods, the limitations of each method and a guide to the main literature without attempting to give an exhaustive reference list in this rapidly developing field.

We will be mainly concerned with spin 1/2 and fermion systems on a lattice although very important and successful work has been done for instance in the Quantum Hall effect[4]. A numerical simulation of course cannot provide a theory but can give valuable intuition about the properties of a model. Such properties are for instance the symmetry of the ground state, existence of hidden symmetries, hints about exact states, qualitative behavior of correlations (e.g. ferromagnetic vs. antiferromagnetic) and if scaling to infinite system is possible information about long range order, predominant correlations etc.

The organization of this chapter is as follows: we will first give examples of different systems to obtain an idea about the size of systems that can be studied using exact diagonalization. Then we will demonstrate the use of symmetry to reduce the size of matrices to be diagonalized, give some ideas about organizing the code and quantities that can be calculated. Next we will briefly discuss numerical methods to diagonalize matrices and we will close with some examples where finite size scaling to infinite system is possible, which will also motivate the discussion in the next chapter about simulation of larger systems using Quantum Monte Carlo methods.

DIMENSION OF HAMILTONIAN MATRIX

In this section we will look at the dimension of the Hamiltonian matrix corresponding to different systems without taking into account any particular symmetry properties of the lattice. In this case the dimension is identical for a one or higher dimensional lattices or one with arbitrary connectivity and only depends on the number of lattice sites and of course the number of particles present. Let us call the dimension D. For technical reasons a boundary value of $D \sim 1000$ exists; matrices smaller than 1000 by 1000 can be "exactly" diagonalized and the full spectrum of eigenvalues and eigenvectors can be obtained, while for matrices with $D > 1000$, iterative methods must be used (Lanczos, Davidson) which yield only the low lying eigenvalues and eigenstates. Of course this boundary is dictated by the capacity of present machines used.

In terms of increasing complexity:

- Spin 1/2 systems where the z-component of the total spin, S^z is conserved:
The dimension of the matrix is given by:

$$D = \frac{N(N-1)\cdots(N-M_\uparrow+1)}{M_\uparrow!}$$

where N is the number of lattice sites and $M_\uparrow, M_\downarrow$ is the number of up, down spins. The maximum dimension occurs for $S^z = 0$, ($M_\uparrow = M_\downarrow = N/2$). For example for $N = 4$, $M_\uparrow = N/2$ the basis set consists of 6 states: ↑↑↓↓, ↓↑↑↓, ↓↓↑↑, ↑↓↓↑, ↑↓↑↓, ↓↑↓↑.

The prototype system of this case is the Heisenberg Hamiltonian[5]:

$$H = \sum_{i,j} J_x S_i^x S_j^x + J_y S_i^y S_j^y + J_z S_i^z S_j^z$$

where $\vec{S}$ are Pauli spin matrices. Keeping in mind that $S_i^x S_j^x + S_i^y S_j^y \sim S_i^+ S_j^- + S_i^- S_j^+$ we see that the number of up and down spins is conserved.

From the expression for D we obtain that for $N = 12$, $D = 924$ and the size of systems that can by typically be studied, even after using the symmetry of the lattice are not much larger than $N = 12$

- Spin 1/2 systems where the z-component of the total spin, S^z is not conserved:
A typical example for this case is the Ising model in a transverse magnetic field:

$$H = \sum_{i,j} J_{ij} S_i^z S_j^z + \sum_i h S_i^x$$

The dimension of the matrix in this case is given by $D = 2^N$ and already for $N = 10$, $D = 1024$.

Note: in terms of the dimension of the basis set there is a direct correspondence between spinless fermion systems and spin systems, a spinless fermion corresponding for instance to an up spin and an empty site to a down spin. This formal equivalence is more explicit in one dimensional systems through the Jordan-Wigner transformation.

- Hubbard type models:

In this case there are four possible states per site, empty, spin up, spin down, doubly occupied. If S^z is conserved, then the dimension D is given by:

$$D = \frac{N(N-1)\cdots(N-M_\uparrow+1)}{M_\uparrow!} \frac{N(N-1)\cdots(N-M_\downarrow+1)}{M_\downarrow!}$$

and is maximum for half-filled band ($M_\uparrow = M_\downarrow = N/2$). The standard Hamiltonian for this class of systems is the Hubbard model[2]:

$$H = -t \sum_{i,j\sigma}(c_{i\sigma}^\dagger c_{j\sigma} + H.c.) + U \sum_i n_{i\uparrow} n_{i\downarrow}$$

where $c_{i\sigma}(c_{i\sigma}^\dagger)$ are annihilation(creation) operators of a fermion with spin $\sigma = \uparrow, \downarrow$, $n_{i\sigma} = c_{i\sigma}^\dagger c_{i\sigma}$.
Here already for $N = 6$, $D = 400$ and for $N = 8$, $D = 4900$ which gives a boundary for exact diagonalization for systems with $N < 10$.

- Large U Hubbard models: the $t - J$ model

This model, extensively studied recently in connection with the mechanism for high temperature superconductivity, is obtained in the limit of large U where doubly occupied sites are only taken into account as virtual states and an effective Hamiltonian is derived in first order in t/U:

$$H = -t \sum_{i\sigma} c_{i\sigma}^\dagger c_{i+1\sigma} + h.c. + t^2/U \sum_i (\vec{\sigma}_i \cdot \vec{\sigma}_{i+1} - 1)$$

where the sum over sites excludes the empty sites ($\vec{\sigma}_i$ are Pauli spin matrices) This Hamiltonian has the complexity almost of a spin chain system (with conserved S^z) and fairly large systems $N \sim 16$ have been studied[1].

- Finally it becomes quite evident that for more complicated models like the Kondo lattice or Anderson lattice the dimension of the basis set rapidly increases and only rather small system $N \sim 4-6$ sites have been studied even after full use of the symmetries of the problem.

USE OF SYMMETRY

The symmetry properties of the Hamiltonian can be used to reduce the dimension of matrices to diagonalize. Of course the cost is the more complicated structure of the matrices and the larger number of them to be diagonalized.

To demonstrate the application of the method we will take as an example a one dimensional system with translational symmetry and periodic boundary conditions. Then the total momentum of the system is a good quantum number.

We can implement the translational symmetry as follows: when a projection operator $\sum_n e^{ikn} T_n$, $n = 1,...N$ is applied on any basis state it creates a special symmetry state. T_n is is the translation operator by n sites. Special symmetry states then belonging to different irreducible representations (different k) do not mix under the action of the Hamiltonian.

A simple example demonstrating the idea is the following: consider a chain of four spins; the basis set consists of 6 states ↑↑↓↓, ↓↑↑↓, ↓↓↑↑, ↑↓↓↑, ↑↓↑↓, ↓↑↓↑. From this set we can construct two independent special symmetry states;

$$|a> \sim \sum_n e^{ikn} T_n | \uparrow\uparrow\downarrow\downarrow >$$

$$|b> \sim \sum_n e^{ikn} T_n | \uparrow\downarrow\uparrow\downarrow >$$

Then only the state $|a>$ is present in the representation k=1 and 3 while both states $|a>, |b>$ are in $k = 0$ and 2. Therefore our initial problem of diagonalizing a 6 by 6 matrix has been reduced to the diagonalization of 4 matrices, two of dimension 1 and two of dimension 2.

The price we have to pay is that the construction of a matrix element $<a|H|b>$ involves the calculation of N^2 terms (actually N terms if we use the property that the Hamiltonian commutes with the translation operator). Therefore reduced size of matrices to diagonalize results to more complicated Hamiltonian and usually there is a tradeoff to be made between size and complexity.

A quantum number that can be calculated without resorting to the use of symmetry is the total spin of every state in a rotationally invariant Hamiltonian (total S commutes with the Hamiltonian). This can be done in two ways; the first is to add a fictitious term in the Hamiltonian λS^2 with λ large enough so that the different multiplets split in separate bands, identify then the S of the band and then subtract the $\lambda S(S+1)$ from the energies.

A second (trivial but more time consuming) way is to diagonalize the Hamiltonian matrix for different values of total S^z. Then (for spin 1/2 system) in the $S^z = N/2$ spectrum only states with $S = N/2$ are present; for $S^z = N/2 - 1$ only states with $S = N, N - 1$ and so on.

STRUCTURE AND WHAT TO CALCULATE

There are many ways to organize a code, depending on the problem and the technical means available. Here we will briefly describe a possible organization of the code, best suited for relatively small size systems, and give some ideas about the quantities that can and have been calculated in the literature.

A typical code can be split in three parts; the first creates the basis set of states $|i>, i = 1, \ldots D$ and if symmetry is to be implemented finds the parents states and their cycle on which to apply the projection operator.

In the second part the matrix elements for the different operators composing the Hamiltonian $<i|T|j>, <i|U|j>$ are calculated and also for any other interesting operator O, $<i|O|j>$ (we take as an example the Hubbard Hamiltonian).

In the third part the operators composing the Hamiltonian are multiplied by their respective couplings $H = -t <T> +U <U>$ and the resulting total Hamiltonian matrix diagonalized.

We find it useful to store the matrix elements calculated separately so that if different couplings t, U need to be studied then the different operators composing the Hamiltonian can be recalled, multiplied by the couplings and the total Hamiltonian matrix diagonalized. The space requirements for storing the values and positions of the nonzero elements of the operators are of the $O(D)$ as most matrices are sparse. This procedure is in general worth to implement as it is usually costly to calculate the matrix elements every time a new set of couplings is needed.

Once the Hamiltonian has been diagonalized and the set of eigenvalues ϵ_n, eigenfunctions $|n>$ and their expansion in terms of the original basis set determined $|n> = \sum_i a_{n,i}|i>$, then it is easy to calculate the expectation value of any operator O for any eigenstate $|n>$ using $<n|O|n> = \sum_{i,j} a^*_{n,i} a_{n,j} <i|O|j>$. The operator O can also be a non-equal time correlation.

About the quantities that can be calculated from the ground state energies for different number of particles: the gap for charge excitations[6] given by $\Delta = \mu_+ - \mu_-$ where $\mu_+ = E_0(N+1) - E_0(N)$ is the chemical potential for adding a particle while $\mu_- = E_0(N) - E_0(N-1)$ for adding a hole; the "binding energy" for a pair in a given background of M particles $E_b = E(M+2) + E(M) - 2E(M+1)$ which corresponds to the difference in energy for two particles being in the same cluster or far apart.

Because finite size effects are very important in these studies (a gap of the order $1/N$ always exists in a finite system) it is very important to study these quantities as a function of system size and extrapolate whenever possible to infinite system. As we saw from the dimension counting earlier this is often very difficult for systems in higher than one dimension.

If the full spectrum is available (for $D < 1000$) thermodynamic quantities can be calculated[7], as

$$<O> = \frac{\sum_n <n|O|n> e^{-\beta \epsilon_n}}{Z}$$

Also the finite frequency conductivity,

$$\sigma(\omega) \sim \sum_n \delta(\epsilon_n - \epsilon_0 - \omega) \frac{|<0|j|n>|^2}{(\epsilon_n - \epsilon_0)}$$

which for finite systems consists of a set of lines at the positions of the excited states. There are attempts to smooth out this set of discrete lines either by fitting a Lorentzian shape or by introducing a phase factor in the Hamiltonian which corresponds to a change of boundary conditions and produces a continuous spectrum. The spectral function can also be calculated using a method based on the Lanczos algorithm.[8]

The same idea[9] of changing the hopping from site to site by a phase factor from $t \to te^{i\phi}$ corresponds in a one dimensional ring with periodic boundary conditions to applying a nontrivial vector potential. The change in ground state energy $\partial^2 E_0/\partial \phi^2$ for $\phi \to 0$ gives the effective hopping t^*/t. This can be simply seen in the case of a free particle where the spectrum $E_k = -2t\cos k$ changes to $E_k = -2t\cos(k+\phi)$ which trivially gives $t^*/t = 1$.

DIAGONALIZATION METHODS

As was mentioned in the first section there is a technical boundary in the dimension of the basis set at $D \sim 1000$. For matrices of dimension less than 1000 by 1000 standard library routines can be used to diagonalize the matrices and obtain the complete set of eigenvalues and eigenvectors. As an example we mention the EISPACK[10] and NAG libraries. For large mainframe machines the time required for a diagonalization of a 1000 by 1000 matrix is typically of the order of 1 minute cpu time. The standard procedure for these diagonalization methods is first to transform the matrix in a tridiagonal form by a succession of similarity transformations and in a second step diagonalize the tridiagonal matrix. The time for these procedures increases as $\sim D^3$.

For matrices larger than 1000 by 1000 there exist different approximate methods which usually give the low lying eigenstates and eigenvectors. One of them is the Davidson method[11] or the different variations of the Lanczos method[12].

FINITE SIZE SCALING

The main limitation of course in the exact diagonalization studies of small systems is the finite size of the studied clusters. The way to extract information about the infinite system is to study a given quantity as a function of system size and after assuming some scaling behavior in $1/N$ extrapolate to infinite system. Below we will give some examples where such a study was performed. The first is the problem of spinless fermions in one dimension interacting through nearest neighbor interaction. This model is equivalent to the Heisenberg spin $1/2$ chain using the Jordan-Wigner transformation. The gap in the excitation spectrum and the critical value above which the gap opens is known from Bethe ansatz analysis. The correct behavior was reproduced[13] using a series of systems up to 12 sites and using as scaling law a polynomial in $1/N$. A similar analysis for the one dimensional Hubbard model[13] gives the correct gap behavior for $U/t > 1$ but the wrong exponent in the singular region $U/t < 1$.

The second example is a model of a particle interacting via a on-site interaction with a bath of spinless fermions. The question in this case is the effective mass of the particle. It can be exactly calculated analytically using again the Bethe ansatz method in the case where the bare mass of the particle is equal to the mass of the particles composing the fermionic bath. It can be reproduced numerically[9] for the whole range of particle-bath interaction using the phase factor trick mentioned in the previous section by studying lattices between 6 and 14 sites and using as a scaling behavior a polynomial in $1/N$ of the same order as the number of systems studied.

CONCLUSION

From the above discussion it becomes clear that the advantages of the exact diagonalization method is that complete information about a model can be obtained for a Hamiltonian of any complexity. The limitations on the size that can be studied are very strict and mainly only one dimensional systems can be extensively studied (with scaling). These prompt the discussion in the second part on Quantum Monte Carlo methods where larger systems can be studied but in general on simpler Hamiltonians and higher temperatures.

REFERENCES

1. X. Zotos, Phys. Rev. **B37**, 5594 (1988); E.Y. Loh, T. Martin, P. Prelovsek, D.K. Campbell, Phys. Rev. **B38**, 2494 (1988); S.A. Trugman, Phys. Rev. **37**, 1597 (1988) and present volume; Y. Hatsugai, M. Imada, N. Nagaosa, preprint; J. Bonca, P. Prelovsek, I. Sega, Phys. Rev. **B39**, 7074 (1989); A. Mistriotis,

H. Büttner, W. Pesch, J. Phys. **CL** 1021, (1988); E. Dagotto, A. Moreo, R. Joynt, S. Bacci, E. Gagliano, preprint; P. Horsch, W. Stephan, M. Ziegler, K. von Szczepanski, preprint

2. H.Q. Lin, J. Hirsch, D.J. Scalapino, Phys. Rev. **37**, 7359 (1988), J.A. Riera, A.P. Young, preprint
3. J.E. Hirsch, S. Tang, E. Loh, D.J. Scalapino, Phys. Rev. Lett. **60**, 1668 (1988); W.H. Stephan, W.v.d. Linden, P. Horsch, preprint
4. F.D.M. Haldane, in The Quantum Hall Effect edited by R.E. Prange and S.M. Girvin (Springer-Verlag, New York,1986)
5. J. Oitmaa, D.D. Betts, Can. J. Phys. **56**, 897 (1978); J. Borysowicz, T.A. Kaplan, P. Horsch, Phys. rev. **B31**, 1590 (1985)
6. E.H. Lieb, F.Y. Wu, Phys. Rev. Lett. **20**, 1445 (1968)
7. H. Shiba, Prog. Theor. Phys. **48**, 2171 (1972)
8. P. Maldague, Phys. Rev. **B16**, 2437 (1977); A.M. Oles, G. Treglia, D.Spanjaard, R. Jullien, Phys. Rev. **B32**, 2167 (1985); E.Y. Loh, D.K. Campbell Synth. Metals **27**, A499 (1988); R. Haydock, V. Heine, M.J. Kelly in Solid State Physics, vol. 35, ed. by H. Ehrenreich, F. Seitz, D. Turnbull,(Academic,N.Y. ,1980); O. Gunnarson, K. Schönhammer, Phys. Rev. **B31**, 4815 (1985); C.A. Balseiro, A.G. Rojo, E.R. Gagliano, B. Alascio, Phys. Rev. **B38**, 9315 (1988)
9. X. Zotos, F. Pelzer, Phys. Rev.**B37**, 5045 (1988)
10. B.T. Smith, Matrix Eigensystem Routines-EISPACK Guide, Lect. Notes in Computer Science, Springer-Verlag (1976)
11. G. Cisneros, C.F. Bunge, Comput. Chem. **8**, 157 (1984); J. Weber, R. Lacroix, G. Wanner, Comp. & Chem. **4**, 55 (1980)
12. H.H. Roomany, H.W. Wyld, L.E. Holloway, Phys. Rev. **D21** , 1557 (1980); E. Dagotto, A. Moreo, Phys. Rev. **D31**, 865 (1985)
13. G. Spronken, R. Jullien, M. Avignon, Phys. Rev. **B24**, 5356 (1981); L. Sneddon, J. Phys. **C11**, 2823 (1978); B. Fourcade, G. Spronken, Phys. Rev. **B29**, 5012 (1984)

B) QUANTUM MONTE CARLO METHODS

M. Dzierzawa, X. Zotos

Institut für Theorie der Kondensierten Materie

Universität Karlsruhe, Karlsruhe, FRG

INTRODUCTION

In the previous section we discussed exact diagonalization studies and concluded that the size of the systems that can be studied are rather limited while complete information on the system can be obtained and the complexity of the Hamiltonian poses no extra problems. A complementary method is the Quantum Monte Carlo method which allows the study of much larger systems (of the order of hundred sites) although the information obtained is more limited and the Hamiltonians so far studied simpler. There is an abundance of methods that have been developed for fermion and spin systems[1]; here we will only describe two methods, the first because it seems to be the most flexible and presently widely used[2-4], the second because it is very recent and the experience on it rather limited[5]. For the interested reader a very complete technical account of both methods has recently been presented[6].

We will describe the two methods in parallel to underline their similar procedure. Their characteristics are the following: in the first, which we will call the determinant method, the chemical potential is fixed (grand canonical ensemble) and finite temperature quantities as well as non-equal imaginary time correlations, susceptibilities, can be calculated. The central quantity to calculate is the partition function:

$$Z = Tr e^{-\beta(H-\mu N)}$$

where the trace is over number of particles and states, and the thermal averages of a quantity O is given by:

$$<O> = \frac{Tr O e^{-\beta(H-\mu N)}}{Z}$$

In the second method, that we will call the projection method, the number of particles is fixed and zero temperature quantities can be calculated e.g. energy, densities or equal time correlation functions. The principle of the method is the following; one starts with a Slater determinant trial function $|\Psi_T>$ and applies the projection operator $e^{-\beta H}$. As any trial function can be expanded in terms of a complete set of eigenfunctions of the Hamiltonian the ground state expectation value of a quantity O is given by:

$$<O> = \frac{<\Psi_T|e^{-\beta H/2}Oe^{-\beta H/2}|\Psi_T>}{<\Psi_T|e^{-\beta H}|\Psi_T>} = <0|O|0> + \mathcal{O}(e^{-\beta(\epsilon_1-\epsilon_0)})$$

For $\beta \to \infty$ in principle the ground state properties can be obtained assuming of course that the ground state has a finite overlap with the trial state. As for a finite system the energy difference $\epsilon_1 - \epsilon_0$ of the excited states from the ground state is $\mathcal{O}(1/N)$ β must be larger than N (unless there is a gap). In practice we find convergence of the results for finite values of β. We can argue that low lying states have the same qualitative properties as the ground state.

We will not attempt to give a very detailed presentation of the methods, which can be found in the literature, but rather a plan of the implementation of the calculation and some practical comments. The typical steps in a Quantum Monte Carlo method are as follows: first using the Trotter decomposition we will separate the kinetic and potential part of the Hamiltonian; then introducing the auxiliary Hubbard - Stratonovich fields we will turn the problem to a one of independent particles moving in a random potential. Next we will discuss that the Green's function is the basic quantity out of which more complicated operators can be calculated and finally we will describe two methods the Langevin and Monte Carlo for performing an importance sampling of the auxiliary Hubbard- Stratonovich fields. We will close the discussion with some comments about the resources necessary for a simulation.

PROCEDURE

a) Trotter decomposition

As our example for the implementation of the method we will again consider the Hubbard model[4-6] on which so much work has recently been done, also used as a prototype for testing the different algorithms. It is described by the following Hamiltonian:

$$H = -t\sum_{i,j\sigma}(c^\dagger_{i\sigma}c_{j\sigma} + H.c.) + U\sum_i n_{i\uparrow}n_{i\downarrow}$$

where $c_{i\sigma}(c^\dagger_{i\sigma})$ are annihilation(creation) operators of a fermion with spin $\sigma = \uparrow, \downarrow$, $n_{i\sigma} = c^\dagger_{i\sigma}c_{i\sigma}$.

In the determinant method we want to calculate the partition function $Z = Tr e^{-\beta(H-\mu N)}$ while in the projection, the quantity $Q = <\Psi_T|e^{-\beta H}|\Psi_T>$. To separate the kinetic energy part T from the interaction part U we write

$$e^{-\beta H} = B_L B_{L-1}...B_l...B_1$$

where $B_l = e^{-\Delta\tau U_l}e^{-\Delta\tau T}$, with $\Delta\tau L = \beta$ and corrections of the order $\Delta\tau^2$. For $\Delta\tau \to 0$ this decomposition is exact. In practice we take $\Delta\tau^2 tU < 0.1$ so that the error remains of the order 1 percent.

We should notice that the Tr in the determinant method can also be thought as a sum over all many particle trial functions which form a complete basis set in every particle number subspace.

b) Hubbard - Stratonovich fields

At this point the operators U are still many body operators as they involve a product of density operators. To obtain a single particle representation we use the identity[3]:

$$e^{-\lambda^2 n_\uparrow n_\downarrow} \sim \int_{-\infty}^{\infty} e^{-\phi^2/2} e^{-\lambda\phi(n_\uparrow - n_\downarrow)} d\phi$$

We should note that the auxiliary field ϕ couples with the spin density $\sigma^z = n_\uparrow - n_\downarrow$ on every site. The operator $e^{-\beta H}$ becomes:

$$e^{-\beta H} \sim \int d\phi e^{-\phi^2/2} \prod_l e^{-\Delta\tau \tilde{U}_l} e^{-\Delta\tau T}$$

where ϕ denotes collectively the fields ϕ_{il} for all space sites i and time slices l. $\tilde{U}_l$ contains now only single particle operators. In the determinant method we can now perform the trace over fermion operators[4] as the expression is quadratic and obtain:

$$Z = det(1 + B_L B_{L-1}...B_1)_\uparrow det(1 + B_L B_{L-1}...B_1)_\downarrow$$

In the projection method we have:

$$Q = det <\Psi_{T\uparrow}|(B_L B_{L-1}...B_1)_\uparrow|\Psi_{T\uparrow}> det <\Psi_{T\downarrow}|(B_L B_{L-1}...B_1)_\downarrow|\Psi_{T\downarrow}>$$

The operators B_l have the following matrix representation:

$$B_l = e^{-\Delta\tau \tilde{U}_l} e^{-\Delta\tau T} \sim \begin{pmatrix} e^{\mp\lambda\phi_{1l}} & 0 & . & . \\ 0 & e^{\mp\lambda\phi_{2l}} & 0 & . \\ . & 0 & e^{\mp\lambda\phi_{3l}} & 0 \\ . & . & 0 & . \end{pmatrix} \cdot \begin{pmatrix} 1 & \Delta\tau t & 0 & . \\ \Delta\tau t & 1 & \Delta\tau t & 0 \\ 0 & \Delta\tau t & 1 & \Delta\tau t \\ . & 0 & \Delta\tau t & . \end{pmatrix}$$

with the minus sign refering to the up spin and the plus to the down spin. We linearized the kinetic energy term as this greatly simplifies the matrix multiplication. Alternative schemes to speed the multiplication and simplify matrix inversions, used later, is the checkerboard decomposition:

$$e^{-\Delta\tau T} \sim \begin{pmatrix} c & 0 & . & . & . \\ 0 & c & s & . & . \\ . & s & c & 0 & . \\ . & . & 0 & c & s \\ . & . & . & s & c \end{pmatrix} \cdot \begin{pmatrix} c & s & . & . & . \\ s & c & 0 & . & . \\ . & 0 & c & s & . \\ . & . & s & c & 0 \\ . & . & . & 0 & c \end{pmatrix}$$

where $c = \cosh(\Delta\tau t)$, $s = \sinh(\Delta\tau t)$. Also while in the linearized version the symmetry of the single particle dispersion $\epsilon_k = -\epsilon_{\pi-k}$ is lost, in the checkerboard decomposition it is preserved.

To make clear the matrix representation of the trial Slater determinant state consider a system of $N = 4$ sites with two spin up and two spin down fermions in the following configuration $\uparrow\downarrow\uparrow\downarrow$. The matrices $|\Psi_{T\uparrow}>$, $|\Psi_{T\downarrow}>$ for the up, down spin fermions then are of dimension N by M ($M = 2, N = 4$) are:

$$|\Psi_{T\uparrow}> = \begin{pmatrix} 1 & 0 \\ 0 & 0 \\ 0 & 1 \\ 0 & 0 \end{pmatrix} ; |\Psi_{T\downarrow}> = \begin{pmatrix} 0 & 0 \\ 1 & 0 \\ 0 & 0 \\ 0 & 1 \end{pmatrix}$$

Then the expression for Q involves the multiplication of the N by M right matrix with the L matrices B_l followed by the left matrix and then taking the determinant of the resulting M by M matrix.

At this point the calculation of the partition function Z or the projection Q turned to a statistical mechanics problem where we have to perform the integration over the $N \cdot L$ auxiliary Hubbard-Stratonovich fields. In contrast though to most statistical mechanics problems which involve local interactions, the action here is highly non-local.

We would like to note that many other possibilities for performing a Hubbard-Stratonovich decomposition exist; for instance in terms of a continuous but bounded one:

$$e^{-\lambda n_\uparrow n_\downarrow} \sim \int_{-\pi}^{\pi} d\theta e^{-\lambda^* \cos\theta (n_\uparrow - n_\downarrow)} \; ; \; I_0(\lambda^*) = e^{\lambda/2}$$

A very convenient one for Monte Carlo sampling as we will see later is in terms of a discrete spin (Ising type) variable[3]:

$$e^{-\lambda n_\uparrow n_\downarrow} \sim \sum_{s=\pm 1} e^{-\lambda^* s(n_\uparrow - n_\downarrow)} \; ; \; \cosh \lambda^* = e^{\lambda/2}$$

c) Green's function

In this section we will demonstrate that the central quantity in the calculation is the Green's function, used to evaluate any other of more complicated correlation and also for updating the Hubbard - Stratonovich fields.

In the projection method the ground state expectation value of an operator O is given by:

$$<O> = \frac{\int d\phi W(\phi) <O>_\phi}{\int d\phi W(\phi)}$$

The expectation value $<O>_\phi$ and the weight $W(\phi)$ at the particular configuration ϕ are defined as:

$$<O>_\phi = \frac{<\Psi_T|B_L...O...B_1|\Psi_T>}{<\Psi_T|B_L...B_1|\Psi_T>} \quad W(\phi) = e^{-\phi^2/2} <\Psi_T|B_L...B_1|\Psi_T>$$

The expectation value $<O>_\phi$ is between single particle states representing fermions moving in a space-imaginary time random potential independently of each other. Therefore for every given random field configuration we can apply Wick's theorem for the operator O and decompose it in a product of single particle Green's functions $g_{ij} = <c_i^\dagger c_j>$. For example if

$$<O> = <n_{i\uparrow} n_{j\uparrow}> = <c_i^\dagger c_i><c_j^\dagger c_j> + <c_i^\dagger c_j><c_i c_j^\dagger>$$

Therefore for every field configuration we have to calculate the quantities g_{ij} for all i, j and combine them to form the different operators we want to study. In the determinant method[4] we similarly define the equal time Green's function at time slice l $G(l) = <c_i c_j^\dagger>_l = (1 + B_l...B_1 B_L...B_{l+1})_{ij}^{-1}$.

A few technical remarks; first in the projection method, we can average over all the positions of the operator O to obtain better statistics;

$$< \Psi_T | B_L ... O ... B_1 | \Psi_T > = (1/L) \sum_l < B_L ... B_l O B_{l-1} ... B_1 >$$

or in the determinant method over all time slices l, $< c_i c_j^\dagger > = (1/L) \sum_l < c_i c_j^\dagger >_l$. Also for each individual field configuration the space and time invariance is lost as the fermions move in a random field so that $< c_i^\dagger c_j > \neq < c_j^\dagger c_i >$ and $< c_i^\dagger c_{i+k} >$ depends on i. The quantities $g_{ii} = < c_i^\dagger c_i >$ are really unphysical[2] as they can also take negative or larger than one values. It is after integrating over all field configurations that the physical quantities and symmetries are recovered.

A very important technical point which arises, and which has hindered calculations at low temperatures (large β) is the following: for large β we have to multiply an ever increasing number of evolution matrices. The matrix elements arising from the multiplication involve scales from $e^{-\beta D}$ to $e^{+\beta D}$, where D is a energy scale related to the hopping t and interaction U and therefore exponentially large and vanishingly small. In an actual numerical calculation this results in saturation of the precision of the machine and the wrong evaluation of the determinant or the inverse of $(1 + B_L...B_1)$. In the projection method this is manifested as the collapse along one direction of the initially orthogonal set of single particle wavefunctions composing the trial function after a few multiplications by the evolution matrices. The remedy to this problem was recently found and is the repeated orthogonalization of the basis set[5,6]. After a certain number of multiplications of the basis set by the evolution matrices B_l the resulting trial function is orthonormalized which amounts to writing it as $|\Psi_T> = |\Psi'> DR$ with D a diagonal matrix including the large difference energy scales and R an orthogonal matrix of dimension M by M (M is the number of fermions). Then the evolution proceeds for a certain number of steps and again the orthogonalization procedure has to be repeated. In the end we have to compute the determinant formed by the product of the left hand trial function and the evolved right hand function, while the trail of the D and R matrices has not to be taken into account as they cancel from the calculation with the denominator.

In the determinant method[4] the same is accomplished by a succesion of modified Gram-Schmidt orthogonalizations; the expression $(1 + B_L B_{L-1}....B_1)$ becomes $(1 + B_L B_{L-1}....U'D'R')$ where $U'D'R'$ is a decomposition of a first group of B operators, with U' an orthogonal, D' a diagonal and R' an upper right triangular matrix. We proceed then with the evolution by multiplying the $U'D'$ matrices with the next block of B matrices, decomposing and so on. Finally we obtain the form $(1 + UDR)$ which we rewrite as $U(U^{-1}R^{-1}+D)R$. After a final Gram-Schmidt orthogonalization of $(U^{-1}R^{-1} + D)$ we obtain the product of an orthogonal, diagonal with the large scale elements and upper triadiagonal matrix, which we can safely invert or take the determinant.

d) Sampling

Up to this point the formulation of the evaluation of the partition function or projection is exact. But the functional integrals over the auxiliary Hubbard - Stratonovich fields can not be done exactly and we must proceed to stochastic methods. Two methods are most extensively used, the Langevin method[7] and variations of the Monte Carlo method[8] for the importance sampling of the fields.

The Monte Carlo method is more appropriate for Ising spins; starting from an initial spin configuration one attempts to flip each spin $s \to s'(=-s)$. If the ratio of weights $r = W(s')/W(s)$ greater than one then the flip is accepted while if it is less than one a random number $0 < \eta < 1$ is created; then for $r > \eta$ the flip is accepted while if $r < \eta$ the flip rejected.

This is done by moving through the lattice time slice by time slice; for every time slice l we attempt to flip the spin on each site, the ratio r depending only on the diagonal element of $G(l)$. Once a flip is accepted then we must correct the Green's function, $G(l) \to G'(l)$, for which a fast algorithm exists[2]. To move from time slice l to $l+1$ we use the relation $G(l+1) = B_{l+1}G(l)B_{l+1}^{-1}$. Still the Green's function must be recalculated from scratch every certain number of time slices because of numerical deterioration.

As the acceptance ratio should be close to 1/2 in order to obtain successsive uncorrelated spin configurations, an alternative sampling method can be used, the heat bath algorithm. In this case a spin is flipped with probability $W(s')/(W(s') + W(s))$.

In the Langevin method, more appropriate for the continuous field representation, a fictitious time t is introduced and the time evolution of the fields ϕ is governed by the Langevin equation:

$$\frac{\partial \phi}{\partial t} = -\frac{\partial V_{eff}}{\partial \phi} + \eta(t)$$

where $V_{eff} = \phi^2/2 - \ln \det <\Psi_T|B_L...B_1|\Psi_T>$ is an effective potential and η a random noise term with δ function correlations. It can be shown that field configurations generated according to the Langevin equation then have a distribution given by the weight $W(\phi)$. Then $<O> = (1/T)\int_0^T dt <O[\phi(t)]>$, $T \to \infty$. As we mentioned earlier, the term $\partial V_{eff}/\partial \phi$ which is the force term, is actually given by the Green's function.

To obtain a sequence of field configurations ϕ we must integrate the Langevin equation[9]. The simulation proceeds as follows; first we discretize the time t and at a given moment, for the field $\phi(t)$, we calculate the Green's functions g_{ij} and use them to calculate first the force term and then the new field configuration $\phi(t + \Delta t)$. As we are interested to sample the quantity $<O[\phi(t)]>$ over uncorrelated field configurations $\phi(t)$, the correlation time being of the order of $1/\Delta t$, we do not measure at every time step. In contrast, in the Monte Carlo sampling it is sufficient to sweep one to two times through the lattice between measurements.

Several approximations done in analytical calculations can be represented by tuning the sampling procedure. If we turn the noise term η off and let the fields ϕ evolve according to the Langevin equation we end up in a minimum of the effective potential which corresponds to a mean field solution. Actually for the Hubbard-Stratonovich decomposition of the previous section it corresponds to a mean field factorization $\sigma^z\sigma^z = (n_\uparrow - n_\downarrow)(n_\uparrow - n_\downarrow) \to \sigma^z <\sigma^z>$. A second possibility is to restrict the sampling over fields which are independent of the the imaginary time slice l, corresponding to the static approximation extensively used in the literature[10].

One important issue that arises in fermionic system simulations is the sign of the determinant. It turns out that for some field configurations (actually strongly space-time dependent) $W(\phi)$ is negative and therefore cannot be used as a probability. A simple example can demonstrate the cases when negative signs can appear: consider a one dimensional system with periodic boundary conditions (a ring) and a trial function where the particles are at certain positions $c_1^\dagger c_2^\dagger ... c_{M-1}^\dagger c_M^\dagger |0>$. The matrices B_l composed of a potential and kinetic part (in the linearized version) are then applied on this right trial function; the potential part gives only a weight factor for every configuration while the kinetic part moves the particles on the chain without interchanging them though. If however we have periodic boundary conditions it is possible that a particle is removed say from site 1 and placed on site N; $(c_N^\dagger c_1) c_1^\dagger c_2^\dagger ... c_{M-1}^\dagger c_M^\dagger |0>$. As we must keep the standard ordering of operators a negative sign will appear if the number of the $M-1$ particles is odd. Therefore we see that on a ring with an even number of particles of given spin negative signs can appear due to permutation of fermions around the ring. In two dimensions this of course cannot be avoided as there are always many paths for interchanging particles.

To partly solve the negative sign problem we reformulate the expression:

$$<O> = \frac{\sum_\phi W(\phi) <O>_\phi}{\sum_\phi W(\phi)} = \frac{\sum_\phi |W(\phi)| <sO>_\phi}{\sum_\phi |W(\phi)| <s>_\phi}$$

where s is the sign of the weight for the configuration ϕ. Then the weight is always positive but the expectation value of the sign must also be calculated. This in most cases poses no problem except when the average sign is approximately zero and we have to divide by a vanishing quantity with a statistical error. This problem seems to be particularly acute in the two dimensional Hubbard model at ~ 80 percent filling[5,6].

CONCLUSION

We will conclude by giving some impression about the size of systems studied; the lattice sizes are typically of the order 8 by 8 (except recent work with the projection method) and temperatures down to $\beta \sim 20$. The computer time[6] for these simulations scales as $\sim N^3$ for the spatial dimension and $\sim L$ for the time one. There are several attempts to develop more efficient algorithms[10] where the time scales linearly with the size of the system ($\sim N$) so that larger systems can be studied.

Between the Langevin and Monte Carlo methods it seems that for the Hubbard model it is easier to obtain good sampling using the discrete spin decomposition and Monte Carlo sampling. However this decomposition is not always possible, like for instance in the case when a phonon field is included. Up to now most studies have concentrated on models with one on site Hubbard interaction. If more interactions are present, a larger number of Hubbard - Stratonovich fields must be introduced and the experience so far on the sampling of many auxiliary fields is rather limited. In this aspect the Monte Carlo techniques are in contrast to exact diagonalization studies where the the complexity of the Hamiltonian produces only little extra weight.

One of the authors (X.Z) would like to thank Prof. F.D.M. Haldane for sharing his experience in exact diagonalization studies and the ISI program on "High T_c superconductivity" in Torino for a fruitful visit. Both authors would like to thank the organizers of the Evora Nato School and acknowledge financial support by the BMFT program No. 13N5501 and the Esprit project MESH No. 3041.

REFERENCES

1. Quantum Monte Carlo Methods, edited by M. Suzuki (Springer-Verlag 1987); D.J. Scalapino, Frontiers and Borderlines in Many-Particle Physics (1988); J.E. Hirsch, R.L. Sugar, D.J. Scalapino, R. Blankenbecler, Phys. Rev. **B26**, 5033 (1982); D. Kung, D. Dahl, R. Blankenbecler, R. Deza, J.R. Fulco, Phys. Rev. **B32**, 2022 (1985)
2. R. Blankenbecler, D.J. Scalapino, R.L. Sugar, Phys. Rev. **D24**, 2278 (1981)
3. J.E. Hirsch, Phys. Rev. **B28**, 4059 (1983)
4. J.E. Hirsch, Phys. Rev. **B31**, 4403 (1985); J.E. Hirsch, H.Q. Lin, Phys. Rev. **B37**, 5070 (1988)
5. S. Sorella, E. Tosatti, S. Baroni, R.Car, M. Parrinelo, Modern Phys. **B1**, 993 (1988); S. Sorella, S. Baroni, R. Car, M. Parinello, Europhys. Lett. **8**, 663 (1989)
6. E.Y. Loh Jr., J.E. Gubernatis, in "Electrons Phase Transitions", ed. W. Hauke, Y.V. Kopaev, North Holland Physics (Elsevier, N.Y.,1990); S.R. White, D.J. Scalapino, R.L. Sugar, E.Y. Loh, J.E. Gubernatis, R.T. Scalettar, Phys. Rev. **B40**, 506 (1989).
7. G. Parisi, Nucl. Phys. **B180**, 378 (1981)
8. K.Binder: in Monte Carlo Methods in Statistical Physics, ed. by K. Binder Topics in Current Physics, vol. 7 (Springer Verlag, New York, 1979)
9. S.R. White, J.W. Wilkins, Phys. Rev. **B37**, 5024 (1988)
10. M. Cyrot, Journal de Phys. **33**, 125 (1972); E.N. Economou, C.T. White, R.R. DeMarco, Phys. Rev. **18**, 3946 (1978)
11. R.T. Scalettar, D.J. Scalapino, R.L. Sugar, D. Toussaint, Phys. Rev. **B36**, 8632 (1987); M. Imada, J. Phys. Soc. Jpn. **57**, 2689 (1988).

HIGH - DENSITY EXPANSION FOR ELECTRON SYSTEMS

M. Bartkowiak[*], P. Münger[†], K.A. Chao[†], and R. Micnas[*]

[*]Institute of Physics, A. Mickiewicz University
Matejki 48/49, PL-60769 Poznań, Poland
[†]Department of Physics, University of Linköping
S-58183 Linköping, Sweden

A diagrammatic technique based on the Wick's theorem for the Hubbard's operators is constructed and applied to perform systematic high-density expansion (HDE) for the extended Hubbard model (EHM) and the spinless fermion (SF) model[1,2,3].

In the loopless approximation for the Green's functions (which has been shown to be equivalent to the Hubbard I equation of motion decoupling scheme) we have determined the band structure of the EHM for $T = 0$. The spectrum consists of two, three or four subbands, depending on the band filling. Within the framework of the selfconsistent first-order approximation, the finite temperature phase diagram of the half-filled EHM in the atomic limit has been derived. It has been shown that the selfconsistent version of the HDE leads to an unphysical jump of the order parameter below the critical temperature. Moreover, this approximation fails to describe finite bandwidth effects on the charge ordering for large enough hopping parameter. From the staggered magnetic susceptibility calculated in the single-loop approximation, the magnetic phase diagram of the half-filled Hubbard model has been determined. The Néel temperature appears to be lower than that obtained in both second and fourth order of the unrenormalized linked cluster expansion[4].

The Horwitz-Callen renormalization scheme and higher order approximations have been applied to the SF model in order to test usefulness of the HDE methods in the theory of narrow-band systems. We have presented temperature dependence of the charge ordering parameter and charge susceptibility, and determined the corresponding finite-temperature phase diagram. Some unphysical features which appear in the selfconsistent first-order approximation schemes have been identified and discussed. The ground state energy of the SF model for half-filling has been calculated to the second and third order in $1/z$. The results are in an excellent agreement with the known exact solutions. Finally, we draw some general conclusions regarding violation of sum rules and internal consistency of the diagrammatic perturbation expansion methods in terms of the Hubbard's operators.

REFERENCES

1. M.Bartkowiak, Int.J.Mod.Phys. **B1**, 1277 (1987).
2. M.Bartkowiak, P.Münger, and R.Micnas, Int.J.Mod.Phys. **B2**, 483 (1988).
3. M.Bartkowiak, P.Münger, and K.A.Chao, Int.J.Mod.Phys. **B2**, 521 (1988).
4. K.Kubo, Prog.Theor.Phys. **64**, 758 (1980).

PERTURBATIVE RESULTS USING THE CUMULANT EXPANSION IN THE ANDERSON LATTICE[#]

Gerardo Martínez[*] and Mario E. Foglio

Instituto de Física "Gleb Wataghin", UNICAMP
13081, Campinas, SP, Brasil

We have developed a method to calculate the free energy and the Green's functions of the periodic Anderson lattice by using the cumulant expansion [1] and a generalized Wick's theorem [2] for the Hubbard operators in the infinite-U limit. Taking the works of Hubbard [3] and Hewson [4] as a starting point we state the rules for calculating the diagrammatic contributions. We sum up some sub-classes of diagrams: the "chains" and the "chains of loops".

The calculation proceeds as follows. The local and extended states are factored apart in the cumulant expansion. For the extended states the usual Wick's theorem is applied whereas for the local states we collect all possible contributions which, afterwards, can be reduced systematically by using the specific commutation rules of the Hubbard operators. Thus (2n)-time ionic propagators appearing in the series expansion can have a definite expression.

Selecting the minimal ("chain") diagrams a gap in the local density of states is opened around the local energy level. This gap is proportional to the effective hybridization of the system, consistent with the saddle point approximation of the slave boson method [5]. The inclusion of two-particle correlations, represented here by chains of loop diagrams, introduces new states within the gap which are present in the mixed valence region. These states are not Kondo-like resonances but look more like impurity states. They could be associated to incoherent scattering of the particles. Further, the exact solution of the atomic limit is compared with our results in the limit of negligible bandwidth. We see that this approximate solution substitutes several groups of delta functions by weighted single deltas at average positions. Detailed calculations of this work will be published separately.

REFERENCES

[#]) Work financed by FAPESP and CNPq.
[*]) Present address: Max-Planck-Institut für Festkörperforschung Heisenbergstr. 1, D-7000 Stuttgart 80, Fed. Rep. Germany
1. R. Kubo, J. Phys. Soc. Jpn. 17, 1100 (1962)
2. D.H.Y. Yang and Y.-L. Wang, Phys. Rev. B10, 4714 (1975)
3. J. Hubbard, Proc. Roy. Soc. A296, 82 (1966)
4. A.C. Hewson, J. Phys. C: Solid St. Phys. 10, 4973 (1977)
5. P. Coleman, Phys. Rev. B29, 3035 (1984);
   D.M. Newns and N. Read, Adv. Phys. 36, 799 (1987)

APPLICATION OF GUTZWILLER'S CORRELATED METHOD TO THE ELECTRONIC EFFECTIVE MASS OF DEGENERATE N-TYPE SILICON

A. Ferreira da Silva

Instituto de Física - UFBa
Campus da Federação
40210 Salvador, Bahia, Brazil

The effective mass m^* of Si:P is calculated for donor concentration N at which experimental data are available. Within the framework of Gutzwiller's variational method for highy correlated system[1], we have performed the calculation following a Brinkman-Rice-Berggren-Ferreira da Silva scheme[2]. Here we have applied a variational treatment which takes into account the many-valley effect of the silicon semiconductor[3]. The m^* obtained is written as

$$m^* = |1 - (U/U_o^2)|^{-1}, \qquad (1)$$

where U is the Hubbard correlation energy and U_o is the average energy without correlation[2]. The results are given in Table 1. The quantitative agreement between our calculation and experiment[4] is certainly satisfactory.

Table 1. The m^* for Highly Doped Samples of Si:P.

N (10^{18}cm^{-3})	Present calculation (m^*)	Experiment (m^*)
96	1.02	1.03
16	1.10	1.11
8.9	1.20	1.13
6.7	1.36	1.20

REFERENCES

1. M. C. Gutzwiller, Phys. Rev. 137:A1726 (1965).
2. W. F. Brinkman and T. M. Rice, Phys. Rev. B2:4302 (1970); K.-F. Berggren, Phil. Mag. 30:1 (1970); A. Ferreira da Silva, Phys. Rev. B38: 10055 (1988).
3. A. Ferreira da Silva, Phys. Rev. Lett. 59:1263 (1987); Phys. Rev. B37:4499 (1988).
4. N. Kobayashi, S. Ikehata, S. Kobayashi, Solid State Commun. 24:67 (1977).

OPTICAL ABSORPTION IN DISORDERED SEMICONDUCTOR SYSTEMS: APPLICATION TO THE CORRELATED PHOSPHORUS-DOPED SILICON

A. Ferreira da Silva and F. de Brito Mota

Instituto de Física - UFBa
Campus da Federação
40210 Salvador, Bahia, Brazil

We have investigated the optical absorption in doped semiconductors via the ground-state $H_2^+ \to H_2^0$ excitation donor-pair transition with application to Si:P. A correlated scheme is used for the H_2- like impurity molecule[1]. The many-valley characteristic of the host silicon is taken into account in the calculation[2]. The absorption coefficient can be written as[3]

$$\alpha = <<G^+(E,R)>> = \int P(R) \, |E - Ei(R) + i0^+|^{-1} dR, \qquad (1)$$

where $<< >>$ is the average Green function. $P(R)$ is the pair distribution function and Ei is the excitation energy. The results are shown in Figure 1. A good agreement compared to the available experimental data is found[4];

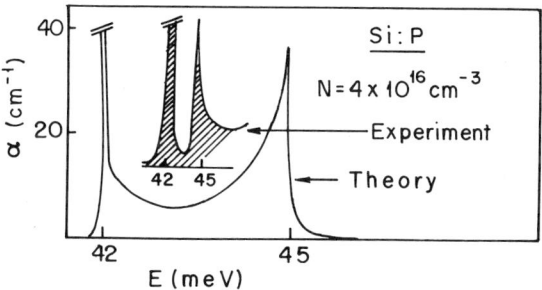

Fig. 1 α as a function of E

REFERENCES

1. A. Ferreira da Silva, J. Phys. C. Solid State Physics, 13:1427 (1980).
2. A Ferreira da Silva, Phys. Rev. Lett. 59: 1263 (1987)
3. E. A. A. e Silva, I. C. da Cunha Lima and A. Ferreira da Silva, Solid State Commun. 61:795(1987).
4. M. Capizzi, G. A. Thomas, F. de Rosa, R. N. Bhatt and T. N. Rice, Phys. Rev. Lett. 44:1019 (1980).

High T_c Superconductivity: Lessons to be Learned from Neutron Scattering

H. Capellmann

Institut Laue Langevin
F-38042 Grenoble Cedex, France
and

Institut für Theoretische Physik C, Technische Hochschule Aachen
D-5100 Aachen, Federal Republic of Germany

1) General remarks

In principle neutron scattering is able to provide all information about microscopic lattice properties and magnetic properties. Due to the nuclear interaction ionic structures

$$S(\mathbf{q}) \propto \sum_{ij} \langle e^{i\mathbf{q}(\mathbf{R}_i^0 - \mathbf{R}_j^0)} \rangle \tag{1}$$

and the dynamics of the ionic displacement-displacement correlation function

$$\sum_{ij} \int e^{i\mathbf{q}(\mathbf{R}_i^0 - \mathbf{R}_j^0)} \langle \mathbf{u}_i(t) \cdot \mathbf{u}_j(0) \rangle e^{i\omega t} dt \tag{2}$$

are obtained ($\mathbf{R}_i^0$ are the average ionic positions, $\mathbf{u}_i$ the displacements).

The magnetic interaction (due to the neutron magnetic moment) allows the measurement of the magnetization-magnetization correlation function, often proportional to the spin-spin correlation function

$$\sum_{ij} \int e^{i\mathbf{q}(\mathbf{R}_i - \mathbf{R}_j)} \langle \mathbf{S}_i(t) \cdot \mathbf{S}_j(0) \rangle e^{i\omega t} dt \tag{3}$$

containing information about magnetic ordering and about fluctuations.

The different correlation functions above can be separated in principle by the use of polarized neutrons and polarization analysis. In practice the study of magnetic excitations can be very difficult because of intensity problems: In high T_c cuprates and even in their magnetic cousins $La_2 CuO_4$ or $YBa_2 Cu_3O_6$ (antiferromagnetic and not superconducting) the density of states for lattice interactions, i. e. the number of phonon modes in the region up to 100 meV, is several orders of magnitude larger than the number of magnetic degrees of freedom in that same low energy region:

In $YBa_2Cu_3O_7$ for example, there are 39 lattice degrees of freedom per formula unit, giving rise to the same number of phonon modes, which may be multiply excited at higher temperatures. Many theoretical models are based on magnetic (spin 1/2) degrees of freedom being associated with Cu-ions within the CuO_2 planes: At best there are only two spin 1/2 two level systems per formula unit (below we shall see that the experimentally established bound is actually much lower in the low energy region).

A direct consequence for theoretical models is that any Fermi liquid type theory, even if based on magnetic interactions, will have to take into account interactions with phonons, simply because their number of degrees of freedom dominates in the low temperature region: Even if magnetic fluctuations were to be responsible for Cooper pair formation, one would have to demonstrate that this mechanism is unaffected by the comparatively large number of phononic excitations in the system.

The other class of theoretical models based directly on phononic mechanisms can to a large extend be tested directly, the phonon spectrum can be studied extensively in practise, provided that large enough single crystals are available.

The third class of models based on electronic excitations and polarizabilities can only be tested indirectly in neutron scattering: Since the neutron does not couple to charge densities directly, electronic charge distributions and excitations cannot be studied directly, but indirectly only through possible effects on nuclear displacements (phonons).

2) Experimental observations

A) Lattice properties

Although neutron scattering has also played a decisive role in structure determinations of high T_c materials, we shall concentrate on excitation spectra.

The first (crude) information comes from phonon densities of states. La_2CuO_4 (the parent compound) and the superconductor $La_{1.85}Sr_{0.15}CuO_4$ were studied first [1]; the general features were: The energy scale of the spectra extended up to 90 meV (setting a scale of $\sim 1000K$), the highest energy modes being associated with oxygen vibrations. No marked differences showed up in the densities of states of La_2CuO_4 and $La_{1.85}Sr_{0.15}CuO_4$. This was in contrast to later observations on superconducting $YBa_2Cu_3O_7$ and nonsuperconducting $YBa_2Cu_3O_6$ and $YBa_2(Cu_{0.9}Zn_{0.1})_3O_7$. [2] The energy scale of excitations again extended roughly up to 1000K (this is a general feature of all high T_c superconducting oxides and related materials, the scale is due to oxygen), but marked differences showed up between the superconductor and its nonsuperconducting reference compounds. These differences concentrated in the highest energy modes, the superconductor being characterized by a softening of these high energy oxygen modes. Similar results were obtained when comparing the nonsuperconducting (and nonmagnetic) $BaBiO_3$ and superconducting $Ba_{1-x}K_xBiO_3$ [3]. The general energy scale is similar to the cuprates (slightly smaller in $BaBiO_3$, of order 70meV). Again the high energy oxygen modes soften markedly when changing the composition by the addition of potassium to achieve high T_c ($\sim 30K$) superconductivity.

These observations are compatible with very strong electron phonon coupling being responsible for

a) softening of high energy oxygen vibrations

b) superconductivity, the high T_c being consistent with the high energy scale of oxygen vibrations, considerably higher than in conventional superconductors.

The assumption about important electron phonon coupling effects to high energy oxygen modes gained strong support recently through measurement of phonon dispersion curves of La_2CuO_4 [4] and $YBa_2Cu_3O_{6+x}$ [5]. A complete investigation of all phonon branches (in the main high symmetry directions) of La_2CuO_4 revealed highly unusual behavior in the highest energy oxygen modes; these are breathing type modes within the CuO_2 planes. This mode is split into two well defined excitations of energy $720 k_B K$ and $1060 k_B K$ (15 and 22 THz) [4]. The appearance of an extra mode, which in principle cannot be explained within the context of a harmonic lattice model alone, points towards a coupling to some other degree of freedom, possibly of electronic origin and consistent with an important electron phonon coupling. Similar observations (although less complete) were made for $YBa_2Cu_3O_{6+x}$ [5].

B) Magnetic properties

The magnetic properties of "high temperature" superconducting Cu-oxides and related (non-superconducting) compounds have been the subject of intense study - both experimental and theoretical - during the last several years.

The parent compound for the superconducting cuprates La_2CuO_4 (which itself is not superconducting) has long range antiferromagnetic order at low temperatures [6], the Néel temperature T_N depending strongly on oxygen concentration. The maximum ordering temperature is of order of 300 K, the ordered moment around 0.6 μ_B. Upon doping with Sr or Ba magnetic order dissapears and superconductivity is observed. In La_2CuO_4 excitations very sharp in q close to the two dimensional magnetic Bragg position have been observed even far above T_N (in the paramagnetic phase [7].* The sharpness in q gave rise to speculations about strong quasi two dimensional short range order (with correlation lengths of several hundred Angstrom far above T_N)[8], although a direct signature of such magnetic short range order (which would be a strong quasi elastic peak corresponding to the relaxation of the short range ordered regions [9]) is missing. Although very sharp in q the experimentally observed response is braod in energy, a very surprising feature stimulating considerable interest and still waiting for an explanation.

Similar features are observed in $YBa_2Cu_3O_6$ which is the equivalent to the LaCu oxide (semiconducting, non superconducting, and antiferromagnetic): The Néel temperature T_N is even higher ($\sim$ 450 K) [10] and again there are excitations very sharp in q (around the twodimensional magnetic Bragg point) and broad in energy above T_N in the paramagnetic phase, without a strong quasi elastic peak being observed [11].

A central question is: what happens to the magnetic Cu moments when changing the composition such that magnetic order disappears and superconductivity occurs - in La_2CuO_4 by substituting part of the La by Sr or Ba, in $YBa_2Cu_3O_6$ by increasing the oxygen content. Two experiments attempting to measure the total magnetic intensity in the low energy region ($\leq$ 30 meV) in well characterized samples in the superconducting regions (away from the transition region from magnetism towards superconductivity) have been published so far. This type of experiment establishes the amplitude of the "magnetic moment" (a detailed discussion of the meaning of "magnetic moment" is given in [12]) important for the thermally accessible low energy region. Both experimetns found no magnetic moments, establishing servere upper bounds for the magnetic intensity. These bounds obtained for $YBa_2Cu_3O_{6.9}$ [13] lead to the author's conclusion, that magnetic excitations can most probably be ruled out as a source for superconductivity. Similarly magnetic fluctuations in the energy region < 40meV were reported to have similarly small intensity in $La_{1.83}Sr_{0.17}CuO_4$ [14].

Close to the borderline in the compositional phase diagram between magnetic and superconducting compositions typically a large amount of local inhomogeneities (on an atomic scale) is present in the samples studied, which makes the interpretation of the data obtained difficult. When approaching the borderline from the magnetic side, neutron diffraction results indicate that the ordered magnetic moment averaged over the crystal at low temperatures decreases (as does the ordering temperature T_N), whereas the amplitude of the moment seen locally by the μ' ons (as observed in μ' on spin rotation experiments) changes only very little [15]. This indicates an inhomogeneous magnetization distribution. In $YBa_2Cu_3O_x$ an obvious source for such inhomogeneities are the Cu-O chains (the Cu1-O4 positions), the O4-position being completely filled for $x = 7$ and empty for $x = 6$ [16]. In the intermediate concentration region some of the copper atoms will have oxygen neighbours, others will not, thus enabling the establishment of random Cu moments in these chains, loosely coupled to the magnetization distribution in the CuO_2 planes. To some extent these Cu moments in the chains can be considered as "impurities", induced by the local disorder in oxygen O4 occupancy.

This picture is suggested by neutron magnetic form factor measurements for intermediate concentration regions [17]: It was shown that an applied field induces ordered Cu moments in the Cu-O chains (for an oxygen concentration of order $O_{6.5}$) whereas no field aligend moments were observed in the CuO_2 planes. The amplitude of the induced moment in the Cu-O chains increased like T^{-1} at low temperatures, consistent with almost free Cu moments in the chains (coupled only very loosely to other moments). A similar indication for Cu moments in the chains for intermediate oxygen concentrations comes from the observation of an altered magnetic structure at low temperatures [18]: In samples with rather high transition temperatures T_N and concentrations around $O_{6.2}$ the dominant magnetic Bragg peaks at high temperatures are (1/2, 1/2, 1) and (1/2, 1/2, 2), the third index being integer (in the c-direction the magnetic unit cell parameter is equal to the chemical lattice unit length). At low temperatures these intensities decrease at the expense of new Bragg peaks, e. g. (1/2, 1/2, 3/2) with half integer indices for the c-direction, indicating a doubling of the corresponding magnetic repeat length. A simple explanation for this behaviour is based on the influence of loosely coupled Cu moments in the Cu-O chains [16], causing a change in the magnetic structure at low temperatures.

These "impurity" Cu moments in the chains are of secondary interest only compared to the magnetic properties of the CuO_2 planes: These planes are certainly responsible for magnetism in the O_6 compound and are carrying the superconductivity in the O_7 concentration range. These CuO_2 planes therefore constitute the more interesting element.

The expected scenario for stable moments with strong planar coupling J and very weak $J_\perp$ perpendicular to the planes is qualitatively well understood [8,9]. Whereas for usual three dimensional systems with coupling J the ordering temperature is of order J, the strictly two dimensional system does not order at all above $T = 0$, but the correlation length increases exponentially fast with decreasing temperature:

$$\zeta_{2d} \sim \exp\{\alpha J/T\} \quad \text{for} \quad T < \alpha J \tag{4}$$

α is a numerical factor of order unity. Because of this exponential rise even a very weak perpendicular coupling $J_\perp$ can lead to high transition temperatures T_c for three dimensional long range order. Crudely the systems order if

$$kT \simeq \zeta_{2d}^2(T) \cdot J_\perp \tag{5}$$

For La_2CuO_4 a ratio of $J_\perp/J \approx 10^{-5}$ has been estimated [8].

Because of the giant short range order above T_N (for $T < J$) and of the rapid temperature variation of ζ_{2d} the 3d order parameter below T_N (the sublattice magnetization and the magnetic Bragg peaks) develops much faster for T decreasing below T_N than in a system with isotropic couplings in all 3 directions, rising to almost its full $T = 0$ value within a small temperature interval below T_N. The exponential form of ζ_{2d} above T_N also has consequences for the intensity distribution of the magnetic correlation function (responsible for magnetic neutron scattering): The full Bragg intensity (for T well below T_N) should for $T > T_N$ (but $T < J$) be contained in a small (q,ω) region centered around the 2d-Bragg point Q_{2d} and $\omega = 0$. The width in q is of order ζ_{2d}^{-1} and the width in ω of order $\Gamma_c \sim v_c \cdot \zeta_{2d}^{-1}$. The reciprocal of the ω - width is a characteristic time in which a short range ordered region of diameter ζ_{2d} may relax, this time being of order $\zeta \cdot v_c^{-1}$, where v_c is a typical velocity with which magnetic fluctuations may spread in the system. Even if this velocity v_c is equal to the high spin wave velocity, Γ_c will be very small due to the exponentially large correlation length ζ_{2d}. Grempel has given a more accurate derivation of the width Γ_c, based on a mode-mode coupling theory [9], finding a value even slightly smaller than estimated above. This has been confirmed by recent computer simulations [19]. The above is a manifestation of an almost obvious phenomenon: Large correlated regions can relax only very slowly as a whole.

The difference to be observed in neutron scattering between real long range order ($\zeta_{2d} \sim \infty$) and gaint short range order is: For long range order a Bragg peak with macroscopic intensity (of order N_{2d} = number of moments in the planes) is observed at the point $(Q_{2d}, \omega = 0)$; for giant short range order this full intensity is spread over a finite region, but both widths in q and ω are of order ζ_{2d}^{-1}. For a given temperature these widths decrease exponentially fast with increasing coupling constant J. If the scenario developed here were applicable the full quasielastic peak should be well within the energy interval covered by our experiment. Outside this small q region propagating spin waves should exist, due to the strong short range order: For wavelengths small compared to ζ, waves can propagate within the short range ordered regions. This intuitive notion was also confirmed by recent computer simulations []. Concerning the intensities to be measured in a neutron scattering experiment: The quasielastic intensity integrated over the small (q,ω) region around $(Q_{2d}, \omega = 0)$ of diameter $(\zeta_{2d}^{-1}, v_c\zeta_{2d}^{-1})$ should be macroscopic: It is equivalent to a Bragg intensity (of order N_{2d}). The intensity integrated over an equally small interval outside the quasielastic region should be much smaller, because a typical spin wave intensity for a fixed q value is only of order 1, much smaller than a Bragg intensity. The scenario described above is actually observed in a number of systems with strong quasi two dimensional couplings of stable moments (e. g. K_2NiF_2 [20]).

The situation observed in $YBa_2Cu_3O_{6+\delta}$ (and in La_2CuO_4 as well), however, differs drastically from this picture.

i) Above T_N no strong quasielastic peak around the 2d Bragg point is present, the integrated intensity in the small (q,ω) region around $(Q_{2d}, \omega = 0)$ is much too small.

ii) The ω dependence of the local magnetization-magnetization correlation function does not contain the strong Bragg-like contribution in the quasielastic region close to $\omega = 0$.

iii) The energy integrated q dependence of the same correlation function is relatively smooth, short range order effects being weak due to the absence of the quasielastic peak.

iv) The total intensity integrated over all q and ω is much too small.

v) The order parameter below T_N varies as a function of decreasing temperature very much like a normal 3-d order parameter observed in systems with similar couplings in all three spatial directions. It does not have the deep rise to almost the full $T = 0$ value within a small region below T_N.

vi) The observed excitations for finite ω above T_N in the region of Q_{2d} do not have the behaviour of antiferromagnetic spin waves: Their intensity does not decrease with ω^{-1} for increasing energy as expected for antiferromagnetic spin waves [7,11].

These discrepancies might be resolved if amplitude fluctuations play an important role in the phase transitions [12]. Such a concept has been discussed entensively in the context of weak itinerant ferro- or antiferromagnetism [21]. In magnetic insulators, however, amplitude fluctuations were thought to play a negligible role.

α) The decrease of the experimentally observed total intensity (point iv above) and the absence of a strong quasielastic peak (point i) may be regarded as a direct signature of a diminished amplitude. If amplitude fluctuations drive the phase transition a spread in energy over a region kT_N should be expected, instead of the much reduced energy region $v_c \zeta_{2d}^{-1}$ present in systems with giant 2d short range order of stable moments. The wider energy region of amplitude fluctuations in connection with the instrumental energy limits may explain the reduced amplitude observed experimentally.

β) Since the magnetization has largely vanished in amplitude above T_N, there exists no region of giant short range order and no quasi elastic peak. Then lowering T from the paramagnetic region requires magnetic amplitude to be built up at the same time when spatial correlations of magnetization densities develope. As a consequence long range order below T_N will build up slower than for the case when regions with preexisting giant 2d short range order only have no lock into a common preferential direction to establish 3d order.

γ) The excitations observed in the narrow ridge close to Q_{2d} have a variation of intensity as a function of energy consistent with amplitude fluctuations and inconsistent with transverse antiferromagnetic spin wave. The inconsistency for the latter (spin waves) is due to the following: In magnetic neutron scattering one measures correlation functions of the type $< S^+ S^- >$ and $< S^z S^z >$. Decomposing the correlation functions $< S^+ S^- >$ into creation and annihilation processes for antiferromagnetic spin waves yields two factors of $1/\omega$: The first from the Bosefactor for $\omega < T$, the second from the "form factor of the antiferromagnetic magnon"[22], which is also $1/\omega$. As a result the discrepancy discussed in [7,11] arises.

The experimental data are consistent with Bose like excitations (yielding one factor of $1/\omega$) but with an energy independent coupling to the neutron, i. e. an energy independent form factor of the excitation in question. Amplitude fluctuations are such excitations [23].

3) Conclusions

Neutron scattering on high T_c oxides and related nonsuperconducting compounds has revealed novel and highly interesting phenomena in lattice, electronic and magnetic properties (the information about electronic anomalies being only indirect via phonon anomalies). Although conclusions on possible mechanisms responsible for high temperature superconducting are still somewhat speculative the following features are suggestive:

The highest energy oxygen vibrations (of breathing type) within the CuO_2 planes are highly anomalous in La_2CuO_4 and $YBa_2Cu_3O_{6+x}$. For finite wavevector these modes split in energy indicating a coupling to some other degree of freedom. This is possibly an electronic excitation pointing towards important electron-phonon coupling effects. Phonon density of states measurements on superconducting compounds, obtained from the parent compounds cited above by substitution of La by Sr and by increasing the oxygen content (i. e. x tending towards 1) show a softening of the highest (anomalous) phonon modes, indicating increased electron-phonon coupling effects. This phenomenon is observed in the $Ba_{1-x}K_xBiO_3$ system as well. It is important for possible consequences on high T_c superconductivity mechanisms, that the phonon modes involved are still of very high energy, setting a temperature scale of 1000K, considerably higher than in conventional superconductors. Incidentally, the anomalies observed in tunneling spectra from point contact spectroscopy [24] of the Eliashberg function $\alpha^2 F(\omega)$ occur at the same energies.

These phenomena are compatible with superconductivity being driven by a coupling to high energy oxygen vibration and polarizabilities in the same energy region.

The study of magnetic properties has revealed antiferromagnetism in the parent compounds La_2CuO_4 and $YBa_2Cu_3O_6$ accompanied by very unusual excitation spectra, sharp in q around the quasi $2-d$ Bragg position, but much broader in energy than an usual picture of stable moments with strong planar coupling and weakly coupled planes would suggest. Furthermore the decrease of the experimentally observed total intensity of magnetic fluctuations in the low energy region suggests that amplitude fluctuations (a signature of unstable moments) have to be considered.

When changing the composition to achieve superconductivity the total magnetic amplitude observed experimentally in the low energy region decreases drastically (except for some "impurity" contributions due to inhomogeneities in intermediate concentration regions). For example in $YBa_2Cu_3O_{6.9}$ this total amplitude is experimentally unobservable and only an upper bound can be established based on experimental error bars. Even this upper bound is extremely low, corresponding to less than 2.5 % of the Cu atoms within the CuO_2 planes (responsible for superconductivity) carrying a spin 1/2 (this corresponds to an energy integration over a region up to 30 meV). The "magnetic moment" apparently is no longer a well defined variable for these high T_c superconductors. Compared to excitations of lattice degrees of freedom, the number of magnetic fluctuations is extremely small (by two orders of magnitude or more in the low energy region).

In this author's opinion this makes a magnetic mechanism responsible for high T_c superconductivity (at least in $YBa_2Cu_3O_7$) unlikely. In any case, even if magnetic interactions were to play an important role, one still has to consider the effect of lattice excitations: Because of their dominance in number alone they dominate the thermodynamic behaviour in most of the temperature region below the superconducting T_c. A consistent theory based magnetic interactions would have to demonstrate that the large number of excited lattice excitations does not destroy the eventual magnetic pairing.

A simpler scenario seems to be based on high energy lattice excitations, possibly coupled to electronic polarizabilities, as a source for superconductivity.

Further experimental and theoretical progress is necessary to clarify this basic problem of high T_c superconductivity.

References

[1] B. Renker et al., Z. Phys. B67, 15 (1987)

[2] B. Renker et al., Z. Phys. B71, 437 (1988)
F. Gompf, B. Renker, E. Gering, Physica C (1988)

[3] C. K. Loong et al., Phys. Rev. Lett. 62, 2628 (1989)

[4] L. Pintschovius et al., to be published

[5] W. Reichardt et al., to be published

[6] D. Vaknin et al., Phys. Rev. Lett. 58, 2802 (1987)
T. Freloft et al., Phys. Rev. B36, 826 (1987)
Y. J. Uemura et al., Phys. Rev. Lett. 59, 1045 (1987)
D. C. Johnston et al., Phys. Rev. Lett. B36, 4007 (1987)

[7] G. Shirane et al., Phys. Rev. Lett. 59, 1613 (1987)
Y. Endoh et al., Phys. Rev. B37, 1443 (1988)

[8] S. Chakravarty, B. I. Halperin, D. R. Nelson, Phys. Rev. Lett. 60, 1057 (1988)

[9] D. R. Grempel, Phys. Rev. Lett. 61, 1041 (1988)

[10] N. Nishida et al., Jap. J. Appl. Phys. 26, L1856 (1987)
J. M. Tranquada et al., Phys. Rev. Lett. 60, 156 (1988)
J. Brewer et al., Phys. Rev. Lett. 60, 1073 (1988)

[11] G. Shirane, ICM proc.
M. Sato et al., Phys. Rev. Lett. 61, 1317 (1988)
P. Burlet et al., Physica C153, (1988)

[12] O. Schärpf et al., Z. Phys. B (to be published)

[13] T. Brueckel et al., Europhys. Lett. 4, 1189 (1987)

[14] A. G. Gukasov, V. P. Plakhty, O. P. Smirnov, I. A. Zobkalo, Sol. State Comm. 69, 497 (1989)

[15] Y. J. Uemura et al., Physica C153, 760 (1988)
J. I. Budnick et al., Europhys. Lett. 5, 651 (1988)

[16] J. J. Capponi et al., Europhys. Lett. 3, 1301 (1987)
A. W. Hewat, J. J. Capponi, C. Chaillout, M. Marezio, Sol. State Comm. 64, 301 (1987)

[17] B. Gillon et al., JCM proc.
J. Schweitzer, privat communication

[18] See G. Shirane, ref. 11

[19] S. Tyc, B. I. Halperin, S. Chakravarty, Phys. Rev. Lett. 62, 835 (1989)

[20] R. J. Birgeneau, H. J. Guggenheim, G. Shirane, Phys. Rev. B1, 2211 (1970)
K. Yamada et al., Phys. Rev. B39, 2336 (1989)

[21] For reviews see: T. Moriya: Spin fluctuations in itinerant- electron Magnetism, Springer series sol. st. sci., 56, Heidelberg (1985); H. Capellmann, ed., Metallic Magnetism, Springer Series, Topics in Current Physics 42, Heidelberg (1987)

[22] S. W. Lovesey, Theory of Neutron Scattering from Condensed Matter, Volume 2: Polarization Effects and Magnetic Scattering, Oxford, Clarendon Press (1986)

[23] H. Capellmann, V. Vieira, Phys. Rev. B25, 3333 (1982)

[24] L. N. Bulaevskii et al., Supercond. Sci. Technol. 1, 205 (1988)

SUPERCONDUCTING NETWORKS IN A MAGNETIC FIELD: EXACT SOLUTION OF THE J^2 MODEL

L.M. Floria[1,2] and R.B. Griffiths[3]

[1] Instituto de Ciencia de Materiales (CSIC-UZ) and Dpto.de Matematica Aplicada, Universidad de Zaragoza. 50009 Zaragoza Spain
[2] Laboratoire Léon Brillouin. CEN Saclay, F-91191 Gif-sur-Yvette Cedex, France
[3] Dpt. of Physics. Carnegie-Mellon University. Pittsburgh PA 15213. USA

INTRODUCTION

The behaviour of micronetworks made of superconducting wires in the presence of a magnetic field has been recently investigated by different groups both experimentally and theoretically. For an updated scope of this field see, for example, Mooij and Schon (1988.)

One of the most interesting aspects of the problem is to find the arrangement of of fluxoid quanta which minimizes the energy of the system. In the case of a periodic network, it is known that arrangements of fluxoid quanta which are commensurate with the underlying network periodicity, lower always the energy. Since modern microlithographic techniques allow the fabrication of non-periodic patterns, a great interest has aroused in the study of networks with geometry "intermediate" between periodic and random.

We will present here a model which is simple enough to be accessible to the mathematics but which, nonetheless, seems to capture some of the essential physics behind the phenomenology of superconducting networks in a field.

In the next section we introduce the J^2 model and, to motivate it, we start with the experiment on fluxoid quantization reported by Little and Parks in the early 60's. The explanation of their results is at the very heart of the J^2 model.

The last section is devoted to the exposition of the exact solution of the model. First, the periodic network is analyzed and it is found that the model exhibit some features which agree with the known phenomenology. Afterwords, the model is solved in a quasiperiodic geometry: the Fibonacci lattice; that requires the introduction of some precise concepts.

The exposition is intended to satisfy the didactic requirements of a summer school more than being strictly rigorous, although all the results shown here are rigorous. A more detailed account of the exact solution of the 1-d J^2 model is being prepared.

THE J^2 MODEL

F. London (1950) predicted that the fluxoid, defined as

$$\int_S H \, dS + c \int_P \Lambda \, J \, dl \qquad (1b)$$

is quantized in units of hc/e in a superconductor. S is a surface bounded by the perimeter P, J is the supercurrent, H is the magnetic field and Λ is the London parameter. Although London's derivation was based on a semi-classical approach (Bohr-Sommerfeld quantization of electronic angular momentum), the validity of the result is by no means limited to that approach: the superconducting order parameter (complex) must be single valued, so that the circulation of the phase gradient (which coincides with the fluxoid in the appropriated units) has to be an integer multiple of 2π:

$$\int \nabla \phi \, dl = \frac{4\pi}{c\phi_0} \int \lambda^2 \, J \, dl + \frac{2\pi}{\phi_0} \int A \, dl = 2\pi n \qquad (2b)$$

where ϕ_0=hc/e is the flux quantum, A is the vector potential, λ is the penetration depth (related to the London parameter through $\Lambda=4\pi\lambda^2/c^2$), and ϕ is the phase of the superconducting order parameter. The terms in (1b) or (2b), i.e.: the magnetic flux and the term involving the persistent current, must adjust in order to add up to an integer.

An experimental demonstration of fluxoid quantization was reported by Parks and Little (1963). They examined the variation with the magnetic field of the transition temperature of a "thin-walled" superconducting cylinder in an applied magnetic field in the axial direction. They observed a periodic variation of the transition temperature ("Little and Parks oscillations") with maxima at every value of the magnetic flux equal to an integer multiple of the flux quantum. Their interpretation of this result is as follows.

Assume that the penetration depth is much greater than the wall thickness, so that the magnetic field in the cylinder is equal to the applied one. As the field is varied the persistent currents must adjust themselves in order to preserve flux quantization:

$$J = \frac{n(hc/e) - \pi r^2 H}{2\pi rc\Lambda} \qquad (3b)$$

where n is the number of fluxoid quanta and r is the radius of the cylinder. The kinetic energy associated to the supercurrents is then:

$$\frac{1}{2}\Lambda J^2 = \frac{h^2}{16\pi^2 e^2 \Lambda r^2}(n - (2e/hc)\Phi)^2 \qquad (4b)$$

where Φ is the magnetic flux. In order to minimize the kinetic energy, the number of fluxoid quanta, n, should vary with the magnetic flux as indicated in figure 1a, and the kinetic energy due to the persistent currents is periodic in the flux (figure 1b). Consequently, the free energy of the superconducting phase must be periodic in the flux. On the other hand, the free energy of the normal phase does not essentially vary with the flux. Therefore, the transition temperature T_c must show periodic variations as the flux increases, as observed.

The important point to keep in mind is that the observed "Little and Parks oscillations" of T_c, as the field varies, are the result of the adjustment of the supercurrents to satisfy flux quantization while minimizing the associated kinetic energy.

Suppose now that, instead of a superconducting ring, one has a network of superconducting wires in a magnetic field. Based on the elegant argument of Little and Parks let us consider a simple model in which there is a persistent current J_{ij} attached to each link joining the vertices i and j, and the energy of the system is given by the sum of the squared currents. Besides, one has the current conservation condition on each vertex and the fluxoid quantization condition on each loop of the network. This is the J^2 model, which can be defined on any planar geometry for the network (periodic, quasi-periodic, fractal, random,...)

We restrict ourselves here to a particular geometry: parallel horizontal lines equally spaced (unit spacing) crossed by vertical lines at positions given by a sequence $\{x_k\}$ which for the present is arbitrary. Assuming that the solution has periodicity of period one in the vertical direction, the currents in the horizontal wires must vanish and one arrives to an effective one-dimensional model, defined by the fluxoid quantization condition on each elementary plaquette:

$$J_{k+1} - J_k = n_k - \alpha_k f \tag{5b}$$

where J_k is the current on the kth vertical wire (in units of $c\phi_0/2\lambda^2$), $\alpha_k = x_{k+1} - x_k$ is the area of that elementary plaquette, n_k is the number (integer) of fluxoid quanta in area α_k, and f is the applied field in units of $\phi_0/$(unit area); note that $f = \sum_k n_k / \sum_k \alpha_k$. The energy of the system is given by the kinetic energy:

$$E = \sum_k J_k^2 \tag{6b}$$

This simple model, eqs. (5b) and (6b), was introduced by Grest et al. (1988). They used simulated annealing techniques on particular finite chains (periodic, quasiperiodic Fibonacci and random Fibonacci) to determine the fluxoid quanta arrangement $\{n_k\}$ which minimizes the energy (6b) for a range of (discrete) values of the field f.

In spite of its apparent simplicity, this model involves infinite-range interactions: any change in one of the n_k's produces changes in all the J_k's. The model is in fact related (Grest et al., 1988) to a 1-d model of charges ($q_k = n_k - \alpha_k f$) with long range interactions. As we will see,

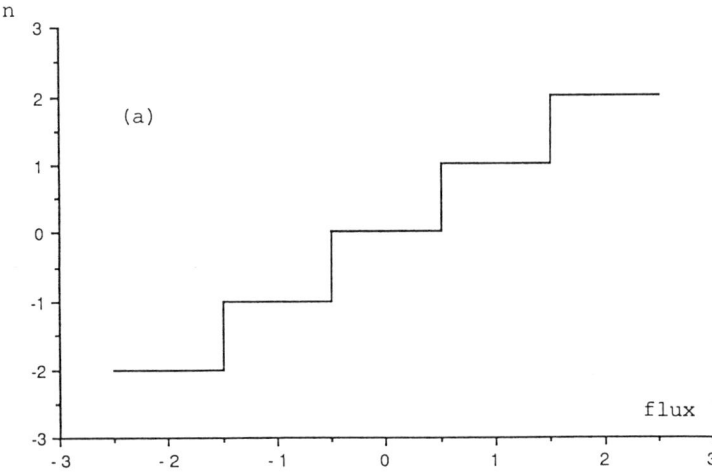
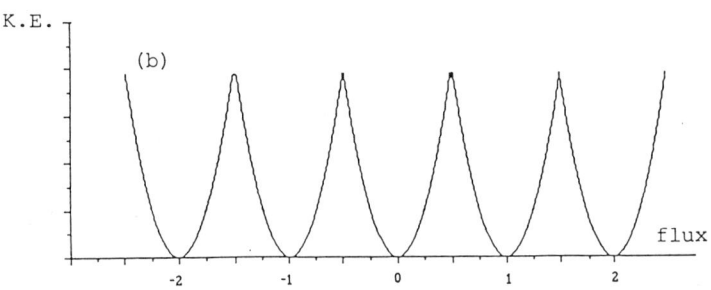

Fig. 1. Little and Parks experiment. (a) The number of fluxoid quanta as a function of the magnetic flux (in units of a flux quantum). (b) The variation of the Kinetic Energy (4b) (arbitrary units) with the magnetic flux.

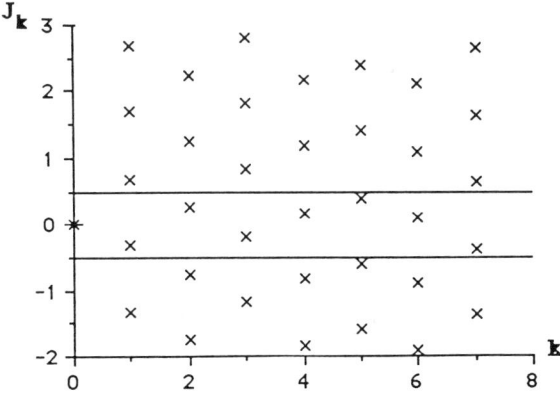

Fig. 2. The possible values of the currents for $J_0=0$ are marked by crosses. The minimum energy configuration is obtained when the values belong to the band between $-1/2$ and $1/2$.

the model shows interesting features and it can give some insight in the physics of superconducting networks in a magnetic field. A major vertue of the model is that one can found its exact solution.

EXACT SOLUTION OF THE J^2 MODEL

A particular state of the 1-d J^2 model can be especified either by the sequence of fluxoid quanta in each elementary plaquette $\{n_k\}$, k=0, ∓1, ∓2,..., or by the sequence of currents $\{J_k\}$ flowing through the wires. For numerical simulation purposes it is perhaps more convenient to use the fluxoid quanta representation, while for getting exact results we have found the currents representation more comfortable. As a major example of the simplicity reached when working in the currents representation, let us obtain the sequence $\{J_k\}$ for the ground state of the model in an arbitrary lattice, especified by the sequence $\{x_k\}$.

Let J_0 be the current in $x_0=0$. From the fluxoid quantization condition (5b), the fractional part of J_1, $Frac(J_1)$, does not depend on n_0, being uniquely determined by $Frac(J_0)$. In the same manner one concludes that for all k, $Frac(J_k)$ is uniquely determined by $Frac(J_0)$ (A way to visualise this simple argument is shown in figure 2.)

Among all possible sequences $\{J_k\}$ which share that same sequence of fractional parts, the one which minimizes the energy is clearly that one satisfying $|J_k| \leq 1/2$, that is,

$$J_k = Frac(J_0 - fx_k + 1/2) - 1/2 \qquad (1c)$$

This formula gives the ground state configuration as a function of the parameter J_0, which should be determined by minimizing the energy of the configuration respect to it. If one represents the sequence $\{J_k\}$ in the unit circle (identifying -1/2 and 1/2), a change in J_0 is just a uniform rotation which does not change the distribution of currents in the circle: J_0 is a sort of harmless parameter. Nonetheless, one has to find the optimum J_0 for which the energy is minimal, and that will depend on the form of the distribution of J_k's.

<u>Periodic Lattice</u>

We will consider first the simplest case of a periodic chain ($\alpha_k=1$), and will obtain the ground state energy per site defined as:

$$E_g = \lim_{N\to\infty} \frac{1}{N} \sum_{k=-N/2}^{N/2} J_k^2 \qquad (2c)$$

From (1c), the sequence of currents in the ground state is:

$$J_k = Frac(J_0 + 1/2 - kf) - 1/2 \qquad (3c)$$

For f irrational, this sequence fills uniform and densely the unit circle (i.e.: the interval (-1/2, 1/2].) The energy is then independent of J_0 and is simply equal to:

$$E_g = \int_{-1/2}^{1/2} J^2 \, dJ = 1/12 \tag{4c}$$

For rational values of f=p/q (p and q relatively primes), the distribution of J_k's consists of q Dirac peaks evenly separated appart a distance 1/q. Now the energy depends on J_0 and the minimum energy is reached when the set of peaks is centered around the value zero. The peaks are then located at $\frac{2r+1}{2q} - \frac{1}{2}$ (r=0,...,q-1) and the energy is easily obtained as:

$$E_g = \frac{1}{q}\sum_{r=0}^{q-1} (\frac{2r+1}{2q} - \frac{1}{2})^2 = \frac{1}{12}(1 - \frac{1}{q^2}) \tag{5c}$$

The fluxoid quanta arrangement $\{n_k\}$ has the form:

$$n_k = \text{Int}((k+1)f + \alpha) - \text{Int}(kf + \alpha) \tag{6c}$$

where Int(x) means integer part of x. For rational f, $\{n_k\}$ is a periodic sequence while for irrational f it is quasiperiodic. The formula (6c) is formally identical to the ground state of models for particles on periodic substrate potentials with certain type of interactions, where the concept of commensurability of the structure respect to the substrate potential plays a major role.

The periodic sequences of fluxoid quanta in the ground state, for rational f, can be quite naturally called "commensurate" superlattices of fluxoid quanta, when the underlying network is periodic, and the results (4c) and (5c) indicate that the fluxoid quanta (when they are allowed) arrange themselves in a commensurate way to minimize their energy. The energy as a function of the field shows then "dips" at those fields for which commensurability is allowed, and the more commensurate the superlattice is the deeper the dip is.

This feature of the model agrees qualitatively with the observed (Pannetier et al., 1983 and 1984) variation of the transition temperature with the field in periodic superconducting networks: sharp dips in $\Delta T_c = T_c(0) - T_c(f)$ at every rational value of f=p/q, the more intense the smaller q is.

The Winding Density of a Non-Periodic Sequence

We have seen that the distribution of currents in the ground state (appart from additive constants and the parameter J_0 which merely rotates the distribution) is given by Frac(fx_k) where x_k is the sequence of points in the underlying 1-d lattice.

The calculation of that distribution was trivial in the case of a periodic lattice ($x_k=k$) because only uniform rotations of constant angle f are involved. That is not anymore the case when $\{x_k\}$ is a non-periodic sequence and one would like to devise some method, as general as possible, to determine a closed form of the ground state current's distribution for any given geometry.

First of all, let us define precisely the quantity we are interested in. Let $\{z_n\}$ be a sequence of real numbers and let $\{\zeta_n\}$ the sequence of their fractional parts, i.e.: $\zeta_n=\mathrm{Frac}(z_n)$, $0\leq\zeta_n<1$. Consider now a finite piece consisting of N subsequent poits of the sequence $\zeta_j: j=m, m+1,\ldots, n=m+N-1$. Now let us call $R_N(\zeta) = \frac{1}{N}\times$(number of j's such that $\zeta_j<\zeta$). One realizes that $R_N(0)=0$, $R_N(1)=1$ and R_N is non-decreasing. If it is the case that one has a limit:

$$R(\zeta) = \lim_{N\to\infty} R_N(\zeta) \tag{7c}$$

then, let define the winding density of the sequence $\{z_n\}$ as the derivative $r(\zeta)=R'(\zeta)$.

Example: for a periodic sequence of irrational period f, $R(\zeta)=\zeta$ and the winding density is a constant function of value 1. If the period is rational f=p/q, the R function consists of horizontal plateaus with q finite vertical jumps of height 1/q, so that the winding density consist of q Dirac peaks.

The distribution of currents in the ground state of the 1-d J^2 model is given by the winding density of the sequence $z_n = f x_n$ where $\{x_n\}$ is the given lattice.

Let us derive an important relation satisfied by the Fourier transform of the winding density. We can extend $r(\zeta)$ (which is defined in the unit interval) to a periodic function $r(y)$, and as such it can be expanded as a Fourier series:

$$r(y) = \sum_{n\in\mathbb{Z}} r_n \exp(i2\pi ny) \tag{8c}$$

where

$$r_n = \int_0^1 r(y) \exp(-i2\pi ny)\, dy \tag{9c}$$

and, since n is an integer and $r(y)$ is periodic, the last integral may be over any interval of unit length. The normalization condition is

$$r_0 = \int_0^1 r(y)\, dy = 1 \tag{10c}$$

Given the way $r(\zeta)$ has been defined, one can imagine the integral defining r_p as the limit:

$$r_p = \lim_{N\to\infty} \frac{1}{N}\sum_{j=m}^{m+N-1} \exp(-i2\pi p\zeta_j) = \lim_{N\to\infty} \frac{1}{N}\sum_{j=m}^{m+N-1} \exp(-i2\pi p z_j) \tag{11c}$$

where z_j can replace ζ_j because p is an integer. If one defines the Fourier transform of the sequence $\{z_n\}$ as:

$$\sigma(k) = \lim_{|N-M|\to\infty} \frac{1}{1+N-M}\sum_{n=M}^{N} \exp(-ikx_n) \tag{12c}$$

we have from (11c) that

$$r_p = \sigma(2\pi p) \tag{13c}$$

This formula allows to calculate the winding density of $\{z_n\}$, provided that the Fourier coeficients (12c) of the sequence are at hand.

The Fibonacci Network

We will show now the results concerning the ground state of the 1-d J^2 model on the Fibonacci lattice. This lattice has been for a long time favored by the experts in the field of incommensurate structures. It also interests us because numerical simulations on this model with this particular geometry have been reported recently (Grest et al., 1988 and Chaikin et al., 1988.) On the other hand, experimental results on the variation of ΔT_c versus field have also been reported (Behrooz et al., 1987 and 1988) for Fibonacci networks, i.e.: networks made of wires periodically spaced in the y direction and following the Fibonacci lattice in the perpendicular (x) direction.

The Fibonacci lattice $\{x_k\}$ has the following inter-site distances:

$$\alpha_k = x_{k+1} - x_k = 1 + \tau^{-1}(\text{Int}((k+1)\tau^{-1}) - \text{Int}(k\tau^{-1})) \qquad (14c)$$

where $\tau=(\sqrt{5}+1)/2$ is the golden mean, a most remarkable irrational. Looking at (14c) one sees that α_k takes on two values, either 1 or $1+\tau^{-1}$ but the sequence has no repeating pattern because the number τ^{-1} inside the argument of the integer part functions is irrational. Nevertheless the sequence contains certain degree of regularity: it is quasiperiodic. In very simple terms it means that if one places oneself somewhere in the lattice and picks up a finite piece of it, one has to move only a finite distance appart in order to find exactly the same piece again, whatever the size of the piece be and no matter where was one placed; the bigger the size chosen, the farther it will be found again.

The lattice has an average lattice constant $a=1+\tau^{-2}$, and one can look at it as if over a periodic spacing of $\underline{a}$ one adds a distortion given by a certain periodic function $h(y)$, in the following sense:

$$x_n = na + h(n\omega) \qquad (15c)$$

where $\omega=\tau^{-1}$, $h(y)=-\tau^{-1}y$ $(-1/2<y\leq 1/2)$, and $h(y)=h(1+y)$. Structures built in this way, with h a periodic function and ω irrational, are said to posses a "hull function" h. This is a central concept introduced by Aubry (1983), and quite important results in the field are based on it.

From (8c),(13c),(12c) and (15c) one obtains that the winding density of the sequence $\{-fx_k\}$ is:

a) $r(y) = 1$, if $f(1+\tau^{-2})$ is not of the form $(n_0+m_0\tau^{-1})/p_0$, $(n_0, m_0, p_0$ integers.)

b) $r(y) = \frac{1}{\kappa}(1 + \text{Int}(\kappa/2+yp_0) + \text{Int}(\kappa/2-yp_0))$, if $f(1+\tau^{-2})=(n_0+m_0\tau^{-1})/p_0$

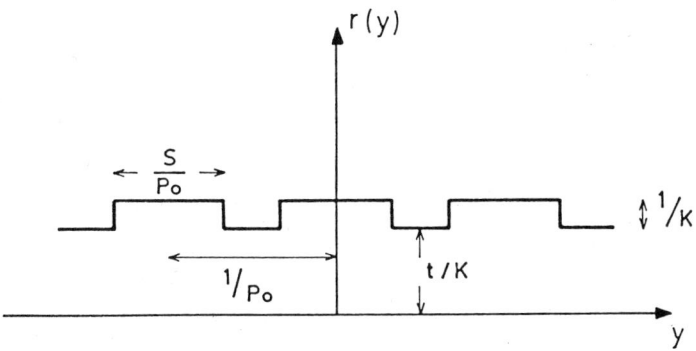

Fig. 3. Winding density of the sequence $-fx_k$ (x_k being the Fibonacci lattice) for a value of f of the type b) (see text).

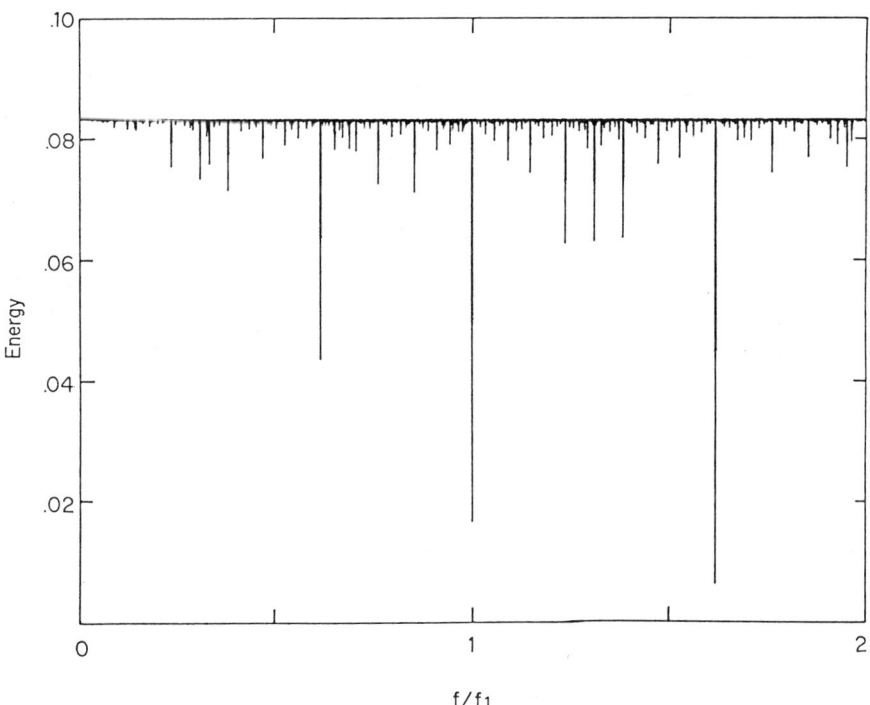

Fig. 4. Ground state energy versus field of the 1-d J^2 model on the Fibonacci lattice.

This winding density is schematically represented in figure 3. It is a square wave of period $1/p_0$ and the width of the hills differs from the width of the valleys. It is, indeed, a simple form for the ground state currents distribution. The ground state energy is now given by:

$$E_g = \min_z \int_{-1/2}^{1/2} y^2 \, r(y-z) \, dy \tag{16c}$$

where the parameter z is related trivially to J_0. For fields of the form a) above, $r(y)$ is constant and the integral in (16c) does not depend on z, being equal to 1/12. For fields of the form b), the optimal z is one for which the p_0 hills of r in the unit interval are centered around the origin. The ground state energy in that case is:

$$E_g = \frac{1}{12\kappa}(t + s(p_0^2 + s^2 - 1)) \tag{17c}$$

where $t = \text{Int}(\kappa)$, and $s = \text{Frac}(\kappa)$. In figure 4 E_g is plotted as a function of the field. We see that for the Fibonacci lattice (in fact, for at least a certain class of non-periodic lattices) there are a countably infinite set of values of the field for which the fluxoid quanta can arrange themselves in a structure which lowers the energy below the background of 1/12. This set of numbers: $(n_0 + m_0 \tau^{-1})/p_0$ has a structure similar in many respects to the rationals.

The experimental results previously mentioned (Behrooz et al., 1986 and 1987) on Fibonacci networks show a variation of ΔT_c exhibiting sharp dips at certain values of the field in agreement with the deepest dips of the E_g for the simple 1-d J^2 model.

The similarity of the results obtained for the Fibonacci lattice with those for the periodic lattice suggests that for at least a certain class of non-periodic lattices, some generalization of the concept of commensurability could be of relevance. A detailed study of this issue will be presented elsewhere.

ACKNOWLEDGEMENTS

One of us (LMF) wants to acknowledge D. Gonzalez for continuous encouragement, S. Aubry and A.R. Bishop for their kind hospitality at Saclay and Los Alamos respectively, and G. Abramovici for reading the manuscript. This work has been partially supported by the National Science Foundation (Grant DMR-8613218), the European Community (Contract SC10036F) and CAI-CONAI (Programa Europa).

REFERENCES

Aubry S., 1983, J. Phys. (Paris) **44**, 147.
Aubry S. and Le Daeron P., 1983, Physica D **8**, 381.
Aubry S., Godreche C. and Vallet F., 1987, J. Phys. (Paris) **48**, 327.
Behrooz A., Burns M., Deckman H., Levine D., Whitehead B. and Chaikin P.M., 1986, Phys. Rev. Lett. **57**, 368.

Behrooz A., Burns M., Deckman H., Levine D., Whitehead B. and Chaikin P.M., 1987, Phys. Rev. B 35, 8396.
Chaikin P.M., Behrooz A., Itzler M.A., Wilks C., Whitehead B., Grest G.S. and Levine D., 1988, Physica B 152, 113.
Grest G.S., Chaikin P.M. and Levine D., 1988, Phys. Rev. Lett. 60, 1162.
Pannetier B., Chaussy J. and Rammal R., 1983, J. Phys. (Paris) 44, L853.
Pannetier B., Chaussy J. Rammal R. and Villegier J., 1984, Phys. Rev. Lett. 53, 1845.

Vortex Dynamics in Networks of Josephson Junctions

Ulrich Eckern

Institut für Theorie der Kondensierten Materie
Universität Karlsruhe
Karlsruhe, Federal Republic of Germany

1 Introduction

Since the proposal by Josephson[1] in 1962 that a tunnel Junction, i.e. two superconductors weakly coupled through an oxide barrier, should show a zero-voltage supercurrent due to the tunneling of Cooper pairs, this system has been studied with unending enthusiasm. In particular, initiated by Leggett,[2] the last decade has seen a remarkable activity in a field which is often called Quantum Mechanics of Macroscopic Variables;[3] experiments on small Josephson junctions at low temperature are found to be in excellent agreement with theoretical predictions.

As a matter of introduction, recall that the relevant variable for a tunnel junction is the order parameter phase difference φ, whose *classical* dynamics is determined by the following equations:

$$\hbar\dot\varphi = 2eV \; ; \;\; C\dot V + V/R + I_J \sin\varphi = I_T , \tag{1}$$

where V, C, R, I_J, and I_T denote the voltage, the capacitance, the resistance, the critical current, and the external current, respectively. Alternatively:

$$M\ddot\varphi + \eta_s \dot\varphi = -\partial U/\partial \varphi . \tag{2}$$

Here the potential is given by

$$U(\varphi) = -E_J \cos\varphi - f \cdot \varphi , \tag{3}$$

where $E_J = \hbar I_J/2e$ is called the Josephson energy, $f = \hbar I_T/2e$, $M = (\hbar/2e)^2 C$, and $\eta_s = (\hbar/2\pi) \cdot R_0/R$; $R_0 = \pi\hbar/2e^2 \sim 6.5k\Omega$ denotes the unit of resistance.

Leaving aside for a moment the dissipative term, it is clear that the above equation can be "derived" from the Hamiltonian

$$\mathcal{H}_0 = Q^2/2C + U(\varphi) \tag{4}$$

by imposing the commutation relation $[Q,\varphi] = -2e \cdot i$, where Q is the charge difference across the junction. Phenomena related to the fact that Q and φ have, in principle, to be considered operators and not classical variables, become experimentally accessible in small junctions and for low temperature. More precisely, define the charging energy $E_C = e^2/2C$ and the parameter δ by

$$\delta^{-1} \sim \langle(\Delta\varphi)^2\rangle \sim (E_C/E_J)^{1/2} , \qquad (5)$$

which also appears as exponent in a typical WKB calculation; note that E_J is the potential barrier, and $\hbar\omega_0 \sim (E_C E_J)^{1/2}$ the characteristic frequency. In the sixties, typical parameters have been such that $\delta \sim 10^6$, which allowed indeed to treat φ as a classical varible. However, very recently,[5] (networks of) tunnel junctions have been deviced with $\delta \sim 1$; thus it can be expected – and is observed – that quantum phenomena are very pronounced in such a system.

Of course, given a classical equation like (2), it is not possible to derive the correct quantum mechanics of φ. In particular, note that R^{-1} is often a sum of two contributions, namely

$$R^{-1} = R_t^{-1} + R_\Omega^{-1} , \qquad (6)$$

where R_t is the strongly temperature dependent resistance due to quasiparticle tunneling, and R_Ω denotes an external shunt resistor. In order of magnitude, one has

$$R_t(T) \sim R_N \cdot \exp(\Delta/kT) , \qquad (7)$$

where $R_N = R_t(T_c)$, and Δ is the magnitude of the superconducting gap, taking the two superconductors to be identical for simplicity. As it turns out, in the presence of dissipation, the theoretical formulation proceeds most conveniently within the path integral formulation.[6] As a result,[7,8] an effective action is derived which contains, besides a term which follows from $\mathcal{H}_0$, two additive contributions describing the shunt resistor[7] and the quasiparticle tunneling,[8] respectively. While the former leads to a quadratic form, it is found that the latter has an important 4π-periodicity,[9] which reflects the fact that the microscopic process is the tunneling of single electrons,[10] in contrast to the continuous change of the charge in the presence of an external shunt. Note also that $\dot{\varphi}/R_t$ is the low frequency limit of a more general expression,[8] which incorporates the creation of quasiparticles for frequencies $\omega > 2\Delta/\hbar$.

In the following chapters, I wish to discuss some aspects of the low temperature behavior of a two-dimensional network of Josephson junctions, mainly based on the appropriate generalization of the Hamiltonian (4). In particular, I wish to pursue the dual description in terms of vortices, which is known to be adequate in the classical limit.[11] Most of the material to be presented is based on a recent publication with Albert Schmid, in which further details and references can be found.[12]

2 Description of the Model

I consider a system of weakly coupled superconducting grains which form a two-dimensional square lattice. The relevant variables are the phases $\{\varphi_{\vec{l}}\}$, where $\vec{l} = (l_x, l_y)$ labels the lattice sites (the lattice constant is taken to be unity), and the corresponding charges $\{Q_{\vec{l}}\}$:

$$[Q_{\vec{l}}, \varphi_{\vec{l}'}] = -2e \cdot i\, \delta_{\vec{l},\vec{l}'} . \qquad (8)$$

The Hamiltonian of the network is given by ($f = 0$)

$$\mathcal{H}_0 = \frac{1}{2} \sum_{\vec{l},\vec{l'}} Q_{\vec{l}} (\mathcal{C}^{-1})_{\vec{l},\vec{l'}} Q_{\vec{l'}} - E_J \sum_{\vec{l},\vec{\mu}} \cos(\varphi_{\vec{l}+\vec{\mu}} - \varphi_{\vec{l}}) , \qquad (9)$$

where $\vec{\mu} = (1,0)$ and $(0,1)$; $\mathcal{C}$ denotes the capacitance matrix. Its Fourier transform in the long wave-length limit is given by

$$\mathcal{C}(\vec{q}) = c_0 + C q^2 , \qquad (10)$$

where c_0 and C denote the ground and nearest-neighbor capacitance, respectively. In passing, I remark that C is the sum of a geometric (C_0) and a quasiparticle part,[8] namely

$$C = C_0 + C_{qp} , \quad C_{qp} = 3\alpha e^2 / 16\Delta , \qquad (11)$$

where I defined $\alpha = R_0/R_N$ as a dimensionless measure of the coupling between the grains (which is *not* a measure of the dissipation in the junctions). For example, $E_J = \alpha \Delta / 2$ in the low temperature limit. For distances large compared to the lattice spacing, the Coulomb interaction in real space ($r = |\vec{l} - \vec{l'}|$) follows from (10) to be given by

$$\mathcal{C}^{-1}(r) = (2\pi C)^{-1} \cdot K_0(r/\lambda) , \qquad (12)$$

which identifies $\lambda = (C/c_0)^{1/2}$ as the screening length. Thus peculiar features can be expected in the limit $c_0 = 0$, which means in particular $\lambda \gg N$, where N is the size of the system (assuming N^2 grains). For the junctions described in Ref. 5, however, λ is presumably of the order one, i.e. the Coulomb interaction is of short range.

From the Hamiltonian (9), the corresponding Euclidean action S_0 is derived in the usual way.[6] Due to the extended nature of the vortices to be studied below, especially in the quantum regime, it is sufficient (for some purposes) to consider the *continuum limit*, which is defined by the approximation

$$\varphi_{\vec{l}+\vec{\mu}} - \varphi_{\vec{l}} \simeq (\vec{\mu} \cdot \nabla) \varphi_{\vec{l}} . \qquad (13)$$

Henceforth, I replace $\vec{l}$ by $\vec{r}$. Defining in addition $m = \hbar^2 c_0 / 4e^2$, the action is given by

$$S_0 = \frac{1}{2} \int d\tau \int d^2 r \left[m \dot{\varphi}^2 + M(\nabla \dot{\varphi})^2 + E_J (\nabla \varphi)^2 \right] . \qquad (14)$$

Note that for finite temperature, the τ-integration is restricted to $0 \ldots \hbar \beta$, $\beta = (kT)^{-1}$, with periodic boundary conditions (see also Refs. 9 and 10). In addition, the external current contribution is

$$S_T = - \int d\tau \int d^2 r (\vec{f} \cdot \nabla) \varphi . \qquad (15)$$

Finally, the dissipative contribution corresponds to a term in the action which is nonlocal in time; note that in the continuum limit, the special features of the quasiparticle part are not apparent. The result is

$$S_D = \frac{1}{8} \int d\tau d\tau' \int d^2 r \, A_s(\tau - \tau') \left[\nabla \varphi(\vec{r}, \tau) - \nabla \varphi(\vec{r}, \tau') \right]^2 . \qquad (16)$$

Here I also neglected an unimportant contribution;[12] $A_s(\tau - \tau')$ is given by

$$A_s(\tau - \tau') = \frac{\hbar \alpha_s}{\pi^2} \frac{1}{(\tau - \tau')^2} \tag{17}$$

with $\alpha_s = R_0/R$ (recall that $\eta_s = (\hbar/2\pi)\alpha_s$). The above expressions apply in the continuum limit, in the *adiabatic limit* of small frequencies ($\hbar\omega \ll 2\Delta$), and for low temperature. The total action

$$S = S_0 + S_T + S_D \tag{18}$$

is the basis of the following considerations (except for some final remarks).

Ignoring vortices for a moment, it is obvious that the response and the fluctuations of the system are characterized by the quantity

$$\begin{aligned} D^{-1}_{\vec{q},\omega} &= [D_0^{-1} + \frac{1}{2} A_s \nabla^2]_{\vec{q},\omega} \\ &= m\omega^2 + [M\omega^2 + E_J + \eta_s |\omega|]q^2 , \end{aligned} \tag{19}$$

from which the "spin-wave" dispersion $\omega_{\vec{q}}$,

$$\omega_{\vec{q}}^2 = \frac{(cq)^2}{1 + (cq)^2/\omega_0^2} , \tag{20}$$

is easily identified. Here I defined $c^2 = E_J/m$ and $(\hbar\omega_0)^2 = 8E_J E_C$; note that $\lambda = c/\omega_0$, and that $\omega_{\vec{q}} = \omega_0$ for $m = 0$ ($\lambda = \infty$).

3 Statics and Dynamics of Vortices

The transformation to the vortex picture is achieved by separating from $\varphi(\vec{r},\tau)$ a contribution which incorporates explicitly the vortex configuration. Thus I put

$$\varphi = \varphi^V + \varphi^S \tag{21}$$

and choose φ^V to be the appropriate solution of $\nabla^2 \varphi^V = 0$, namely

$$\varphi^V(\vec{r},\tau) = \sum_j e_j \arctan \frac{y - y_j(\tau)}{x - x_j(\tau)} , \tag{22}$$

where $\vec{r}_j(\tau) = (x_j(\tau), y_j(\tau))$ is the center of the jth vortex. Note that φ^V is a multiple-valued function, which increases by $2\pi e_j$ ($e_j = \pm 1$) as one goes around the jth vortex; e_j will also be called the charge of the vortex, and I impose at once the charge neutrality condition, $\Sigma_j e_j = 0$. Inserting the ansatz (21) into the equation of motion, $D^{-1}\varphi = 0$, I obtain for the spin-wave part φ^S the following linear inhomogeneous equation:

$$D^{-1}\varphi^S = (m - M\nabla^2)\partial_\tau^2 \varphi^V , \tag{23}$$

which leads to

$$\varphi^S_{\vec{q},\omega} = -i\omega(m + Mq^2) D_{\vec{q},\omega} [\partial_\tau \varphi^V]_{\vec{q},\omega} . \tag{24}$$

Inserting the total φ into (18) leads to an effective action in terms of the vortex coordinates $\{\vec{r}_j\}$, which I denote by $\mathcal{A}$ (and $\mathcal{A}_0$, etc.). The following relations are useful:

$$[\nabla \varphi^V]_{\vec{q}} = 2\pi i \sum_j e_j \frac{(q_y, -q_x)}{q^2} e^{-i\vec{q}\cdot\vec{r}_j} ; \qquad (25)$$

$$[\partial_\tau \varphi^V]_{\vec{q}} = 2\pi i \sum_j e_j \frac{q_x \dot{y}_j - q_y \dot{x}_j}{q^2} e^{-i\vec{q}\cdot\vec{r}_j} . \qquad (26)$$

In addition, it is clear that in some expressions a large wave-vector cutoff is necessary, for example, as in the following quantity (see below):

$$V(\vec{r}) = E_J \int \frac{d^2 q}{q^2} (e^{i\vec{q}\cdot\vec{r}} - 1) e^{-q/q_c} , \qquad (27)$$

where $q_c = (2\pi)^{1/2}$ as required by the original lattice model. Correspondingly,

$$\int d^2 q \to \int d^2 q \, e^{-q/q_c} = 2\pi q_c^2 = (2\pi)^2 . \qquad (28)$$

For the sake of transparency, I present the main features by concentrating on simple cases.

(i) $m = 0$ ($\lambda = \infty$)

In this case, in which the spin-wave spectrum has only an optical part, it is possible to neglect φ^S. With $\varphi \simeq \varphi^V$, it is straightforward to derive the following results:

$$\mathcal{A}_0 = \frac{1}{2} \int d\tau \left[\mathcal{M} \sum_j \dot{\vec{r}}_j^2 + \sum_{i,j} e_i e_j V(\vec{r}_i - \vec{r}_j) \right] \qquad (29)$$

$$\mathcal{A}_T = -2\pi \sum_j e_j \int d\tau (f_y x_j - f_x y_j) \qquad (30)$$

$$\mathcal{A}_D = \sum_{i,j} e_i e_j \int d\tau d\tau' \, A_s(\tau - \tau') \, F(\vec{r}_i(\tau) - \vec{r}_j(\tau')) \qquad (31)$$

Here the vortex mass is given by $\mathcal{M} = 2\pi^2 M$, $V(\vec{r})$ is the (logarithmically increasing) interaction potential familiar from the classical limit,[11] and I have taken the external current, represented by $\vec{f} = (f_x, f_y)$, to be space independent. Furthermore, $F(\vec{r})$ is given by

$$F(\vec{r}) = -\frac{V(\vec{r})}{4 E_J} = \frac{\pi}{2} \ln \frac{1 + (1 + 2\pi r^2)^{1/2}}{2} . \qquad (32)$$

Ignoring the dissipative term $\mathcal{A}_D$, (29) and (30) correspond to the following equation of motion:

$$\mathcal{M} \ddot{\vec{r}}_i = \vec{\mathcal{F}}_i - 2\pi e_i \, \hat{z} \times \vec{f} , \qquad (33)$$

where $\hat{z} = (0,0,1)$, and $\vec{\mathcal{F}}_i$, representing the force due to the other vortices, is of the standard form:[15]

$$\vec{\mathcal{F}}_i = -e_i \sum_{j \neq i} e_j (\nabla V)_{\vec{r}_i - \vec{r}_j} = -2\pi E_J \, e_i \, \hat{z} \times (\nabla \varphi^V)' \,. \tag{34}$$

Here the prime, $(\nabla \varphi^V)'$, indicates that the ith term has to be omitted in (22); also, I used the large distance result, $V(\vec{r}) \simeq -2\pi E_J \ln r$, to arrive at the last equality. Note that a vortex moves *perpendicular* to the total current (external plus the current due to the other vortices).

In order to derive the dissipative contribution to (33), analytic continuation procedures are required.[8] Here I give only the final result, which follows from (31), namely

$$\mathcal{M} \ddot{\vec{r}}_i + 2e_i \sum_j e_j \int dt' \, A_s^R(t-t') \, (\nabla F)_{\vec{r}_i(t) - \vec{r}_j(t')} = \dots \,, \tag{35}$$

where $A_s^R(t-t')$ is the Fourier transform of $A_s^R(\omega) = A_s(\omega \to -i\omega + 0)$. Retaining only the term $i = j$, and in addition expanding for small distances, I find

$$\mathcal{M} \ddot{\vec{r}}_i + \tilde{\eta} \dot{\vec{r}}_i = \dots \,, \tag{36}$$

where $\tilde{\eta}$ is given by $\tilde{\eta} = 2\pi^2 \eta_s = \pi \hbar \alpha_s$, which is not unexpected in view of the result $\mathcal{M} = 2\pi^2 M$.

(ii) $M = 0$, $\eta_s = 0$

In this case, it is rather cumbersome to follow the above procedure. However, it is obvious from (14) that the problem is symmetric with respect to space and time (for zero temperature), which allows for the following elegant argument.[16] Defining $\tilde{x} = (x, y, c\tau)$, the corresponding gradient, $\tilde{\nabla}$, and the "magnetic field" $\tilde{\vec{h}} = \tilde{\nabla} \varphi$, the action (14) can be written as follows:

$$S_0 = \frac{E_J}{2c} \int d^3x \, (\tilde{\vec{h}})^2 \,. \tag{37}$$

Also, introduce the charge and the current density by

$$\rho = \sum_j e_j \delta(\vec{r} - \vec{r}_j) \,, \quad \vec{j} = \sum_j e_j \dot{\vec{r}}_j \delta(\vec{r} - \vec{r}_j) \,. \tag{38}$$

Thus the continuity equation $\partial_\tau \rho + \nabla \cdot \vec{j} = 0$ transforms into $\tilde{\nabla} \cdot \tilde{\vec{j}} = 0$, where $\tilde{\vec{j}} = (\rho, \vec{j}/c)$. Finally, it is obvious that

$$D^{-1} \varphi = 0 \;\to\; \tilde{\nabla} \cdot \tilde{\vec{h}} = 0 \,. \tag{39}$$

Choosing $\tilde{\vec{h}}$ such that $\tilde{\nabla} \times \tilde{\vec{h}} = 2\pi \tilde{\vec{j}}$, it follows from the analogy with the magnetostatic problem that $\mathcal{A}_0$ is given by

$$\mathcal{A}_0 = \frac{\pi E_J}{2c} \int d^3x \, d^3x' \, \frac{\tilde{\vec{j}}(\tilde{x}) \cdot \tilde{\vec{j}}(\tilde{x}')}{|\tilde{x} - \tilde{x}'|} \,. \tag{40}$$

In terms of the vortex coordinates, this result reads

$$\mathcal{A}_0 = \frac{\pi E_J}{2c} \sum_{i,j} \int d\tau d\tau' \frac{\dot{\vec{r}}_i(\tau) \cdot \dot{\vec{r}}_j(\tau') + c^2}{\{[\vec{r}_i(\tau) - \vec{r}_j(\tau')]^2 + c^2(\tau - \tau')^2\}^{1/2}}. \quad (41)$$

The expression (40) is particularly appealing since it is of the same form as the magnetostatic energy of unimodular line currents (in three dimensions) flowing along the trajectories of the vortex centers. I only remark that some progress in the description of the phase transition of this model, which is expected to be of the 3D-XY type, in terms of vortex loops has been made recently.[17]

Of course, for a static configuration, i.e. $\vec{r}_j(\tau)$ independent of τ, one may integrate with respect to time and recover the potential given in (29). More interesting, however, is the fact that (41) also includes a contribution which represents dissipation in the vortex motion due to decay into spin-waves. In particular, taking again the $i = j$ term and expanding for small velocities, $|\vec{r}_j(\tau) - \vec{r}_j(\tau')| \ll c|\tau - \tau'|$, one obtains after Fourier transformation[12,13]

$$\mathcal{A}_0 \to \frac{1}{2} \sum_j \int \frac{d\omega}{2\pi} |\omega| \eta(\omega) |\vec{r}_j(\omega)|^2, \quad (42)$$

where $\eta(\omega) = \pi m |\omega| \ln(1/|\omega|)$ means a frequency dependent friction (which is called *subohmic* since $\eta(\omega) \to 0$ for $\omega \to 0$; this feature is connected with the fact that the vortices are coupled to a *two-dimensional* system of acoustic vibrations). It turns out, however, that the friction represented by $\eta(\omega)$ is weak in the sense that it leads only to logarithmic corrections to the free acceleration of a vortex by an external current.[12]

(*iii*) $M = 0$, $\eta_s \neq 0$

I wish to remark that, in all cases, the evaluation of $\mathcal{A}$ can be simplified by the fact that $[\nabla \varphi^S]_{\vec{q}}$ is perpendicular to $[\nabla \varphi^V]_{\vec{q}}$. Concerning the dissipation related to η_s, it follows from (19) that a characteristic frequency can be defined by

$$\hbar \omega_s = \hbar E_J/\eta_s = 2\pi E_J/\alpha_s = \pi \Delta \alpha / \alpha_s. \quad (43)$$

Assuming that $\alpha = R_0/R_N \gg \alpha_s$, it seems possible to expand the result (24) with respect to η_s. In particular, it follows that the expression $\dot{\varphi}^2 + c^2(\nabla \varphi^S)^2$ has only a correction in second order. As a result, the contribution $\sim \eta_s$ to $\mathcal{A}$ is precisely given by (31), and $\mathcal{A}_0$ retains the form given in (41) in this order.

Finally, I briefly discuss the case where both M and m are different from zero. As a rule, it seems that the acoustic vibrations implied by a finite value of m dominate at large distances in space and time. This means that the most important consequence of a finite M is the effective mass $\mathcal{M}$ of a vortex. Besides the kinetic energy term in (29), the effective action is thus given by (41), (30), and (31), at least as long as α_s is small. A particular case is the limit in which $\lambda = (M/m)^{1/2}$ is larger than the sample dimension. Then it is possible to put $m = 0$ and take $\mathcal{A}_0$ as given by (29). Presumably, the last case is mostly of theoretical interest.

4 Vortex Dynamics in Other Systems

Without going into great detail, I summarize in this section the *classical* equations of motion which are often used to describe vortex dynamics in superfluid ^{4}Helium and in superconductors. In order to simplify the notation, I denote the coordinate of the vortex under consideration by $\vec{r}_0(t)$, and define $\vec{v}_s'$ to be the local superfluid velocity at $\vec{r}_0$, however, *excluding* the velocity field of this particular vortex. Also, $e_0 = \pm 1$ is the sign of the vorticity and, having especially films in mind, d is the film thickness, and ρ_s the superfluid mass density integrated across the film; finally, m^* denotes the Helium mass and twice the electron mass, respectively.

(i) Superfluid Helium Films

The equation of motion is obtained by balancing the Magnus force against the viscous drag due to interactions with thermal excitations and with the substrate (see Vinen,[15] and also Ref. 18):

$$B\dot{\vec{r}}_0 + B'e_0\hat{z} \times \dot{\vec{r}}_0 = e_0\rho_s \frac{2\pi\hbar}{m^*}\hat{z} \times (\dot{\vec{r}}_0 - \vec{v}_s'), \qquad (44)$$

where B and B' are phenomenological coefficients. Of particular importance is the form of the rhs of (44), which leads to the conclusion that for $B = B' = 0$, the vortex rides along with the local superfluid velocity. It seems that the Magnus force as given in (44) is intimately related to the Galileil invariance of the Helium system, and the above conclusion is believed to hold at low temperature and for a translation invariant substrate. Generally, the vortex moves in a direction characterized by the angle θ (defined such that $\theta = 0$ if the vortex moves *perpendicular* to $\vec{v}_s'$), which is given by

$$\tan\theta = (2\pi\hbar\rho_s/m^* - B')/B . \qquad (45)$$

(ii) Superconducting Films

The theory of vortex dynamics in superconductors has been first discussed in detail by Bardeen and Stephen,[19] and is essentially based on the idea that a vortex has a normal core of radius $\sim \xi$, where ξ is the coherence length. (For an overview, see Tinkham's book.[20]) Thus dissipation is due to ordinary Ohmic losses, i.e. through electrons scattering at static impurities or defects, in the normal region. Their result reads:[19]

$$\eta(\dot{\vec{r}}_0 - \kappa_H \vec{v}_s') = \rho_s \frac{2\pi\hbar}{m^*}\hat{z} \times (-\vec{v}_s'), \qquad (46)$$

where I have taken $e_0 = 1$; the friction constant η is given by

$$\eta = \frac{R_0}{R_N}\frac{\hbar}{\xi^2} = R_0 \sigma_N d \frac{\hbar}{\xi^2} \qquad (47)$$

with σ_N the normal state (Drude) conductivity. In addition, κ_H is given by

$$\kappa_H = H/H_{c2}, \quad H_{c2} = \phi_0/2\pi\xi^2, \qquad (48)$$

where H is the magnitude of the magnetic field (which is applied perpendicular to the film), and ϕ_0 the flux quantum. (However, shortly thereafter it was argued[21] that $\kappa_H = 1$ should be the correct result.) Note that η is essentially independent of temperature.

Using that at low temperature the integrated superfluid density is given by $2\rho_s/m^* = nd$, where n is the total electron density, the Hall angle is found to be given by the same expression as in the normal state, namely

$$\tan\theta = \omega_c\tau \sim \kappa_H \frac{\Delta}{\epsilon_F}\frac{\ell}{\xi}, \qquad (49)$$

where ω_c is the cyclotron frequency, τ the elastic scattering time, $\ell = v_F\tau$ the mean free path, ϵ_F the Fermi energy, and I used that $\xi \sim \hbar v_F/\Delta$, which applies for $\ell \gg \xi$. Note that the Hall angle is very small, except for extremely pure samples, since $\Delta \sim 10^{-4}\epsilon_F$. It seems that the debate about the Hall angle was somehow inconclusive (see, however, Ref. 22), which also may be related to the fact that pinning of vortices by impurities[23] – neglected above – plays experimentally a far more important role;[20] on the other hand, the theory of the viscosity as well as of nonlinear phenomena is well developed.[24]

In comparison, I wish to emphasize what I believe is an important difference between Helium and superconductors, which is apparent from (44) and (46). In the former case, the vortex moves in the direction of the local $\vec{v}'_s$ in the limit of *vanishing* viscosity, $B \to 0$, with a velocity smaller than $\vec{v}'_s$ provided B' is finite. In the latter case, however, the limit of an extremely clean metal corresponds to a very strong viscosity, i. e. formally $\eta \to \infty$. In this limit, the vortex moves in the direction of the local $\vec{v}'_s$, with a velocity smaller than $\vec{v}'_s$ if $\kappa_H < 1$. It appears that the dynamics of vortices is fundamentally different in the two systems.

Finally returning to the model of a network of Josephson junctions studied in the preceding sections, I emphasize that I do not find any indication of a finite θ. Rather, the vortices are found to move *perpendicular* to the local current, which seems somehow surprising since the viscosity is small at low temperature. Presently, I can only speculate that the underlying lattice structure (which means e.g. that the vortices do not have a normal core) is responsible for this result.

5 Conclusion

In these notes, I presented selected aspects of the statics and the dynamics of a network of Josephson junctions, within the framework of the dual description in terms of vortices. In this approach, it is straightforward to make contact with the classical (high temperature) limit results.[11] I remark that the lattice model (for $M = 0$) has been studied recently by extensive Monte Carlo simulations, which show a reentrant behavior upon decreasing the temperature, as well as a first order transition within the superconducting region.[25] However, for example, the effect of a finite nearest-neighbor capacitance ($M \neq 0$) on the phase diagram, especially at zero temperature, is still an open question.[26,10] In addition, I wish to mention that recent experiments on granular[27] as well as on continuous[28] superconducting films have shown several unexpected features, pointing towards the normal state film resistance as the (most?) important parameter.

As a word of caution, I emphasize that I restricted myself to discuss the adiabatic limit of the general model,[12] i.e. the limit of small frequencies

$\hbar\omega \ll 2\Delta$. In this case, the parameter R_N or $\alpha = R_0/R_N$, where R_N equals the normal state sheet resistance (for a square sample) in the absence of shunt resistors, enters only indirectly through E_J and the quasiparticle capacitance. However, the adiabatic condition is, at best, only marginally satisfied when E_C and E_J are of the order of a few Kelvin. For example, for $\alpha_s = 0$, a vortex is freely accelerated by an external current until its energy is large enough to create quasiparticles,[12] i.e. is larger than 2Δ. This suggests that for a detailed understanding of the dynamics at low temperatures, the full nonlinear model has to be considered, which unfortunately is a formidable problem. On the other hand, I emphasize that high frequencies probe the "normal" part of the current-voltage characteristic, which possibly may explain the significance of the normal state resistance.

Acknowledgement

Financial support by the Deutsche Forschungsgemeinschaft through a Heisenberg fellowship is gratefully acknowledged.

References

1. B. D. Josephson, Phys. Letters **1**, 251 (1962); Advan. Phys. **14**, 419 (1965).
2. A. J. Leggett, Suppl. Progr. Theor. Phys. **69**, 80 (1980).
3. For a recent review and further references, see: A. J. Leggett, Jap. J. Appl. Phys. **26**, Suppl. 26-3, 1986 (1987).
4. P. W. Anderson, in *Lectures on the Many Body Problem*, edited by E. R. Caianiello (Academic, New York, 1964), Vol. 2, p. 113.
5. L. J. Geerligs, M. Peters, L. E. M. de Groot, A. Verbruggen, and J. E. Mooij, Phys. Rev. Lett. **63**, 326 (1989).
6. R. P. Feynman and A. R. Hibbs, *Quantum Mechanics and Path Integrals* (McGraw-Hill, New York, 1965).
7. A. O. Caldeira and A. J. Leggett, Phys. Rev. Lett. **46**, 211 (1981); Ann. Phys. (N. Y.) **149**, 374 (1983).
8. V. Ambegaokar, U. Eckern, and G. Schön, Phys. Rev. Lett. **48**, 1745 (1982); Phys. Rev. B **30**, 6419 (1984); A. I. Larkin and Yu. N. Ovchinnikov, Phys. Rev. B **28**, 6281 (1983).
9. F. Guinea and G. Schön, Europhys. Lett. **1**, 585 (1986); J. Low Temp. Phys. **69**, 219 (1987).
10. For a recent review see, for example: U. Eckern and G. Schön, in *Festkörperprobleme/Advances in Solid State Physics*, edited by U. Rössler (Vieweg, Braunschweig, 1989), Vol. 29, p. 1.
11. V. L. Berezinskii, Zh. Eksp. Teor. Fiz. **61**, 1144 (1972) [Sov. Phys. JETP **34**, 610 (1972)]; J. M. Kosterlitz and D. J. Thouless, J. Phys. C **5**, L124 (1972); **6**, 1181 (1973).
12. U. Eckern and A. Schmid, Phys. Rev. B **39**, 6441 (1989); see also Ref. 13.
13. A. I. Larkin, Yu. N. Ovchinnikov, and A. Schmid, Physica B **152**, 266 (1988).
14. Compare Eq. (2.4) of Ref. 12.
15. Y. B. Kim and M. J. Stephen, in *Superconductivity*, edited by R. D. Parks (Dekker, New York, 1969), Vol. 2, p. 1107; W. F. Vinen, *ibid*. p. 1167.

16. V. N. Popov, *Functional Integrals in Quantum Field Theory and Statistical Physics* (Reidel, Boston, 1983), Sec. 21.
17. S. R. Shenoy, Phys. Rev. B (1989), to be published.
18. V. Ambegaokar, B. I. Halperin, D. R. Nelson, and E. D. Siggia, Phys. Rev. B **21**, 1806 (1980).
19. J. Bardeen and M. J. Stephen, Phys. Rev. **140**, A1197 (1965).
20. M. Tinkham, *Introduction to Superconductivity* (McGraw-Hill, New York, 1975), Sec. 5.
21. P. Nozières and W. F. Vinen, Philos. Mag. **14**, 667 (1966).
22. J. Bardeen and R. D. Sherman, Phys. Rev. B **12**, 2634 (1975).
23. A. Schmid and W. Hauger, J. Low Temp. Phys. **11**, 667 (1973).
24. See, for example: A. Schmid, Phys. Kondens. Materie **5**, 302 (1966); A. I. Larkin and Yu. N. Ovchinnikov, Pis'ma Zh. Eksp. Teor. Fiz. **23**, 210 (1976) [JETP Lett. **23**, 187 (1976)]; Zh. Eksp. Teor. Fiz. **73**, 299 (1977) [Sov. Phys. JETP **46**, 155 (1977)].
25. L. Jacobs, J. V. José, M. A. Novotny, and A. M. Goldman, Phys. Rev. B **38**, 4562 (1988); L. Jacobs and J. V. José, Physica B **152**, 148 (1988); J. Choi and J. V. José, Phys. Rev. Lett. **62**, 1904 (1989).
26. S. Chakravarty, S. Kivelson, G. T. Zimanyi, and B. I. Halperin, Phys. Rev. B **35**, 7256 (1986); R. A. Ferrell and B. Mirhashem, *ibid.* **37**, 648 (1988); S. E. Korshunov, Europhys. Lett. **9**, 107 (1989); W. Zwerger, *ibid.* **9**, 421 (1989).
27. B. G. Orr, H. M. Jaeger, A. M. Goldman, and C. G. Kuper, Phys. Rev. Lett. **56**, 378 (1986); H. M. Jaeger, D. B. Haviland, A. M. Goldman, and B. G. Orr, Phys. Rev. B **34**, 4920 (1986).
28. D. B. Haviland, Y. Liu, and A. M. Goldman, Phys. Rev. Lett. **62**, 2180 (1989).

COMPARISON OF EFFECTIVE MODELS FOR CuO$_2$ LAYERS IN OXIDE SUPERCONDUCTORS

A. Ramšak and P. Prelovšek

J. Stefan Institute and Department of Physics
University of Ljubljana
61111 Ljubljana, Yugoslavia

Several effective models, as derived from a general two-band Hubbard model for CuO2 layers in oxide superconductors are studied. In particular we compare the Hubbard model with the hole-spin models (unsymmetrized and symmetrized) and a generalized effectice single-band (t-J) model.[1] Our exact diagonalization results for a finite CuO2 chain of 4 cells show that spectral properties of a single mobile quasiparticle in a general Hubbard model can be well reproduced by reduced effective models. This remains valid even in the mixed valence regime, where the straightforward derivation of the hole-spin model is not possible due to the breakdown of the perturbation expression using hopping parameter in the Hubbard model as a small parameter.

In figure are presented the lowest lying branches for a system with a single additional hole on a chain of 4 cells for different models: a)Two band Hubbard model; b)Hole-spin model; c)Symmetrized hole-spin model; d)Kondo-lattice model; e)Generalized t-J model.

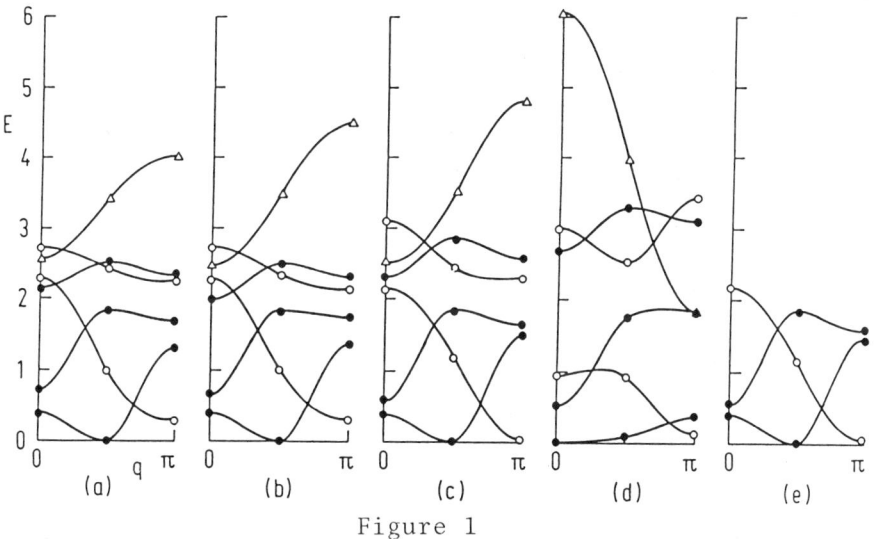

Figure 1

REFERENCE

[1] A. Ramšak and P. Prelovšek, Phys. Rev. B **40**, 2239 (1989).

CHARGE DENSITY WAVES IN QUASI-ONE-DIMENSIONAL SYSTEMS

Aleksa Bjeliš

Department of Theoretical Physics
Faculty of Sciences, P.O.B. 162
41001 Zagreb, Croatia, Yugoslavia

INTRODUCTION

The search for new superconducting materials and novel types of superconductivity has been, and still is, the main challenge in the investigation of low dimensional systems. Besides the great achievements made by discoveries of superconductivity in organic chain compounds and layered perovskites, this activity has also highlighted another type of instability which characterizes particularly quasi-one-dimensional conductors. This is the stabilization of the periodic deformation in the crystal lattice, accompained by the formation of the electronic charge density wave (CDW), and the openning of the gap at the Fermi level.

This instability was predicted by Peierls[1] in the early days of the modern condensed matter physics, and has been for many years a curious example of spontaneous symmetry breaking in the many electron system coupled to the lattice. Fröhlich[2] made a further step forward by conjecturing that the Peierls state with the incommensurate modulation is degenerate in phase and possesses a particular type of coherence (quantum rigidity) which makes possible a collective charge transport without dissipation. That was one of the first microscopic proposition for the explanation of superconductivity. There was however no way to extend the model to realistic three-dimensional systems, i.e. to make a necessary step towards the understanding of perfect diamagnetism in the superconducting state. The microscopic theory of superconductivity went in another direction[3] well-known by now.

The revival of interest for the Peierls state and the Fröhlich mechanism of collective transport was initiated by the experimental investigations on the Krogmann salt $K_2Pt(CN)_4Br_{0.3} \cdot H_2O$ (KCP)[4] and the organic salt TTF-TCNQ.[5] Like in the preceeding theoretical developments, the modulations and Peierls gaps in the single-particle excitations observed respectively in structural and spectroscopic measurements were the first well established evidences for the new state. The first clear-cut experiments showing collective transport were performed a few years later.[6-8]

The parallel theoretical investigations led to, among others, two important conclusions. At first, it was realized that the coupling to the lattice is not necessary for the stabilization of CDW. Utilizing more

Applications of Statistical and Field Theory Methods to Condensed Matter
Edited by D. Baeriswyl et al., Plenum Press, New York, 1990

involved theoretical techniques, it was shown that the CDW state is one of possible ground states for interacting quasi-one-dimensional electron gas,[9-11] beside the spin density wave and superconducting state. Later on, it became apparent that the electron-electron interactions are dominant in TTF-TCNQ and similar materials,[12] as well as in the first organic superconductors belonging to the family of Bechgaard salts.[13]

The second theoretical conclusion invokes the sensitivity of the CDW coherence to the presence of impurities, or, more generally,[14-16] to any kind of disturbances which affect the translational symmetry. In contrast to the superconducting condensate, the Fröhlich condensate is not able to move frictionlessly in real systems. This conclusion was confirmed by the first experimental evidences of Fröhlich transport.[6] The measurements on the transition metal trichalcogenide $NbSe_3$ showed that the small, finite threshold field E_T has to be applied in order to depin the CDW and to put it into the state of collective motion. This nonlinear regime remains dissipative in the most of known materials. Still, the recent low temperature data on the blue bronze $K_{0.3}MoO_3$[17,18] suggest that above much higher threshold fields the CDW motion may become rigid and very coherent, ressembling to the perfect Fröhlich transport.

The text which follows covers only some aspects of the vast field of CDW physics developed in about twenty last years.[19] In the consideration of microscopic aspects of CDW instability and of the ground state (first Section) the particular attention is devoted to the characteristic energy (i.e. temperature) scales, to the role of interchain couplings and to the effects of finite ionic mass, i.e. of nonadiabatic electronic response to the lattice dynamics.

The second Section starts with the discussion of the coherence of the Fröhlich condensate, and of the collective phason excitations. The roles of the Coulomb interaction and impurities on the phason dispersion, and the mechanism of depinning in the regime of weak impurities are also shortly discussed.

In the third Section we consider topological defects in the CDW, distinguishing between the static ($E<E_T$) and dynamic ($E>E_T$) regimes. In particular, it is emphasized that the nucleation of dynamic dislocation lines is closely connected with the microscopic generation of phase slippages (PSs). The latter process is the possible source of narrow band noise (NBN), which is the specific property of the CDW in the collective motion, observed in the most of the known systems.[19]

CHARGE DENSITY WAVE ORDER

Materials which exhibit the CDW ground state are as a rule highly anisotropic, i.e. built as sequences of weakly coupled chains or planes. The Fermi surfaces in such systems are not only anisotropic, but also have a particular, so called nesting shape. This means that finite portions on opposite sides of surface almost coincide after the translation by a wave vector $\vec{Q}$ (Fig.1). The number of such portions (i.e. the number of components in the corresponding star of wave vectors, Fig.1) depends on the crystal symmetry. In quasi-one-dimensional systems the Fermi surface consists of two open sheets, and the star reduces to one pair of wave vectors, $\pm\vec{Q}$, or even to a single wave vector in the case of half-filled band, in which $-\vec{Q} = \vec{G} + \vec{Q}$, where $\vec{G}$ is a vector of reciprocal lattice.

The nesting property of Fermi surface is crucial for the stabilization of CDW state. It enables the condensation of macroscopic number of coherent electron-hole pairs with the common wave vector $\vec{Q}$. The

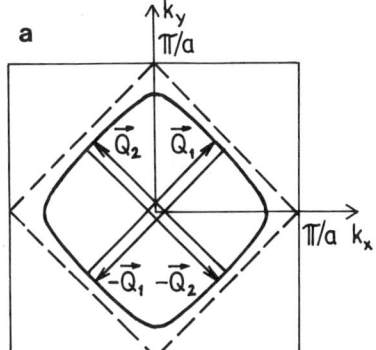 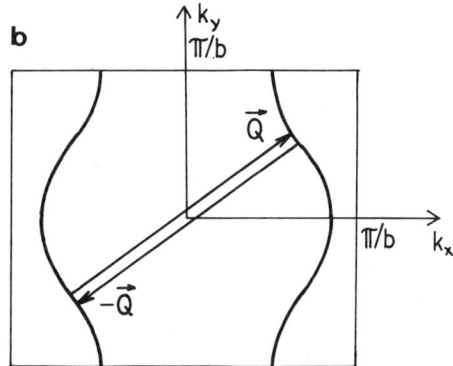

Fig. 1. The nesting Fermi surfaces for (a) the two-dimensional band close to the half-filling and (b) the quasi-one-dimensional band. The star of wave vectors comprises four and two wave vectors respectively.

anomalous scattering leading to this condensation involves generally both Coulomb electron-electron and electron-phonon interactions. These interaction are respectively instantenous and retarded. Furthermore, they are also anisotropic, i.e. intrachain couplings are stronger than interchain ones. The degree of the interaction anisotropy and of the phonon nonadiabaticity, together with the level of anisotropy of the electron spectrum are microscopic parameters which essentially determine the electronic correlations above the critical temperature, and the properties of the ordered state below it. Let us roughly schematize the characteristic regimes, particularly pointing out those which are relevant for the actually best known CDW materials.

CDW instability

The first distinction to be made is that between the high temperature regime of one-dimensional critical fluctuations and the low temperature regime of three-dimensional long range order. In the former regime defined provisionally by $T > t_\perp$, the transverse dispersion in Fig.1. is effectively flattened by temperature. The probability of excitation of of virtual electron-hole pairs with the momentum $2k_F$ is then logarithmically singular [$\sim \log(E_F/T)$] and independent of the transverse momenta. The Fermi energy E_F has here the role of electron-band cut-off. On the other hand, the virtual electron-electron Cooper pairs have just the same properties, if the bare interaction is instantenous. Thus both channels have to be treated on the same footing. The logarithmic (so called parquet[9]) contributions in all orders comprise not only terms from each channel (usually called random phase approximation (RPA) and ladder series respectively for electron-hole and electron-electron processes), but also the mixed terms. There are few methods[9-11] which enable the calculation of various correlation functions with all logarithmic processes taken into account. The conclusions are usually parametrized in terms of bare coupling constants for the forward ($k \simeq 0$) and backward ($k \simeq 2k_F$) scatterings, g_2 and g_1 respectively. The CDW correlation function at the wave number $2k_F$ and in the static limit ($\omega = 0$) behaves like[20,21]

$$P(2k_F, \omega=0) \sim (g_2-g_1/2)^{-1} (E_F/T)^{2g_2-g_1} . \qquad (1)$$

These correlations dominate over superconducting ones for $2g_2-g_1 > 0$.

327

The degeneracy of RPA and ladder channels is the consequence of a particular property of instantenous interactions. They do not introduce any cut-off in the scatterings, so that the integration limits in all virtual processes are defined by the electron bandwidth ($\sim E_F$). The cut-off for the backward electron-electron interaction via phonons is however finite and of the order of Debye energy $\hbar\omega_D \simeq \hbar\omega_0(2k_F) \ll E_F$. The corresponding forward scattering is negligible for acoustic phonons. The low energy cut-off $\hbar\omega_D$ intervenes in all *inelastic* electron-electron scattering via $2k_F$ phonons.[22] In particular, each virtual formation of electron-electron pair in a given parquet diagram carries now the factor $\log(\hbar\omega_D/T)$. On the other hand, the quasistatic virtual formation of electron-hole pairs, i.e. the whole series of RPA diagrams (Fig.2) retains the same logarithmic factors $\log(E_F/T)$ as in the non-retarded limit. In other words, the crystal lattice can follow the electron-hole propagation, i.e the $2k_F$ fluctuations in the electronic charge density, provided that the latter are slow enough on the phonon time scale ω_D^{-1}.

Fig. 2. RPA series of diagrams for the phonon propagator.

The phonon retardation thus leads to the breaking of degeneracy of electron-electron and electron-hole blocks in parquet diagrams. It is then instructive to distinguish between ranges of temperatures higher or lower than the Debye temperature.[22-24] For $T < \hbar\omega_D$ all parquet diagrams remain logarithmically anomalous. The problem is then even more complex than that with instantenous coupling, since now two cut-offs are present. The situation is much simpler for $T > \hbar\omega_D$ (more precisely for $2\pi T > \hbar\omega_D$).[22,24] Then only the pure RPA series remains anomalous. One has the Migdal theorem[25] satisfied, i.e. the vertex corrections in RPA diagrams (Fig.2) are of the order $\hbar\omega_D/E_F \sim (m_{el}/M)^{1/2}$, where m_{el} and M are the effective electron mass and the ionic mass respectively. Consequently only CDW one-dimensional fluctuations are developed. They are accompanied by the softening of the $2k_F$ phonon mode(s) coupled to the charge fluctuations. This simple adiabatic regime may persist down to low temperatures, i.e. it may involve the cross-over to the three-dimensional fluctuations and the onset of long-range order. This is obvious if the latter is above the Debye temperature. However, even this conditions is not necessary, since below the mean-field temperature the pseudo-gap of the width Δ_0 (see eq. (30)) develops in the electron spectrum,[26,27] so that the condition $2\pi T > \hbar\omega_D$ is gradually replaced by the condition $\Delta_0 > \hbar\omega_D$ as the temperature decreases.

The critical temperature of three-dimensional ordering, T_p, is quite high[19] in most of CDW sistems, so that the condition $2\pi T_p > \hbar\omega_D$ seems to be well satisfied. The best known examples are trichalcogenic and tetrachalcogenic compounds and blue bronzes. Moreover, in these systems the electron-phonon coupling is stronger than the Coulomb electron-electron interaction. The latter then cause only inessential corrections to the logarithmically anomalous RPA polarizability,[24] as well as to the critical temperature of three-dimensional order, T_p, and to the accompanying order parameter. The opposite limit of CDW instability due to the Coulomb interactions, and the lattice deformation driven by a relatively weak electron-phonon coupling is realized in e.g. the organic compound TTF-TCNQ. The corresponding power law of the one-dimensional fluctuations and the scales relevant for the critical temperature in this limit are discussed in detail in Ref. 12.

The summation of RPA series of diagrams (Fig.2) gives the well known result for the softening of the Kohn phonon.

$$\omega^2(k,\vec{k}_\perp) = \omega_0^2(k,\vec{k}_\perp)\left\{\left[1 - \lambda\ P(k,\vec{k}_\perp)\right] + \right.$$
$$\left. + 4\lambda\ g_{1\perp}\cos(\phi + \vec{k}_\perp\vec{b})\ \log^2(E_F/T)\right\} \quad (2)$$

and for the corresponding elastic contribution to the free energy

$$F^{(2)} = \int d^3k\ \omega^2(k,\vec{k}_\perp)\ |u(k,\vec{k}_\perp)|^2 \quad , \quad (3)$$

where $u(k,\vec{k}_\perp)$ is the Fourier component of the $2k_F$-modulated lattice displacement, ω_0 is the bare phonon frequency, λ is the electron-phonon coupling constant, and b is the interchain distance.

The $k_\perp$-dependences are intentionally indicated in eq.(2) since they are essential for the development of transverse correlations and the eventual onset of three-dimensional order. These dependence come from three physical sources. One of them is the interchain backward electron-electron coupling[28] $g_{1\perp}$, with the lowest order correction to the electron-hole bubble shown in Fig.3b and included into eq.(2). This term favors the phase shift $\phi = \pi$ between CDWs on neighboring chains. E.g. for the rectangular chain lattice the wave vector of transverse modulation would be $\vec{Q}_\perp = (\pi/b, \pi/c)$.

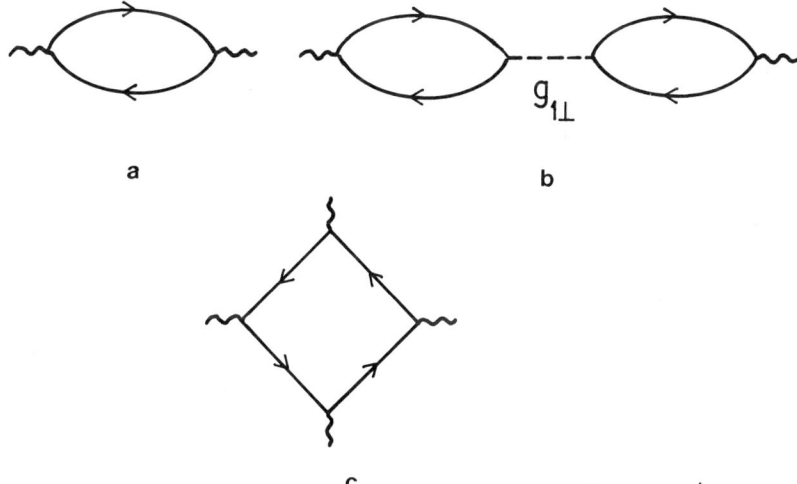

Fig. 3. (a) Quadratic contribution to the free energy coming from the electron-hole (bubble) polarizability; (b) the lowest order correction to the phonon self-energy due to the electron-electron interchain coupling; (c) the fourth-order contribution to the free energy.

The further cause of transverse correlations are ionic elastic forces which result in the finite dispersion of the bare phonon frequency $\omega_0(\vec{k})$.

The wave vector $\vec{Q}_\perp$ favorized by this mechanism depends on the details in the transverse dispersion $\omega_0(\vec{k}_\perp)$.

Finally the $\vec{k}$ - dependence of the bubble electron polarizability $P(\vec{k},\omega)$ (Fig.3a) is due to the transverse band dispersion, which, as already noted, favors the best possible nesting of Fermi surfaces. The usual model tight-binding dispersion, appropriate again for the rectangular chain lattice, is given by[29,30]

$$E_\pm(k,k_\perp) = \pm v_F(k \pm k_F) + 2t_b \cos(k_\perp b) + 2t'_b \cos(2k_\perp b) \,. \tag{4}$$

The first term is the longitudinal dispersion linearized around the Fermi wave numbers $\pm k_F$. The second term comes from the electron hopping between neighboring chains. The third term is the one which prevents the perfect nesting with the wave vector $(2k_F, Q_\perp = \pi/b)$. It could be identified as the hopping between the next neighboring chains. This hopping is however very weak. In fact, the more important contribution comes from the higher, quadratic term in the longitudinal expansion around $\pm k_F$. The latter can be conveniently expressed as the effective transverse term[30] with

$$t'_b = \left[t_b^2 \cos(k_F a)\right] \Big/ \left[4 t_a \sin^2(k_F a)\right] \,, \tag{5}$$

where t_a is the longitudinal tight binding bandwidth. The dispersion in the other transverse direction is not explicitly written in eq.(4). The two transverse directions are usually not equivalent. E.g. in blue bronzes which have the layered crystal packing, the electron spectrum is almost two dimensional, with the open Fermi surfaces due to the stacking structure of the layers.[31]

The enumerated interchain couplings determine, sometimes in a competitive way, the optimum value of the transverse CDW wave vector, $\vec{Q}_\perp$. The even more important outcome is however the strength of transverse correlations. It is expressed in terms of characteristic correlation length $\xi_{0\perp}$, obtained by expanding the right-hand side in eq.(3) around the minimum at $(Q_\parallel, Q_\perp)$, where $Q_\parallel$ is equal or close to $2k_F$. This expansion gives the form of Kohn anomaly in the dispersion of soft phonon above the critical temperature. One gets[24]

$$\xi_{0\perp} = b\, g_{1\perp} \log^2(E_F/T) \,, \tag{6a}$$

$$\xi_{0\perp} = b\, (C_\perp/C\lambda)^{1/2} \tag{6b}$$

and

$$\xi_{0\perp} = b\, t_b/T \tag{6c}$$

for Coulomb, lattice and hopping interchain coupling respectively. $C_\perp$ and C are transverse and longitudinal elastic constants respectively. For further considerations it usually suffices to keep the largest of the coefficients (6a,b,c). Note that the Coulomb and hopping transverse correlation lengths increase by lowering temperature, while that due to lattice coupling remains approximately constant. The overwhelming contribution to the correlation length in the longitudinal direction comes from the expansion of the polarizability $P(k,\vec{Q}_\perp)$ and is given by

$$\xi_0 \simeq v_F/T \,. \tag{7}$$

The magnitude of transverse correlation length $\xi_{0\perp}$ determines the way in which the system passes by lowering the temperature from one-dimensional to three-dimensional (or two-dimensional) fluctuations. Two regimes, $\xi_{o\perp} > b$ and $\xi_{o\perp} < b$ should be distinguished from this point of view. Thus, if in the temperature range

$$T_P^o \simeq E_F \exp(-1/2\,\lambda) \tag{8}$$

in which the renormalized phonon frequency (2) approaches zero, the transverse correlations are strong enough ($\xi_{o\perp}(T_P^o) > b$), the critical fluctuations preceeding the phase transition are three-dimensional, although anisotropic. E.g. if the transverse fluctuations are due to the electron hopping (eq.6c), this condition is realized if $T_P^o \lesssim t_\perp$. It is then sufficient to undertake the standard mean-field procedure in the whole temperature range. The critical temperature depends on the degree of nesting, i.e. the larger is t_b', the lower is T_P .[30,32] This dependence is shown in Fig.(4). Similarly, the critical temperature is of the order of T_P^o if one of other two mechanisms from eq.(6) is stronger than the nesting

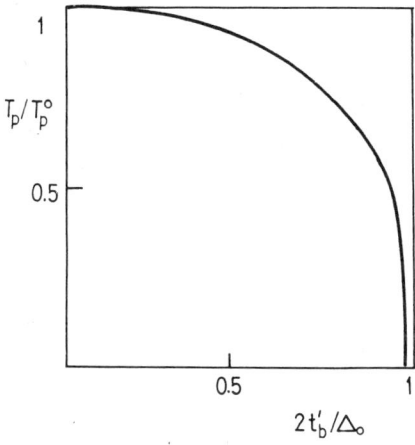

Fig. 4. Mean-field critical temperature T_P vs imperfect nesting parameter t_b' (Eq.(4)).

mechanism. As is seen from Fig.(4), $T_P = T_P^o$ for $t_b' = 0$. Then the law of corresponding states is fulfilled, i.e. $2\Delta_o / T_P \simeq 3.5$, where Δ_o is the Peierls gap in the electron spectrum at T=0. This gap does not change even for finite value of t_b' up to $t_{b,cr}' \simeq \Delta_o$, while for $t_b' > t_o'$ the conducting state remains stable down to T = 0. Thus for finite t_b' the ratio Δ_o / T_P is larger than 3.5 and increases together with t_b' .

In the opposite limit of weak transverse correlations, $\xi_{o\perp}(T_P^o) < b$, the fluctuations at T^o are still one-dimensional. The RPA approach, eqs. (2,3), and even its extension to the parquet summation, are then inadequate, since they lead to the instability at a finite temperature[9] (T_P^o in eq.(8)), in contradistinction with the general property of one- -dimensional thermodynamic systems with short-ranged interactions.[33] Instead of going diagrammatically beyond these mean-field schemes, it is more convenient to pass from the original electron-phonon Hamiltonian to the Landau expansion of the corresponding free energy. The next term in this expansion is of the fourth-order in deformation $u(\vec{k})$, Fig.3c, and reads as

$$F^{(4)} \simeq \frac{0.1\,\lambda\,C}{2\,(T_P^o)^2\,n_F} \int d^3k\,|u(\vec{k})|^4 , \tag{9}$$

where n_F is the density of states at the Fermi level.

Eqs.(3) and (9) complete the Landau model, since the further terms are usually marginal for the treatment of critical fluctuation. The same expansion applies also to the already mentioned anisotropic three-dimensional regime $\xi_{0\perp} > b$. The critical fluctuation in the latter regime are present only in the rather narrow temperature region close to T_P^0. Due to the anisotropy this regime is somewhat widened with respect to that of standard three-dimensional Landau model.[24] On contrary, for $\xi_{0\perp} < b$ the critical range persists from T_P^0 down to low temperatures, as is known from exact treatments of the one-dimensional Landau model.[34-36] Well below T_P^0 the fluctuations include predominantly the phase of the order parameter,[36] while the amplitude is well fixed at its mean-field value at $T = 0$, Δ_0. The long-range order of phase, i.e. the complete three-dimensional order, is established below the critical temperature[37]

$$T_{3d} \simeq n_F (T_P^0)^2 \xi_{0\perp} / b \qquad (10)$$

as follows from the mean-field[37] and scaling[36] treatments of transverse correlations in eqs. (3) and (9). In particular, for hopping correlations (6c)

$$T_{3d} \simeq t_b \exp(-1/2\lambda) . \qquad (11)$$

This result does not take into account the nesting details parametrized by e.g. t_b' in eq.(4). Note that the cut-off t_b appears in eq.(11) instead of eq.(8). Thus, the law of corresponding state is replaced by $\Delta_0/T_P \simeq E_F/t_b >>$ $>> 3.5$. It is obvious from the above considerations that the deviations from the law corresponding states in two opposite limits, Fig.4 and eqs. (10) and (11), have different physical origins.

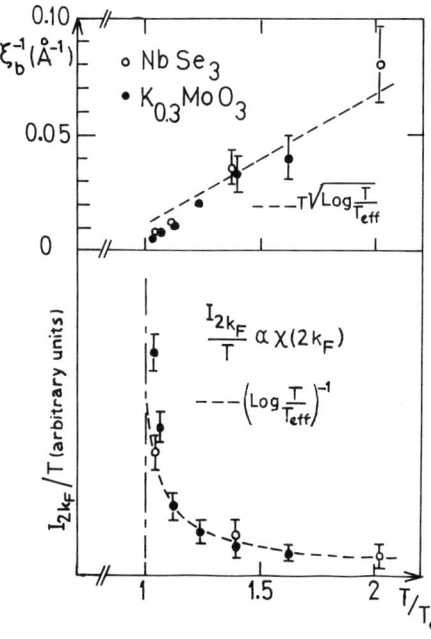

Fig. 5. Temperature dependence of the width and the height of the diffusive $2k_F$-line above the critical temperature for $NbSe_3$ and $K_{0.3}MoO_3$ (after Ref.38).

The materials like NbSe$_3$ and K$_{0.3}$MoO$_3$ are likely to be anisotropic three-dimensional electron-phonon systems. The most direct evidence for this comes from the X-ray data on diffusive $2k_F$-lines above the critical temperature, shown in Fig.(5)[8]. The intensity of this scattering $I(k,T)$ is proportional to $T/\omega^2(k)$. The temperature dependence of the height and the width of $2k_F$-line is in a good agreement with the RPA result (2). The width is expressed in terms of the temperature dependent coherence length at $T > T_p$

$$\omega^2(k) = \omega^2(2k_F)\left[1 + \xi_\parallel^2(k - 2k_F)^2\right] \tag{12}$$

The RPA expression for $\xi_\parallel$ is

$$\xi_\parallel = \xi_0 / \log(T/T_P^0) \tag{13}$$

as follows from eq.(3). The fits in Fig. (5) suggest that the critical temperature T_p is slightly above T_p^0 (T_c and T_{eff} respectively in Fig.5), i.e. $T_p^0 \simeq 0.93\, T_p$[38]. The three-dimensional ordering at a temperature higher than T_p^0 is triggered by the transverse adjustement of the CDW phase, as can be easily concluded from the mean-field treatment of the interchain coupling with $\xi_{0\perp} > b$.

Ordered state

The properties of the CDW ordering below the phase transition can be inferred to a great extent from the already formulated Landau model, defined by eqs. (3) and (9). Here we make a complementary step by considering the eigenstates of electron-phonon Hamiltonian. This will give an additional insight into the Peierls-Fröhlich ground state, in particular into its symmetry and quantum properties. In the adiabatic limit, analogous to the RPA used previously, this procedure leads again to the free energy which reduces to the Landau functional at temperatures that are not far below T_p (i.e. for $T_p - T < T_p$), and even more gives qualitatively the same description in terms of collective phase and amplitude fluctuations in the temperature range below T_p.

Before formulating the Hamiltonian, it is convenient to introduce the electron field operators for the left ($-k_F$) and right ($+k_F$) Fermi surface from Fig.1,

$$\psi(\vec{r}) = \exp(i\vec{Q}\vec{r})\,\psi_+(\vec{r}) + \exp(-i\vec{Q}\vec{r})\,\psi_-(\vec{r}), \tag{14}$$

and to pass to Fourier components

$$\psi_\alpha(\vec{r}) = \int d^3k\, \exp(i\vec{k}\vec{r})\,\psi_\alpha(\vec{k}), \qquad \alpha = \pm \tag{15}$$

The Hamiltonian reads as

$$H = 2\left[\frac{a}{2\pi}\right]\int dk \sum_\alpha \left[\alpha \hbar v_F k + E_F/2\right]\psi_\alpha^+(k)\,\psi_\alpha(k) +$$

$$+ \left[\frac{a}{2\pi}\right]\int dp\,\hbar\omega_0\left\{b_{Q+p}^+ b_{Q+p} + b_{-Q-p}^+ b_{-Q-p} + 1\right\} +$$

$$+ 2(\sqrt{2}\,g)\left[\frac{\hbar}{2\rho\omega_0}\right]^{1/2}\left(\frac{a}{2\pi}\right)^2 \int dp \int dk \qquad (16)$$

$$\left[\left[b_{Q+p} + b^+_{-Q-p}\right]\psi^+_+(k)\,\psi_-(k-p) + \left[b^+_{Q+p} + b_{-Q-p}\right]\psi^+_-(k-p)\,\psi_+(k)\right],$$

where $g^2 \equiv \lambda E_F \omega_0^2 M/n_0$, and $n_0 = 2ak_F/\pi$ is the number of electrons per unit cell. The longitudinal electron dispersion is again linearized around the Fermi level, and the transverse dispersion is not explicitly written for simplicity. The latter can be easily included into the discussion which follows. The phonon operators b_Q, b^+_Q are in the standard way connected with the Fourier components of lattice displacement u and momentum p;

$$u_{Q+p} = (\hbar/2M\omega_0)^{1/2}\left[b_{Q+p} + b^+_{-Q-p}\right], \qquad (17a)$$

$$p_{Q+p} = i(\hbar M\omega_0/2)^{1/2}\left[b^+_{-Q-p} - b_{Q+p}\right], \qquad (17b)$$

and $u_{-Q-p} = u^+_{Q+p}$, $p_{-Q-p} = p^+_{Q+p}$.

In the classical approximation which is legitimate in the limit $M \to \infty$, the lattice kinetic energy in eq.(16) commutes with the rest of the Hamiltonian depending only on the displacements (17a). The self-consistent diagonalization of the Hamiltonian (see below) then leads to the Peierls' result. Together with the opening of the gap $\Delta_0 = |\Delta|$ at the Fermi level, the lattice acquires the sinusoidal deformation determined by

$$u_{Q+p} = (\Delta/g)\cdot\delta_{p,0} \qquad (18)$$

Taking into account the finiteness of the ionic mass M, i.e. the quantum aspect of the lattice, it appears appropriate to represent the phonon modes Q and -Q through coherent states.[39] Let us therefore introduce the variational trial state

$$|G_{ad}\rangle = |el\rangle|ph\rangle, \qquad (19)$$

with

$$|ph\rangle \equiv |ph;z_Q,z_{-Q}\rangle \equiv \exp(-|z_Q|^2 - |z_{-Q}|^2)\cdot\exp(z_Q b^+_Q + z_{-Q} b^+_{-Q})|0\rangle_{ph}, \qquad (20)$$

where $|0\rangle_{ph}$ is the phonon vacuum. The aim is to calculate the ground state energy as

$$E_g = \min \frac{\langle G_{ad}|H|G_{ad}\rangle}{\langle G_{ad}|G_{ad}\rangle} \qquad (21)$$

where the minimization will lead to the self-consistent values of parameters z_Q and z_{-Q}. The first step is to perform the averaging $\langle ph|H|ph\rangle$. One gets the effective electron Hamiltonian

$$\langle ph|H|ph\rangle = \frac{a}{\pi}\int dk \left\{ \sum_\alpha \left[\alpha\hbar vk + E_F/2\right] \psi_\alpha^+(k)\psi_\alpha(k) + \right.$$

$$+2g\left[\frac{\hbar}{M\omega_0}\right]^{1/2}\left[(z_Q + z_{-Q}^*)\psi_+^+(k)\psi_-(k) + (z_Q^* + z_{-Q})\psi_-^+(k)\psi_+(k)\right]\right\} +$$

$$+ \hbar\omega_0\left[|z_Q|^2 + |z_{-Q}|^2 + 1\right] . \qquad (22)$$

The electronic part of the problem is the same as in the pure classical approximation, and can be diagonalized by e.g. the transformation of the Bogoliubov type

$$a_-(k;\Delta) = \alpha\psi_-(k) - \beta\exp(-i\phi)\psi_+(k)$$
$$b_-(k;\Delta) = \beta\psi_-(k) - \alpha\exp(-i\phi)\psi_+(k) \qquad (23)$$

where

$$\alpha^2 = \frac{1}{2}\left\{1 + \frac{\hbar vk}{\sqrt{(\hbar vk)^2 + |\Delta|^2}}\right\} ,$$

$$\beta^2 = \frac{1}{2}\left\{1 - \frac{\hbar vk}{\sqrt{(\hbar vk)^2 + |\Delta|^2}}\right\} \qquad (24)$$

and

$$\Delta \equiv |\Delta|\exp(i\phi) \equiv \sqrt{2}\, g\, u = g\left[\frac{\hbar}{M\omega_0}\right]^{1/2}(z_Q + z_{-Q}^*) . \qquad (25)$$

The effective Hamiltonian (22) reduces to

$$\langle ph|H|ph\rangle = \frac{a}{\pi}\int dk \left[-\sqrt{(\hbar vk)^2 + |\Delta|^2}\,(a_-^+a_- - a_+^+a_+) + \right.$$

$$\left. + \frac{E_F}{2}(a_-^+a_- - a_+^+a_+)\right] + \hbar\omega_0\left[\frac{M\omega_0}{2\hbar g^2}|\Delta|^2 + |z_Q - z_{-Q}^*|^2 + 1\right] , \qquad (26)$$

where $a_\pm \equiv a_\pm(k;\Delta)$. The electron part of the trial function (19) is the Fermi sea with the semiconducting gap $|\Delta|$,

$$|el(\Delta)\rangle = \prod_{k_{occ}} a_-^+(k;\Delta)|0\rangle_{el} , \qquad (27)$$

335

where $|0\rangle_{el}$ is the electron vacuum. The corresponding energy (21) is given by

$$E_g = \min\left\{-\frac{a}{\pi}\int_{-k_F}^{k_F} dk \left[\sqrt{(\hbar vk)^2 + |\Delta|^2} - E_F/2\right] + \hbar\omega_0 \left[\frac{M\omega_0 |\Delta|^2}{2\hbar g^2} + \frac{1}{2}|z_Q - z^*_{-Q}|^2 + 1\right]\right\}. \quad (28)$$

The minimization with respect to $z_Q - z^*_{-Q}$ gives

$$z_Q = z^*_{-Q} = \frac{1}{2g}\sqrt{\frac{M\omega_0}{\hbar}} |\Delta| \exp(i\phi) \quad (29)$$

and one obtains the classical ground state energy. The remaining minimization with respect to Δ gives the already mentioned result for the Peierls gap

$$|\Delta| \equiv \Delta_0 \simeq 2E_F \exp(-1/\lambda). \quad (30)$$

The phase ϕ is an arbitrary constant. The condensation energy is

$$E_g = -\frac{n_0 |\Delta|^2}{4 E_F}, \quad (31)$$

as follows from eq. (28)

The arbitrariness of phase Φ leads to the degeneracy in the ground state, introduced variationally (and therefore still approximately) by the assumption (19). The degenerate set of states can be conventionally written, after eqs. (27) and (29), as

$$|G_{ad};\phi\rangle = \exp(-|z_Q|^2)\exp\left[|z_Q|(e^{i\phi} b^+_Q + e^{-i\phi} b^+_{-Q})\right]|0\rangle_{ph} \cdot |el;\phi\rangle, \quad (32)$$

with

$$|z_Q| = \frac{1}{2g}\sqrt{\frac{M\omega_0}{\hbar}} \Delta_0,$$

and

$$|el;\phi\rangle = \prod_{k<0}\left[\beta(k) - e^{i\phi}\alpha(k)\psi^+_-(k)\psi_+(k)\right]\left[\beta(k) - e^{-i\phi}\alpha(k)\psi^+_+(-k)\psi_-(-k)\right] \cdot$$

$$\left[\prod_{k<0}\psi^+_+(k)\psi^+_-(-k)\right]|0\rangle_{el}. \quad (33)$$

The electron part of the wave function is written in the form similar to the ground state in the BCS theory of superconductivity. The first and second [...]-brackets in eq. (33) represent the electron-hole pairs with

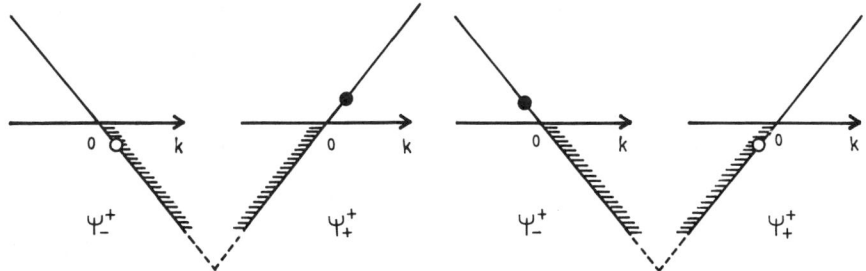

Fig. 6. Electron-hole pair with the wave number $2k_F$ (left) and $-2k_F$ (right). The dispersion curve is linearized around the Fermi level, represented by the horizontal lines. See also eqs. (14), (16), (36) and (37).

the wave vectors $2k_F$ and $-2k_F$ respectively (Fig.6). These pairs have a common, k-independent phase ϕ, and are therefore coherent. This is a property analogous to the coherence of Cooper pairs in the BCS-state. The phase has a meaning of the macroscopic quantum variable for the Peierls-Fröhlich state (32). Putting in another way, the phase is a good quantum number, so that the states (32) with different values of ϕ are orthogonal,

$$\langle G_{ad};\phi' | G_{ad};\phi \rangle = \delta(\phi' - \phi) . \tag{34}$$

This is the consequence of the orthogonality of the electronic parts (33). Note that the corresponding phonon coherent states themselves are not orthogonal, i.e.

$$\langle ph;\phi' | ph;\phi \rangle = \exp\left[-4|z_Q|^2 \sin^2(\phi' - \phi) \right] . \tag{35}$$

The relation (35) contains two quantities which describe the Peierls order, namely the gap parameter Δ, and the lattice displacement u. Indeed, it is easy to verify that the latter quantity is the mean value of the displacement operator (17a) in the state (32). The third useful quantity which characterizes the electron CDW itself is its amplitude, defined as the mean value of the $(\pm 2k_F)$-components of electron density,

$$\rho_{\pm} = \langle \sum_k \psi_{\mp}^+(k)\psi_{\pm}(k) \rangle = -\frac{n_0 \Delta_0}{2\lambda E_F} \exp(\pm i\varphi) . \tag{36}$$

The quantity closely related to $\rho_{\pm}$ is the number of electron-hole pairs, defined as

$$N_{\pm} = \langle \hat{N}_{\pm} \rangle \equiv \langle \sum_{k>0} \psi_{\pm}^+(k)\psi_{\pm}(k)\left[1 - \psi_{\mp}^+(k)\psi_{\mp}(k)\right] \rangle =$$

$$= \langle \sum_{k>0} \left[\psi_{\pm}^+(k)\psi_{\pm}(k)\right] \psi_{\mp}^+(k)\psi_{\mp}(k) \rangle = \frac{n_0 \Delta_0}{4 E_F} . \tag{37}$$

The Peierls-Fröhlich state can be interpreted as the condensate with the macroscopic occupation of these pairs.

It is worthwhile at this stage to point out the differences of this state with respect to the condensate of Cooper pairs. The Peierls-Fröhlich condensate involves electron-hole pairs excited above the conducting Fermi sea, represented by the last bracket in the eq.(33), while the superconducting condensate involves all electron-electron pairs above the "empty" vacuum. Consequently, the state (33) conserves a total number of particles (band electrons), in contrast to the BCS-state, in which this number fluctuates, and being twice the number of Cooper pairs, represents the quantity conjugated to the phase ϕ. The macroscopic numbers which scale with the volume V and fluctuate are just N_+ and N_-. Note that N_+ and N_- in eq.(37) are defined as numbers per unit cell). The corresponding mean variances scale as the square root of volume,

$$\langle (\hat{N}_\pm - N_\pm)^2 \rangle^{1/2} = \left(\frac{\pi \Delta_0 n_0}{8 E_F} \right)^{1/2}, \tag{38}$$

i.e. the relative mean fluctuations vanish as $V^{-1/2}$. Thus, some linear combination of $\hat{N}_+$ and $\hat{N}_-$ might have the role of variable conjugated to the phase ϕ.

The further, perhaps far more important difference with respect to the superconducting state, is the phonon coherence, expressed through the static displacement u, eq.(25), and through the corresponding macroscopic occupation of $\pm Q$ phonon modes. The mean number of phonons per unit cell is given by

$$n_{\pm Q} \equiv \langle \hat{n}_{\pm Q} \rangle = \langle b^+_{\pm Q} b_{\pm Q} \rangle = |z_Q|^2 = M \omega_0 |u|^2 / 2\hbar. \tag{39}$$

The total mean number again scales with the volume, now through the mass M. The corresponding mean variances are

$$\langle (\hat{n}_{\pm Q} - n_{\pm Q})^2 \rangle^{1/2} = \sqrt{2 n_{\pm Q}}. \tag{40}$$

Thus, some linear combination of $\hat{n}_Q$ and $\hat{n}_{-Q}$ might be a proper choice for a phonon variable conjugated to the macroscopic phase ϕ.

The state (32) is evidently the adiabatic one, since the phase and the amplitude of the lattice deformation, introduced as the variational parameters in eqs.(19) and (20), entered also in the many-body electron wave function (33). Thus, electrons acquire their electron-hole coherence by following adiabatically the lattice. This is the consequence of the variational scheme (21). It is approximate since $|G_{ad}\rangle$ is not an eigenstate of the Hamiltonian (16), even after the mean-field step by which only the phonon modes with wave vectors $\pm Q$ are retained.

The less restrictive variational procedure would start from the set of states $|z_Q, z_{-Q}; el(\Delta_e)\rangle$ in which generally $z_Q \neq z^*_{-Q}$ and, in addition. the CDW parameter Δ_e is free. The Hamiltonian projects a given state of this type to other states with the same value of Δ_e. In other words, the off--diagonal matrix elements

$$\langle z'_Q, z'_{-Q}; el(\Delta_e) | H | z_Q, z_{-Q}; el(\Delta_e) \rangle \tag{41}$$

with $z'_Q \neq z_Q$ and $z'_{-Q} \neq z_{-Q}$, are finite. All such elements were neglected in the procedure (21). The more rigorous procedure involves the diagonalization of the problem (41). Let us mention that the matrix elements (41) comprise two types of effects. The linear combinations of the states for which the relation (25) between the electron and lattice order is not fulfilled, allow for the inclusion of electron nonadiabaticity. On the other hand the combinations of states with $z_Q \neq z^*_{-Q}$ take into account, even if the adiabatic condition (25) holds, the internal quantum degrees of freedom of lattice, i.e. of two coupled phonon modes $\pm Q$ retained in the eq.(41). The latter effect is not present for a dimerized CDW for which $Q = G/2$ and only one mode remains.

The effects of quantum lattice fluctuations on the dimerized[40,41] and other[42] Peierls ground states at T=0 were considered numerically[40] and analytically.[41,42] Some of results are summarized in Fig.7. With the increase of bare phonon frequency ω_0 the Peierls displacement (i.e. the amplitude of ion staggering) gradually decreases towards zero. The equivalent result for the critical temperature T_P was obtained in the renormalization group analysis of the path integral approach to the electron-phonon problem.[42] The Monte Carlo simulations[40] even suggest that the dimerization entirely disappears above some critical value of ω_0/E_F of the order of unity, and for strong enough electron-phonon coupling, $\lambda \gtrsim 1$. The strong reduction of Peierls order at $\omega_0/\Delta_0 \gtrsim 1$ (which due to large value of λ corresponds to $\omega_0/E_F \gtrsim 1$ in Fig.7) is caused by the tendency of electrons to form polaronic states which now can be partially and "antiadiabatically" followed by the lattice. It was argued[43] that in this regime the superconducting state might become stable. As was already mentioned, the similar conclusion follows from the analysis of parquet sum at finite temperatures.

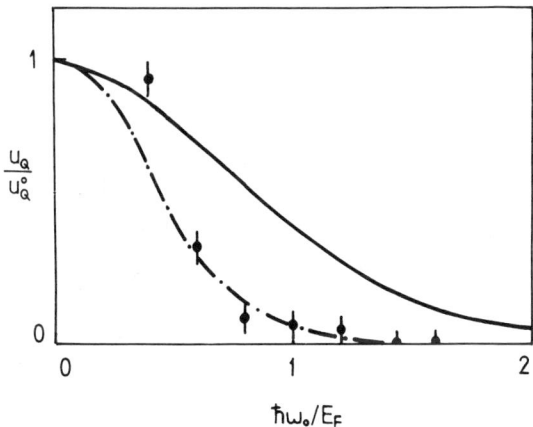

Fig. 7. Few theoretical predictions on the dependence of Peierls displacement as a function of bare phonon frequency. Dots, full-line and dashed-dotted line are results of Refs. 40, 41 and 42 respectively.

Polaronic, bipolaric or, more commonly, localized charge states are not suppressed even in the limit of very strong electron-phonon coupling, as could be expected from the electron-phonon model (10) with the lattice taken as a continuum. In this regime the discreteness of crystal becomes important. The analysis of discrete electron-phonon Holstein model led to the conclusion that for $\lambda/n_0 \gtrsim 20d$, where d is the dimension of the system the CDW state is unstable with respect to the states of short-range ordered charge defects.[43]

Having in mind the results mentioned above, one comes to the conclusion that the standard Peierls-Fröhlich CDW order occurs in the systems in which the Debye frequency is small enough and the electron-phonon coupling is not extremely strong. It is reasonable to situate blue bronzes, trichalcogenides and tetrachalcogenides in this physical regime.

COHERENCE, COLLECTIVE EXCITATIONS AND PINNING

The phase degeneracy of the state (32) is the property of ideal electron-phonon system (16), which possesses the the continuum translational symmetry, and is free of any perturbation which spoils this symmetry. The immediate consequence of this degeneracy is the superfluid rigidity of the Fröhlich condensate. By putting the CDW in a stationary motion with a constant velocity

$$\phi = \dot{\phi} t , \quad v_{CDW\parallel} = \dot{\phi}/Q , \qquad (42)$$

one induces a current

$$j_{CDW} = e n_0 v_{CDW} \qquad (43)$$

at T=0. Thus, the number of collective carriers in the Fröhlich superfluid is equal to the number of band electrons. As is discussed in detail elsewhere,[44] this number varies with temperature, i.e. decreases towards zero as $T \to T_p$. Furthermore the homogeneous compression (or dilatation) of CDW,

$$\phi = \phi' x \qquad (44)$$

induces the variation of band filling and the change of the electron charge density

$$\rho_{CDW} = n_0 \phi'/Q . \qquad (45)$$

The relations (43) and (45) are also valid for local spatial and temporal variations of phase, providing that the latter are slow enough on the respective scales $\xi_\parallel$ and $\hbar/\Delta$. The corresponding equation of continuity for the charge in the condensate is the simple consequences of the equality $\partial^2 \phi/\partial x \partial t = \partial^2 \phi/\partial t \partial x$.

The motion (42) and the deformation (44) of CDW, both increase the free energy of the system. The former gives a finite kinetic term in which the lattice contribution is dominant due to $M \gg m_{el}$. It is given by

$$F_{kin} \simeq \frac{n_F}{\lambda \omega_0^2} \int |\Delta|^2 (\dot{\phi})^2 d^3r . \qquad (46)$$

The phase deformation (44), as well as more general deformations comprising transverse gradients, lead to the finite elastic contribution which coincides with the gradient terms in the Landau expansion (3), in which $|\Delta(T)|$ is fixed and temperature is extended below T_p, down to T = 0. Eqs. (3) and (46) then reduce to the standard problem of the elastic medium with the acoustic oscillations, so called phasons. Their dispersion is[14]

$$\omega_{ph}^2(\vec{q}) \simeq \frac{m_{el}}{m^*} v_F^2 \left[q^2 + \left(\frac{\xi_\perp}{\xi_\parallel} \right)^2 q_\perp^2 \right] . \tag{47}$$

The Peierls transition breaks the translational symmetry, and $\omega_{ph}(\vec{q})$ is the corresponding Goldstone mode. Being the combination of lattice inertia and CDW elasticity, the dispersion (47) is determined by the effective CDW mass of mixed origin

$$m^* \simeq m_{el} \frac{4 |\Delta|^2}{\lambda \omega_0^2} . \tag{48}$$

The values of m^* in real systems is quite large ($\sim 10^2 - 10^3 m_{el}$), so that the phason velocity is at least one order of magnitude smaller than the Fermi velocity.

The Goldstone mode in the BCS superconductor is the electronic plasmon. Its frequency is raised to a finite value due to the long range Coulomb interaction between electrons. To see how the electron-electron interaction affects the phason mode, let us recall that the phase deformations (44) violate the local charge neutrality. The contribution to the free energy coming from the induced Coulomb forces is given by[45]

$$F_{Coul} = \int d^3r \int d^3r' V_{scr}(\vec{r} - \vec{r}') \left(\frac{n_0}{Q_\parallel} \right)^2 \frac{d\phi(\vec{r})}{dx} \frac{d\phi(\vec{r}')}{dx'} , \tag{49}$$

where $V_{scr}(\vec{r}-\vec{r}')$ is the electron-electron electrostatic potential screened by other (non-CDW) charges in the system. The phason dispersion (47) is replaced by[45,46]

$$\omega_{ph}^2(\vec{q}) = v_{ph\parallel}^2 q_\parallel^2 + v_{ph\perp}^2 q_\perp^2 + \frac{v_{ph}^2}{E_F} \frac{4\pi e^2 n_0 q_\parallel^2}{\varepsilon_{\parallel\infty} q_\parallel^2 + \varepsilon_{\perp\infty} q_\perp^2} , \tag{50}$$

where $\varepsilon_{\parallel\infty}$ and $\varepsilon_{\perp\infty}$ are respectively longitudinal and transverse dielectric constants in the high frequency limit $\omega \gg \omega_{ph}$. If the system is as a whole a dielectric, i.e. if there are no free carriers which remain noncondensed, the dielectric constant has a standard semiconducting form

$$\varepsilon_\infty = 1 + \frac{4\pi n_0 e^2}{m_{el} |\Delta|^2} . \tag{51}$$

The phason frequency is then raised[14] to $\sim \lambda^{1/2} \omega_0$ in the long wavelength limit $q_\perp = 0$, $q_\parallel \to 0$.

If the system contains normal conducting electrons which do not participate in the CDW order, the dielectric constant ε_∞ has a large metallic value, so that the last term in the eq. (50) is negligible, and phasons remain acoustic. In other words, conducting electrons screen almost perfectly CDW charge fluctuations, as long as the latter are not too fast.

The expression (49) indicates that the phason response to the dc field ($q_\parallel = 0$) is gapless even in the dielectric regime (51). The Coulomb interaction (49) does not prevent the perfect Fröhlich superfluidity, no matter how weak the external field is. Perturbations which do not have the continuous translational invariance of the Hamiltonian (16) however act just in this direction. E.g. the interaction of the CDW with impurities gives the phase dependent contribution to the free energy[15,16,47]

$$F_{imp} = \sum_i \int d^3r\, V(\vec{r} - \vec{r}_i) \frac{en_0}{E_F} |\Delta| \cos\left[\vec{Q}\vec{r} + \varphi(\vec{r}_i)\right], \qquad (52)$$

where $V(\vec{r}-\vec{r}_i)$ is the potential of the impurity at the position $\vec{r}_i$. If this potential is weak enough, the CDW does not accomodate its phase locally at each impurity site, but instead deforms in an averaged way on a scale larger than the mean distance between neighboring impurities. This scale is characterized by the Lee-Rice length[47] which represents the mean dimension of domain inside which the CDW phase is still coherent. In the three-dimensional system it is given by

$$\ell_{LR} \simeq \frac{v_F^2 k_F^2}{b^2 V^2 n_i}, \qquad (53)$$

where n_i is the density of impurities. Due to the lack of coherence the phason oscillations with wavelengths larger than ℓ_{LR} cannot propagate as acoustic waves. Instead, the phason dispersion has a gap of the order

$$v_{ph}\ell_{LR}^{-1} \simeq \omega_{ph}(q \simeq 0) \qquad (54)$$

in the long wavelength limit. Correspondingly, for the external electric field smaller than the threshold field[15,47]

$$E_T \simeq \frac{Q_\parallel E_F}{n_0 e \ell_{LR}^2} \sim n_i^2 \qquad (55)$$

the CDW does not move coherently, but remains pinned to impurities. The result (55) follows from the simple balance of elastic energy (3), the pinning energy (52) and the coupling to the electric field

$$F_{field} = Q_\parallel^{-1} \int d^3r\, e n_0 \varphi E. \qquad (56)$$

TOPOLOGICAL DEFECTS

The above procedure neglects the variations of CDW amplitude $|\Delta(x)|$. This is consistent since ℓ_{LR} is much larger than the correlation lengths $\xi_\parallel$, $\xi_\perp$ which define the characteristic scale on which amplitude variations become more favorable than phase variations. The former are present even at low temperatures if there are strong local constraints, including strong impurities in eq.(52). Without such constraints the amplitude fluctuations are only thermodynamically activated. This mechanism becomes important at temperatures close to T_p, i.e. for $|\Delta(T)| < T_p$. The following discussion will be limited to low temperatures, with the particular emphasis on configurations in which $\Delta(x)$ locally vanishes. The static ($E < E_T$) and dynamic ($E > E_T$) regimes will be distinguished.

Topological defects in the static CDW

Let us start from the hypothetical one-dimensional ordered system with the phase difference π imposed along the segment L. If $L < \xi_\parallel$ the CDW forms an amplitude kink (Fig.8a). The well known example of this type is realized in Peierls systems with $Q_\parallel = \pi/a$ in which the strong constraint comes from the lattice discreteness.[48,49] It forces the CDW to have nodes on the ionic sites. The CDW "phase" can have only two values, as is shown in Fig.8b. The order parameter is one-componenet, i.e. it is completely determined by the value of dimerization amplitude. The amplitude kinks are formed in regions which join two opposite bond-antibond alternations (Fig. 8b). Microscopically, these topological objects are local electronic states with the energy level in the middle of Peierls gap, and with the fractional charge e/2 per spin.[48,49] The rest of electronic charge is delocalized in the Bloch states which are only smoothly perturbed on the much larger scale defined by $\xi_\parallel$.

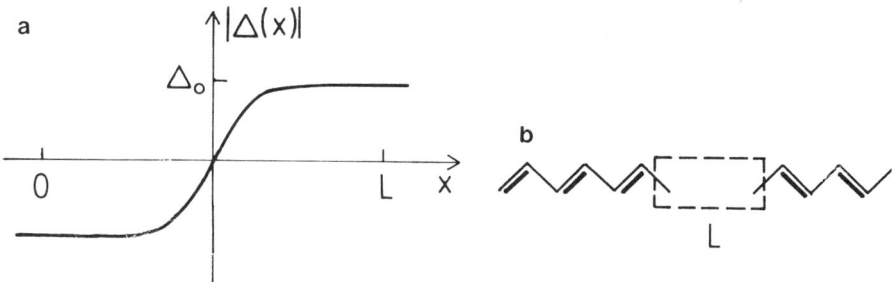

Fig. 8. The π amplitude kink (a), and the kink in the dimerized Peierls system (b).

The similar effects are in principle possible also for higher order commensurabilities of CDW with the lattice. The topological defects are then the phase kinks in which the amplitude does not pass through zero. Still, one has localized electronic states with the charge fractionalization e/μ, where $\mu = 2\pi/Q_\parallel a$. These defects are however hardly realizable in real systems, although in some of them the Peierls wave number $Q_\parallel$ is very close to the commensurate value $\pi/2a$. The reason is that the lattice constraints are rather weak, and introduce the anisotropy of the relative order $(\Delta_0/E_F)^{\mu-2}$ in the phase dependence of the free energy.[50] The corresponding scale L in the Fig.8 is not small enough, and the resulting kinks are too smooth to be distinguished from other (e.g. phason) fluctuations of the order parameter. In other words, the topological aspect of such defects ceases to to be significant, since the energy barriers for their annihilation are not high enough and can be shadowed by e.g. thermodynamic fluctuations.

The interchain interactions also act against the kink stability.[32] Since the CDW should have a finite transverse coherence, the kink on a given chain has to be extended to a domain wall. The kinks are however charged, so that the walls are energetically unfavorable due to the repulsive Coulomb forces.

Another type of topological defects in threedimensional long-range ordered systems with the complex order parameter are vortices.[51] In that sense the CDW condensate is again analogous to other superfluids. The additional particular property of the CDW is the $2k_F$- modulation of electronic density and ionic displacements. Being in a periodic pattern, CDW vortices can be also interpreted as dislocation lines,[47,52,53] in analogy with the dislocation lines in usual plastic media. The "CDW--crystal" in the continuum limit (3) is still somewhat specific. It may be be visualized as the array of equidistant planes which are perpendicular to the chain direction. The "lattice constant" is equal to the Peierls wavelength π/k_F. The Burgers vector for such crystal is always directed along the chains and has the elementary value π/k_F. It designates a change by one longitudinal wave-front along the circulation loop which encircles the dislocation line.

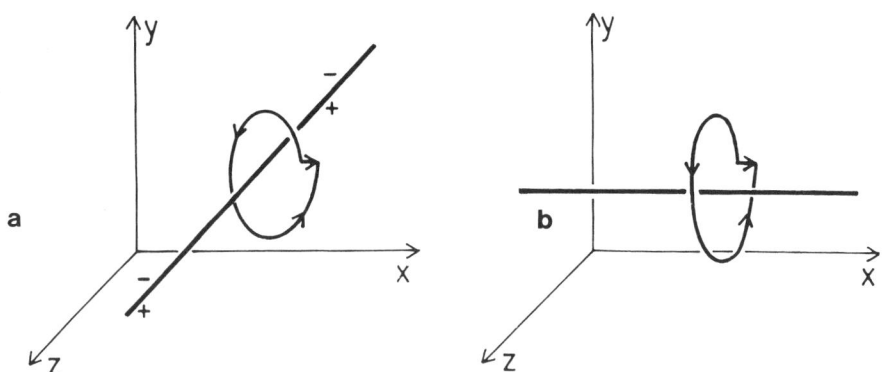

Fig. 9. Edge (a) and screw (b) dislocation lines. Chains and Burgers vector are directed along x-axis.

The properties of CDW dislocation lines on a large scale follow from the phase dependent elastic part of the free energy (3). Thus the equation of motion for dislocation lines is same as that for phasons. The former are the particular solutions of the phason differential equation, not included in the plane wave set. The solution for static edge and screw straight dislocation lines (Fig.9) are respectively

$$\phi_{edge} = \tan^{-1}\left[\frac{\xi_\parallel}{\xi_\perp}\frac{y}{x}\right] \qquad (57a)$$

and

$$\phi_{screw} = \tan^{-1}\left(\frac{z}{y}\right) \qquad (57b)$$

where z is the longitudinal coordinate, and y and z are the transverse coordinates.

Two remarks concerning these solutions are in order. Firstly, the elastic energy (3) and the coordinate dependences in eqs.(57) are highly anisotropic due to $\xi_\parallel \gg \xi_\perp$. The energy per unit length of the edge dislocation line is therefore larger than that of the screw one, the ratio being of the order $\xi_\parallel/\xi_\perp$.[53] Secondly, taking into account the relation (45), one concludes that the screw dislocation line is neutral, while the edge dislocation line is associated to the dipolar charge distribution, as indicated in Fig.9. This reasoning can be extended to dislocation loops which are also particular solutions of the differential equation mentioned above. The pure edge loop is charged in a way shown in Fig.10a, while the combined edge-screw loop from Fig.10b is neutral.

The above properties of CDW dislocation have a direct influence on their stability (i.e. metastability) in the static regime. The dislocation prefers to decrease its energy, and to eliminate as much as possible the Coulomb repulsion induced by charge redistributions. In other words it will tend to have mostly a screw alignment. Thus, provided that there are no external strains, or that these strains are not strong enough, the dislocation line created by some process of injection or by thermal activation, aligns longitudinally (Fig.9b), and climbs towards the nearest boundary due to the interaction with its image. Similarly, the created loop will orient itself in a screw-edge position, and deform into two long screw lines joined by short edge bridges, as shown in Fig.10b. These two lines finally annihilate each other.

There are various possible sources of local strain which may prevent the above self-annihilation of dislocations. For example, the chain

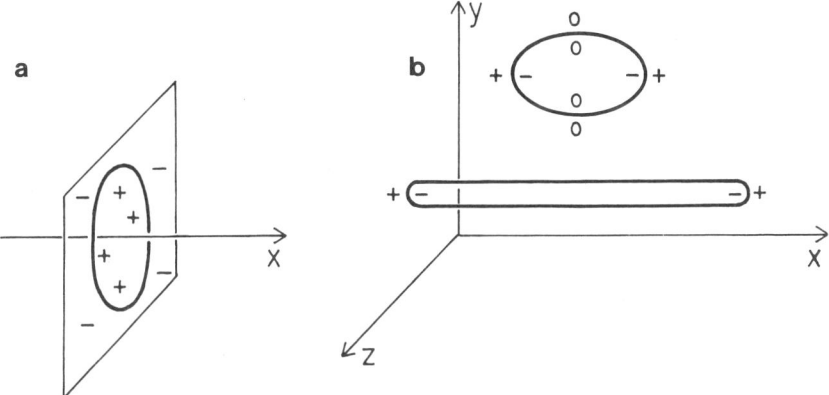

Fig. 10. (a) Edge dislocation loop; (b) Edge-screw dislocation loop before and after the reduction to two long screw dislocation lines.

discreteness acts against the transverse motion of screw dislocation lines.[53] This Peierls-Nabarro mechanism of dislocation pinning is particularly effective when $\xi_\perp$ is small, i.e. of the order of or smaller than the interchain distance. Furthermore, dislocation lines and loops can be trapped by strong point or more extended defects like impurities, crystal defects, contacts etc. E.g. the mechanism of CDW pinning on strong and charged impurities (52) may proceed via the formation of small edge loops which neutralize the impurity charge.[54] Similarly, strong neutral impurities may be encircled by screw-edge loops. Such local configurations are complemetary to the Lee-Rice domains (53) in the limit of weak impurities.

The controlled way to introduce local strains is to apply the electric field, which has the role of external stress as is seen from eq. (56). Let us assume that the phase is fixed at the edges of segment of the length L. The aim is to determine the critical electric field, i.e. the lowest field necessary to create a dislocation loop. The answer follows from the balance of the energy of loop and the gain of energy due to the polarization of CDW charge along the segment (Fig.11). The estimation of the critical electric field for the creation of smallest loop of dimensions $\xi_\perp \times \xi_\perp$ (Fig.10a) or $\xi_\perp \times \xi_\parallel$ (Fig.10b) is[53]

$$E^C_{DL} \simeq (\text{cte})\, \Delta/eL \ . \tag{58}$$

The field E^C_{DL} is not only the critical field for the nucleation of dislocation loop, but also the threshold field for the onset of collective CDW motion. This equivalence is not obvious within present static considerations. It however follows from a more complete approach which takes into account dissipative processes in the CDW condensate.[55] A particular model of this type will be considered in the rest of the paper. The application of this model to the problem in Fig.11 give the numerical result[56]

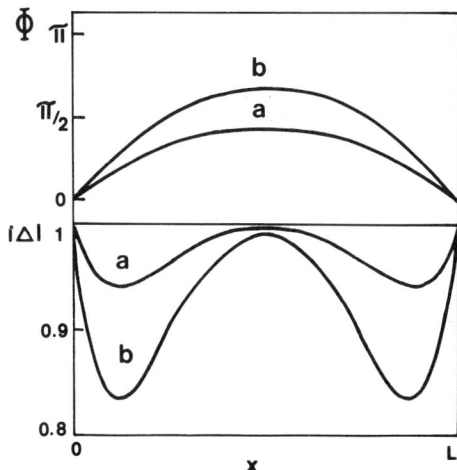

Fig. 11. The deformation of CDW phase and amplitude between two fixed ends at 0 and L, for two values of the electric field [E(a) < E(b)] below the threshold field (59).

$$E^c_{PS} \simeq \frac{\Delta}{e\,\xi_\parallel} \left[\frac{\xi_\parallel}{L}\right]^{1.23} . \tag{59}$$

The results (58) and (59) predict the decrease of E^c with the increase of L, with power laws which are slightly different and experimentaly hardly distinguishable. The segments of Fig.11 are realized in the experiments by e.g. putting electrical contacts at the distance L. The segments have to be short enough, in order to get values of E^c which are larger than the bulk threshold field due to the depinning from e.g. weak impurities (eq.(55)). The numerous measurements[57] can be well interpreted by expressions (58) and (59). Characteristic values of $E^c L$ vary for various materials and experimental setups within the range from 0.1 to few milivolts. It should be mentioned that the prefactors in eqs. (58) and (59) depend on the details in the contact design, and in particular on the possible preexistence of trapped dislocation loops, as mentioned previously.

The preexistence of dislocation loops might be also important for the understanding of the bulk depinning in the presence of strong localized impurities, with threedimensional configurations which are complicated generalization of that in Fig.11. This possibility was considered already in the early work of Lee and Rice[47], but the complete analysis is still lacking. One possible process of depinning in the external field could proceed through the accumulation of the charge in the edge dislocation loops centered at impurities. This accumulation leads to the increase of the dimensions of loops. At some critical dimension, i.e. at some threshold external field, the loops start to blow up spontaneously and are released by the impurities, initiating so the collective CDW transport.

Dynamic topological defects

The phason mode introduced in eqs. (47), (50) and (54) is dispersive, without any damping. This is not the case in real materials. In most of them the phason is overdamped,[19] at least at higher temperatures, e.g. at $T \gtrsim 40$ K in $K_{0.3}MoO_3$.[58] Thus, the classical equation of motion for the phase which follows from eqs.(3), (52) and (56) contains also the viscous term:

$$\frac{d^2\phi}{dt^2} + \frac{1}{\tau}\frac{d\phi}{dt} + \xi_\parallel^2 \frac{d^2\phi}{dx^2} + \xi_\perp^2 \frac{\partial^2\phi}{\partial \vec{r}_\perp^2} - \frac{eE}{m^*} + \frac{\delta F_{imp}}{\delta\phi} = 0 \tag{60}$$

Among various microscopic scattering processes which may contribute to the τ^{-1} term in eq.(60), two mechanisms appear to be particularly important, and are roughly of the same order. These are the phason scattering on thermal phonons[59] with $\tau^{-1} = \tau_{ph}^{-1}$, and the single particle scattering of electrons involved in the electron-hole condensate,[58] with

$$\tau^{-1} = \tau_{el}^{-1} (m_{el}/m^*) .$$

The total τ may be strongly temperature dependent, in particular it may increase exponentially as the temperature decreases.[60]

The static regime $E < E_T$ is overdamped if $\tau^{-1} >> \omega_{ph}(q \to 0) \equiv \omega_{pin}$, where ω_{pin} is estimated in eq.(54). Then the inertial term in eq. (60) is negligible. It is also irrelevant for $E > E_T$. In the weak pinning limit (53) and well above the threshold field, $E \gg E_T$, the last term in eq. (60) can be also neglected. This equation than has a simple stationary solution

$$\phi = E \tau t / m^* , \qquad (61)$$

which corresponds to the previously introduced state of motion (42), and gives the relation of Ohmic type between the collective current j_{CDW} (43) and the external electric field.

The presence of the dissipative parameter τ in eq.(61) indicates that the CDW flow is far from being a superfluid one. In fact, there are experimental[19] and theoretical[61] indications that the conductivity which follows from eqs.(43) and (61) cannot be larger than the Ohmic conductivity of the same system, were it remained in the metallic state. The further important property of the CDW flow is its nonstationarity, present at least locally in real systems. The eq.(61) does not express it explicitly, since it does not include the effects of strong defects like sample boundaries, contacts etc., and of inhomogeneities of the electric field which are again present e.g. in the vicinity of contacts. Furthermore in the limit of strong impurities, the corresponding term in the eq.(60) is not expected to be negligible in the whole range of experimentally available electrical fields, so that the flow is not stationary even in the bulk. In order to show how the nonstationary diffusive flow leads to the production of a new type of dislocation lines, we limit the further discussion to its properties in the vicinity of extended defects (obstacles) or field inhomogeneities. Like in the static regime, they are easier to handle than the strong but localized defects with characteristic dimensions smaller than $\xi_\parallel$ and $\xi_\perp$.

The obstacles cause a local slowing or even a stoppage of collective CDW motion, or, putting in a different way, an accumulation of longitudinal and/or transverse strains which would increase limitlessly. Note that in the case of longitudinal strain, this corresponds, due to the eq. (45), to the accumulation of CDW charge. Obviously there should be some process in which this strain accumulation has to be eliminated, enabling in particular the evacuation of the charge via the quasi-particle (i.e. Ohmic) channel. For example, at current contacts the collective current has to be fully converted into the Ohmic one. Gor'kov[55] proposed that this conversion proceeds through successive changes of number of wave fronts by one, i.e. by a successive release of one electron per chain. These slippages of phase by 2π in turn may be realized only through local and temporary collapses of CDW amplitude, i.e. through the formation of temporary amplitude kinks.

Microscopic models for phase slip (PS) processes were developed[55,62] for semimetallic ("gapless") systems in which the CDW charge converts into the conducting electrons. The formation of PSs is a diffusive process, with the characteristic diffusion time still much shorter than the time interval $\pi v_{CDW}/k_F$ in which one wave front sweeps one Peierls wavelength, π/k_F. Due to this, the charge conversion can indeed proceed by releasing one by one electron per chain.

The PS diffusion was analysed in the limits of "dirty" ($\tau_{el}^{-1} \gg \Delta, T_p$)[55] and "clean" ($\tau_{el}^{-1} \ll \Delta, T_p$) systems, where τ_{el} corresponds to the scattering of electrons on impurities. Both approaches lead to the qualitatively same descriptions of PSs. In the "dirty" limit which will be discussed in more details here, the problem can be reduced to the time-dependent equation of Landau type

$$\dot{\Delta} + iE\Delta - \Delta + |\Delta|^2\Delta - \nabla^2\Delta = 0 \quad , \tag{62}$$

with $\Delta \equiv |\Delta|\exp(i\phi)$. The corresponding expression for the current is given by

$$j \simeq \frac{\sigma T_P}{\xi_\parallel e}\left[E - \eta E |\Delta|^2 - \eta |\Delta|^2 \dot{\phi}\right] \tag{63}$$

The first term in the latter expression is the contribution of normal electrons, the second term is the effective reduction of the ohmic part due to the shallow gap in the electron spectrum, and the third term is the normal CDW contribution. The coefficient in front of this term, $\eta \equiv (T_P \tau_{el})^2$ is smaller than unity, as specified by the model. The length and time scales in expressions (62) and (63) are $\xi_\parallel$ ($\xi_\perp$) and $(T^2 \tau_{el})^{-1}$ respectively. The units for $|\Delta|$ and E are T_P and $T_P/e\xi_\parallel$ respectively. Note that the unit for the electric field is extremely large for usual values of T_P and $\xi_\parallel$ mentioned previously. Thus, the characteristic experimental electric fields are of the order $E \sim 10^{-4}$.

The differential equation (62) does not contain the term coming from presumably weak impurities, which cause the relaxation τ_{el}. Here again, like in the previous discussion of eq.(60), it was assumed that the electric field is much stronger than the bulk threshold field, so that this term is negligible. On the other hand, the strong obstacles are not included into the equation (62), but will be introduced as boundary conditions. This makes the problem (62) completely defined. Note that eq. (62) represents two coupled real differential equations for e.q. the amplitude $|\Delta|$ and the phase ϕ of the order parameter. After neglecting completely the variations of the amplitude from its equilibrium value, the remaining equation for phase coincides with eq. (60) in the limit $E \gg E_T$.

Longitudinal PS diffusion

The first situation to be considered is that of CDW stopped at a transversally extended barrier. More precisely, it is assumed that the dimensions of the barrier are much larger than $\xi_\perp$. Since $\xi_\perp \ll \xi_\parallel$, the PS diffusion is then almost entirely longitudinal, i.e. the problem is reduced to an almost one-dimensional one, so that the transverse variations in eq. (62) can be completely ignored.

The one-dimensional problem (62) is still the nonlinear parabolic equation for the complex variable. In the limit $E \ll 1$ which is of physical interest, the only efficient method is the numerical analysis.[56,63] It gives the solutions which are of limit cycle type, and represent the periodic repetition of PSs in time, with the period equal to $2\pi/E$. As shown in Fig.12, the PS is a rather sharp diffusive pulse, as announced previously. It is preceeded by the slow diffusion of shallow minimum in $|\Delta(x)|$. The characteristic length scale for this diffusion is

$$x_{PS} \simeq \xi_\parallel \left[E/(|\Delta|/\xi_\parallel e)\right]^{-0.28} \gg \xi_\parallel \quad . \tag{64}$$

The oscillating contribution to the current coming from the above solution is, due to the sharpness of pulses, highly multiharmonic. The multiharmonicity is the common property of the most of NBN data.[19,64-68]

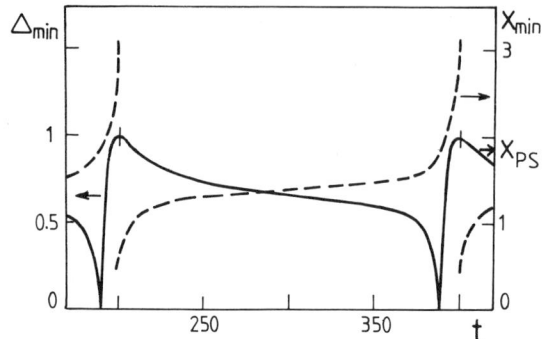

Fig. 12. The value of the minimum in $|\Delta(x)|$ (full line) and its position (dashed line) as the functions of time. The position of the phase slippage is indicated by the arrow on the x_{min}-axis.

Some measurements[66] also indicate the locality of NBN sources, which is the starting assumption of the present model. When extended to external electric fields with a finite ac component with the frequency ω_{ext}, the numerical analysis[63] of eq. (62) shows the existence of resonances at harmonic and subharmonic ratios of $\omega_{ext}/\omega_{CDW}$, which are observed in numerous measurements.[19,65,67,68] The presence of subharmonic resonances is the direct consequence of the multiharmonicity of PS pulses.

The eq. (63) also gives the ratio between the oscillating current density and the NBN frequency ω_{CDW} (equal to E in dimensionless units)

$$j_{CDW}/\omega_{CDW} = \eta \frac{2\pi^2 n_0 e}{k_F} \qquad (65)$$

In contrast to the usual expression,[19] the above result contains the factor $\eta < 1$ on the r.h. side. This is the consequence of the diffusive process responsible for PS generation. Eq. (65) may explain the deviation from the ideal ratio [eq.(65) with $\eta = 1$] observed in some experiments.[19,31,64-68]

The above results were obtained for a semi-infinite system with only one end fixed. Applying the field to the finite segment with both ends fixed, one comes to the problem already specified by eq.(58) and Fig.11. The limit-cycle solution are then possible only at fields larger than the critical field $E^c(L)$ (eq.(58)). Below this field the CDW acquires a static deformed configuration with the polarized phase, Fig.11a, and with minima in the amplitude $|\Delta(x)|$ close to each end, Fig.11b. At $E > E^c(L)$ the PS diffusion from two ends interfere. This interference leads to the interesting effects in the finite external ac field,[63] also observed experimentally.[68] The lock-in close to resonant harmonic and subharmonic values of external frequency then occur in the finite frequency bands, with the accompanying side wings of negative resistance in the current--field characteristics.

Dynamic dislocation lines

The space variation of the order parameter $|\Delta|\exp(i\phi)$ in the PS region is shown in Fig.13. If the phase difference between the right and the left side of PS center is, say, $\Delta\phi = 0$ immediately before the PS, it will be $\Delta\phi = 2\pi$ immediately after the PS. At the very moment of PS the amplitude kink in Fig.13.b is accompanied by the local jump in ϕ equal to π within the short scale of the order of $\xi_\parallel$. Qualitatively, this is just the structure shown in Fig.8. The rest of the phase difference from π to 2π is extended over the much longer distance of the order of x_{PS} (eq.(64)). Taking into account Figs. 12 and 13, the one-dimensional PS can be interpreted as an instantaneous topological defect which contains the amplitude kink. In the same manner the PSs in front of an perfectly flat transverse obstacle with the boundary condition

$$\Delta(x=0, \vec{r}_\perp, t) = \Delta_0 \tag{66}$$

occur simultaneously at all chains. The resulting topological defect is the transverse layer containing an instataneous planar domain wall of simultaneous amplitude kinks.

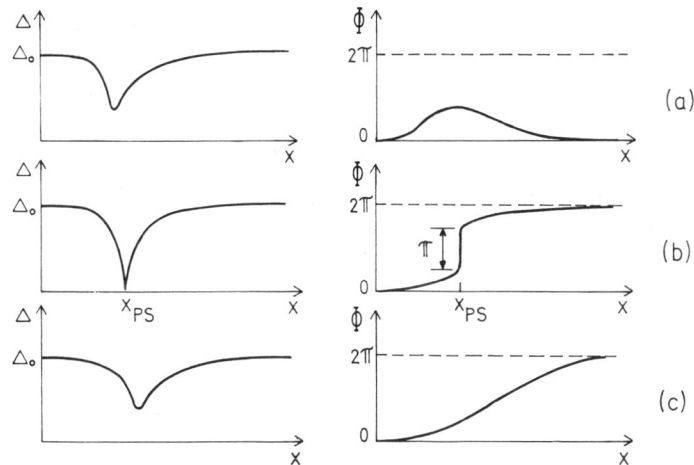

Fig. 13. The schematic x-dependence of CDW amplitude and phase before (a), at (b) and after (c) the moment of the phase slippage.

Let us now suppose that the boundary condition has same transverse dependence

$$\Delta(x=0, \vec{r}_\perp, t) = \Delta_0(\vec{r}_\perp) \tag{67}$$

and ask what kind of topological defects occur due to the PS generation. The problem (62,67) is now three-dimensional, and the numerical analysis becomes too cumbersome. However, in contrast to dynamical properties, the topological aspects of the solution are not sensitive to the time scale. The furhter discussion can be therefore limited to the range of large fields, $E \gg 1$, in which one looses some physical aspects of PS diffusion, e.g. the multiharmonicity. However, in this asymptotic limit the eq.(62) is solvable in an approximate analytic way.[55] Namely, due to the dominance of the term E Δ, the differential equation can be linearized in the vicinity of boundary. The particular solution in this region is static and decreases exponentially as x increases. It can be written in the form

$$\Delta_L(x,\vec{r}_\perp) \equiv R_L(x,\vec{r}_\perp) \exp[i\phi_L(x,\vec{r}_\perp)] ,\qquad (68)$$

with the amplitude $R_L(x,\vec{r}_\perp)$ and the phase $\phi_L(x,\vec{r}_\perp)$ uniquely determined by the boundary condition (67). Far from the boundary the solution of eq.(62) is the stationary flowing CDW, which has to end with the kink at $x = 0$ in order to adjust the requirement that the CDW stops at $x = 0$;

$$\Delta_R(x,t) = \tanh(x/\sqrt{2}) \exp(iEt) .\qquad (69)$$

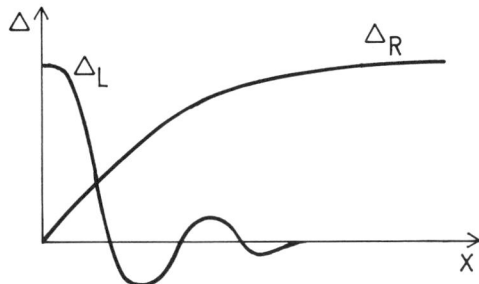

Fig. 14. The matching of two solutions, given by eqs.(68) and (69), of the eq.(62) in the limit $E \gg 1$.

The complete solution is the superposition of these two (Fig.14),

$$\Delta(x,\vec{r}_\perp,t) = \Delta_L(x,\vec{r}_\perp) + \Delta_R(x,t) .\qquad (70)$$

The PS center is defined by the equation

$$\Delta(x,\vec{r}_\perp) = 0 ,\qquad (71)$$

i.e. by

$$\tanh(x_{PS}/\sqrt{2}) = R_L(x_{PS}, \vec{r}_{\perp PS}) \qquad (72a)$$

and

$$E\, t_{PS,n} = (2n+1)\pi + \phi_L(x_{PS}, \vec{r}_{\perp PS}) .\qquad (72b)$$

$\vec{r}_{PS} \equiv (x_{PS}, \vec{r}_{\perp PS})$ and $t_{PS,n}$ are the PS position and moment respectively. The points of simultaneous PSs as a function of time follow from the inverted second equation (72b). One gets the periodic succession of curves $\vec{r}_{PS}(t + 2\pi n/E)$, which move in such a way that at each chain there is one PS per period $2\pi/E$. The first equation (72a) gives the distance of the PS center from the boundary at the given point $\vec{r}_\perp$ of the transverse cross section.

The curves $\vec{r}_{PS}(t)$ have the topological properties of dislocation lines, whose shapes and dynamics depend on the "morphology" of boundary, eq.(68). One general (a) and one particular (b) pattern of dislocation lines at a given moment are shown in Fig.15. The latter follows from the boundary condition

$$\Delta_0(\vec{r}_\perp) = \exp(i\, a\, r_{\perp 1}), \qquad (73)$$

and represents equidistant straight lines which climb in the transverse direction with the velocity $v_\perp = E/a$. Note that in the limit $a \to 0$ (i.e. $v_\perp \to \infty$) the boundary condition (73) reduces to the uninform condition (66), i.e. the solution in Fig.15.b reduces to the previous frontal domain wall of simultaneous PSs in the plane $x = x_{PS}$. The pattern in Fig.15.b has the same topological properties as the pattern of vortex lines invoked phenomenologically in Ref.(51). The latter were based on the model used here in the discussion of static dislocation lines (57). The present solution (72) however does not have the form (57), but is characterized by the previously discussed space and time scales, shown in Figs.12 and 13.

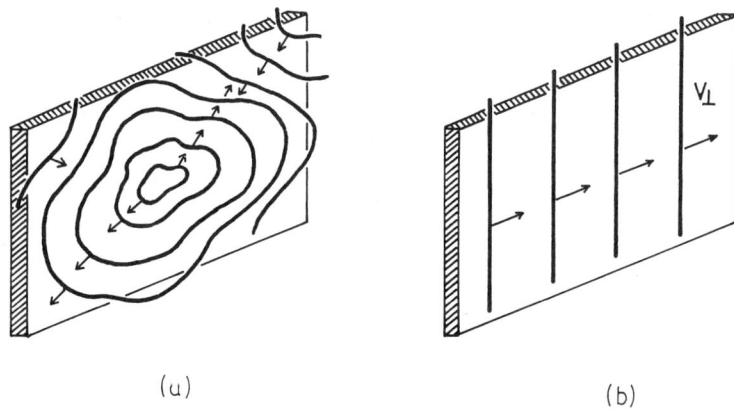

(a) (b)

Fig. 15. Dynamic dislocation lines in front of the transverse barrier for the general and particular boundary conditions, given by eqs. (68) and (73) respectively.

Dynamical dislocation lines and loops from Fig.15 are particular configurations corresponding to the rather restrictive choice of obstacles, defined by eq. (67) and (73). In the real systems one could expect boundary conditions which inforce also stoppages of CDW at lateral (parallel to chains) and more complex barriers. Furthermore, the dynamical dislocations are generated also due to nonuniform electric field $\vec{E}(\vec{r})$ which causes[69] the local accumulation of strain and, consequently, the PS diffusion. The latter possibility is particularly interesting since it may be realized far from any strong "mechanical" barrier, and leads to the slow periodic generation of dislocations which move along the system. This motion carries entropy and may be responsible for the additional heat transport observed at $E > E_T$ in $Rb_{0.3}MoO_3$.[70] Furthermore, finite temperature gradients may lead to similar effects, in particular to the division of sample into domains separated by regions in which PSs and dislocations are perpetually generated.

In the latter two cases of the continuous changes of electrical field and/or temperature one has weak local variations of the parameters in eq.(62), so that the numerical approach is again unavoidable. This analysis is presented elsewhere.[71] The only point to be mentioned here is that characteristic time scales for such phenomena, $|\partial E / \partial \vec{r}| \cdot \xi_\perp$ and $|\delta E / \partial x| \cdot \xi_\parallel$, are much smaller than E. The former scale could correspond to the broad band noise (BBN), usually present in real systems above E_T.[19] This time scale determines the dynamics of dislocation lines in the bulk, i.e. of the nonuniformity of the bulk CDW flow. This flow influences the fast PS generation at rigid barriers through the matching shown in eq.(70) and Fig.14, and ultimately causes the broadening of the coherent current oscillations (i.e. NBN). One comes to the conclusion that the BBN and the finite widths of NBN lines both originate from the nucleation of dynamical dislocations. More precisely, the NBN linewidths come from slow temporal, and not spatial, modulations in the PS diffusion, in accordance with the experimental observations.[72]

ACKNOWLEDGEMENTS

The author acknowledges useful discussions with S. Barišić, C. Bourbonnais and D. Jelčić. The work is partially supported by Federal Yugoslav Program No.P-233 and Yu-US Collaboration Program, No. DOE-JF-738.

REFERENCES

1. R. E. Peierls, Ann. Phys. $\underline{4}$, 121 (1930) and "Quantum Theory of Solids", p.108, Oxford University Press (1955).
2. H. Fröhlich, Proc. Royal Soc. A $\underline{223}$, 296 (1954).
3. J. Bardeen, L. N. Cooper and J. R. Schrieffer, Phys. Rev. $\underline{108}$, 1175 (1957).
4. For the review see B. Renker and R. Comes in "Low Dimensional Cooperative Phenomena", p.235, ed. H. J. Keller, Plenum Press (New York) (1975).
5. For the review see A. J. Heeger in "Highly Conducting One-Dimensional Compounds", p.69, ed. J. T. Devreese, R. P. Evrard and V. E. Van Doren, Plenum Press (New York) (1979).
6. P. Monceau, N. P. Ong, A. M. Portis, A. Meerschaut and J. Rouxel, Phys. Rev. Lett. $\underline{37}$, 602 (1976).
7. J. Dumas, C. Schlenker, J. Marcus and A. Buder, Phys. Rev. Lett. $\underline{50}$, 757 (1983).
8. R. C. Lacoe, H. J. Schulz, D. Jérome and L. Johannsen, Phys. Rev. Lett. $\underline{55}$, 2351 (1985).
9. Yu. A. Bychkov, L. P. Gor'kov and J. E. Dzyaloshinskii, Zh. Eksp. Teor. Fiz. $\underline{50}$, 738 (1966) [Sov. Phys. JETP $\underline{23}$, 489 (1966)].
10. J. Solyom, Adv. in Physics $\underline{28}$, 201 (1979) and references therein.
11. V. J. Emery, in "Highly Conducting One-Dimensional Solids", p.247, ed. J. T. Devreese, R. P. Evrard and V. E. van Doren, Plenum Press (New York) (1979).
12. S. Barišić and A. Bjeliš, in "Theoretical Aspects of Band Structures and Electronic Properties of Pseudo One-Dimensional Solids", p.49, Ed. by H. Kamimura, Riedel, Dordrecht (1985).
13. S. Barišić and S. A. Brazovskii in "Recent Developments in Condensed Matter Physics", vol.1, ed. J. T. Devreese, Plenum (New York) (1981); V. J. Emery, R. Bruinsma and S. Barišić, Phys.Rev.Lett. $\underline{48}$, 1039 (1982).
14. P. A. Lee, T. M. Rice and P. W. Anderson, Solid State Commun. $\underline{14}$, 703 (1974).
15. K. B. Efetov and A. I. Larkin, Zh. Eksp. Teor. Fiz. $\underline{72}$, 2350 (1977) [Sov. Phys. JETP $\underline{59}$, 1057 (1977)].
16. H. Fukuyama and P. A. Lee, Phys. Rev. $\underline{B17}$, 535 (1978), P. A. Lee and H. Fukuyama, ibid. $\underline{B17}$, 542 (1978).
17. L. Mihaly and G. X. Tessema, Phys. Rev. $\underline{B33}$, 5858 (1986).
18. G. Mihaly and P. Beauchêne, Solid State Commun. $\underline{63}$, 911 (1987).
19. For general reviews see, e.g., P. Monceau, Electronic Properties of Inorganic Quasi One-Dimensional materials II, p.139, Edited by P. Monceau, Riedel, Dordrecht (1985); G. Grüner, Rev. Mod. Phys. $\underline{60}$, 1129 (1988).

20. I. E. Dzyaloshinskii and A. I. Larkin, Zh. Eksp. Teor. Fiz. 61, 791 (1971) [Sov. Phys. JETP 34, 422 (1972)].
21. N. Menyhard and J. Sólyom, J. Low Temp. Phys. 12, 529 (1973).
22. L. P. Gor'kov and J. E. Dzyaloshinskii, Zh. Eksp. Teor. Fiz. 67, 397 (1974) [Sov. Phys. JETP 40, 198 (1975)].
23. S. Barišić, Lecture Notes in Physics (Springer) 65, 85 (1977).
24. S. Barišić, Electronic Properties of Inorganic Quasi One-Dimensional Compounds, p.1, Ed. P. Monceau, Riedel, Dordrecht (1985).
25. e.g. A. A. Abrikosov, L. P. Gor'kov and I. E. Dzyaloshinskii, Quantum Field Methods in Statistical Physics, Prentice-Hall, Engelwood Clifft, (1963).
26. A. Bjeliš and S. Barišić, J. Physique Letters 36, 169 (1975).
27. S. A. Brazovskii and I. E. Dzyaloshinskii, Zh. Eksp. Teor. Fiz. 71, 2338 (1976) [Sov. Phys. JETP 44, 1233 (1976)].
28. K. Šaub, S. Barišić and J. Friedel, Phys. Lett. 56A, 302 (1976).
29. B. Horovitz, H. Gutfreund and M. Weger, Phys. Rev. B12, 3174 (1975).
30. K. Yamaji, J. Phys. Soc. Jpn. 51, 2787 (1982).
31. C. Schlenker and J. Dumas, in "Crystal Chemistry and Properties of Materials with Quasi-One-Dimensional Structures", p.135, ed. J. Rouxel, Riedel Publ. Comp. (1986).
32. S. A. Brazovskii, L. P. Gor'kov and A. G. Lebed, Zh. Eksp. Teor. Fiz. 83, 1198 (1982). [Sov. Phys. JETP 56, 683 (1983)].
33. See e. g. L. D. Landau and E. I. Lifchitz, "Physique Statistique", Editeur Mir, Moscou (1967).
34. D. J. Scalapino, M. Sears and R. A. Ferrell, Phys. Rev. B6, 3409 (1972).
35. G. Toulouse, N. Cimento 23B, 234 (1974).
36. K. Uzelac and S. Barišić, J. Physique 36, 1267 (1975), and J. Physique Lett. 38, 47 (1977).
37. K. B. Efetov and A. I. Larkin, Zh. Eksp. Teor. Fiz. 66, 2290 (1974) [Sov. Phys. JETP 39, 1129 (1974)].
38. J. P. Pouget, in "Charge Density Waves in Solids", eds. L. P. Gor'kov and G. Grüner, Elsevier (to be published).
39. See e. g. J. W. Negele and H. Orland, "Quantum Many-Particles Systems", Addison-Wesley Publ., New York (1988).
40. J. E. Hirsch and E. Fradkin, Phys. Rev. Lett. 49, 402 (1982).
41. H. Zheng, D. Feinberg and M. Avignon, Phys. Rev. B39, 9405 (1989).
42. C. Bourbonnais and L. G. Caron, J. Physique 50, 2751 (1989).
43. S. Aubry, G. Abramovici, D. Feinberg, P. Quemerais and J.-L. Raimbault, Lecture Notes in Physics, to be published (1989).
44. K. Maki, this Volume.
45. K. B. Efetov and A. I. Larkin, Zh. Eksp. Teor. Fiz. 72, 2350 (1977) [Sov. Phys. JETP 45, 1236 (1977)].
46. S. Barišić and I. Batistić, J. Physique Lett. 46, 819 (1985).
47. P. A. Lee and T. M. Rice, Phys. Rev. B19, 3970 (1979).
48. W. P. Su, J. R. Schrieffer and A. J. Heeger, Phys. Rev. B22, 2099 (1980).
49. S. A. Brazovskii, I. E. Dzyaloshinskii and N. N. Kirova, Zh. Eksp. Teor. Fiz. 81, 2279 (1981) [Sov. Phys. JETP 54, 1209 (1981)].
50. S. A. Brazovskii, I. E. Dzyaloshinskii and S. G. Obukhov, Zh. Exsp. Teor. Fiz. 72, 1550 (1977) [Sov. Phys. JETP 45, 814 (1977)].
51. N. P. Ong and K. Maki, Phys. Rev. B32 (1985); K. Maki, Physica 143B, 59 (1986).
52. J. Dumas and D. Feinberg, Europhys. Lett. 2. 555 (1986).
53. D. Feinberg and J. Friedel, J. Physique (Paris) 49, 485 (1988).
54. S. Barišić and I. Batistić, J. Physique 50, 2717 (1989).
55. L. P. Gor'kov, Pis'ma Zh. Eksp. Teor. Fiz. 76 (1983) [JETP Lett. 38, 87 (1983)]; Zh. Eksp. Teor. Fiz. 86, 1818 (1984) [Sov. Phys. JETP 59, 1957 (1985)].
56. I. Batistić, A. Bjeliš and L. P. Gor'kov, J. Phys. (Paris) 45, 1049 (1984).

57. J. C. Gill, Solid State Commun. __44__. 1041 (1982); P. J. Yetman and J. C. Gill, Solid State Commun. __62__. 201 (1987); G. Mihaly, Gy Hutiray and L. Mihaly, L. Phys. Rev. __B28__. 4896 (1983); M. Prester, Phys. Rev. __B32__, 2621 (1985); D. V. Borodin, S. V. Zaitsev-Zotov and F. Ya Nad', Zh. Eksp. Teor. Fiz. __90__, 318 (1986); __93__, 87 (1987); [Sov. Phys. JETP __63__, 184 (1986); __66__, 793 (1987)]; M. C. Saint-Lager, P. Monceau and M. Renard, Synth. Metals __29__ (1989).
58. G. Mihaly, Physica Scripta T__29__, 67 (1989).
59. J. Takada, M. Wong and T. Holstein, Phys. Rev. __B32__, 4639 (1985).
60. P. B. Littlewood, Solid State Commun. __65__, 1347 (1988).
61. L. P. Gor'kov and E. N. Dolgov, Zh. Eksp. Teor. Fiz. __77__, 396 (1979). [Sov. Phys. JETP __50__, 203 (1979)].
62. S. N. Artemenko, A. F. Volkov and A. W. Kruglov, Zh. Eksp. Teor. Fiz. __91__, 1536 (1986) [Sov. Phys. JETP __64__, 906 (1987)].
63. D. Jelčić, A. Bjeliš and I. Batistić, Phys. Rev. __B38__, 4045 (1988).
64. R. M. Fleming and G. G. Grimes, Phys. Rev. Lett. __42__, 1423 (1979).
65. P. Monceau, J. Richard and M. Renard, Phys. Rev. __B25__, 918, 931 (1982).
66. N. P. Ong and G. Verma, Phys. Rev. __B27__, 4495 (1983).
67. A. Zettl and G. Grüner, Solid State Commun. __46__, 501 (1983); Phys. Rev. __B29__, 755 (1984).
68. R. E. Thorne, W. G. Lyons, J. W. Lyding, J. R. Tucker and J. Bardeen, J. Phys. Rev. __B35__, 6348, 6380 (1987).
69. A. Bjeliš, Physica Scripta T__29__, 62 (1989).
70. G. Mihaly, G. Kriza and G. Grüner, Europ. Lett. __9__, 163 (1989); L. Forró, A. Janossy, M. Raki and C. Ayache, to be published.
71. D. Jelčić and A. Bjeliš, to be published.
72. S. Bhattacharya, J. P. Stokes, M. G. Higgins and R. A. Klemm, Phys. Rev. lett. __59__, 1849 (1987).

RECENT DEVELOPMENTS IN CHARGE DENSITY WAVE SYSTEMS

Pierre Monceau

Centre de Recherches sur les Très Basses Températures, CNRS, BP 166 X
38042 Grenoble-Cédex, France

1. INTRODUCTION

Collective transport phenomena are among the most fascinating properties in solid state physics. The best known example is superconductivity where the energy gap in the excitations at the Fermi level, as found by BCS[1], does not prevent conductivity. This is so because the interaction involved does not require a specified reference frame and because Cooper pairs can be built either in states "k and -k" or "k+κ or -k+κ". The latter state leads to a uniform velocity such as :

$$mv_s = \hbar \kappa$$

However in 1954, before BCS, Fröhlich[2] proposed a model in a jellium approximation in which a sliding charge density wave (CDW) could lead to a superconducting state. It is now well recognized that in systems of restricted dimensionality the interaction between ions and electrons, the so-called electron-phonon interaction, leads to structural instabilities at low temperature. According to the relative strength of several electron-electron couplings, the modulated ground state can be a CDW or, if the spin orientation is concerned, a spin-density wave (SDW).

The CDW instability was first predicted by Peierls[3]. As is known from band theory, every Brillouin zone constitues a locus of discontinuity for the electronic energy. If, in a one-dimensional electronic system with a Fermi vector of k_F, a periodic lattice distortion of wave-vector $2k_F$ is introduced, the band structure will be modified because of the new periodicity. A new Brillouin zone appears at $|k_F|$ and so, each occupied electronic energy for $|k| < k_F$ decreases, giving rise to a new ground state of the system characterized by a charge density wave (CDW) with wave vector $q = 2k_F$. The occupied electronic states are Bloch wave functions with the superlattice periodicity :

$$\psi_k = \exp(ikr) \sum_n V_{k,n} \exp(inqr)$$

and consequently the electronic density has Fourier components with wave-vectors $\pm nq$, especially for the fundamental ones $\pm q$:

$$\rho_{el} = \rho_0 + 2\rho_q \cos(qr+\phi) + ... \qquad (1)$$

where ρ_0 is the uniform electron density and $2\rho_q$ the charge modulation amplitude. The phase, ϕ, specifies the position of the CDW relative to the lattice ions. The local electron charge density is partially neutralized by a concomitant displacement of each ion to a new equilibrium position, the displacement of the nth ion, initially at nr_0, being

$$u_n = u_0 \sin(nqr_0 + \phi) \tag{2}$$

Since a gap, Δ, is opened at the Fermi level, the CDW state has an energy lower than the metallic state. This structural transition (a Peierls transition) occurs if the crystal distortion is energetically favourable when compared to the gain energy caused by the gap formation.

CDW formation has been observed in two-dimensional layered compounds, namely transition metal dichalcogenides, the Fermi surface of which approximates to a cylinder with nearly parallel faces. Thus a large fraction of states on the Fermi surface are connected by the same vector $q = 2k_F$ (the nesting condition). Nevertheless, the low-temperature ground state remains metallic. In contrast, for a strictly one-dimensional conductor, the Fermi surface consists of two parallel planes, so that all states are connected by the same q. The energy gap removes the whole Fermi surface, and the low temperature ground state is insulating.

In the strictly one-dimensional case no long range order can be established because of fluctuations and there is no phase transition at any temperature. In practice, however, with pseudo one-dimensional conductors, we can identify a characteristic temperature, called the Peierls transition temperature below which a lattice distortion occurs and the condensed state can be described by an order parameter. The latter can be defined either in terms of the electron density modulation as $\rho_q \exp(i\phi)$ (see equation 1) or in terms of the lattice distortion which is proportional to ρ_q.

The opening of a gap below the Peierls transition temperature is reminiscent of semiconductors, but the essential feature of a CDW is that its wavelength, $\lambda_{CDW} = 2\pi/2k_F$, is controlled by the Fermi surface dimensions and is generally unrelated to the undistorted lattice periodicities, i.e. the CDW is incommensurate with the lattice. Consequently the crystal no longer has a translation group and in contrast to semiconductors, the phase, ϕ, of the lattice distortion is not fixed relative to the lattice but is able to slide along q. This phenomenon is easy to understand if we recognize that if the lattice is regular, no position is energetically favoured and no locking results. In more theoretical terms : if we think of the CDW as resulting from an electronic interaction via the lattice phonons, this interaction is the same in every galilean frame, provided that the frame velocity is small compared with the sound velocity (in which case the interaction would be strongly modified). CDW condensation may thus arise in any set of galilean frames with uniform velocity, v, giving in the laboratory frame an electronic current density,

$$J = -n_0 e v \tag{3}$$

where n_0 is of the order of the electron number density condensed in the band below the CDW gap. This Fröhlich mode is a direct consequence of translation invariance. In practice, as shown by Lee, Rice and Anderson[4] this translation invariance is broken because the phase, ϕ, can in fact be pinned to the lattice, for example by impurities or by a long-period commensurability between the CDW wavelength and the lattice or by Coulomb interaction between adjacent chains. Oscillations of the pinned CDW are expected to produce a large low-frequency ac conductivity and a large dielectric constant. An applied dc electric field, however, can supply the CDW with an energy sufficient to overcome the pinning, so that above a threshold field, the CDW can slide and carry a current. Unfortunately damping prevents superconductivity. This extra conductivity associated with the collective CDW motion, called Fröhlich conductivity, has been observed[5] for the first time in 1976 and since this time, an intense experimental and theoretical activity has been devoted to the understanding of the properties of this collective transport mode.

In this article emphasis will be essentially made on new developments in the field. For a complete survey (including an exhaustive list of references), recent review articles[6] and conference proceedings[7] are available. The organization of the paper is as follows : in Part 2 the general properties of the sliding CDW will be shortly reviewed. Then Part 3 will describe the observation of the CDW structure in real space by scanning tunneling microscopy. Boundary conditions, size effects and non local effects associated with the growth of CDW dislocations will be analysed in Part 4. The end of the paper will be devoted to the properties of the CDW at low and very low temperatures with respect to the CDW damping (Part 5), the possible new type of transport at helium temperatures with no apparent friction (Part 6) and finally the description of the CDW ground state as a disordered medium as revealed by specific heat measurements below 1 K (Part 7).

2. DESCRIPTION OF THE SLIDING CDW STATE

2.1 - Materials

Up to now three families of inorganic compounds exhibit non-linear transport properties at any temperature below Peierls transitions namely transition metal trichalcogenides as $NbSe_3$, TaS_3 with monoclinic or orthorhombic structures, NbS_3, molybdenum oxides called blue bronzes $K_{0.30}MoO_3$ and $Rb_{0.30}MoO_3$ and halogened transition metal tetrachalcogenides as $(NbSe_4)_2I$, $(NbSe_4)_{10}I_3$, $(TaSe_4)_2I$. Without going into details the structure of these compounds can be described in chains of trigonal prisms stacked on the top of each other with a cross-section close to an isosceles triangle in the case of $NbSe_3$, of layers of infinite chains of MoO_6 octahedra with K ions separating the layers in the case of $K_{0.3}MoO_3$ and of parallel $TaSe_4$ chains with iodine atoms lying between them in the case of $(TaSe_4)_2I$. Except $NbSe_3$ which remains metallic at low temperature all the other compounds exhibit a semiconducting behaviour below the Peierls transition temperature. According to the compounds, this transition temperature is distributed between 330 K for NbS_3 down to 59 K for $NbSe_3$. The wavelengths of the CDW distortion appear to be incommensurate, very often near of four lattice distances, along the chain direction. A temperature dependence of the CDW wavelength has only been detected in orthorhombic TaS_3 and blue bronze with an apparent commensurability at low temperature. Non linearity has also been detected in the organic material TTF-TCNQ in a limited range of temperature between 54 K and 49 K where a unique CDW develops on the TCNQ stacks. Similar behaviour is also expected in SDW materials as recently found in $(TMTSF)_2NO_3$.[8]

2.2 - General properties of the non-linear state

The properties of the new current-carrying state can be summarized as follows :

- The d.c. electrical conductivity increases above a threshold field E_T.
- The conductivity is strongly frequency dependent in the range of 100 MHz-a few GHz.
- Above the threshold field, noise is generated in the crystal which can be analysed as the combination of a periodic time dependent voltage and a broad noise following a 1/f variation.
- Interference effects occur between the ac voltage generated in the crystal in the non-linear state and an external rf field.
- Hysteresis and memory effects are observed, principally at low temperature.

Fig. 1 shows a typical variation of the electrical conductivity (normalized to the ohmic value) as a function of the reduced electric field. The V(I) characteristic deviates from the ohmic law above a critical current I_T which leads to a threshold field defined as $E_T = RI_T/\ell$ where R is the resistance of the sample, ℓ : the distance between voltage electrodes. E_T varies typically from a few mV/cm in $NbSe_3$ to a few tenths of a volt for the other compounds. It is experimentally found that E_T strongly increases when T is lowered. Several phenomenological laws have been tried to fit the $\sigma(E)$ variation. One particularly has been largely used which is reminiscent of some kind of Zener-tunneling process such as :

$$\sigma(E) = \sigma_a + \sigma_b (1 - \frac{E_T}{E}) \exp(-\frac{E_0}{E}) \qquad (4)$$

with σ_a : the ohmic conductivity, $E_0 = kE_T$ with k between 2 and 5. The so small activation field E_0 precludes a single electron process because the gap which can be derived from such a Zener expression is several orders of magnitude lower than the thermal energy kT. When $E \to \infty$ the conductivity saturates to the value $\sigma_a+\sigma_b$ which is of the order of the metallic conductivity extrapolated from above the critical temperature. E_T is also seen to largely increase when the crystals are doped with impurities. Other defects can also play a role as dislocations, grain boundaries and in small size crystals the surface and/or the contacts. Pinning by the contacts has been revealed by measuring the increase of the threshold when the length is reduced ($\ell < 100$ µm) (see Part 4). Careful experiments have been recently reported[9,10] on specimens with small cross-section (< 50 µm^2) : E_T increases with decreasing cross-sectional area A and E_T is found to correlate with the ratio of the crystal circumference C to the cross section C/A.

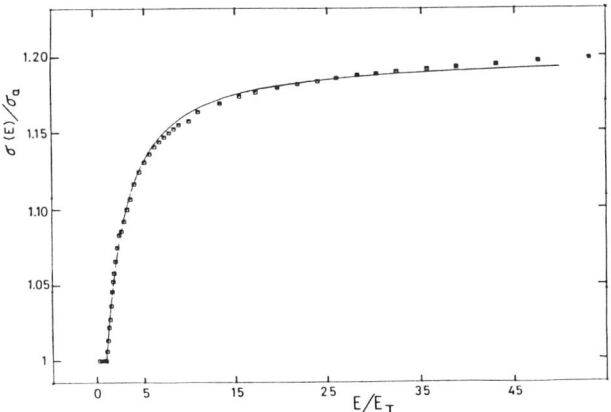

Fig. 1. Variation of the normalized conductivity of NbSe$_3$ of T = 128 K as a function of the electric field (normalized to the threshold value). The curve is the fit with Eq. 4 with $\sigma_b/\sigma_a = 0.26$, k = 1.03.

The low field ac conductivity shows a strong increase in the frequency range of 100 MHz-1 GHz and a saturation at a value close to the dc one for infinite fields. This behaviour is described in terms of a harmonic oscillator response due to the oscillations of the pinned CDW mode[11]. For NbSe$_3$ and TaS$_3$ the response is overdamped. Recent measurements in the range of 10-100 GHz have revealed the inertial term and have allowed to estimate the CDW effective mass, the pinning frequency and the damping constant. However at low temperatures the description with a single oscillator fails and a distribution of pinning frequency has to be taken into account.

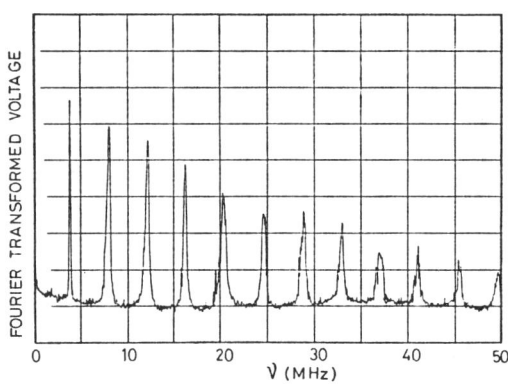

Fig. 2. Fourier transformed voltage spectra as a function of frequency for a NbSe$_3$ sample (T = 42 K) in the non-linear state.

When E passes beyond E_T, a time-dependent voltage is generated in the crystal which can be studied with a spectrum analyser. Besides a broad band noise the frequency dependence of which varies as 1/f, the Fourier transformed voltage spectra, as shown in Fig. 2 for NbSe3, reveal a fundamental frequency and many harmonics. The fundamental frequency appears at E_T and increases with the current applied to the sample.

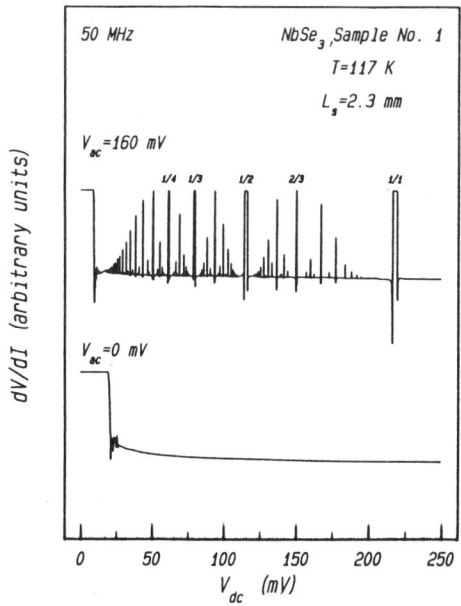

Fig. 3. Differential resistance dV/dI as a function of the applied dc voltage in a presence of a 50 MHz. The 1/5, 1/4, 1/3, 2/5, 1/2, 2/3 and 1/1 peaks are completely mode locked (from ref. 12).

Steps can be observed[12] in the dc V(I) characteristics if a rf current is superposed to a dc current exceeding the critical one. Such frequency synchronization is expected in non-linear phenomena such as Josephson junctions (Shapiro steps) for rf frequencies ω_{ac} in the immediate vicinity of harmonics and subharmonics eigenfrequencies of the phenomena. Fig. 3 shows the differential resistance dV/dI as a function of the dc V_{dc} bias measured in presence of a 50 MHz ac field. Constant-current steps are observed when the internal coherent CDW oscillation frequency ω_n is mode-locked to the applied ac frequency ω_{ac} such as $\omega_n/\omega_{ac} = p/q$ (p,q integers). It is to be noticed that during synchronization, the sample recovers nearly the ohmic differential conductivity (complete mode-locking) implying (see below) that the whole sample is oscillating nearly coherently.

Because of the strong interaction of the CDW with impurities, it is unlikely that the CDW can be described by a unique ground state and many metastable states have to be taken into account. Deformation of the CDW phase can be induced by a current and by temperature. The time scale for the metastable states to relax to the ground state can be very broad depending of the materials and of the temperature ; it is found that this decay time increases strongly at low temperature (see Part 5).

2.3 - <u>Current models for non-linear transport phenomena in CDW systems</u>

Bardeen[13] was the first to interpret the non-linear conductivity in the CDW systems described above as the Fröhlich conduction induced by the CDW motion. The major part of the theoretical work has been carried in the incommensurate case considering the extra conductivity as due to a collective motion of the CDW phase.

However the existence of a threshold field, E_T, leads to a problem. As for the vortex pinning in type II superconductors a true threshold can only be accounted if we introduce some elasticity of the phase : a random distribution of impurities or dislocations is the more probable cause of such a pinning force. But by its interaction with a completely rigid lattice the force summation will be random leading to a $V^{1/2}$ resultant (V being the volume ~ number of impurities), and to E_T going to zero in the thermodynamic limit. To the contrary, some elasticity allows for a deformation of the phase, and a finite second order effect due to the individual pinning forces. A very popular model by Lee and Rice[14] shows that the phase coherence between distant points tends to zero with the distance if an arbitrarily small elasticity is introduced in the random pinning problem :

$$<[\rho(\vec{r}+\vec{L})-\rho(\vec{r})-\vec{q}\vec{L}]^2>$$

extrapolating to zero when L increases. This is due to the accumulation of small phase disturbances over a great number of disturbing centers. They defined a "domain size" such that the phase deviation to ideality is of the order of π : the domain gives an internal random summation, but in actual problems L is only a few microns and any measurement is concerned with a many domain problem. *If it can be assumed that the domains* act independently, the result will be the sum over a great number of domains and the total pinning force be proportional to the total length : E_T is independent of the length. One of the consequences of this model is an E_T dependence with the square of the impurity concentration.

A domain is associated with an equivalent mass, some dissipative mechanism, Γ, (thermalization of the phase motion by the phonon bath), a net charge, and a resultant pinning force which is of course periodic in ϕ, since if ϕ is increased by 2π each impurity sees the same charge distribution in its vicinity. This leads to an equation of motion[15,16] :

$$\phi'' + \Gamma\phi' + \omega_p^2 \sin\phi = q \frac{eE}{M^*} \tag{5}$$

where E is the applied field, $q = 2\pi/\lambda_{CDW}$, ω_p the pinning frequency and M^* the Fröhlich mass.

This simple equation leads to at least qualitative explanations of many CDW phenomena :
- A threshold field defined by : $E_T = (\lambda_{CDW}/2\pi) \times (M^*\omega_p^2/e)$.
- If $E = E_0\cos\omega t$ and $E_0 << E_T$ a linearization of the $\sin\phi$ term gives a linear response theory, leading to a good agreement with the complex conductivity measured in low fields : $\sigma(\omega)$ and $\varepsilon(\omega)$ (overdamped oscillator).

For a dc field E higher than E_T, the "$\sin\phi$" force term gives rise to a velocity modulation at a fundamental frequency, ν, and its harmonics which can be considered as the origin of the ac voltage generated in these systems. It has to be noted that the λ_{CDW} assumed periodicity for the force (where λ_{CDW} is the CDW wavelength), means that the fundamental frequency is linked to the mean CDW velocity by :

$$v_{CDW} = \lambda_{CDW}\nu \tag{6}$$

Therefore according to Eq.3 the extra current carried by the CDW into motion is given by :

$$J_{CDW} = n_0 e v_{CDW} = n_0 e \lambda_{CDW}\nu \tag{7}$$

A consequence of the classical equation of motion (Eq.5) is that for E slightly higher than E_T, the extra d.c. current varies as :

$$J_{CDW} \sim (E-E_T)^{1/2}$$

However, experimental results show a nearly 3/2 power law. Some attempts have been made to explain the regime near E_T such as the calculation of Fisher[17] establishing some analogy between the vicinity of E_T and the critical behaviour of a second order phase transition, leading to the 3/2 exponent.

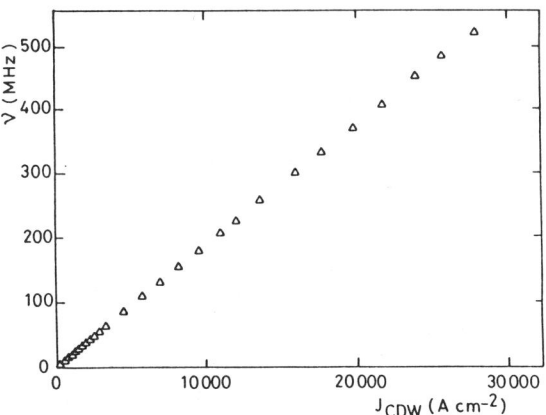

Fig. 4. Variation of the current J_{CDW} carried by the CDW as a function of the fundamental frequency measured in the Fourier-transformed voltage for an orthorhombic TaS_3 sample at T = 127 K. The slope $J_{CDW}/\nu = ne\lambda_{CDW}$ leads to the number of electrons condensed below the CDW gap.

According to Eq.7 the slope of J_{CDW}/ν is a measurement of the number of electrons condensed below the CDW gap. The extra-current J_{CDW} is measured from the non-linear V(I) characteristics. Fig. 4 shows the linear relationship between J_{CDW} and ν for an orthorhombic TaS_3 sample which is still valid with a CDW current density of 30 000 A/cm². The number of electrons deduced from the ν/J_{CDW} slope is for any CDW compound of the order of the electron concentration in the bands affected by the CDW condensation as it can be calculated from band structures or from chemical bonds. This result is thought to be the proof of the Fröhlich conductivity : when the field overcomes the threshold one, the electrons, which were trapped below the CDW gap, coherently participate to the electrical conductivity.

If, in Eq.5, E is the superposition of a dc field $E > E_T$ and a small ac field $E_o\cos\omega t$, the non-linear $\sin\phi$ term gives a frequency linking between the sliding CDW wave and the applied ω_{ac}, if the eigenfrequency is near of ω_{ac}. During this synchronization the wave velocity is independent of the continuous field, and therefore the differential conductivity equals the linear ohmic value.

3. CDW STRUCTURE BY SCANNING-TUNNELING MICROSCOPY

The modulation of the ion positions can be detected by X-ray, neutron or electron diffraction measurements. Superlattice spots appear near the main Bragg spots that correspond to the unmodulated structure. Measurements of the inverse separation of these superlattice spots give the CDW wavelength. In real space, images of the CDWs have been obtained using high resolution electronic diffraction. This method is very well suited to study defects in the CDW lattice induced, for instance, by electron irradiation.

However the more promising technique is by scanning-tunneling microscopy[18] (STM). According to the simple tunneling theory, when a small amplitude voltage bias V is applied, the tunneling current between two surfaces separated of d is :

$$I = \alpha V \exp[-\sqrt{\phi} d]$$

where ϕ is the extraction potential. For $\phi \sim 4$ eV, a change in d of 1 Å yields a change in I of one order of magnitude. Thus the STM microscope consists of a small tip sweeped along the surface to be analysed. Keeping the tunneling current I constant, the z deflection is recorded as a function of x and y which generates a three-dimensional image of the surface : z(x,y). The electronic structure of the electrodes has been taken into account by Tersoff and

Hamman[19]. Modelizing the tip as a spherical potential well, they found the tunneling current to be :

$$I = \alpha \, V \, D(E_F) \, \rho(r_0, E_F)$$

where $D(E_F)$ is the density of states (DOS) of the tip at the Fermi level, and $\rho(r_0,E_F)$ the DOS of the surface at the Fermi level and at the position r_0 of the tip. Then the STM yields an image of the atomic arrangement at the surface by following the spatial modulation of the DOS at the tip position.

As far as CDWs are concerned, the modulation of the conduction electron density at the CDW wavelength is easily detected by the STM and the z deflection is related to the fraction of electrons transferred into the CDW condensate. The layered transition metal dichalcogenides have been intensively studied by the STM technique at nitrogen[20] and helium[21] temperatures. The STM images reflect the amplitude of the CDW charge modulation -weak in the case of 2H-NbSe$_2$, and very strong in the case of 1T-TaSe$_2$. The two independent CDWs in NbSe$_3$ have very recently been revealed[22] by STM. The CDW modulations are localized on different types of chains as predicted by band-structure calculations. The sliding CDW has also been studied with the STM method[23]. By keeping fixed the tunneling tip position, sharp peaks in the Fourier-transformed spectra of the tunneling current have been found when the current bias exceeds the threshold value for CDW depinning. When sliding the hill and valley of the CDW modulates the distance d between the tip position and the surface. Then the tunneling current is modulated with a time period $\tau = \lambda_{CDW}/v$ with v the CDW velocity as in the case of the narrow band noise (see Eq. 6). This technique also shows that the depinning is inhomogeneous through the cross-section and that the CDW starts to slide at the surface.

Finally STM images reveal the importance of the crystal defects on the CDW charge modulation. The range of perturbation induced by a given defect and the redistribution of CDW contours around impurities located at the surface are now accessible by this technique which can be very useful for studying the pinning effects in CDW dynamics.

4. BOUNDARY AND NON-LOCAL EFFECTS

As explained in Part 2, the threshold electric field E_T for CDW depinning results from the competition between the elastic energy of the CDW and the pinning energy provided by impurities randomly distributed in volume. However this model does not take into account the boundary conditions. At the electrodes where the CDW velocity vanishes, the condensate CDW current should be converted in quasi particles and this process can only occur at places where the Peierls gap Δ is zero.

By analogy with type II superconductors Maki[24] has described vortices around which the phase rotates of 2π. If one draw (Fig. 5) the planes corresponding to $\phi = 2n\pi$ (n algebraic integer), these vortices look like edge dislocations. Around the dislocation core the phase gradient is gigantic : the core is a normal area with $\Delta = 0$. If one supposes for example that A is fixed by the end of the sample, if a vortex is created at B and climbs to B' every line at the right of B has been translated by one wavelength. A continuous flow of vortices may assure a transition between a moving part BC and a static one AB. The electrons injected at the electrode travel as excitations to the vortex core where they can condense easily ($\Delta = 0$), and when two electrons per chain have been condensed the core is translated to the next chain ...

If Maki has given an equivalent description of the Abrikosov-Gor'kov vortices in superconductors, Gor'kov[25] and Batistic et al.[26] have treated the equivalent of the superconducting weak links. Neglecting the transverse variations for a very thin film of

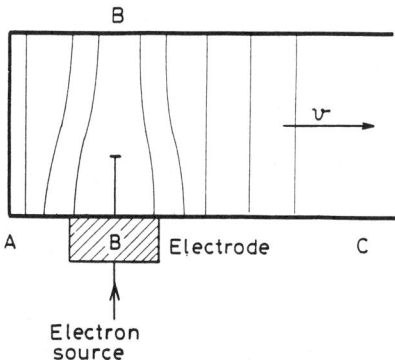

Fig. 5. Schematic topological defect in the CDW lattice to accomodate the phase gradient between parts in the sample in which the CDW moves and is at rest.

CDW sample it was shown that phase slippage centers are necessary to accomodate the electrons arriving (or leaving) from an electrode.

The strong deformations of the CDW at the electrodes are revealed in size effects measurements by measuring the threshold field as a function of the distance between electrodes. The threshold voltage sharply increases when ℓ is reduced. V_T then can be written as

$$V_T = E_P L + V_0 \tag{8}$$

where the first term results from the bulk pinning and V_0 has been interpreted as the potential necessary for the nucleation of a vortex or a phase slip center. The importance of thermodynamic fluctuations near T_P in the initiation of CDW dislocations has been established[10,27] but for $T < T_P$, V_0 is sizeable and typically 0.2 mV to 0.5 mV for NbSe$_3$ at T = 40 K and 1 mV for TaS$_3$ at T = 120 K. However such a small value for V_0 discredits a nucleation process. Indeed the nucleation of a loop the minimal size of which being the transverse amplitude coherence-length $\xi_\perp$ carries a charge given by $Q = \pi\xi_\perp^2 \lambda_{CDW} n_0 e$. Then the electrical source should supply the extra energy: $W_e = QV_0$. The nucleation will be favourable if W_e is at least larger than the condensation energy given by $2\pi\xi_\perp\xi_{//}E_c n_0$ with $\xi_{//}$ the amplitude coherence-length along the chains, E_c the condensation energy per electron. So V_0 should be larger than $2\xi_{//}E_c/\lambda$; with the well accepted values for NbSe$_3$, V_0 is found to be ~ 20 mV, two orders of magnitude larger than the experimental value. As in elastic theories for metals, a smoother mechanism is the growth of bubbles from preexisting dislocations acting for instance as Frank-Read sources. In this case, the reduction in energy delivered by the external source for enlarging the loops is $2\pi\xi_\perp/D$ where D is the diameter of the loop. With the measured V_0, D is typically 500 Å to 1000 Å much less than the transverse dimensions of the sample (~ 10^4 Å).

Then V_0 can be differently interpreted. A recent critical state model has been developed[28] in which, in addition to the bulk pinning, the growth of dislocation loops acting as Frank-Read sources are hindered by pinning from impurities. V_0 is the consequence of this irreversible pinning suffered by dislocation loops.

The growth of dislocation loops should appear at any velocity discontinuity to release the charge accumulation. Depinning under inhomogeneous conditions can be achieved by applying a thermal gradient along the sample length or when several independent sources deliver current in different segments of the same crysal. In the latter case two types of configuration can be used as shown in Fig. 6 : in Fig. 6a, a current source supplies i on a segment ℓ with typically 25 µm < ℓ < 80 µm whereas the whole crystal of length L ~ 1 mm is fed by a constant current source delivering I.[29] In Fig. 6b, segment L and ℓ are now apart and separated by a distance d. Near the electrodes, the distortion of the CDW can

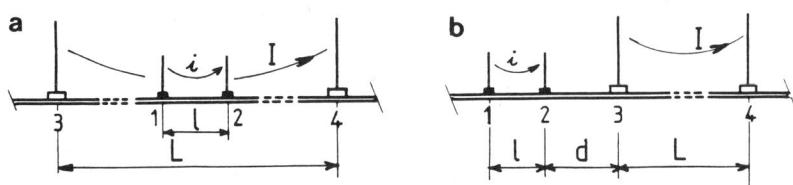

Fig. 6. Schema for CDW depinning when two independent current sources deliver current in different segments of the same crystal. Typically L~1 mm, 20 μm<ℓ<80 μm, the width of the contacts=8 μm. In a) segments L and ℓ are overlapping, in b) L and ℓ are apart and separated of d.

spread outside the segment where the sample is in the non-linear state and this configuration 6b allows the study of non-local effects.[30]

When the segments L and ℓ are overlapping (Fig. 6a) the depinning can be studied with the conjunction of I and i currents.[29] The fundamental frequency equivalent to the CDW velocity (see Eq. 6) is obtained from the Fourier transformed time-dependent voltage. First $I > I_T$ is applied and the whole sample is in the non-linear state with velocity v_L. Then i is increased in the segment ℓ. The extra force brought by i will slightly increase the CDW velocity between 3 and 4 which indicates the long range CDW coherence. Then for i beyond a value i_0 dislocation loops under electrodes 1 and 2 are activated and the CDW coherence is broken along the sample : two frequencies appear in the Fourier spectra, a CDW velocity in the segments 1-3 and 4-2 and an independent CDW velocity for the inner segment 1-2. In reality the two electrodes 1 and 2 are not equivalent and the breaking under 1 and 2 occurs for slightly different i values.

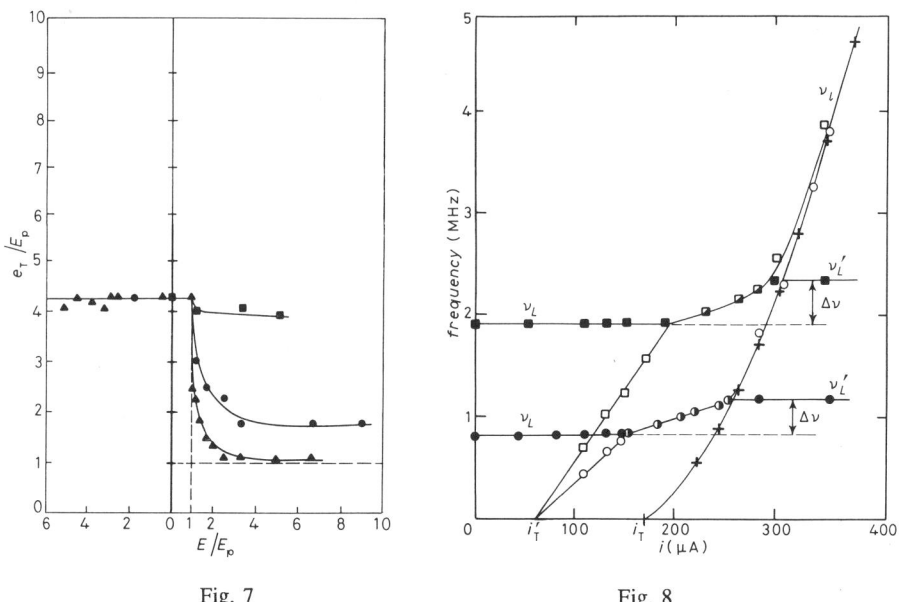

Fig. 7   Fig. 8

Fig. 7. Variation of the threshold field e_T of a small segment ℓ (50 μm) at T = 42 K as a function of the electric field on the neighbouring segment L with d = 0 (▲), 30 μm (●) and 60 μm (■). Electric fields are normalized with the threshold from impurity bulk pinning, E_p.

Fig. 8. Variation of the fundamental narrow-band noise frequency v_ℓ (○, □) in a segment (50 μm) of a NbSe$_3$ single crystal at 42 K as a function of the applied current i for two different constant currents above threshold in a neighbouring segment L (500 μm) inducing a narrow-band noise frequency v_L (●, ■) in L. The locking regime for each v_L is shown by ◉ and ◪.

With L and ℓ segments apart (Fig. 6b), the depinning of the small segment ℓ can be studied according to the state -linear or non-linear- of the segment L.[30] The threshold of ℓ alone is $e_T = E_p + V_o/\ell$ whereas the threshold of L is $E_T \approx E_p$ since L is $\gg \ell$. Fig. 7 shows that e_T decreases when the neighbouring segment L starts to slide. This effect is only detectable for d < 100 µm and for I and i with the same polarity. When d = 0 (adjacent segments) e_T decreases sharply to reach the volumic value E_p. Thus the depinning on the adjacent segment suppresses the boundary effects on the segment ℓ revealed by V_o/ℓ.

The CDW velocities can also be studied from the noise analysis. The current I in L is kept constant and the CDW velocity in I is then v_L. Fig. 8 shows the CDW velocities in L and ℓ as a function of the current i in ℓ. When i is increased beyond i_T^+, the CDW starts to move in ℓ with a velocity v_ℓ. When v_ℓ is equal to v_L, the two segments lock together ; similarly to the case described in Fig. 6a, the further increase of i speeds up the unique velocity in L+ℓ. Then for higher value of i coherence breaking occurs, the CDW velocity in L staying constant at a value $v_L' > v_L$ whereas v_ℓ continues to increase with i. The velocity diagram in Fig. 8 has been totally explained with the model of dislocation pinning at the electrodes.[30] These experiments show that the irreversible pinning forces are always paid by the segment with the largest velocity. However the most spectacular result is the long range CDW coherence as manifested by the variation of v_L in the long segment : although I has been kept absolutely constant, when the boundary conditions are changed at one electrode by a modification of the electrical condition of the adjacent segment, the CDW velocity on a distance of 1 mm or more is as a whole increased from v_L to v_L'. The CDW velocity does not suffer any small local variations along the sample length but the CDW coherence is quasi infinite.

5. DAMPING AND SCREENING EFFECTS

The equation of motion of the CDW with a single degree of freedom has been phenomenological derived in Eq. 5 as a particle moving in a periodic pinning potential. The dielectric constant arising from the oscillations of the CDW is then :

$$\varepsilon(\omega) \sim \frac{\Omega_p^2}{\omega_p^2-\omega^2+i\Gamma\omega} \tag{9}$$

with $\Omega_p^2 = 4\pi n e^2/M^*$. In the low frequency limit the dielectric constant follows a relaxational response

$$\varepsilon(\omega) = \frac{\varepsilon(0)}{1+i\omega\tau} \tag{10}$$

with $\tau = \Gamma/\omega_p^2$. However in order to fit the experimental dielectric response, internal degrees of freedom should be included through a distribution of relaxation times[31] and :

$$\varepsilon(\omega) \sim \frac{\varepsilon(0)}{\left[1+(i\omega\tau)^{1-\alpha}\right]^\beta} \tag{11}$$

where α and β characterize the width of distribution of relaxation times and the skewness, respectively. The conductivity $\sigma(\omega)$ can be deduced from Eq. 9 as follows :

$$\sigma(\omega) = \frac{ne^2}{i\omega M^*} \frac{\omega^2}{\omega_p^2-\omega^2-i\omega\Gamma}$$

$$= \frac{ne^2}{\Gamma M^*} \left[\frac{\Gamma^2\omega^2}{(\omega_p^2-\omega^2)^2 + \Gamma^2\omega^2} - \frac{i\Gamma(\omega_p^2-\omega^2)}{(\omega_p^2-\omega^2)^2 + \Gamma^2\omega^2}\right] \tag{12}$$

As the CDW is pinned, $\sigma(\omega)$ contributes only at finite frequencies.

Experimentally it was found that the dielectric relaxation time τ and the static dielectric constant $\varepsilon(0)$ exhibit an Arrhenius temperature dependence.[31,32] Also in conductivity measurements the amplitude of extra conductivity σ_{CDW} at a given $E/E_T > 1$ decreases exponentially when the temperature is reduced with the same activation energy as the normal free carrier conductivity.[33] Thus the viscous forces or the friction of the CDW seems to diverge when T is reduced and the dissipation mechanism must imply normal carrier dissipation. These results have been explained by taking into account screening effects of the CDW deformations and long range Coulomb interactions.[34,35] The charged CDW deformations are electrostatically coupled to normal electrons and now the total current should be written as :

$$j = \varepsilon \dot{E} + \sigma E + j_{CDW}$$

where the first term is the displacement current with ε the dielectric constant, the second one the linear ohmic current where σ is thermally activated as $\exp(-\Delta/kT)$ and the last term the extra CDW current. The relevant parameter is now $\omega\varepsilon/\sigma$ or ω/ω_1 ($\omega_1 = \sigma/\varepsilon$).

- At low frequencies, or at relatively high temperature, the conduction electrons are able to screen the CDW deformations. This back flow current induces a ohmic dissipation which accounts for the enhanced damping. The effective damping was found[35] to be :

$$\Gamma_{eff} = \Gamma_o + \Gamma_1 \left[1+(\omega/\omega_1)^2\right]^{-1}$$

Γ_o is the damping which originates from phason-phason or phason-phonon scattering. This damping varies as T^2 and should vanish[36] at low T. However if some damping comes from scattering with impurities, then Γ_o should remain finite and temperature independent.

- When $\omega\varepsilon/\sigma \gg 1$ at low temperature, the few normal electrons are no more able to screen the CDW deformations and there are Coulomb interactions of the CDW with itself. These Coulomb-Coulomb interactions introduce a high frequency plasmon mode in the CDW excitations.

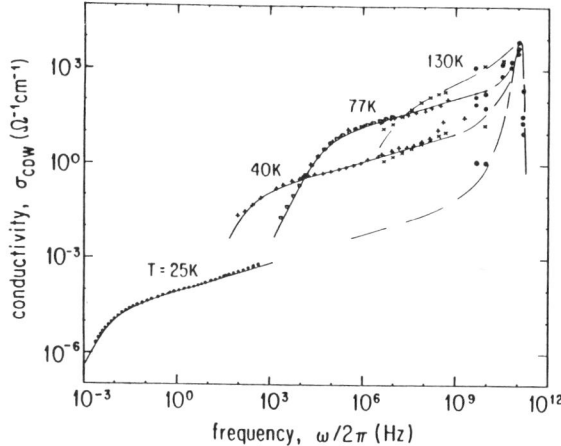

Fig. 9. Variation of the CDW conductivity as a function of frequency at several temperatures (from ref. 37).

Thus the ac response show two modes : an overdamped low frequency mode dominant at high temperature which strongly interact with normal carriers and a high frequency (microwave range) underdamped mode dominant at low temperatures[37]. Fig. 9 shows the variation of the conductivity of $K_{0.3}MoO_3$ as a function of the frequency at several temperatures. The underdamped pinned mode around 100 GHz emerges sharply below 40 K. The frequency of this pinned mode is in good agreement with either inelastic neutron scattering[38] or far infrared measurements.[39] Littlewood[35] recently brought the likely explanation for the splitting in two modes of the pinned phason excitations : longitudinal and transverse modes of the CDW should be distinguished. Longitudinal modes couple to an electrostatic potential and are strongly screened by interaction with free carriers whereas transverse modes couple to electromagnetic radiation and are unscreened and consequently much less damped. The mixing between longitudinal and transverse character of the pinned mode is assumed to result from the non uniform pinning due to the disordered nature of the CDW. However other interpretation of these modes have also been proposed from a strong pinning theory of CDW dynamics.[40]

Screening effects modify also strongly the CDW excitation dynamics. As shown by Lee-Rice-Anderson[4] the coupled electron-phonon mode which leads to the Kohn anomaly at high temperature is split into two different modes below the Peierls transition : an optical mode Ω_+, and an acoustic mode Ω_-, with the following frequency dependencies :

$$\Omega_+^2 = \lambda \omega_Q^2 + \frac{1}{3} \frac{m^*}{M^*} v_F^2 |q-Q|^2 \quad (13)$$

$$\Omega_-^2 = \frac{m^*}{M^*} v_F^2 |Q-q|^2 \quad (14)$$

with $Q = 2k_F$, λ : the electron phonon coupling constant, ω_Q the bare phonon frequency at high temperature, m^* the band mass of the electrons. Ω_+ is an amplitude mode and Ω_- is a phase mode or phason. From Eq. 14, the velocity of this mode is

$$C_0 = \left(\frac{m^*}{M^*}\right)^{1/2} v_F \quad (15)$$

In the case where the effects of impurities are neglected ($\omega_p = 0$), the dispersion law of the phase mode in the case of screening has been calculated by several authors.[41-45] Longitudinal and transverse modes have to be analysed separately : a longitudinal compression of the CDW implies a charge redistribution leading to long range Coulomb forces which consequently at T = 0 raises the Ω-phase mode to a finite frequency calculated by Lee, Rice and Anderson[4] to be $(1.5 \lambda)^{1/2} \omega_Q$. On the contrary a transverse shear only implies a dephasing of the CDW between adjacent chains but not a charge redistribution which preserves the acoustic character of the phase mode. With Coulomb force interactions, the dispersion relation Eq. 14 becomes[45], with x the longitudinal chain axis :

$$\Omega_-^2 = \left(c_0^2 + \frac{\Omega_p^2 q_x^2}{\varepsilon_z q_x^2 + q_0^2}\right) q_x^2 \quad (16)$$

with two limits

$$\Omega_-^2 = \left(c_0^2 + \frac{\Omega_p^2}{q_0^2}\right) q_x^2 \quad \text{for} \quad q_x^2 \ll \frac{q_0^2}{\varepsilon_z} \quad (17a)$$

and

$$\Omega_-^2 = \frac{3}{2} \lambda \omega_Q^2 + c_0^2 q_x^2 \quad \text{for} \quad q_x^2 \gg \frac{q_0^2}{\varepsilon_z} \quad (17b)$$

where ε_z is the screening dielectric constant, $\varepsilon_z = 1 + (2/3) \Omega_p^2/\Delta^2$ with $\Omega_p^2 = 4\pi n e^2/M^*$ and $q_0^2 = 4\pi e^2 n_{qp}/T$ the square of the Thomas-Fermi screening wave number of the quasi-particle gas with the concentration n_{qp}. This quasi-particle concentration is thermally activated. The cross-over between both regimes (Eqs 17a and 17b) results from the comparison between the phase mode length and the Thomas-Fermi screening length. When $|q_x| \ll q_0 \varepsilon_z^{-1/2}$ the phase mode length is much larger than the Thomas-Fermi screening length, there is no local charge density in the scale of a phase mode length and the excitations remain acoustic. However the phason velocity is seen to become stiffer. On the other hand when $|q_x| \gg q_0 \varepsilon_z^{-1/2}$ quasi particle are unable to screen any charge fluctuation in the scale of the phase mode length and the spectrum shows an optical mode with a frequency which for $q \to 0$ agrees with Lee-Rice-Anderson result.

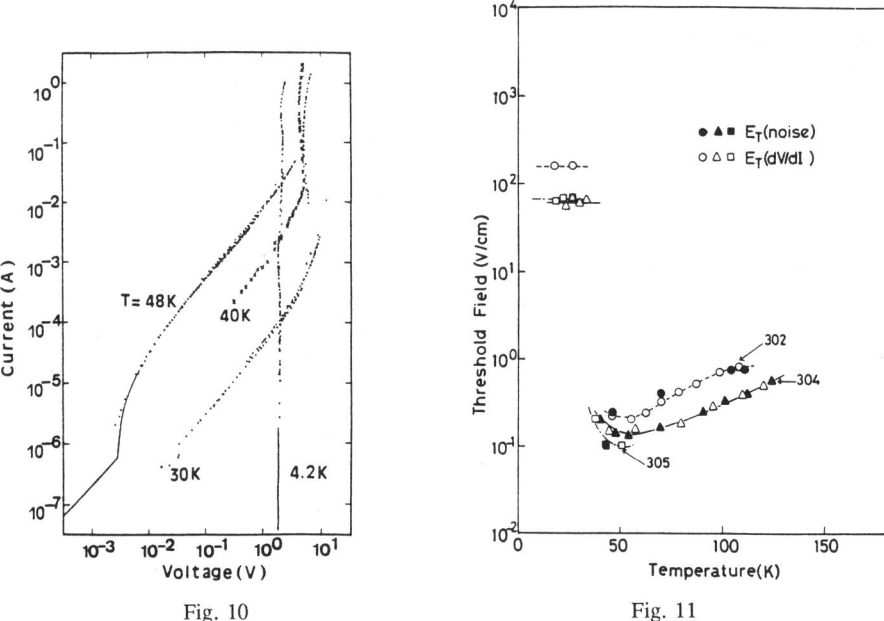

Fig. 10.
Fig. 11.

Fig. 10. Total current as a function of applied voltage for a crystal of $K_{0.3}MoO_3$ at several temperature. At T = 48 K V_T is ~ 3 mV and at 4.2 K the switching occurs at $V_T^* = 2V$ (from ref. 48).

Fig. 11. Variation of E_T and E_T^* as a function of temperatue for three different samples of $K_{0.3}MoO_3$ (from ref. 46).

6. DEPINNING OF THE CDW AT HELIUM TEMPERATURE

Following the analysis of Part 5, the CDW current is expected to vanish at very low temperature due to the huge enhancement of the damping. However in $K_{0.3}MoO_3$ at helium temperature an abrupt increase of the current by several orders of magnitude occurs above a threshold voltage in the range of 10 V/cm - 100 V/cm.[46-48] Fig. 10 shows the total current as a function of the applied voltage for a crystal of $K_{0.3}MoO_3$ at several temperatures. At 4.2 K for a threshold $V^* = E^*/L$, a switching appears between the insulating state at low voltage to a highly conducting state in which the current increases apparently with a zero differential resistivity as :

$$\sigma_{CDW} \propto \frac{j}{E_T}$$

Two threshold fields have then to be considered : E_T corresponds to the depinning of the CDW at high temperature and its variation with temperature is shown in Fig. 11, and E_T^* which abruptly appears below 40 K such as $E_T^* \sim 10^3 E_T$. Both depinning processes are overlapping in a small temperature range as seen in Fig. 10. The sharp (I-V) characteristics at low temperature is now commonly associated with CDW depinning although earlier measurements on the same kind of compounds were explained by impact ionization.[49] This low temperature non-linear state is characterized by large broad noise, periodic current oscillations the frequency of which is increasing with the current[46,50-52], intermittency, negative differential resistance region.[52] However the linear relation between J_{CDW} and the fundamental frequency shows, if CDW motion is involved (Eq. 7), that only a small part (~ 1 %) of the cross-section is in the non-linear state.[46,52] Below E_T^*, the polarization (defined as P = 2ed where e is the electric charge and d the displacement of the CDW) resulting from bipolarity voltage pulses shows a divergent behaviour[53] when E_T^* is approached and for a given $E < E_T^*$, a time dependence following a stretched-exponential form.[47] On the contrary, when unipolar pulses are applied, the polarization is reversible[54] and linear with E up to E^*. This reversible polarization has been attributed[54] to the rigid displacement of the CDW while the remanent polarization would reflect configurational change between weakly CDW domains induced by the reversal of the voltage step. Thus at low temperatures in this scheme, the internal CDW excitations are frozen and the CDW moves rigidly. If so, the classical model with a single degree of freedom should apply. From Eq. 5, the threshold field is :

$$E_T = \frac{\lambda_{CDW}}{2\pi} \frac{M^* \omega_p^2}{e}$$

and from Eq. 9 the dielectric constant in the low frequency range is :

$$\varepsilon_{\omega \to 0} = \frac{4\pi n e^2}{M^* \omega_p^2}$$

which leads to

$$\varepsilon_{\omega \to 0} E_T = 2ne\, \lambda_{CDW} = cst \tag{18}$$

This simple relation is model independent. Eq. 18 was shown to hold in the high temperature range and in the new low temperature state. Moreover, at high temperature, following the Bardeen model[13] the non-linear conductivities $\sigma(\omega)$ and σ_{dc} can be scaled such as :

$$\sigma(\omega/\omega_T) = \sigma(V/V_T)$$

with
$$V_T = \alpha\, \omega_T$$

Typical values of α were 1 mV/MHz which yield typical values of ω_T in the megahertz range i.e in the range where dielectric relaxation takes place.[55]

Thus the image which emerges from the above analysis is the following : at high temperature when screening of CDW deformations occurs, the CDW depinning is associated with the low frequency mode in the ac response. When T is reduced and when the long range Coulomb forces become effective, the low frequency mode is shifted at very low frequencies as discussed in Part 5 and shown in Fig. 9 ; the CDW is more rigid and the unique mode associated with the sliding CDW is the pinned mode in the 100 GHz range. The maximum of the dc conductivity reached in this sliding state is given by Eq. 12 when $\omega \to 0$ i.e $\sigma_{max} = ne^2/\Gamma M^*$. The only damping effect now is the coupling of phasons with the phonon bath, Γ_o which, as explained in Part 5, vanishes at T = 0. However finite damping still occurs for instance at the electrodes in the CDW-normal electron conversion process which excludes any superconductivity state by CDW sliding.

Although very appealing, the discussion above may however have several important drawbacks. The first one is the very particular temperature dependence of E_T and E_T^* with an apparent discontinuity around 30-40 K. In other CDW systems, insulating at low temperature as TaS_3, it seems that there is an unique threshold field the value of which is increasing continuously when T is reduced. Since negative differential resistance has been observed[52], breakdown avalanche and non uniform current paths are also probably favoured and filamentary conduction may take place. That might also be connected with the so small cross-section of the sample participating to the non-linear state. When a voltage is applied at low temperature, other excitations can also be created as solitons or kinks on single chains. Such local excitations have been invoked for the explanation of non-linear behaviour at helium temperature of other one-dimensional systems as TTF-TCNQ[56], or TaS$_3$.[57] Finally the CDW ground state is far from equilibrium and many metastable states are present which relaxe over a very long time as it will be now presented in the last part.

7. LOW ENERGY EXCITATIONS OF CDW METASTABLE STATES

The CDW ground state results from the interaction between the CDW elasticity with the randomly distributed impurities. Due to this randomness, the E = 0 ground state comprises many metastable states which are defined as local deformations of the pinned CDW phase. These metastable states have been essentially characterized through the polarization they induce when an electric field is applied. However, as in other disordered materials as glasses, spin glasses or polymers, the metastable states are expected to contribute to the thermodynamical properties at very low temperature.

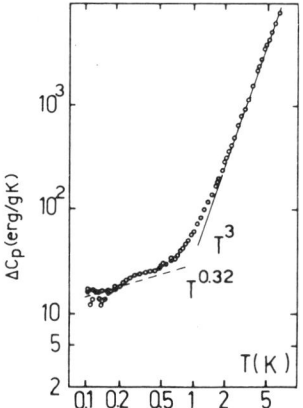

Fig. 12. Specific heat of orthorhombic TaS$_3$ (after substraction of a nuclear hyperfine contribution) as a function of temperature.

Fig. 12 shows the specific heat of TaS$_3$ as function of temperature.[58] Beyond the regular phonon contribution in T^3, an excess specific heat is measured below 1 K with a $T^{0.32}$ law. Such an excess contribution T^ν has been obtained in (TaSe$_4$)$_2$I with $\nu = 0.22$[59], ~ 1 in NbSe$_3$ [60] and ~ 0.6 in K$_{0.3}$MoO$_3$.[60]

Moreover, in the temperature range where these excitations are detectable, the thermal relaxation does not follow an exponential decay. The technique of measurements is a transient heat pulse technique in which the specific heat is calculated from the decay of the temperature increment after a heat pulse as $\Delta T(t) = \Delta T_0 \exp{-t/\tau}$ with $\tau = C_p R_\ell$ (R_ℓ being the thermal resistivity of the thermal link to the cold sink). For heat pulse duration of ~ 1 s, $\Delta T(t)$ was shown to decrease with a stretched exponential variation. But the recovering towards the equilibrium depends of the duration during which the thermal increment ΔT_0 has been applied. These aging effects bind even more the CDW systems with the other disordered compounds.[61] Fig. 13 shows the variation of $\Delta T(t)/\Delta T_0$ as a function of time during which the temperature increment ΔT_0 has been applied : 1 s, 5 hours and 13 hours. The energy relaxation needs longer time when the perturbation ΔT has been applied longer. Similarly to spin glasses, this time plays the role of "waiting time".

The curves in Fig. 13 correspond to the following fits :

$$\Delta T \propto (t/\tau)^{-\alpha} \exp[(-t/\tau)^{\beta}]$$

with $\tau = 28$ s, $\alpha = 0.14$, $\beta = 0.4$ for the shorter waiting time (1 s) ; $\tau = 260$ s, $\alpha = 0.04$, $\beta = 0.42$ for a waiting time of 5 hours ; $\tau = 350$ s, $\alpha = 0.035$, $\beta = 0.43$ for a waiting time of 13 hours. Then for short time, the relaxation is described by a power law $t^{-\alpha}$ with $\alpha = 0.04$. For a given waiting time (1 s) $\Delta T(t)$ has been measured for several temperatures and $\beta(T)$ and $\tau(T)$ deduced : $\beta \sim 0.3$ at the lowest temperature and reaches 1 at $T \sim 0.6$ K while τ shows an Arrhenius variation with an activation energy of 0.3-0.4 K.

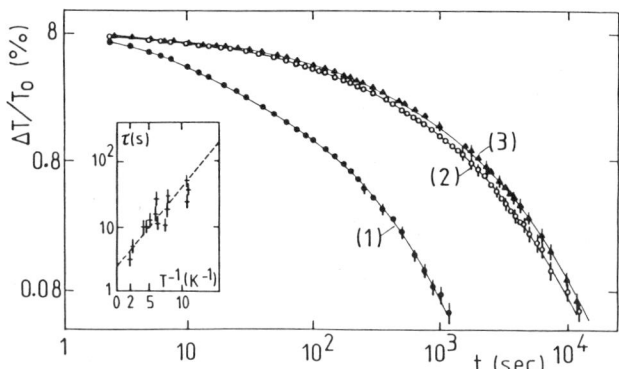

Fig. 13. Variation of $\Delta T/\Delta T_o$ as a function of time in a log-log plot when the heat increment ΔT_o is applied 1 s in (1), 5 hours in (2) and 13 hours in (3). The curves are the fits corresponding to Eq. 19. The inset shows the temperature dependence of the time constant τ in the case of pulse heat flow.

Thus at low temperature, the relaxation processes occur over a very broad distribution. The CDW metastable states can be described as a frozen landscape of potential wells and hills with some height and depth for inhibiting the evolution of the system in phase space. The time necessary for the system to jump from one metastable state to another is thermally activated as :

$$\tau = \tau_0 \exp(W/kT)$$

The small activation energy (0.3-0.4 K) measured in the energy relaxation reflects the very small barrier height between metastable states. A very small perturbation in energy (as small as 10^{-6} eV) allows the system to explore many neighbouring states. These results have to be examined in connection with those reported in Part 6 especially concerning the time response of the polarization induced when $E < E_T^*$ is applied. The CDW degrees of freedom are not frozen at low temperatures and excitations between many metastable states on a very long time scale are detectable.

Finally other experiments have shown these metastable states : at low temperature magnetoresistance at liquid helium temperature in $NbSe_3$ exhibits large amplitude Shubnikov-de Haas oscillations which reveals a rather simple Fermi surface. As approximated to an ellipsoïd, the volume occupied by this pocket is about 0.1 % of the Brillouin zone leading to a carrier concentration of $\sim 10^{18}$ cm^{-3} very close to the number obtained by the

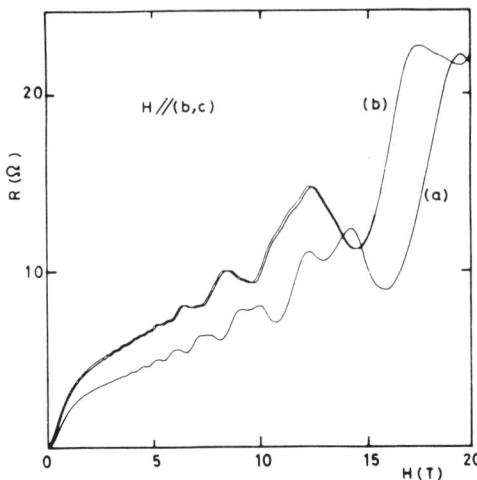

Fig. 14. Shubnikov-de Haas oscillations in the magnetoresistance of NbSe$_3$ at T = 1.8 K (a) when the sample is first cooled from room temperature, (b) when the CDW has been depinned and repinned at helium temperatures.

Hall constant. The period of quantum oscillations is slightly sample dependent varying for instance between 0.36 MG to 0.3 MG when H is applied // to the (b,c) plane. But when the CDW is depinned and repinned at low temperatures, the period of the oscillations decrease and in the case H // (b,c) is always 0.29 MG. Fig. 14 shows the effect of the depinning-repinning process on the magnetoresistance of NbSe$_3$ at 2 K.[62] A shift in the extrema of the Shubnikov-de Haas oscillations means that the Fermi surface has been modified during the depinning-repinning process. The volume of the Fermi surface pocket has been reduced by $\sim$ 10 % which would increase the CDW wave vector of 10^{-4} (there are $\sim 10^{21}$ e/cm^3 in the band affected by the CDW). One can think that when cooled through the Peierls transition, the CDW is away from its equilibrium state and that the depinning-repinning operation has released local distortion towards a better equilibrium state.

Broadening of the superlattice spots in directions parallel and perpendicular to the chain axis of $K_{0.3}MoO_3$ has also been observed[63] when a pulse higher than E_T^* is applied. This perturbation remains when the field is removed.

8. CONCLUSIONS

This new collective conducting state induced by a CDW sliding is now well established and it has been found in different families of quasi one-dimensional materials. The general properties of this state are more or less well analysed although in the recent years, new unexpected properties have been discovered. However in spite of all these efforts most of the fundamental questions remain unsolved. Progress in crystal growth quality and a better characterization of defects either in the crystal structure or in the CDW structure will be crucial for further developments in the field. The central point for understanding the non-linear properties remains to know how the CDW phase slides and especially the role of phase dislocations. These dislocations are necessary at the electrodes for the CDW condensate -normal current conversion and these processes are probably at the origin of at least a part of the observed periodic noise-. But the nature of the pinning -strong pinning or weak pinning interaction- in the volume still remains unsettled. However as recently shown, impurity pinning can be mixed with other pinning origins as surface pinning or pinning at the electrodes. Consequently any reliable study of pinning effect should separate these different contributions. The switching in the I(V) characteristic of $K_{0.3}MoO_3$ at helium temperature shows a new conducting state with a transition from an overdamped disordered motion to a underdamped motion. Experiments on other CDWs insulating at helium temperatures have to be performed in order to establish definitively the generality of this behaviour. Then very interesting would be the scaling between the threshold E_T^* and the pinning energy proportional to ω_p^2. Role of contacts which seems to strongly affect the value of E_T^* needs also to be investigated. However aging effects in energy relaxation measurements below 1 K have revealed the strong disordered nature of the CDW state which allows analogy with other disordered materials as spin glasses. Metastable states are separated by energy barriers which extend to arbitrary low values. Thus these results have to be consistently analysed with those which indicate the non-deformability of the CDW for $E < E_T^*$. New techniques have been recently used as NMR and tunneling in the sliding state. STM microscopy might be very useful for studying locally pinning effects but inherent in the technique only at the surface. The similar collective mode has been shown to exist in a few SDW systems. Other SDW conductors have to be studied and the interaction of impurities on the threshold field better established.

Acknowledgements - I whould like to thank K. Biljakovic, T. Chen, O. Laborde, J.C. Lasjaunias, M. Renard, J. Richard, and M.C. Saint-Lager with whom the experimental work has been conducted in collaboration, F. Levy and A. Meerschaut for providing samples, A. Bjelis, R. Currat, G. Grüner, K. Maki and M. Renard for useful discussions and American Physical Society and Pergamon Press for permission to reproduce Fig. 3, 37 and Fig. 10, 11 respectively.

REFERENCES

1. J. Bardeen, L.N. Cooper and J.R. Schrieffer, Phys. Rev. **108**, 1175 (1957).
2. H. Fröhlich, Proc. Royal Soc. **A223**, 296 (1954).
3. R.E. Peierls, Ann. Phys. **4**, 121 (1930) and *Quantum Theory of Solids*, Oxford University Press, p. 108 (1955).
4. P.A. Lee, T.M. Rice and P.W. Anderson, Solid State Commun. **14**, 703 (1974).
5. P. Monceau, N.P. Ong, A. Portis, A. Meerschaut and J. Rouxel, Phys. Rev. Lett. **37**, 602 (1976).
6. *Electronic Properties of Inorganic Quasi One-Dimensional Compounds*, Part I and II, ed. P. Monceau (D. Reidel, Dordretch, 1985) ; *Crystal Chemistry and Properties of Materials with Quasi One-Dimensional Structures*, ed. J. Rouxel (D. Reidel, Dordretch, 1986) ; *Low Dimensional Electronic Properties of Molybdenum Bronzes and Oxides*, ed. C. Schlenker (D. Reidel, Dordretch, 1989) ; G. Grüner and A. Zettl, Phys. Rep. **119**, 117 (1985) ; C. Grüner, Rev. Mod. Phys. **60**, 1129 (1988) ; J.C. Gill, Contemp. Phys. **27**, 37 (1986) ; J.C. Gill, Physica Scripta T **25**, 51 (1989) ; *Charge Density Waves in Solids*, Modern Problems in Condensed Matter Science Series, to be published, ed. L.P. Gor'kov and G. Grüner (Elsevier, Lausanne).
7. *Proceedings of the International Conference on Charge Density Waves in Solids*, held in Budapest in August 84, *Lecture Notes in Physics*, ed. Gy Hutiray and J. Solyom (Springer-Verlag, Berlin) Vol. 217 (1985) ; *Proceedings of the International Conference on Physics and Chemistry of Quasi One-Dimensional Conductors*, held at Lake Kawaguchi, Japan in May 1986, ed. S. Tanaka and K. Uchinokura, Physica **143B** (1986) ; *Low-Dimensional Conductors and Superconductors*, ed. D. Jérôme and L.G. Caron, NATO ASI Series **B155** (1987).
8. S. Tomic, J.R. Cooper, D. Jérôme and K. Bechgaard, Phys. Rev. Lett. **62**, 462 (1989).
9. P.J. Yetman and J.C. Gill, Solid State Commun. **62**, 201 (1987).
10. D.V. Borodin, S.V. Zaitsev-Zotos, and F.Ya Nad, Zh. Eksp. Teor. Fiz. **93**, 1394 (1987), Soviet Phys. JETP **66**, 793 (1987).
11. D. Reagor, S. Sridhar and G. Grüner, Phys. Rev. B **34**, 2212 and 2223 (1986).
12. R.E. Thorne, J.S. Hubacek, W.G. Lyons, J.W. Lyding and J.R. Tucker, Phys. Rev. B **37**, 10055 (1988).
13. J. Bardeen, Phys. Rev. Lett. **42**, 1498 (1979) : **45**, 1978 (1980) ; Phys. Rev. Lett. **55**, 1010 (1985) ; Physica B **143**, 14 (1986) ; Physica Scripta T **27**, 136 (1989).
14. P.A. Lee and T.M. Rice, Phys. Rev. B **19**, 3970 (1979).
15. G. Grüner, A. Zawadowski, and P.M. Chaikin, Phys. Rev. Lett. **46**, 511 (1981).
16. P. Monceau, J. Richard and M. Renard, Phys. Rev. Lett. **45**, 43 (1980) ; Phys. Rev. B **25**, 931 (1982).

17. D.S. Fisher, Phys. Rev. Lett. **50**, 1486 (1983).
18. G. Binning and H. Rohrer, Rev. Mod. Phys. **59**, 615 (1987).
19. J. Tersoff and D.R. Hamman, Phys. Rev. B **31**, 805 (1985).
20. C.G. Slough, W.W. McNairy, R.V. Coleman, B. Drake and P.K. Hansma, Phys. Rev. B **34**, 994 (1986).
21. B. Giambattista, A. Johnson, W.W. McNairy, C.G. Slough and R.V. Coleman, Phys. Rev. B **38**, 3545 (1988).
22. C.G. Slough, B. Giambattista, A. Johnson, W.W. McNairy and R.V. Coleman, Phys. Rev. B **39**, 5496 (1989).
23. K. Nomura and K. Ichimura, Solid State Commun. **71**, 149 (1989).
24. N.P. Ong, G. Verma and K. Maki, Phys. Rev. Lett. **52**, 663 (1984); N.P. Ong and K. Maki, Phys. Rev. B **32**, 6582 (1985).
25. L.P. Gor'kov, Pis'ma Zh. Eksp. Teor. Fiz. **38**, 76 (1983); JETP Letters **38**, 37 (1983).
26. I. Batistic, A. Bjelis and L.P. Gor'kov, J. Physique **45**, 1059 (1984).
27. J.C. Gill, J. Phys. C **19**, 6589 (1986).
28. M. Renard and M.C. Saint-Lager, to be published.
29. P. Monceau, M. Renard, J. Richard and M.C. Saint-Lager, Physica **143B**, 64 (1986).
30. M.C. Saint-Lager, P. Monceau and M. Renard, Europhys. Lett. **9**, 585 (1989).
31. R.M. Fleming in *Low-Dimensional Conductors and Superconductors*, ed. D. Jérôme and L.G. Caron, NATO ASI Series **B155**, 433 (1987).
32. J.R. Tucker, W.G. Lyons, J.H. Miller, Jr., R.E. Thorne and J.W. Lyding, Phys. Rev. B **34**, 9038 (1980).
33. R.M. Fleming, R.J. Cava, L.F. Schneemeyer, E.A. Rietman and R.G. Dunn, Phys. Rev. B **33**, 5450 (1986).
34. L. Sneddon, Phys. Rev. B **29**, 719 (1984).
35. P. Littlewood, Phys. Rev. B **36**, 3108 (1987); Synth. Metals **29**, F531 (1989).
36. S. Takada, K.Y.M. Wong and T. Holstein, Phys. Rev. B **32**, 4639 (1985).
37. G. Mihaly, T.W. Kim and G. Grüner, to be published and G. Mihaly, Physica Scripta, to be published.
38. C. Escribe-Filippini, J.P. Pouget, B. Hennion and M. Sato, Synth. Metals **19**, 931 (1987).
39. H.K. Ng, G.A. Thomas, and L.F. Schneemeyer, Phys. Rev. B **33**, 8755 (1986).
40. W.G. Lyons and J.R. Tucker, Phys. Rev. B **38**, 4303 (1988).
41. P.A. Lee and H. Fukuyama, Phys. Rev. B **17**, 542 (1978).
42. Y. Kurihara, J. Phys. Soc. Jpn **49**, 852 (1980).
43. S. Barisic, Mol. Cryst. Liq. Cryst. Liq. Cryst. **119**, 413 (1985); in *Low-Dimensional Conductors and Superconductors*, ed. D.Jérôme and L.G. Caron, NATO ASI Series **B155**, 395 (1987).
44. Y. Nakane and S. Takada, J. Phys. Soc. Jpn **54**, 977 (1985).
45. K.Y. Wong and S. Takada, Phys. Rev. B **36**, 5476 (1987).
46. A. Maeda, T. Furuyama and S. Tanaka, Solid State Commun. **55**, 951 (1985).
47. L. Mihaly and G.X. Tessema, Phys. Rev. B **33**, 5858 (1986).
48. G. Mihaly and P. Beauchene, Solid State Commun. **63**, 911 (1987).
49. W. Fogle and J.H. Perlstein, Phys. Rev. B **6**, 1402 (1972).
50. A. Maeda, M. Notomi, K. Uchinokura and S. Tanaka, Phys. Rev. B **36**, 7709 (1987).
51. A. Maeda, T. Furuyama, K. Uchinokura and S. Tanaka, Solid State Commun. **58**, 25 (1986).
52. S. Martin, R.M. Fleming and L.F. Schneemeyer, Phys. Rev. B **38**, 5733 (1988).
53. T. Chen, L. Mihaly and G. Grüner, Phys. Rev. Lett. **60**, 464 (1988).
54. G. Mihaly, T. Chen and G. Grüner, Phys. Rev. B **38**, 12740 (1988).
55. J.H. Miller, J. Richard, J.R. Tucker and J. Bardeen, Phys. Rev. Lett. **51**, 1592 (1983).
56. M.J. Cohen and A.J. Heeger, Phys. Rev. B **16**, 688 (1977).
57. S.K. Zhilinskii, M.E. Itkis, I.Yu Kal'nova, F.Ya Nad' and V.B. Preobrazhenskii, Zh. Eksp. Teor. Fiz. **85**, 362 (1983), Soviet Phys. JETP **58**, 211 (1983).
58. K. Biljakovic, J.C. Lasjaunias, P. Monceau and F. Levy, Europhys. Lett. **8**, 771 (1989).
59. K. Biljakovic, J.C. Lasjaunias, F. Zougmoré, P. Monceau, F. Levy, L. Bernard and R. Currat, Phys. Rev. Lett. **57**, 1907 (1986).

60. K. Biljakovic, K. Hasselbach, J.C. Lasjaunias, unpublished.
61. K. Biljakovic, J.C. Lasjaunias, P. Monceau and F. Levy, Phys. Rev. Lett. **62**, 1512 (1989).
62. J. Richard, P. Monceau and M. Renard, Phys. Rev. B **35**, 4533 (1987).
63. L. Mihaly, K.B. Lee and P.W. Stephens, Phys. Rev. B **36**, 1793 (1987).

SPIN DENSITY WAVES IN ORGANIC CONDUCTORS

Kazumi Maki

Department of Physics
University of Southern California
Los Angeles, CA 90089-0484

RESUMÉ

In the following we discuss two aspects of the spin density wave (SDW) state in organic conductors like TMTSF salts (Bechgaard salts) and DMET salts at low temperatures. As a model we take an anisotropic Hubbard model and we study the properties of the model within mean field theory. In the first part we describe the collective transport associated with the sliding motion of the SDW (the Fröhlich conduction), which shares a number of similarities with the related transport in the charge density wave. So far only a few experiments have been reported on the Fröhich conduction in the SDW. But these results are quite encouraging. In the second part we describe the field induced spin density wave (FISDW) in the related systems, which appears only in a strong magnetic field (H>4T) normal to the best conducting plane. The same model describes semi-quantitatively the observed phase diagram and other properties for H ≤ 8T, while the theory appears to fail mysteriously to describe a class of observed phenomena beyond H = 8T.

INTRODUCTION

We associate usually the modern development in physics of low dimensional systems with a very suggestive paper by Little[1]. Although his suggestion has not been realized so far, the concerted effort to find high transition temperature superconductors in low dimensional systems lead to discovery of organic conductors like TTF-TCNQ, $(TMTSF)_2PF_6$ and $(TMTSF)_2C\ell O_4$ and inorganic conductors like $NbSe_3$ and TaS_3^2. Besides the fact that both $(TMTSF)_2C\ell O_4$ and $(TMTSF)_2PF_6$ under pressure exhibit superconductivity at low temperatures with T_c = 1.2K, they become SDW or FISDW under other conditions, which should exhibit the Fröhlich conduction.

For modelling these systems there has been an early tendency to formulate the system within one dimensional model[3], where the fluctuation is of paramount importance. On the other hand there are accumulating evidences that the system should be formulated as the two or three dimensional where the unnesting parameter (which does not exist in the one dimensional model) will play the crucial role in interpreting the FISDW. Further we believe that mean field theory will provide the basis for a quantitative analysis for these systems. It may appear to be somewhat surprising why mean field theory applies to the low dimensional systems. We argue that they are indeed the three dimensional systems though they are strongly anisotropic. For example the conductivity along the chain direction (we shall denote this as the a direction) and those for other directions can be different by orders of magnitude. As a typical example we may assume for $(TMTSF)_2ClO_4$.

$$\sigma_a/\sigma_b \sim 10^2, \quad \sigma_b/\sigma_c \sim 10^3$$

As a model we take an anisotropic Hubbard model as introduced by Yamaji[4]

$$H = \sum_{\alpha p} \varepsilon(p) C^+_{p\alpha} C_{p\alpha} + V \sum_q n_{q\uparrow} n_{-q\downarrow} \tag{1}$$

where

$$\varepsilon(p) = -2t_a \cos ap_1 - 2t_b \cos bp_2 - 2t_c \cos cp_3 - \mu \tag{2}$$

with $t_a/t_b/t_c \simeq 10/1/0.03$

and μ is the chemical potential and $C^+_{p\alpha}$ is the electron creation operator with momentum $\vec{p}$ and spin α. Further for definiteness we assume that the electron band is about 3/4 filled. In this circumstance the Fermi surface consists of two warped sheets extending in the y, z directions. When t_b and t_c are much smaller than t_a, the almost perfect nesting of two Fermi surfaces is achieved by choosing the nesting vector $Q = (2p_F, \pi/b, \pi/c)$.

Within mean field theory we can diagonalize the Hamiltonian (1) as[5]

$$H = \sum_p \psi_p^+ (\xi(p)\rho_3 + \eta(p) + \Delta \rho_1\sigma_3) \psi_p + 2 U^{-1} |\Delta|^2 \tag{3}$$

where

$$\xi(P) = \frac{1}{2}(\varepsilon(p) - \varepsilon(p+Q)) = 2t_a \sin ap_F \sin a(p_1 - p_F) - 2t_b \cos bP_2$$

$$\eta(p) = \frac{1}{2}(\varepsilon(p) + \varepsilon(p+Q)) = -2t_a \cos ap_F(1 - \cos a(p_1 - p_F)) - \delta\mu$$

$$= \varepsilon_0 \cos(2bp_2) \tag{4}$$

with

$$\varepsilon_0 = -\frac{1}{4} t_b^2 \cos ap_F (t_a \sin^2 ap_F)^{-1} \tag{5}$$

and Δ is determined from the self-consistent equation

$$\Delta = U \langle n_{Q\uparrow} \rangle = \frac{1}{2}\bar{U} 2\pi T \sum_\omega \left\langle \frac{\Delta}{[(\omega + i\eta)^2 + \Delta^2]^{1/2}} \right\rangle \tag{6}$$

Here $\langle\ \rangle$ means the average over p_2, $\bar{U} = UN_0$, ρ_i's are the Pauli matrices operating on the spinor space formed by the right-going and the left-going electrons and N_0 is the electron density of states at the Fermi surface per spin. We have also neglected the t_c term for simplicity.

The thermodynamics of the present model is well known.[5,6] For example the transition temperature is given by

$$1 = \frac{1}{2} \bar{U} \, 2\pi T \sum_\omega (\omega_n^2 + \varepsilon_0^2)^{-1/2} \tag{7}$$

Since this equation has the same form as the gap equation in a BCS superconductor, the $\varepsilon_0 = \Delta(T_c/T_{c0})$ where $\Delta(T/T_c)$ is the energy gap of the BCS theory at finite T. In particular T_c vanishes when $E_0 = \Delta_{00}$. The ε_0 dependence of T_c is shown in Fig. 1a togăther with the jump in the specific heat normalized by the one for $\varepsilon_0 = 0$.

It may be useful to rewrite Eq(6) as

$$1 = \frac{1}{2} \bar{U} \int_\Delta^{\varepsilon_f} dE \sqrt{\frac{dE}{E^2 - \Delta^2}} \langle \tanh(\frac{1}{2T}(E-\eta)) \rangle \tag{8}$$

we see then immediately that

$$\Delta_0(\varepsilon_0) = \Delta_{00} \tag{9}$$

the energy gap at T = 0K is independent of ε_0. The condensation energy at T = 0K is also given by

$$E_G = -\frac{1}{2} N_0 (\Delta_0^2 - \varepsilon_0^2) \tag{10}$$

Therefore, though Δ_0 is independent of ε_0, the condensation energy is depleted as ε_0 increases. In most of TMTSF salts with the SDW ground state, the SDW transition temperature T_c decreases when pressure is applied. Following Yamaji[4] this decrease in T_c is interpreted as due to increase in ε_0 under pressure. As T_c decreases the SDW encounters the superconductivity around P ~7~8 kbar. Then the superconductivity in general invades the SDW region[3,7], which is not surprising since the condensation energy for the SDW is almost gone when $\varepsilon_0 \simeq \Delta_0$. We stress that this kind of behavior cannot be described with the one dimensional model.

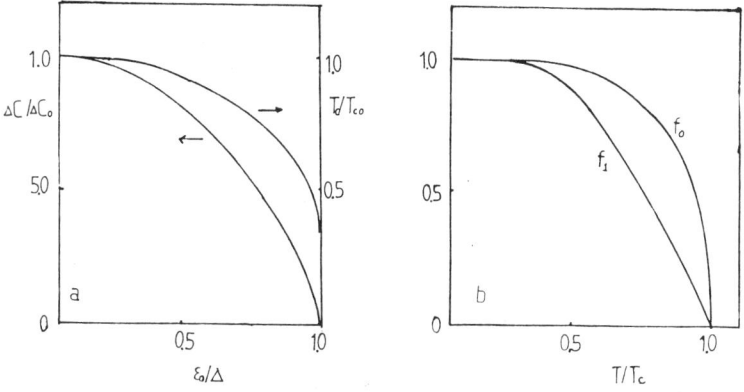

Fig. 1. a) The normalized transition termperature T_c/T_{c0} and the normalized jump in the specific heat are shown as functions of ε_0/Δ_0. b) The condensate densities f_0 (the dynamic limit) and f_1 (the static limit) as functions of the reduced temperature.

PHASON DYNAMICS

The SDW is characterized by appearance of the order parameter

$$\vec{S}(\vec{x}) = 2\Delta U^{-1} \hat{n} \cos(\vec{Q}\vec{x} + \phi) \tag{11}$$

where $\hat{n}$ is a unit vector and ϕ is the phase of the order parameter. In the present model fluctuations in $\hat{n}$ and ϕ constitute the Goldston modes of the SDW's the spin wave and the phason. Both are gapless. However in the real system the spin-spin dipole interaction and the spin orbit scattering break the rotation symmetry of $\hat{n}$, while the impurities and the crystalline defects break the translational symmetry of the SDW though the pinning potential due to impurities is calculated only recently.[9,10] In the following we shall concentrate on the sliding motion of the SDW (i.e the phason).

Following Brazovskii and Dzyaloshinskii[11] we introduce a pseudo unitary transformation (chiral gauge transformation)

$$\psi(\vec{x},t) \rightarrow e^{\frac{1}{2}i\phi\rho_3\sigma_3}\psi(\vec{x},t) = U\psi(\vec{x},t) \tag{12}$$

Then from the definition of the order parameter (6), which is rewritten as

$$\Delta(\vec{x},t) = U\frac{1}{4}Tr(\psi(\vec{x},t)\rho_+\sigma_3\psi(\vec{x},t)) \tag{13}$$

we see that ϕ corresponds to the phase of the order parameter. When ϕ is independent of $\vec{x}$ and t, U commutes with H. The system is invariant for global chiral gauge transformation, though this invariance will be broken in the presence of impurities. (i.e. the pinning centers) Now let us consider the local chiral gauge transformation where ϕ depends slowly on $\vec{x}$ and t. In this case the transformation is no longer unitary and it generates two additional terms in the Hamiltonian density

$$\Delta H = \bar{U}^1 H U + \bar{U}^1 i \frac{\partial}{\partial t} U = \frac{1}{2} v \psi_\alpha^+(\vec{x})\psi_\alpha(\vec{x})\frac{\partial \phi}{\partial x} - \frac{1}{2}\psi_\alpha^+ \rho_3 \psi_\alpha \frac{\partial \phi}{\partial t}$$

$$= (2e)^{-1} v\, n(\vec{x}) \frac{\partial \phi}{\partial x} - (2ev)^{-1} j(\vec{x}) \frac{\partial \phi}{\partial t} \tag{14}$$

where $n(\vec{x})$ and $j(\vec{x})$ are the charge density and the current operators. Assuming that ΔH is a small correction to H, we first find the effective Hamiltonian H_ϕ, which describes the phason dynamics as

$$H_\phi = \frac{1}{4} N_0 f[(\frac{\partial \phi}{\partial t})^2 + v^2(\frac{\partial \phi}{\partial x})^2] \tag{15}$$

and corresponding relations

$$n_{CDW} = -enfQ^{-1} \frac{\partial \phi}{\partial x} \tag{16}$$

$$J_{CDW} = enfQ^{-1} \frac{\partial \phi}{\partial t} \tag{17}$$

where f is a complicate function [5,9] of ω and q. Here ω and q are the frequency and the wave vector associated with ϕ and n is the electron density and $Q = 2p_F$. In the limit $\omega > vq$ and $\omega < vq$, f takes the familiar form

$$f = f_0 \simeq \int_0^{2\pi} \frac{d\phi}{2\pi} \int_0^\infty d\chi \, \text{sech}^2\chi \, \text{th}[\tfrac{1}{2}\beta(\Delta \text{ch}\chi - \eta)] \tag{18}$$

and

$$f = f_1 \simeq 1 - \tfrac{1}{2}\beta\Delta \int_0^{2\pi} \frac{d\phi}{2\pi} \int_0^\infty d\chi \, \text{ch}\chi \, \text{sech}^2[\tfrac{1}{2}\beta(\Delta \text{ch}\chi - \eta)] \tag{19}$$

respectively where $\eta = \varepsilon_0 \cos 2\phi$.

These f_0 and f_1 are called the dynamic and the static limit of the condensate. Further they are presented often as associated with Eq(16) and Eq(17) respectively.[11,12] However, as our present derivation shows, the same f function should appear in both Eqs (16) and (17), so that the conservation of the CDW charge is guaranteed for all temperatures;

$$\frac{\partial n_{CDW}}{\partial t} + \frac{\partial j_{CDW}}{\partial x} = 0 \tag{20}$$

Some limiting behaviors of f_0 and f_1 are useful. In the vicinity of $T = 0K$ we obtain

$$f_0 = 1 - 2\sum_{n=1}^\infty (-1)^{n+1} \tilde{K}(n\beta\Delta) I_0(n\beta\varepsilon_0) \tag{21}$$

$$f_1 = 1 - 2\beta\Delta \sum_{n=1}^\infty (-1)^{n+1} n K_1(n\beta\Delta) I_0(n\beta\varepsilon_0) \tag{22}$$

where $K_1(z)$ and $I_0(z)$ are modified Bessel functions and $\tilde{K}(z)$ defined by

$$\tilde{K}(z) = \int_0^\infty d\chi \, \text{sech}^2\chi \, e^{-z \text{ch}\chi} \tag{23}$$

On the other hand in the vicinity of $T = T_c$ we obtain

$$f_0 \simeq \pi\beta\Delta \sum_{n=1}^\infty (-1)^{n+1} NL(n\beta\varepsilon_0) + f_1 \tag{24}$$

$$f_1 \simeq 2\pi T\Delta^2 \sum_{n=0}^\infty (\omega_n^2 - \tfrac{1}{2}\varepsilon_0^2)(\omega_n^2 + \varepsilon_0^2)^{-5/2} \tag{25}$$

where ω_n is the Matsubara frequency and

$$L(z) = \frac{2}{\pi}\int_0^{\pi/2} dQ \, e^{-z\sin Q} \tag{26}$$

Further in the limit $\varepsilon_0 = 0$, both f_0 and f_1 are evaluated numerically and shown in Fig. 1b. We note that except at $T = 0K$ where $f_0 = f_1 = 1$, we have $f_0 > f_1$. Further as T approaches T_c, f_0 vanishes like $\left(1 - \frac{T}{T_c}\right)^{1/2}$, while f_1 vanishes like $\left(1 - \frac{T}{T_c}\right)$. Second, the coupling to an external electric field E along the chain direction is obtained from

$$e n f Q^{-1} V(x) \frac{\partial \phi}{\partial x} \approx - e n f Q^{-1} \phi E \qquad (27)$$

where $V(x)$ is the external potential. Third the pinning potential is obtained in the presence of nonmagnetic impurities as[5,9]

$$V_{pin} = - (\frac{\pi}{2} N_0 V_2)^2 \Delta(T) \tanh(\frac{1}{2}\beta \Delta(T)) \cos(2\vec{Q}\cdot\vec{x} + 2\phi(\vec{x}))$$

$$\times \sum_i \delta(\vec{x} - \vec{x}_i) \qquad (28)$$

where $\vec{x}_i$ are the impurity site and V_2 is the Fouriers component of the impurity potential with wave vector Q. In contrast to the case of a charge density wave (CDW), the lowest order term is the second order in V_2. Further considering the fact that the SDW transition temperature of most organic conductors is around 10K while the CDW transition temperature is around 100 ~ 200K, we expect that the pinning potential in a SDW is smaller by a factor 10^{-2} than that in a CDW if the impurity concentration in two systems are of the same order of magnitude. This prediction appears to be born out in a recent observation of the nonohmic conductivity in (TMTSF)$_2$NO$_3$ by Tomic et al.[13] Further the observed temperature dependence of the threshold field E_T appears to be consistent with E_T calculated within the present model.[14] In particular E_T is almost independent of T for $T/T_c \leq 0.5$.

THESHOLD ELECTRIC FIELD

Now let us consider the threshold electric field. It is now well established that a number of CDWs in quasi-one dimensional compounds like NbSe$_3$ etc exhibit the Fröhlich conduction[15]; the sliding motion of the CDW. One of the hallmark of the Fröhlich conduction is nonohmic conduction. The electric conductivity is ohmic as long as $E_2 < E_T$ the threshold field. In a typical CDW E_T is of the order of 10^{-2} ~ 10V/cm. The threshold electric field in a CDW has been considered by Fukuyama and Lee[16] and Lee and Rice[17] within a phenomenological model. We shall repeat here their analysis within our microscopic model, which enable us to predict the temperature dependence of E_T. Following Fukuyama and Lee[16] we consider the strong pinning and the weak pinning limit separately. The first limit applies when the impurity scattering is strong and/or the impurity concentration is rather low. In such a situation a CDW chooses the low energy configuration by minimizing ϕ locally at each impurity site, while ϕ between two impurities change linearly with distance interpolating two limiting values. In this circumstance the elastic energy associated with ϕ is proportional to $L^{-2} \propto n_i^{2/3}$ where n_i is the impurity concentration. In the strong pinning limit the elastic energy is negligible in comparison with the potential energy.

Then comparing the potential energy to the force exerted by an electric field on ϕ in Eq(27), we obtain

$$E_T^S(T) = \frac{Q}{e}(\frac{n_i}{n})(\pi N_0 V_2)^2 \Delta(T) \tanh(\frac{\beta}{2}\Delta(T)) f_1^{-1} \qquad (29)$$

where we replace Eq(28) with it's spatial average. Here we took the static limit of the f function as the phase ϕ is dominated by the spatial variation due to the pinning $q \sim L^{-1}$ where $L = (n_i/n)^{-1/3}$ in the strong pinning limit. The superscript S in Eq(29) means the strong pinning limit. $E_T^S(T)$ is proportional to n_i. The temperature dependence of $E_T^S(T)/E_T^S(0)$ is evaluated numerically and shown in Fig. 2. Unlike E_T in a CDW, E_T in a SDW exhibits only weak temperature dependence. E_T is almost constant up to $T = 0.5\ T_c$ then increases monotonically to $1.33\ E_T(0)$ at $T = T_c$. In particular $E_T(T)$ does not diverge at $T = T_c$.

In the weak pinning limit, on the other hand, a single impurity cannot pin the SDW.[16] According to Fukuyama and Lee[16] impurities in the domain of the size L^D will collectively pin the SDW. Here D is the spatial dimension of the SDW. The pinning potential due to the impurities in the single domain is estimated as

$$\bar{V}_{pin} = - (\tfrac{\pi}{2} N_0 V_2)^2 \Delta(T) \tanh[\tfrac{\beta}{2}\Delta(T)](n_i L^D)^{\tfrac{1}{2}} \tag{30}$$

where we replaced the impurity average by a random walk in two dimensional plane (i.e.., the complex plane) with $n_i L^D$ steps. On the other hand the collective pinning costs the elastic energy

$$E_{e\ell} = \tfrac{1}{4} N_0 f_1 \bar{v}^2 L^{-2} \tag{31}$$

where $\bar{v}^2$ is the average of the square of the Fermi velocities in

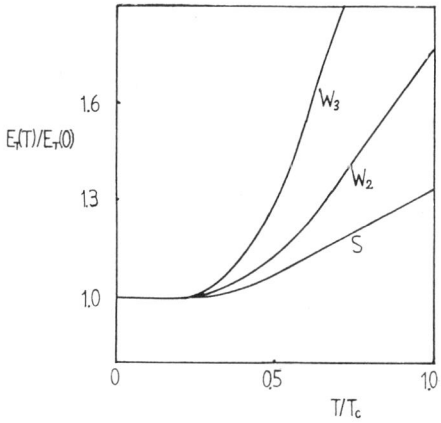

Fig. 2. The temperature dependence of the threshold field in a SDW (with $\varepsilon_0 = 0$) or in a FISDW as functions of the reduced temperature T/T_c. S, W_2, W_3 mean the strong pinning limit, the weak pinning limit (D = 3) and the weak pinning limit (D = 3).

different direction. For D = 2 we have to replace $\bar{v}^2$ by $v\, v^\perp$ for example. The total pinning energy density is minimized for

$$L = \left(N_0 f_1 \bar{v}^2 / D n_i^{1/2} (\tfrac{\pi}{2} N_0 V_2)^2 \Delta(T)\tanh[\tfrac{\beta}{2}\Delta(T)] \right)^{2/4-D} \tag{32}$$

with

$$E_{pin} = - \tfrac{(4-D)}{4D} N_0 f_1 \bar{v}^2 L^{-2} \tag{33}$$

Then the threshold field is given by

$$E_T^W(T) = \frac{(4-D)}{4D} \frac{Q}{e}(\frac{n_i}{n})^{2/4-D}[D(\frac{\pi}{2} N_0 V_2)^2 \Delta(T)\tanh[\frac{\beta}{2}\Delta(T)]f_1^{-1}]^{4/4-D} \quad (34)$$

In particular we obtain

$$E_T^W(T)/E_T^W(0) = (E_T^S(T)/E_T^S(0))^{4/4-D} \quad (35)$$

In general the temperature dependence of $E_T(T)$ in the weak pinning limit is stronger than the one in the strong pinning limit. We plot $E_T^W(T)$ for $D = 2$ and 3 in Fig. 2 togather with $E_T^S(T)/E_T^S(0)$.

MICROWAVE CONDUCTIVITY

Following Lee, Rice and Andersen[8] the microwave conductivity is easily obtained as[18]

$$\sigma(\omega) = \frac{e^2 n}{m} \{\Gamma_n^{-1}(1-f) + i\omega f [\omega(\omega + i\Gamma_p) - \omega_p^2]^{-1}\} \quad (36)$$

where f is the f function, ω_p is the pinning frequency and Γ_n and Γ_p are the normal and the phason damping constant. Except the damping term Γ_p the second term in Eq(36) is read out of H_ϕ by making use of the relation (17). In a model where quasi-particle scattering is due to randomly distributed impurities[18] Γ_n and Γ_p are given by

$$\Gamma_n = 4T(1-f)[\int_\Delta^\infty dz \, \text{sech}^2[\tfrac{1}{2}\beta z](z^2-\Delta^2)[\Gamma z^2 \mp_2 \Gamma'\Delta^2]^{-1}]^{-1} \quad (37)$$

$$\simeq \Gamma_2(1-f)[f(\Delta)]^{-1}$$

and

$$\Gamma_p = \frac{\Delta^2 \Gamma}{T} f^{-1}\int_G^\infty dz \, \text{sech}^2(\tfrac{1}{2}\beta z)\frac{\Gamma_2 z^2}{(z^2-\Delta^2)}[\Gamma_2 z^2 + \Gamma'\Delta^2]^{-1}$$

$$\simeq 2\Gamma_2(\Delta/T)f^{-1}f(\Delta) \, f(-\Delta)[\ln(\frac{4\pi\Delta}{\Gamma}) + 2\sqrt{\frac{\Gamma'}{\Gamma_2}}\tan^{-1}\sqrt{\frac{\Gamma'}{\Gamma_2}}] \quad (38)$$

here $\Gamma = \Gamma_1 + \tfrac{1}{2}\Gamma_2$, $\Gamma' = \Gamma_1 - \tfrac{1}{2}\Gamma_2$ and Γ_1 and Γ_2 are the forward and back scattering rate due to impurities an $f(\Delta) = (1 + e^{\beta\Delta})^{-1}$ is the Fermi distribution function. At $T = T_c$ $\Gamma_n = 2\Gamma_2 = \tau_{tr}^{-1}$ the inverse of the transport life time in the normal state. We have also a relation $\Gamma_p < \Gamma_n$ at all temperatures.

It is an important question to decide which limit of the f function should be used in Eq(36). For example in the dc experiment it is clear that we have to use $f = f_1$ as long as E is not too large compared with E_T, since in this limit ϕ is dominated by the spatial distortion. However when the SDW starts sliding ϕ acquires an intrinsic time dependence related to the narrow band noise frequency. If we assume that the Fukuyama-Lee-Rice coherence Length L is of the order of 10μm as in a relatively pure sample of $NbSe_3$, the characteristic frequency $\omega_c = vL^{-1} = $ 10 GHz. Therefore in this case the nonOhmic conductivity is described by f_1 as long as the narrow band noise frequency is less than ω_c. A similar rule applies also for the microwave response. For the microwave frequency $\omega < \omega_c$ $f = f_1$ has to be used, while for $\omega > \omega_c$ $f = f_0$ has to be used.

Finally the pinning frequency ω_p may be expressed in terms of the pinning potential given in Eq(28). For simplicity let us consider the strong pinning limit. Then ω_p is given by[18]

$$\omega_p^2 = \omega_p^{*2} f_0^{-1} \tag{39}$$

with

$$\omega_p^{*2} = 2n_i N_0 (\pi V_2)^2 \Delta(T) \tanh\left(\frac{\Delta(T)}{2T}\right) \tag{40}$$

It is important to discriminate two cases $\omega_p^{*2} \ll \Delta_0^2$ and $\omega_p^{*2} \gg \Delta_0^2$ where Δ_0 is the energy gap at T = 0K

In the first case $(\omega_p^{*2} \ll \Delta_0^2)$ the low frequency absorption in the normal state is shifted into a resonance absorption around $\omega = \omega_p^* [f_0]^{-1/2}$. In this situation the quasi-particle absorption at $\omega = 2\Delta$ is almost invisible, since the Drude tail at $\omega = 2\Delta$ is reduced by a factor of $(\Gamma_n/2\Delta)^2$, which is rather small. This situation is shown in Fig. 3a. On the other hand when $\omega_p^* \gg 2\Delta_0$, ω_p in Eq(39) is solved as

$$\omega_p \approx 2\Delta\left[1 - \left(\frac{\pi}{2}\left(\frac{2\Delta}{\omega_p^*}\right)^2\right)^2\right]^{1/2} \tag{41}$$

Now ω_p is very close to 2Δ but slightly less than 2Δ. Further it is easily shown that the residue of this pole in Eq(36) is extremely small $\sim (2\Delta/\omega_p^*)^6$. Therefore we don't expect to see the sliding mode in this limit.

A few words an Γ_1 an Γ_2 are in order. Unlike to a BCS superconductor the impurity has destructive effect on a SDW. For example the SDW transition temperature is suppressed from the one in the pure system by[18]

$$-\ell n \frac{T_c}{T_{c0}} = \psi\left(\frac{1}{2} + \frac{\Gamma}{2\pi T}\right) - \psi\left(\frac{1}{2}\right) \tag{42}$$

where T_{c0} is the transition temperature in the pure system and $\psi(z)$ is the di-gamma function. Here we consider the case $\varepsilon_0 = 0$ for simplicity. In particular Eq(42) predicts that T_c is completely suppressed when $\Gamma = 1/2\Delta_0$. Therefore in most of the SDW we require that $\Gamma/\Delta_0 \ll 1$. This condition leads to the small Drude tail at $\omega = 2\Delta$ as described above. Therefore unlike the case of a superconductor or that of a CDW (here the large phason mass $m^*/m \sim 10^3$ helps) it is difficult to observed the quasi-particle structure at $\omega = 2\Delta$ in a SDW.

SOUND PROPAGATION

In a brilliant series of experiments Brill and co-workers[20] and other[21] have shown that the CDW in TaS_3 exhibits strong electromechanical effect; the elastic constants depend strongly on an applied electric field in the chain direction. We don't go into phenomenological models proposed to account for this phenomenon, since we believe that they are inadequate. In our opinion the simplest way to look into the elastic constant is to study the sound propagation.[9,22] For this purpose we introduce the standard electron-phonon interaction in the present model. In the presence of the electron-phonon interaction the ionic potential is screened by the conduction electrons. In a SDW, as in a CDW, the ionic potential is then screened by both the quasi-particle and the phason. However when the SDW is pinned, the phason cannot participate in the screening. Therefore when the temperature is decreased through the SDW transition temperature the sound velocity increases in general with lowering temperature as observed in a SDW of (TMTSF)$_2$PF$_6$ by Chaikin et al[23]. Now when an electric field E exceeding E_T is applied, the SDW is depinned. Then the phason starts contributing in the screening of the ionic potential resulting in the decrease of the sound velocity.[9,22] However the general behavior of the elastic constant is more complicated than this simple picture may suggest. For example the shear moduli observed in TaS$_3$ and NbSe$_3$ exhibit no structure near or at the CDW transition temperature though they decrease in an electric field E exceeding E_T.[24,25] We now understand the difference in these behaviors as due to the diffusion pole in correlation functions.[26] When $\omega \ll Dq^2$ where

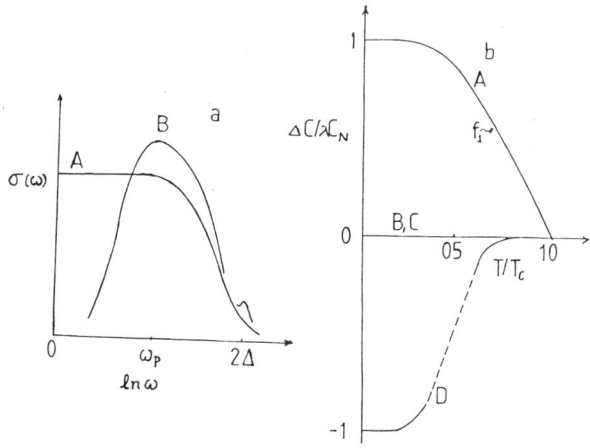

Fig. 3. a) The frequency dependent conductivity as function of the frequency ω. A and B mean $T > T_c$ and $T \ll T_c$ respectively.
b) The changes in the sound velocity $\Delta C = C - C_n$ in a SDW are shown as function of the reduced temperature. A, B, C, and D mean the pinned SDW ($\omega \ll Dq^2$), the pinned SDW ($\omega \gg Dq^2$) the unpinned SDW ($\omega \ll Dq^2$) and the unpinned SDW ($\omega \gg Dq^2$) respectively.

ω and q are the frequency and the wave vector and D is the diffusion constant, what we have described in the above is true, since in this limit both the quasi-particle and the phason participate in the screening. Further the sum of these two contributions is exactly the same as the quasi-particle term in the normal state implying that there will be no change in the elastic constant at the transition temperature when the SDW is completely unpinned.[27] On the other hand when $\omega \gg Dq^2$ the contribution of the quasi-particle term becomes negligible. Therefore in the present limit the elastic constant does not change through the SDW transition when the SDW is pinned. However, when the SDW is unpinned by an electric field the elastic constant decreases, since now the phason does contribute to the screening. Furthermore at low temperatures the amount of change in the elastic constant is independent of the ratio ω/Dq^2, but depends only on λ the electron-phonon coupling constant. We shall summerize the main result here.[26]

1. Due to the diffusion pole, the nature of the sound propagation depends sensitively on whether $\omega/Dq^2 \ll 1$ or $\omega/Dq^2 \gg 1$. Here D is the diffusion constant in a SDW and given by

$$D = \begin{cases} D_0(1+f_2-f_1)^{-1} & \text{for } T \simeq T_c \\ D_0(2\Gamma_2)\Gamma^{-1}\alpha^{2/3}(2T/3\pi\Delta)^{1/2} & \text{for } T \simeq 0 \end{cases} \qquad (43)$$

and $D_0 = v^2/2\Gamma_2$ is the diffusion constant in the normal state and

$$f_2 \simeq \frac{\pi\Delta^2}{\Gamma_2 T} f(\Delta)f(-\Delta)\left(1 + \frac{\Gamma}{2\pi\Delta}\left[\ln(\frac{4\pi\Delta}{\Gamma})+\frac{1}{2}\right]\right) \qquad (44)$$

2. When $\omega/Dq^2 \ll 1$, both the quasi particle and the phason contribute to the screening of the ion potential. Further the sum of two terms is identical to the quasi particle term in the normal state implying that the sound velocity is unchanged at $T = T_c$ when the SDW is unpinned ($C/C_0 = 1 - \lambda$, where λ is the electron-phonon coupling constant and C_0 is the bare phonon velocity). In this limit the attenuation coefficient decreases rapidly below $T = T_c$

$$\alpha/\alpha_N = (1-f_1)^2 - f_2^2 \qquad \text{for } T \simeq T_c \qquad (45)$$

3. When $\omega/Dq^2 \ll 1$ and when the SDW is pinned the sound velocity is given by

$$C/C_0 = 1 - \lambda(1-f_1) \qquad (46)$$

which increases as the temperature decreases through $T = T_c$. Similarly the attenuation coefficent is given by

$$\alpha/\alpha_n \simeq 2f(\Delta) + f_2 \qquad (47)$$

which has a broad peak immediately below $T = T_c$. The present result describes both the sound velocity and the attenuation coefficient in a SDW of $(TMTSF)_2PF_6$ observed by Chaikin et al.[23]

4. When $\omega/Dq^2 \gg 1$, on the other hand, the quasi particle contribution becomes negligible. Then the sound velocity does not exhibit any feature at $T = T_c$ when the SDW is pinned ($C = C_0$). Further the attenuation coefficient is given by

$$\alpha/\alpha_N = (1-f_1)D_0/D \qquad (48)$$

which decreases monotonically below $T = T_c$.

5. In the same limit when the SDW is unpinned by an electric field, the sound velocity decreases to

$$C/C_0 = 1 - \lambda[1+(\omega f_2/D_0 q^2)^2]^{-1} \qquad (49)$$

and

$$\alpha = 2\lambda(\omega f_2/D_0 q)[1+(\omega f_2/D_0 q^2)^2]^{-1} \qquad (50)$$

while in the normal state $C_n = C_0$. Therefore at low temperatures $[\omega f_2/D_0 q^2 \ll 1]$, the change in the sound velocity when the SDW is depinned

$$C_{pin} - C_{unpin} = \lambda C_0 \qquad (51)$$

is independent of ω/Dq^2 and only depends on λ. In particular this will provide a unique way to measure to electron-phonon coupling constant in a SDW. The temperature dependence of the sound velocity in the pinned and unpinned SDW are sketched as functions of temperature for $\omega/Dq^2 \ll 1$ and $\omega/Dq^2 \gg 1$ in Fig. 3b.

FIELD INDUCED SPIN DENSITY WAVES

The field induced spin density wave (FISDW) in $(TMTSF)_2ClO_4$ was discovered in early eighties by Kwak et al[28] as strong anomaly in magnetoresistance at low temperatures (T ≲ 1K) when a magnetic field perpendicular to the most conducting plane (i.e. parallel to the c' axis) is increased beyond 4 Tesla. Later the Hall resistance is shown to exhibit the corresponding anomaly[29-34]. Further the existence of a cascade of new SDW phases is confirmed by the thermodynamic measurement[35]. The latter is shown in Fig 4a. From these experiments we can now construct the phase diagram as shown in Fig. 4b. So far eight distinct SDW phases are identified. Also it is known that $(TMTSF)_2PF_6$ under high pressure (P ~ 7 ~ 8k bar) where the ground state in the absence of magnetic field becomes superconducting, exhibits a similar phase diagram.[37]

In the remaining part of this lecture we describe the theoretical attempt to interpret these phases within the same model. Roughly speaking we can describe the observed phase diagram for H < 10T semi-quantitatively. Further we predict that these FISDWs exhibit the Fröhich conduction as in the ordinary SDW, which we have described in the preceeding sections. Some of these predictions appear to be confirmed experimentally. On the other hand if we look the region beyond H = 10T, recent experiment on $(TMTSF)_2ClO_4$ exhibit the unexpected reentrant behavior;[36,38] the last FISDW which appears around H = 7.5T is destroyed when H exceeds 26T. Further the nature of the last phase (as we shall see we associate this particular phase with the N = 0 state) appears to be quite different from what expected from the theory. In short we cannot

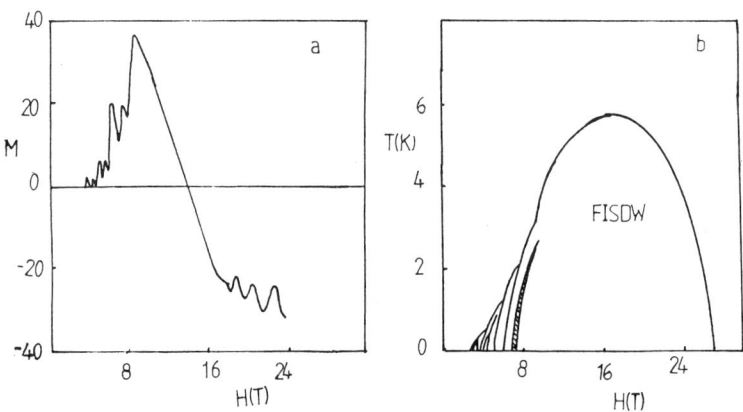

Fig. 4. a) The magnetization observed in $(TMTSF)_2ClO_4$ as function of magnetic field.
b) The phase diagram of $(TMTSF)_2ClO_4$. The FISDW is completely suppressed beyond H = 26T. (Ref. 36)

describe the FISDW in $(TMTSF)_2ClO_4$ in a magnetic field beyond H = 7.5T. We don't understand why the present model suddenly breaks down in describing the high field behavior of the FISDW. Certainly there are a few proposals, which attempt to save the theory. However, unfortunately none of them looks convincing. In the following we shall describe the theory, which is now called the standard model.

STANDARD MODEL

Let us consider the Hamiltonian (1) in a strong magnetic field H perpendicular to the a-b plane. Then the effect of the magnetic field is incorporated into the theory by replacing p_2 by p_2-eHx, where we took the a axis as the x direction. Then within mean field theory the quasi-particle Green's function G obeys;[39]

$$[i\omega-\varepsilon(\vec{p}-e\vec{A})]G(x,x')-\Delta F(x,x')=\delta(x-x')$$

$$[i\omega-\varepsilon(\vec{p}-\vec{Q}-e\vec{A})]F(x,x')-\Delta G(x,x')=0 \quad (52)$$

where $\vec{A}=(0,Hx,0)$

Then making use of an approximate expression of $\varepsilon(p)$ which can be found from Eqs(4) and (5)

$$\varepsilon(p) = v(p_1-p_F)-2t_b\cos(bp_2)-\varepsilon_0\cos(2bp_2) \quad (53)$$

we can simplify Eq (52) as

$$[i\omega-\xi\rho_3-\Delta e^{i(\phi(x)-\phi(0))}\rho_1]\hat{g} = 1 \quad (54)$$

$$\hat{g} = \begin{pmatrix} g(x,x') & \bar{f}(x,x') \\ f(x,x') & \bar{g}(x,x') \end{pmatrix} \quad (55)$$

and g and f are related to G and F by

$$G(x,x') = \exp[i(\phi(x)-\phi(x'))]g(x,x')$$

$$F(x,x') = \exp[i(\phi(x)-\phi'(x'))]f(x,x') \quad (56)$$

with

$$\phi(x) = v^{-1}\int_\infty^x dx(2t_b\cos b(p-eHx)+\varepsilon_0\cos 2b(p-eHx))$$

$$= -(vbeH)^{-1}\{2t_b\sin b(p_2-eHx)+\frac{1}{2}\varepsilon_0\sin 2b(p_2-eHx)+\text{const}\} \quad (57)$$

and $\phi'(x) = \phi(x)\Big|_{p_2 \to p_2+q_y+\pi/b}$ \quad (58)

Here we made use of the transformation introduced by Gor'kov and Lebed[40] and the generalized nesting vector introduced by Héritier, Montambaux and Lederer[41]

$$\vec{Q} = (2p_F+q_x, \pi/b+q_y, \pi/c) \quad (59)$$

where q_x and q_y are small extra momentum shifts. Finally the Φ function is expressed in terms of $\phi(x)$ and $\phi'(x)$ as

$$\Phi(x) = q_x(x-x') + \phi(x) - \phi'(x)$$

$$= q_x(x-x') + \beta\cos k(x-x_0) - \alpha\sin 2k(x-x_0) \quad (60)$$

with

$$\alpha = (H_0/H)\cos bq_y, \quad \beta = (4t_a H_0/\varepsilon_0 H)\sin(1/2 bq_y)$$
$$H_0 = evb/\varepsilon_0 \text{ and } k = ebH \quad (61)$$

We note that Eq(54) is similar to the Gor'kov equation for a BCS superconductor except the x dependent phase factor on Δ; if we can neglect this x dependence, the thermodynamics of the FISDW will be identical to that of a BCS superconductor. It is not difficult to see that $\exp[i\phi(x)]$ is periodic in x and is decomposed as

$$e^{i\phi(x)} = e^{iq_x(x-x_0)} \sum_{n=-\infty}^{\infty} I_n e^{ink(x-x_0)} \quad (62)$$

with

$$I_n = I_n(\alpha,\beta) = i^n \sum_\ell J_\ell(\alpha) J_{n-2\ell}(\beta) \quad (63)$$

and $J_\ell(z)$ is the Bessel function. As shown in Fig. 5a, I_n depends on H. In the high field limit ($H \geq H_0$) I_0 is the largest and further as H increase I_0 approach unity. As the magnetic field is decreased, I_0

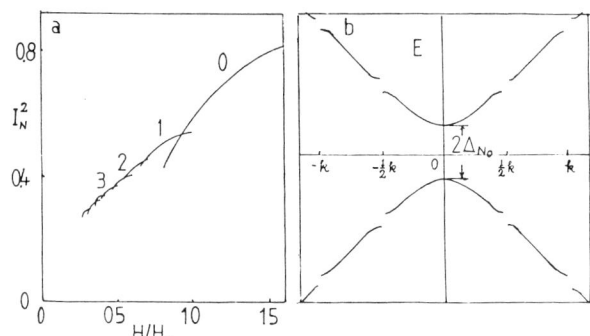

Fig. 5. a) The maximum value of I_N^2 calculated within the present model where we took $4t_a/\varepsilon_0 = 60$.
b) the quasi-particle spectrum of FISDW. The spectrum develops a series of energy gaps at momentum $p = p_F \pm 1/2 Nk$.

decreases with field while I_1 increases with decreasing field. Then I_1 crosses I_0 and becomes the largest. As the magnetic field is further decreased I_2, I_3, etc become successively the largest.

Due to the periodic behavior of $\Delta(x) = \Delta \exp[i\phi(x)]$ the quasi-particle is scattered by $\Delta(x)$ into new states with the momentum shifted nk with n integer. This gives rise to a series of energy gaps at

$p_1 = p_F \pm nk$, in the quasi-particle energy spectrum (see Fig. 5b) with the nth gap proportional to I_n. Then when I_N is larger than all other I_n's the Fermi surface moves into at the center of the Nth gap, which maximizes the condensation energy. This is achieved by choosing $q_x = Nk$. An observation of this multiple gap structure in the electron density states will provide a definitive proof on the correctness of the present model. The present conclusion is reached by several people independently.[42,43]

In order to make further progress we neglect all I_n's but the largest I_n (the single gap approximation or SGA). Then as stated already Eq(54) becomes formally the Gor'kov equation for a BCS superconductor.

For example the transition temperature of the FISDW is given by

$$T_c(H) = 1.14 \, \varepsilon_F e^{-(\bar{U}I_N^2)^{-1}} \qquad (64)$$

Further the thermodynamics is the same as in a BCS superconductor. The energy gap, the condensation energy and the magnetization at $T = 0K$ are given by

$$\Delta_0(H) = 1.57 \, T_c(H) \qquad (65)$$

$$E_G(H) = -\tfrac{1}{2} N_0 \Delta_0^2(H) \qquad (66)$$

$$M(H) = -\frac{\partial}{\partial H} E_G(H) = N_0 \Delta_0^2(H) \frac{\partial}{\partial H} \left(\bar{U} I_N^2(H)\right)^{-1} \qquad (67)$$

Above T_c, E_0 and M are evaluated numerically for appropriate parameters[41] for $(TMTSF)_2ClO_4$ [$t_a = 2843K$ and $t_b = 265K$] and shown in Figs 6, 7a ~ b. By choosing ε_F and $\bar{U}$ as adjustable parameters we can fit the observed $T_c(H)$ and a cascade of the FISDW's quite well for $4T < H < 10T$. For $H > 10$ T the theory predicts still increasing $T_c(H)$ with H, while the observed $T_c(H)$ takes the maximum value 6K around $H \sim 16$ T then $T_c(H)$ starts decreasing. On the lower field side ($H < 4T$) the theory predicts an infinite series of the SDW transitions, while so far eight distinct FISDWs are identified. However introduction of finite mean free path ℓ of the order of $10^3 A^\circ$ can eliminate easily the FISDWs at low temperatures since the quasi-particle scattering in general destroys the SDW. Further we

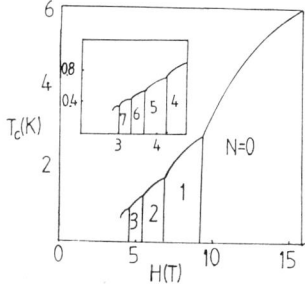

Fig. 6. The FISDW transition temperature predicts an infinite series of the FISDW

note that M derived in the present model describes qualitatively the observed field dependence of the magnetization except the behaviour H beyond 8T. However the observed magnetization has extended shoulders in the low field side of each peak, while the theory predicts a series of large jumps.. A similar agreement is obtained for the jump in the specific heat though the observed jump[35] in the specific heat is always larger by a factor 2 ~ 6 than the one expected from the BCS theory.

We summarize therefore that the standard model describes qualitatively and sometimes semi-quantitatively the thermodynamics of the FISDW observed in $(TMTSF)_2ClO_4$ for $H \leq 7.5$ T. However, Beyond $H = 7.5$ T the theory appears to lose it's predictive power mysteriously. We shall come back to these problems later.

TRANSPORT PROPERTIES

We shall describe in this section the electron spin resonance and the magnetotransport only since the other properties are very similar to the ordinary SDW. The spin in the FISDW is given by

$$\langle \vec{S}(\vec{x}) \rangle = 4\bar{U}^{-1}\Delta \hat{n} \cos(\vec{Q} \cdot \vec{x} + \phi(\vec{x})) \tag{68}$$

where $\hat{n}$ is a unit vector lying in the a-b plane and ϕ is the phase of the order parameter. In the absence of symmetry breaking energy the ground state energy is independent of $\hat{n}$ and ϕ; the ground state has two Goldstone modes the spin wave and the sliding SDW mode. However the spin-spin dipole energy and the spin-orbit scattering break the rotation symmetry of spin while impurities and crystalline defects break the translational symmetry of the SDW. Now the spin dynamics is described by a set of equations similar to the Leggett equation in superfluid ^{3}He. In particular the spin wave consists of two modes the longitudinal and the transversal and corresponding resonance frequencies are given by[44]

$$\omega^2 = \Omega_\ell^2(T) \tag{69}$$

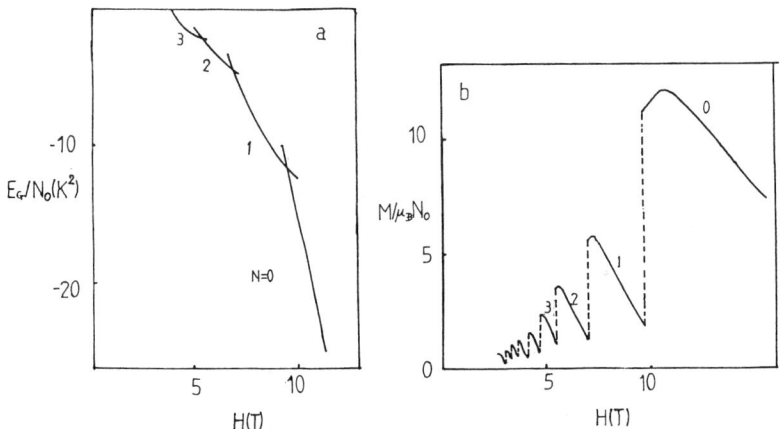

Fig. 7. a) The ground state energy as function of magnetic field.
b) The magnetization at T = 0K as function of magnetic field

$$\omega^2 = (\mu_B H)^2 + \Omega_t^2(T) \tag{70}$$

respectively where μ_B is the Bohr magneton and $\Omega_\ell(T)$ and $\Omega_t(T)$ are proportional to $\Delta(T)$. However since $\mu_B H \sim 10^3 \text{GHz}$ in the present circumstance one needs the submilli wave technique to detect the transverse model. On the other hand the longitudinal mode lies in the more conventional frequency range as in the ESR in an ordinary SDW. However in order to detect the longitudinal mode an oscillating magnetic field has to be applied parallel to the static magnetic field.

The sliding SDW mode can be seen in nonohmic conduction, microwave response and electromechanical effect as in an ordinary SDW. Indeed the phason dynamics is described by the phase Hamiltonian as decribed in Eqs(15), (27) and (28). We shall describe here the magnetotransport which is unique to the FISDW. For example the Hall resistance at low temperatures (T < 0.5K) exhibits a series of plateau like features[33,34], which remind us the quantum Hall effect in the two dimensional electron system. Indeed these plateaus are commonly interpreted in analogy to the quantum Hall effect.[41,45] However, it is important to remember that the plateau value does not correspond to the exactly quantized value. Further the resistance in the chain direction ρ_{xx} never vanishes though at low temperatures ρ_{xx} becomes much smaller than ρ_{xy} (the Hall resistance). We have proposed an alternative model which describes a variety of the observed features semi-quantitatively.[46] First, up to the first order in t_b the electric currents in the a-b plane are given by[46]

and
$$j_x = e\{v + 2t_b a \cot(ap_F)\cos k(x-x_0)\} \tag{71}$$

$$j_y = -2et_b b \sin k(x-x_0) \tag{72}$$

Second, the conductivity tensor is calculated within the standard method;

$$\sigma_{xx} = \sigma_0 \{1 - f_0 P + K \cot^2(ap_F)\Gamma_n \Gamma_n' \Omega^{-2}\}$$

$$\sigma_{yy} = \sigma_0 K \left(\frac{b}{a}\right)^2 \Gamma_n \Gamma_n' \Omega^{-2}$$

$$\sigma_{xy} = -\sigma_{yx} = \sigma_0 K \left(\frac{b}{a}\right) \cot(ap_F)\Gamma_n \Omega^{-1}(1 - f_c) \tag{73}$$

where $\Omega = v b e H$ and

$$f_c = (2\Delta)^2 \int_\Delta^\infty dz (z^2 - \Delta^2)^{-1/2} (\Omega^2 + 4\Delta^2 - 4z^2)^{-1} \tanh\left(\frac{1}{2}\beta z\right) \tag{74}$$

and Γ_n and Γ_n' are two relaxation rates with $\sigma_0 = e^2 n/m\Gamma_n$. Γ_n has been already given in Eq(37), while Γ_n' is given by

$$\Gamma_n' \simeq \beta \int_G^\infty dz\ \text{sech}^2\left(\frac{1}{2}\beta z\right)\frac{z^2}{z^2 - \Delta^2}\left(\Gamma_2 + \Gamma\frac{\Delta^2}{z^2 - \Delta^2}\right)$$

$$\simeq 4\Gamma_2 f(\Delta)\left\{1 + \frac{\pi}{2\Gamma_2 T}\Delta^2 f(-\Delta)\right\} \tag{75}$$

Finally P is the pinned portion of the SDW. When $E < E_T$ we take P = 1. As the electric field exceeds E_T, the SDW starts sliding the P decreases to zero for large E. We note that the nonOhmic conductivity appears in

σ_{xx}, while σ_{yy} and σ_{xy} does not depend on E.

From Eq(73) the corresponding resistivity tensor is found as

$$\rho_{xx} = \rho_0 D_0^{-1}$$

$$\rho_{yy} = \rho_0 [K (\tfrac{b}{a})^2 \Gamma_n \Gamma_n']^{-1} \Omega^2 [1 - f_0 P + K\cot^2(ap_F) \Gamma_n \Gamma_n' \Omega^{-2}] D_0^{-1}$$

$$\rho_{xy} = -\rho_{yx} = \rho_0 \tfrac{a}{b}\cot(ap_F) \Omega \Gamma_n'^{-1}(1-f_c) D_0^{-1} \tag{76}$$

$$D_0 = 1 - f_0 P + K \cot^2(ap_F)(\Gamma_n/\Gamma_n')(1-f_c)^2 \tag{77}$$

Here we neglected the higher order terms in $(\Gamma_n'/\Omega)^2$, which are small. In the normal state when $\Delta = 0$, the present result agrees with the expression of magnetoresistance obtained by Cooper et al.[47]

Compared with the early result[46] we include the relaxation rates Γ_n and Γ_n', which are obtained from a microscopic model. Especially at low temperatures when $1-f_0$ becomes exponentially small (i.e. in the pinned SDW) Eq(76) is further simplifies as

$$\rho_{xx} \approx \left(\tfrac{e^2 n}{m} K \cot^2(ap_F)\right)^{-1} \Gamma_n' (1 - f_c)^{-2} \tag{78}$$

$$\rho_{xy} \approx \left(\tfrac{e^2 n}{m} K \cot(ap_F)\right)^{-1} \Omega (1 - f_c)^{-1} \tag{79}$$

where f_c is now given by

$$f_c = (2\Delta)^2 \Omega^{-1} [(2\Delta)^2 + \Omega^2]^{-1/2} \sinh^{-1}(\Omega/2\Delta)$$

The present result describes fairly well the early magnetoresistance measured by Oshima et al.[31] However, the calculated Hall resistance does not exhibit plateaus, though it exhibits a series of steps as observed. The discrepancy is most serious for the N = 0 state, which extends for H≥ 7.5T. However, the N = 0 state cannot be explained in analogy to the Quantum Hall effect either, since then we expect $\rho_{xy} = \infty$ in contrary to the observation. Further Osada et al[48] observed a nonOhmic resistance in the N = 0 state. However, the result is difficult to interprete within the present model. They observed the nonOhmic behavior only in ρ_{xx} and not in ρ_{xy}, while the present theory predicts that the nonOhmicity of ρ_{xx} and ρ_{xy} should be same, as it is controlled by D_0 in Eq(77). Further the observed ρ_{xx} increases with E contrary to our expectation. A possible interpretation of their result is that they measured ρ_{yy} instead of ρ_{xx}, then we expect something similar to what they saw happens, since ρ_{yy} increases with E. Further the nonOhmicity of ρ_{yy} is much stronger than those of ρ_{xx} and ρ_{xy}.

REMAINING PUZZLES

We have seen that mean field treatment of anisotropic Hubbard model describes semi-quantitatively the thermodynamics (the phase diagram) and the transport properties of the SDW in (TMTSF)-salts like $(TMTSF)_2PF_6$ and $(TMTSF)_2C\ell_4$. In particular it is crucial to incorporate the quasi two dimensionality into the theory. Further this quasi tow dimensionality guarantees the validity of the mean field approach unlike in the case of the one dimensional model.

A similar model applied to the CDW of NbSe$_3$ not onl describes the observed electron tunneling density of states[49,50] quite well but also resolves a puzzle associated with a large value of $2\Delta/k_B T_c \sim 11 \sim 13$. Indeed the quasi two dimensional model tells that the energy gap seen by the tunneling is not 2Δ but $2(\Delta + \varepsilon_0)$ where ε_0 is the unnesting energy introduced in Eq(5) when this identification is made, we can interprete[51] the pressure dependence[52] of the CDW transition temperatures of NbSe$_3$ and the temperature dependence of the tunneling density of states within the mean field theory. Further we have shown that the same model describes semi-quantitatively the field induced spin density waves (FISDW) in (TMTSF)$_2$ClO$_4$ and (TMTSF)$_2$PF$_6$ under high pressure in a large transverse magnetic field (4T < H < 7.5T), though many theoretical predictions have not been tested experimentally. In an ordinary SDW recent observations of the nonOhmic conduction[13] in the SDW of (TMTSF)$_2$NO$_3$ and the microwave measurement[53] in the SDW of (TMTSF)$_2$PF$_6$ rend support to our prediction. On the other hand we have difficulties in interpreting the nonOhmic conductance[48] observed in the N = 0 state of (TMTSF)$_2$ClO$_4$ as already discussed.

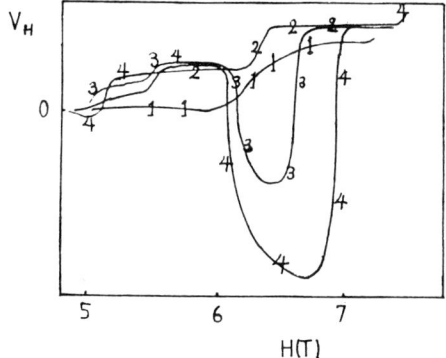

Fig. 8. The Hall voltage in (TMTSF)$_2$ClO$_4$. Note the dependence on the cooling rate between 30K and 4.2K. 1, 2, 3, 4 are the rate at 2×10^{-3}, 1.6×10^{-3}, 0.7×10^{-3} and 10^{-4} K/s. (Ref. 33)

In spite of this impressive array of successes, the present model appears to fail describing the high magnetic field behavior (H > 7.5T) of the FISDW in (TMTSF)$_2$ClO$_4$. It is pity that the most of experiments have been done on the unique compound (TMTSF)$_2$ClO$_4$ and we can not judge at the present moment how general or universal this puzzling failure of our theory is. In any case the experimental results on (TMTSF)$_2$ClO$_4$ indicate that this failure is associated with the N = 0 state and beyond (H > 8 T). We shall summarize the prominent puzzles

1) Why the SDW transition temperature is reentrant? As the magnetic field is increased beyond H > 10T, $T_c(H)$ saturates around 6K for H ~16 ~

17T and T_c vanishes around H = 26T.

2) Nature of the normal state beyond H ~ 25T. Though the normal state beyond H ~ 25T is reached continuously from the normal state in higher temperatures which is metallic, the resistance of the low temperature phase is of activated form. Is it still metallic with strong electron localization?

3) Mystery of the N = 0 state. Our theory predicts still increasing Hall resistivity with H, while a rather extended plateau spanning whole stable region of the N = 0 state (7.5 T < H < 26T) is observed. Further ρ_{xy} in the N = 0 state is very close to 3 (or 2) times of ρ_{xy} in the N = 1 state. Therefore a few questions on this result. Why such an extended plateau? It is related to the fractional quantum Hall state?

4) Negative Hall resistance. When a sample of $(TMTSF)_2ClO_4$ is super-relaxed (i.e. cooled extremely slowly through the anion ordering temperature 25K by say 1K/5 hours) the Hall resistance exhibits a series of large negative dips[32] (see Fig. 8), which depend on the cooling rate. What means the super relaxed SDW? What is origin of the negative Hall effect? Is it related to an appearance of the small electron like pocket in the Fermi surface?

In summary a large amount of both experimental and theoretical works are still required in order to understand the above puzzles.

ACKNOWLEDGEMENTS

The work reported here are written while I am staying at Max-Planck Institute für Festkorperforschung at Stuttgart and at International Centre for Theoretical Physics at Trieste. I wish to thank their hospitalities. This work is supported in by National Science Foundation under grant No. DMR 86-11829.

REFERENCES

1. W.A. Little, Phys, Rev. A134 1416 (1964)
2. For an overview on these systems see: Proceeding of the international conference on Science and Technology of Synthetic metals. Santa Fe. June 26 - July 2, 1988 in Synthetic Metals 27 (1988) - 29 (1989).
3. D. Jérome and H. Schulz, Adv. in Physics 31 299 (1982); D. Jerome, F. Creuzet and C. Bourbonnais, Phys. Scripta T27 130 (1989)
4. K. Yamaji, J. Phys. Soc. Jpn 51 2787 (1982)
5. K. Maki and A. Virosztek, Phys. Rev. B (to be published)
6. H. Hasegawa and H. Fukuyama, J. Phys. Soc. Jpn 55 3978 (1986)
7. K. Yamaji, J. Phys. Soc. Jpn 52 1361 (1983)
8. P.A. Lee, M.T. Rice and P.W. Anderson, Sol. Stat. Commun. 14 703 (1974)
9. A. Virosztek and K. Maki, Phys. Rev B 27 2028 (1988)
10. I. Tüttö and A. Zawadowski, Phys. Rev. Lett 60 1442 (1988)
11. S.A. Brazovskii and I.E. Dzyaloshinskii, Soviet-Phys. JETP 44 1233 (1976)
12. see for example H. Fukuyama and H. Takayama in "Electronic Properties of Inorganic Quasi-One Dimensional Materials, I p41, edited by P. Monceau (Reidel, Dordrecht 1985) and K. Maki, ibid p125.
13. S. Tomić, J.R. Cooper, D. Jérome, and K. Bechgaard, Phys. Rev. Lett 62 2446 (1989)
14. K. Maki and A. Virosztek, Phys. Rev. B 39 9640 (1989)

15. For general reviews see P. Monceau in "Electronic Properties of In organic Quasi-One Dimensional Material II. p 139 Edited by P. Monceau (Reidel, Dordrecht 1985); G. Grüner and A. Zettl, Phys. Rep 119 117 (1989)
16. H. Fukuyama and P.A. Lee, Phys. Rev. B17 535 (1978)
17. P.A. Lee and T.M. Rice, Phys. Rev. B19 3970 (1979)
18. K. Maki and A. Virosztek, Phys. Rev. B39 2511 (1989)
19. S. Takada, J. Phys. Soc, Jpn 53 2193 (1984)
20. J.W. Brill and W. Roark, Phys. Rev. Lett 53 846 (1984); J.W. Brill, W. Roark and G. Minton, Phys. Rev. B33 6831(1986)
21. G. Mozurkewich, P.M. Chaikin, W.G. Clark, and G. Grüner, Sol. State Commun 56 421 (1985)
22. K. Maki and A. Virosztek, Phys. Rev B36 2910 (1987)
23. P.M. Chaikin, T. Tiedje, an A.N. Bloch, Sol. State Commun 41 739 (1982)
24. X.D. Xiang and J.W. Brill, Phys. Rev. B36 2969 (1987)
25. X.D. Xiang and J.W. Brill, Phys. Rev. B39 1290 (1989)
26. A. Virosztek and K. Maki, Phys. Rev (submitted)
27. Y. Nakane and S. Takada, J. Phys. Soc. Jpn 54 977 (1985)
28. J.F. Kwak, J.E. Schirber, R.L. Greene, and E.M. Engler, Phys. Rev. Lett. 46 1296 (1981); J.F. Kwak, Mol. Cryst. Liq. Cryst. 79 111 (1982)
29. P.M. Chaikin, M.-Y. Choi, J.F. Kwak, J.S. Brooks, K.P. Martin, M.J. Naughton, E.M. Engler, and R.L. Greene, Phys. Rev. Lett. 51 2333 (1983)
30. M. Ribault, D. Jérome, J. Tuchendler, C. Weyl and K. Bechgaard, J. Phys Lett (Paris) 44 L 953 (1983)
31. K. Oshima, M. Suzuki, K. Kikuchi, K. Kuroda, I. Ikemoto, and K. Kobayashi, J. Phys. Soc Jpn 53 3295 (1984)
32. M. Ribault, J. Cooper, D. Jerome, D. Mailly, A. Moradpour, and K. Bechgaard, J. Phys. Lett. (Paris) 45, L 935 (1984)
33. M. Ribault, Mol. Cryst. Liq. Cryst. 119 91 (1985)
34. R. V. Chamberlin, M.J. Naughton, X. Yan, L.Y. Chiang, S.Y. Hsu and P.M. Chaikin, Phys. Rev. Lett. 60 1189 (1988)
35. F. Pesty, P. Garoche and K. Bechgaard, Phys. Rev. Lett. 55 2495 (1985); P. Garoche and F. Pesty, J. Magn, Magn. Mater 54-57 1418 (1986)
36. M.J. Naughton, J.S. Brooks, L.Y. Chiang, R.V. Chamberlin, and P.M. Chaikin, Phys. Rev. Lett 55 969 (1985); M.J. Naughton, R.V. Chamberlin, X. Yan, S.Y. Hsu, L.Y. Chiang, M.Ya Azbel and P.M. Chaikin, Phys. Rev. Lett 61 621 (1988)
37. S.T. Hannahs, J.S. Brooks, P.M. Chaikin, L.Y. Chiang, X. Yan, W. Kang and S.H. Bloom, Bull. Ameri Phys. Soc. 34 740 (1989)
38. T. Osada, N. Miura and G. Saito, Physica 143 B 403 (1986)
39. K. Maki, Phys. Rev. B33 4826 (1986); A. Virosztek, L. Chen and K. Maki, Phys. Rev. B34 3371 (1986)
40. L.P. Gor'kov and A.G. Lebed, J. Phys. Lett (Paris) 45 433 (1984)
41. M. Héritier, G. Montambaux and P. Lederer, J. Phys. Lett (Paris) 45 L943 (1984)
42. K. Yamaji, J. Phys. Soc. Jpn 54 1034(1985); Synth. Met. 13 29 (1986)
43. D. Poilblanc, M. Héritier, G. Montambaux, and P. Lederer, J. Phys. C 19 L321 (1986)
44. K. Maki and A. Virosztek, Phys. Rev. B38 2691 (1988)
45. D. Poilblanc, G Montambaux, M. Héritier, and P. Lederer, Phys. Rev. Lett. 60 1189 (1988); M. Ya Azbel, P. Bak and P.M. Chaikin, ibid 59 926 (1987)
46. A. Virosztek and K. Maki, Phys. Rev. B39 616 (1989)
47. J.R. Cooper, M. Miljak, G. Delpanque, D. Jérome, M. Weger, J.M. Fabre, and L. Giral, J. Phys. (Paris) 38 1097(1977)
48. T. Osada, N. Miura, I. Oguro and G. Saito, Phys. Rev. Lett 58 1563 (1987)

49. A. Fournel, J.P. Sorbrier, M. Konczykowski and P. Monceau, Phys. Rev. Lett. 57 2199 (1986)
50. T. Ekino and J. Akimitsu, Jpn, J of Apl. Phys. suppl 26 625 (1987)
51. X.Z. Huang and K. Maki, Phys. Rev. B 40 2575 (1989)
52. A. Briggs, P. Monceau, M. Nuñez-Requeiro, J. Peyrard, M. Ribault and J. Richard, J. Phys. C 13 2117 (1980)
53. T.W. Kim, J.P. Carini, G. Grüner, K. Maki and F. Wudl, Phys. Rev. B (submitted)

POSTER CONTRIBUTIONS

V.M. Castillo, R.C. Pochy, L. Lam, San Jose State University, San Jose, USA: Pattern change in electrodeposits of copper

Alois Würger, ILL Grenoble, France: Temperature dependance of rotational tunnelling

A. Vallat, H. Beck, Institute of Physics, University of Neuchâtel, Switzerland: The frustrated XY model on a fractal lattice

A. Ramsak, P. Prelovsek, J. Stefan Institute, Ljubljana, Yugoslavia: Comparison of effective models for CuO_2 layers in oxyde superconductors

A. Extremera, Faculty of Science, University of Granada, Spain: Unstable behaviour of a superconducting ring containing a Josephson junction

Luiz R. Evangelista, State University of Maringà, Brazil and M. Simoes, State University of Londrina, Brazil: A possible field-theoretical model for nematic-isotropic phase transition in liquid crystals

Arianna Montorsi and Mario Rasetti, Politechnical University of Torino, Italy: A Fermi linearized Hubbard model: Dynamical superalgebra and supersymmetry

Rupak Chatterjee, Physics Department, University of Calgary, Canada: Comments on the magnetic flux quantization in the Hall effect

M. Dzierzawa, Institute of the Theory of Condensed Matter, University of Karlsruhe, FRG: Monte Carlo simulation of the 2-D Hubbard model

Luis Vásquez, and Zhang Fei, Faculty of Science, Universidad Complutense de Madrid, Spain, Yuri Kirshar, Physico-Technical Institute of Low Temperature, Kharkov, USSR, and Boris A. Malomed, Institute for Biological Physics of the USSR, Academy of Sciences, Moscow, USSR: Numerical simulation of the sine-Gordon equation with impulsive force

Gerardo Martinez and Mario E. Foglio: Institute of Physics, Unicamp, Campinas, Brazil: Perturbative results using the cumulant expansion for the Anderson lattice.

D. Gottlieb and M. Lagos, Department of Physics, University of Chile, Santiago, Chile: Magnetically driven lattice instabilities

M. Bartkowiak, Institute of Physics, A. Mickiewicz University, Poznan, Poland: High-density expansion for electron systems

Riccardo Giachetti, Department of Mathematics, University of Cagliari, Italy, and Valerio Tognetti, Department of Physics, University of Florence, Italy: Effective Hamiltonians for the quantum statistical mechanics of nonlinear models

P. Hadley, Department of Physics, University of Stanford, USA, and R. Wiesenfeld, Technical University of Georgia, USA: Attractor crowding in oscillator arrays

S.W. Lovesey, RAL, Didcot, UK, and C.J. Lantwin, Department of Theoretical Physics, Oxford University, UK: Dynamic properties of longitudinally modulated spin systems and incommensurate spin systems in general

Angel Sánchez and Luis Vázques, Department of Theoretical Physics I, Universidad Complutense, Madrid, Spain: The stochastic ϕ^4 atomic chain

U. Eckern, W. Lehr, A. Menzel-Dorwarth, F. Pelzer, and A. Schmid, Institute of the Theory of Condensed Matter, University of Karlsruhe, FRG: Decay of a dissipative object driven by colored noise: the "quasiclassical" Langevin equation

Shepard Smithline, and Karl Fried, James Franck Institute, University of Chicago, USA: The quantum mechanical diffusion of hydrogen on metal surfaces: Isotope effect and the role of surface phonons

U. Zürcher and P. Talkner, Institute of Theoretical Physics, University of Basel, Switzerland: Quantum mechanical harmonic chain attached to heat-baths

Sarah Evans, and G.A. Gehring, Department of Theoretical Physics, University of Oxford, UK: Mixed valent and Kondo lattice behaviour in the Uranium heavy fermion systems

D. Jelcic, A. Bjelis and I. Batistic, Department of Physics, University of Zagreb, Yugoslavia: Charge density wave phase slippages in external electric field: Local generation of "narrow band noise"

CONTRIBUTORS

Barone, P. M. V. B. 79
Bartkowiak, M. 281
Bjeliš, A. 325
Bohr, T. 157
Bruinsma, R. 209

Caldeira, A. O. 27, 79
Capellmann, H. 239,289
Castillo, V. M. 151
Chao, K. A. 281

Dzierzawa, M. 273

Eckern, U. 311
Efetov, K. B. 187
Evangelista, L. R. 117
Extrema, A. 77

Ferreira da Silva, A. 285,287
Floria, L. M. 299
Foglio, M. E. 283

Giachetti, R. 141
Gottlieb, D. 119
Griffiths, R. B. 299

Kleban, P. 83

Lagos, M. 119
Lam, L. 151
Lantwin, C. J. 153
Leggett, A. J. 1,63

Maki, K. 379
Marino, E. C. 121
Martinez, G. 283
Micnas, R. 281
Monceau, P. 357
Münger, P. 281
Mydosh, J.A. 225

Pochy, R. D. 151
Prelovšek, P. 323

Ramšak, A. 323

Sánchez, A. 155
Simões, M. 117

Tagliacozzo, A. 49
Talkner, P. 81
Tognetti, V. 141
Trugman, S. A. 253

Vaia, R. 141
Vázquez, L. 155

Zotos, X. 265,273
Zürcher, U. 81

INDEX

Born-Oppenheimer, 5, 66

Chaos, 157
Charge-density wave, 325, 357
Cold fusion, 63
Conformal invariance, 83
Correlation functions, 86, 99, 106, 121, 216, 289
Critical phenomena, 83, 209

Disorder, 155, 187, 209, 225, 340, 360
Dissipation, 41

Exact diagonalization, 265

Fluctuation-dissipation, 18, 250
Fractals, 151, 157, 304

Harmonic oscillator, 12, 81, 247
Heavy fermions, 225, 239
Hubbard model, 253, 265, 281
Hysteresis, 219

Instanton, 59
Ising model, 100, 122

Josephson junction, 33, 49, 77, 299, 311

Kondo system, 233, 240

Landau-Ginzburg, 128, 211
Liquid crystals, 117
Localization, 187

Lyapunov exponent, 161, 171, 181

Macroscopic quantum tunneling, 1, 27, 49
Measurement theory, 4
Metastability, 27, 219
Magnets, 36, 119, 153, 225, 242

Neutron scattering, 289
Nonlinear σ-model, 132, 192
Nucleation, 46

Path integrals, 38, 121, 141
Peierls state, 325, 357
Phason, 340, 395
Polaron, 256, 339

Quantization, 121, 141
Quantum Monte Carlo, 273

Scaling, 84, 199, 223, 226
Soliton, 121, 144, 155
Spin-density wave, 379
Spin glass, 209, 255
SQUID, 29, 77
Superconductor, 137, 233, 289, 299, 318, 323, 379
Supersymmetry, 187
Surfaces, 101

Transfer matrix, 90
Turbulence, 164

Vortex, 129, 178, 311, 344